International Series in Heat and Mass Transfer

Editors: Arthur E. Bergles and Franz Mayinger

Grigull and Sandner, *Heat Conduction*
(Translated by J. Kestin)

Stephan, *Heat Transfer in Condensation and Boiling*
(Translated by C.V. Green)

Alifanov, *Inverse Heat Transfer Problems*

Oleg M. Alifanov

Inverse Heat Transfer Problems

With 116 Figures

Springer-Verlag
Berlin Heidelberg New York
London Paris Tokyo
Hong Kong Barcelona Budapest

Professor Oleg M. Alifanov
Moscow Aviation Institute
Volokolamskoye Shosse, 4
125871 Moscow
Russia

Translated from the Russian "Identifikacija Processov Teploobmena Letatel'nych Apparatov" (Mashinostroenie, 1979)

ISBN-13:978-3-642-76438-7 e-ISBN-13:978-3-642-76436-3
DOI: 10.1007/978-3-642-76436-3

Library of Congress Cataloging-in-Publication Data
Alifanov, O.M. (Oleg Mikhaĭlovich) [Identifikatsiia protsessov teploobmena letatel'nykh apparatov. English] Inverse heat transfer problems/Oleg M. Alifanov
p. cm.-(International series in heat and mass transfer) Includes bibliographical references and index.
ISBN-13:978-3-642-76438-7

1. Heat-Transmission-Mathematics. 2. Mass transfer-Mathematics. 3. Numerical analysis. 4. Inverse problems (Differential equations) I. Title. II. Series.
QC320.A39813 1994 536'.2-dc20 94-29203

Softcover reprint of the hardcover 1st edition 1994

Typesetting: Macmillan India Ltd., Bangalore, India.
SPIN 10006765 61/3020-5 4 3 2 1 0 - Printed on acid-free paper

Preface

This research monograph presents a systematic treatment of the theory of the propagation of transient electromagnetic fields (such as optical pulses) through dielectric media which exhibit both dispersion and absorption. The work divides naturally into two parts. Part I presents a summary of the fundamental theory of the radiation and propagation of rather general electromagnetic waves in causal, linear media which are homogeneous and isotropic but which otherwise have rather general dispersive and absorbing properties. In Part II, we specialize to the propagation of a plane, transient electromagnetic field in a homogeneous dielectric. Although we have made some contributions to the fundamental theory given in Part I, most of the results of our own research appear in Part II.

The purpose of the theory presented in Part II is to predict and to explain in explicit detail the dynamics of the field after it has propagated far enough through the medium to be in the mature-dispersion regime. It is the subject of a classic theory, based on the research conducted by A. Sommerfeld and L. Brillouin in the early 1900's and described in modern textbooks on advanced electrodynamics. Over the past seventeen years, we have improved the classic theory in several ways. In particular, we have increased the accuracy of many of the approximations, we have applied modern mathematical methods to clarify some of the confusion present in the older theory, we have extended the theory to include the field dynamics in space-time regions that were excluded by the classic theory, we have applied numerical techniques with computers to guide our analysis and to check the accuracy of our analytic results, we have studied the dynamics of a number of special pulses, and we have developed a physical model which provides a simplified quantitative algorithm for obtaining all of the dynamics of the field and explains all of those dynamics in physical terms. All of these advances are included in this book.

Since our primary interest is in the effects of dispersion and absorption on the propagation of the field, we exclude other effects by studying the propagation of a plane wave in a linear, homogeneous, isotropic dielectric. We assume that the plane wave is incident normally on the dielectric which fills a half-space. The electromagnetic field is known for all time on the plane boundary. As in the classic theory, we employ the Lorentz model to describe the interaction between the field and material medium. Our goal is to determine the details of the field behavior within the medium.

Professor V.A. Morozov; his colleagues Dr. E.A. Artyukhin, Dr. V.V. Mikhailov and Dr. S.V. Rumyantsev for their assistance and valuable advice, contributing to improvements in the content of the book; and O.A. Volkova and E.V. Filatova for their help in programming and performing methodical calculations.

The author takes the opportunity to mention an outstanding role of Prof. J. Beck in the appearance and practical application of the first effective approaches to a solution of inverse heat conduction problems which in the main stimulated further investigations in this area and thank him for very useful comments and criticism made on the text of the manuscript thus improving its form and content.

Moscow, January 1994 O.M. Alifanov

Table of Contents

Notations

$a = \lambda/C$	–	thermal diffusivity;
b	–	slab thickness;
C	–	heat capacity per unit volume;
d	–	location of a point of temperature measurement;
l	–	linear dimension;
q	–	heat flux;
S	–	body's inner heat source;
T	–	temperature;
T_w	–	body surface temperature;
T^*(or f), q^*	–	known values of temperature and heat flux, respectively;
x, y, z	–	space Cartesian coordinates;
X	–	moving boundary coordinate;
$\Delta\tau, \Delta x$	–	discretization intervals in time and space coordinate, respectively;
$\Delta\tau_{cr}$	–	critical step in time;
δ	–	error of input data;
θ	–	model temperature (a variable after Kirchoff substantiation);
λ	–	thermal conductivity;
τ	–	time;
τ_m	–	utmost value of the variable;
ν, μ	–	densities of thermal potentials of the simple and double layer, respectively;
σ	–	mean-square error;
$Fo_l = a\tau/l^2$	–	Fourier number;
$\Delta Fo_{cr} = a\Delta\tau_{cr}/l^2$	–	critical step in Fo_l

Mathematical Symbols

A	–	operator, acting in some functional space;
$\boldsymbol{A}_\Delta, \boldsymbol{B}, \boldsymbol{C}$	–	matrices, acting in Euclidean spaces;
A^*	–	operator, adjoint to operator A;
$\boldsymbol{A}_\Delta^T$	–	matrix, transposed to $\boldsymbol{A}_\Delta$;

A^{-1} – operator inverse of operator A;
A_Δ^{-1} – matrix inverse of matrix A_Δ;
A' – Frechet derivative of the operator A;
$\boldsymbol{AB}$ – product of matrices $\boldsymbol{A}$ and $\boldsymbol{B}$;
$C[a, b]$ – space of real-valued functions, continuous on segment $[a, b]$;
$C(G), C$ – space of real-valued functions continuous and bounded in domain G of space E_m;
$C^k(G), C^k$ – set of functions having in domain G derivatives continuous and bounded up to the k-th order, inclusively;
E_m – Euclidean space of m-dimensional real vectors;
$L_2[a, b]$ – space of real-valued functions square integrable on segment $[a, b]$;
$W_2^k[a, b]$ – Sobolev space – class of functions having the square integrable generalized derivatives up to the k-th order inclusively on segment $[a, b]$;
$(\cdot, \cdot)_H$ – inner product in some space H;
$\|\cdot\|_H$ – norm in some space H;
$\|A\|$ – operator norm;
$z = [z_1, z_2, \ldots, z_m]^T$ – vector-column (T-transposition sign);
$z^T = [z_1, z_2, \ldots, z_m]$ – vector-string;
$\{z_i\}_1^l$ – sequence of elements z_i, $i = 1, 2, \ldots, l$;
$\{z : R\}$ – set of all elements z with some property R;
$z^{(k)}(x)$ – the k-th derivative z with respect to x;
$\inf\limits_{u\in G} J[u], \min\limits_{u\in G} J[u]$ – problem notation on minimization of scalar quantity $J[u]$ in set G (inf – the lower bound);
J', J'_u – gradient of functional or function J (with respect to u);
$\subset$ – sign of inclusion;
$\in, \notin$ – signs of belonging, or non-belonging;
$\equiv$ – sign of identity or equality by definition;
$O(h^k)$ – symbol of approximation error of the derivative by difference formula of k-order with respect to approximation step h;
$o(\eta)$ – symbol of infinitesimal quantity of higher order than value η (with $\eta \to 0$):
$\overline{n, m} \equiv n, n+1, \ldots, m$ – sequence of real integers;

$\operatorname{erf}[u] \equiv \dfrac{2}{\sqrt{\pi}} \int\limits_0^u \exp[-y^2]\mathrm{d}y$ – function of errors [172];

$\operatorname{erfc}[u] = 1 - \operatorname{erf}[u]$ – additional function of errors;

$\operatorname{ierfc}[u] \equiv \int\limits_u^\infty \operatorname{erfc}[u]\,\mathrm{d}y = \dfrac{1}{\sqrt{\pi}}\exp[-u^2] - u\operatorname{erfc}[u]$ – integral of function $\operatorname{erfc}[u]$ [172];

$i^k\mathrm{erfc}[u] \equiv \int_u^\infty i^{k-1}\mathrm{erfc}[y]\mathrm{d}y$, $k = 2, 3, \ldots$ – iterated integrals [172];

$I(u)$ – modified Bessel function of the first kind.

In all other cases the notations are explained in the text.

Abbreviations

IHCP – inverse heat conduction problem(s)
IHTP – inverse heat transfer problem(s)
MM – mathematical model

Introduction

Recently a new direction of investigation has developed in the theory and practice of studying heat exchange processes as well as in the thermal design and simulation of thermal conditions of engineering systems. It is based on principles of inverse heat transfer problems. These methods have received special attention in the experimental study of nonstationary heat transfer processes, associated with reentry of space vehicles, rocket launchers, different heat engines, the determination of thermophysical properties of materials, construction and correction of mathematical heat models of engineering systems and in a number of other cases.

What gave rise to the rapid development and broad implementation of the methods based on solving the inverse problems for thermal investigations? The following probably answer this question.

1. The creation of advanced technology in such branches as rocketry, space flights, the aircraft industry, energetics, engine manufacturing, metallurgy, chemical engineering, etc. require solutions to a wide range of thermophysical problems, which in many cases involve units and systems operating close to the thermal limit. At the same time heat transfer experiments can be quite costly and limited to one test, especially when dealing with a full-scale test of machines and equipment. That is why it is extremely important to increase the amount of information from such investigations and to search for new methods of data processing and treatment of measurements in order to increase accuracy of the results. The methods based on solving the inverse heat transfer problems serve these purposes. It turns out that in many cases satisfactory data acquisition can be performed by proper application of inverse heat transfer methods.

Thus, the need to develop reliable methods of inverse problem solving is induced and stimulated through practice.

2. For a long time a purely mathematical aspect of the problem hindered the development of a solution theory for inverse heat transfer problems. These problems, supposedly were considered to be unsolvable and thus of no practical importance. However, the latest achievements in mathematical theory have given rise to the development of effective methods and algorithms for solving a series of inverse heat transfer problems. At present the means exist for the successful solution and effective practical application of inverse heat transfer problems found in many different applied investigations as well as in the experimental development of thermal regimes for engineering parts.

3. Of no small importance in the formation of the methods used for inverse problem solving is the development of computer facilities. 'Manual' techniques can only be used for the most simple problems which is why the methods used for solving inverse problems of heat transfer should be regarded as computer methods. Only with the appearance of powerful computing systems did it become possible to effectively realize and actively introduce into practice experimental studies on heat simulation and design of engineering systems.

It should also be added that inverse problems arising from the innovative and untypical nature of their formulation and the specific character of the mathematics used in this field, are especially attractive to researchers in this new field.

Today, the list of publications on applied questions of investigation and solution of inverse heat conduction problems is extensive. A survey of this literature can be found in various monographs [1, 81, 96, 162]. Note here that J. V. Beck, O. R. Burggraf, N. V. Shumakov, and G. Stolz should rightly be referred to as pioneers in this field, during the period 1957–1964. They suggested the first steps to a solution of boundary inverse heat conduction problems and attracted attention to this problem.

This book can be viewed as an introduction to the applied theory of solving inverse problems of heat transfer. The contents include studies made by the author and his co-workers. In the monograph an attempt has been made to present, in a systematic way, the main principles of formulation, analysis of correctness and solution of inverse problems with respect to the study of unsteady heat transfer processes. Natural liaisons are established and investigated as to the physical essence of the problem and its mathematical abstract form. Attention is paid to the development of approximate analytical, numerical and extremal methods to solve linear and nonlinear problems in restoring heat flux density and temperature fields in solids, bringing these methods to the engineering application stage. Examples occupy quite a considerable part of this book; the majority of methods and algorithms are illustrated using numerical results, characterizing their efficiency. Most universal approaches, such as variational and iterative regularization methods, are described in detail and illustrated by numerical examples.

It should be noted that the author did not intend to make a survey of all methods and algorithms existing at present for solving inverse problems. In particular, some methods developed in the USA are not exhaustively illustrated. Readers are referred to the book by J. Beck, B. Blackwell, C. St. Clair [162], which considers the findings of American authors.

Chapter 1

Statements and Use of Inverse Problems in Studying Heat Transfer Processes and Designing Engineering Units

1.1 Introduction to the problem

1.1.1 Inverse Heat Transfer Problems. What are they for?

Many structures employed in various branches of engineering operate under conditions of strong, often extreme, thermal effects. The general tendency has been the extensive use of heat-loaded engineering objects and the high intensity of heat regimes. At the same time it has been necessary to increase the reliability and service life of goods, whilst reducing the specific consumption of materials. Investigating the processes of heat transfer and providing acceptable heat conditions also occupy an important place in the design and development of production methods related to the heating and cooling of materials as, for example, in continuous steel casting or the different techniques of metal thermal treatment, in glassmaking, foundry work, high-temperature crystal growing out of melt, etc. *The non-stationary state and non-linearity* (considerable, at times) of heat transfer phenomena can be referred to as the special features of heat conditions of modern heat-loaded structures and production methods. These considerably reduce the possibility of using many traditional design-and-theoretical and experimental methods. So it became necessary to develop new approaches to thermophysical and heat engineering studies. Amongst them are methods based on a solution of *inverse problems*, in which it is required, by measurements of the system or process state, to specify one or several characteristics causing this state (in other words, to find not causal-sequential, as in direct problems, but rather sequential-causal quantitative relations). The advantage of these methods is that they help to carry out experimental studies under conditions very similar to full-scale tests or in the operation of the engineering parts. In addition, new information from these investigations makes it possible to speed up experimental work compared to traditional methods and reduces material expense. This is very important for parts used in aviation and rocket-space technology, and it has been in this area of practical applications where the first formulations and methods of solving inverse heat transfer problems have appeared as seen in the next two examples.

The first example is related to the diagnostics of non-stationary heat loads acting on thermo-loaded airframe structures and launching complexes. In particular, a similar question arose when designing a thermoprotection for the first reentry vehicles; the motion in atmosphere being accompanied by considerable and quickly changing thermal loads. The most reliable experimental information by which it could be possible to judge the values of heat loads of such structures are by the data of temperature measurements inside a solid body. The time interval in measuring a heat flux density determined by the time for reaching a melting temperature of the measuring element can run to a fraction of a second. Using this limited and indirect information the heat flux acting on the body may be restored. Similar effects take place when studying the heat action of high-temperature gas jets on the structures of launching devices as well. This is quite an intensive and a short but fast process. In the above cases data processing of temperature measurements will inevitably demand solution of inverse problems, since other methods are inapplicable.

The second example is concerned with the analysis of thermophysical properties of composite heat-protective materials acting with a high-enthalpy gas flow as, for example, when a reentry vehicle is flying in the atmosphere of the Earth or that of any other planet. Thermophysical measurements based on classical approach methods for many materials could only be made at temperatures and heating rates much lower than those realized in reality. To avoid this discrepancy it is possible to simulate the required conditions for model heating on test stands with further results from temperature measurements through the methods used for inverse problems. Thermophysical properties thus obtained correspond to the heating conditions brought near to the natural conditions in which a heat protection operates.

At present the list of practical applications of inverse problem methods has been considerably broadened.

1.1.2 A History of the Question

A retrospective look at the problem of solving *Inverse heat transfer problems* and the applied utilization of corresponding methods demonstrates the rapid development of the scientific trend observed to-day compared to former times. The interest and attention shown by investigators to this problem appeared incidentally. The first formulations and attempts to solve inverse problems should perhaps be related to the determination of the historical climate and thermal conductivity of the earth's ground layer (see Fourier, Poisson and Kelvin in the last century [172]).

Some methods used at present are based on solutions or general approach methods known long ago, an example being solutions of linear problems of heat conduction using Duhamel's integral (1832) with his numerical inversion leading towards a solution of the inverse problem. However, corresponding procedures to determine unsteady heat fluxes appeared much later in works by Mirsepassi, one of the first having been published in 1958 [192], Stolz (1960 [203]), Beck (1962 [158, 159]), Aldoshin, Golosov, Zhuk (1968 [2, 70]), Alifanov (1969[1]

[1]Alifanov, O.M.: Solution of an unsteady-state heat conduction problem for investigation of ablation of heat protection materials. In: Preprints to the All-Union conference of young scientists on heat-and-mass transfer (January 27–30, 1969). Minsk: 1969, p. 187 (in Russian).

[3, 6]) and by other authors. Regularization of heat conditions for solid bodies in the form of the exponential law of temperature change had been established in 1901 by J. Boussinesq. At the same time the basics for the theory of regular regime had been developed by G.M. Kondratiev and later by A.V. Lyikov in the 1940s and 1950s. In 1955 the principle of regular regime had been used by Shumakov to find non-stationary heat fluxes through a successive interval method.[2] Obviously, it is the first "promulgated" technique for solving a boundary inverse problem in heat conduction. Later on it was published in article form [150] and in its complete form with corresponding supplements in a monograph [151].

Note that for a particular case of so-called pseudo-inverse heat conduction problem (see Sect. 2.4) W.H. Giedt in 1955[3] and O.N. Kastelin and L.N. Bronsky in 1956 have published a procedure for its solution which is still applied today.[4]

A solution of the heat conduction problem in the Cauchy generalized formulation presented as an infinite power series had been obtained by Stefan in 1890 [202]. This result can be considered as the first exact solution of a one-dimensional inverse problem with constant coefficients, although not used until the studies of Tyomkin and Burggraf [132, 169] who correspondingly, in 1961 and 1964, obtained similar results by form solutions for a series of other linear inverse problems of heat conduction.

Thus, despite the fact that the necessary preconditions for constructing solutions of inverse problems had already appeared in the last century and at the beginning of this century, practical conclusions have only been drawn quite recently. The most active and stable period for the development of solution methods and their application has been during the last 15–20 years. The causes of such a situation conditioned by the demands of practice, achievements in mathematical theory and progress of computer facilities have been described in the Preface of this book. The mathematical aspect will be dealt with at greater length in the next section.

1.1.3 Methods of Approach to a Solution of Ill-Posed Inverse Problems

Logical formulations for any problem of mathematical physics were introduced by Hadamard in 1902 [180]. Previously it was assumed that if the original mathematical formulation of a problem did not satisfy any of these conditions, it was then of no physical or practical sense, and, consequently, there was no reason to constructing a solution. Gradually, however, the attitude of mathema-

---

[2] Shumakov, N.V.: Methods of experimental study of solid body heating Doctorate thesis, M.: Energy Institute named after G.M. Krzhizhanovsky, 1955.

[3] Giedt, W.H.: The determination of transient temperatures and heat transfer at a gas-metal interface applied to a 40 mm gun barrel, *Jet Propulsion*, 1955, **25** (4), pp. 158–162.

[4] Kastelin O.N., Bronsky L.N.: Heating of a blunt body in the supersonic flow. In: Physical gasdynamics, heat transfer and high temperature gas thermodynamics. M.: Publ. H. of USSR Acad. of Sciences 1962, pp. 233–238.

ticians and physicists towards ill-posed problems began to change. In 1926 T. Carleman made the first attempt to solve an ill-posed problem.[5] In the 1930s new investigations on the determination of the historical climate were made by A.N. Tikhonov. In 1943 he formulated for the first time in a complete form the *conditionally-well-posed statement* for such problems in mathematical physics, assuming a stable solution in the class of functions which were a compact set [136]. This fundamental result, beginning from 1953, runs all through and is further developed in the works by Lavrentiev and Ivanov (see bibliography in [88, 140]). This trend is elaborated upon in an interesting study by John in which he presents a method of solving the heat conduction equation with inverse time.[6]

A weighty mathematical result for these ill-posed problems opening a fruitful direction in mathematical physics and computing mathematics was obtained in 1963 by Tikhonov [137, 138]. His method of *regularization* considerably broadened the bounds of effective practical use of ill-posed problems in various fields of science and technology. Since that time this method has been developed by Tikhonov, Lavrentiev, Ivanov, Ya. Arsenin, Morozov, Bakushinsky, Glasko, Strakhov and many other Soviet mathematicians (see bibliography in [105, 140]). At present we have a harmonious and complete mathematical theory for solving ill-posed problems, as a result of this work.

Most works devoted to the development of the regularization method treat one of its forms, namely *a variational principle* (*method*) of constructing regularizing operators (see Sect. 2.6.4 and Chap. 7).

Other forms are also possible. Among the most universal is a so-called *iterative regularization* which is most effectively realized with the help of nonlinear gradient algorithms. This more general method has been proposed by Alifanov [11, 15] and mathematically grounded together with Rumyantsev [37, 38]. The principles of the method of iterative regularization are considered in Chaps. 6 and 8. Full mathematical aspects are stated.[7]

Of the French workers in the field of ill-posed problem solving is a monograph by R. Lattes and J.-L. Lions, in which they suggest a *quasi-inversion* method specially directed onto the equations with partial derivatives.[8]

Alongside the development of the general theory of ill-posed problems and the construction of regular methods for their solution a process has been observed for the elaboration of stable methods which are effective in practice together with algorithms for solving inverse problems of heat conduction. The initial phase of developing the computational procedures to solve these problems (up to the time when a regularization method appeared in 1963 and later

---

[5] Carleman, T.: *Les fonctions quasi analytiques*. Paris (1926).

[6] John, F.: Numerical solution of the heat equation for preceding times. *Ann. Mat. pura ed appl.*, 4, 40 (1955), 129–142.

[7] Alifanov, O.M.: Artyukhin, E.A.: Rumyantsev, S.V.: Extremal methods of solving ill-posed problems and their application to inverse heat transfer problems. M.: Nauka, 1988. p. 288 (in Russian).

[8] Lattès, R.: Lions, J.-L.: Méthode de Quasi-Réversibilité et Applications. Dunod. Paris 1967. Engl. transl.: Lattès, R.: Lions, J.-L.: The Method of Quasi-reversibility: Applications to Partial Differential Equations. American Elsevier, New York, 1969.

when it attracted the attention of other workers, i.e. somewhere between 1968 and 1970) can be called *a heuristic regularization*. The corresponding methods were known as *direct methods*. In other words, the workers using such algorithms achieved stability and acceptable accuracy of results based mainly on the physical sense and, consequently, upon a physical level. Apart from the above named works to a solution of inverse problems there is that of *trial-and-error* used by Kozdoba [80–82] and the *linear dynamic filtration* method being used by Matsevitiy, Multanovsky and Symbirsky [96, 127, 128]. Rigorous mathematical conditions of algorithm regularization have not yet been formulated in the above approaches. But this does not mean that similar methods are not acceptable. In many cases they are quite workable, effective and able to solve various practical problems.

Alongside heuristic methods, beginning from the end of the 1960s and at the start of the 1970s, rigorous and mathematically substantiated methods for solving inverse heat conduction problems began to appear. The first were enumerated by Tikhonov and Glasko.[9] The majority of them are related to the linear problem formulation and are constructed based upon a variational technique of regularization and, later on, on iterative regularization. To illustrate this, the reader is referred to works on the first [4, 7] and [11, 17, 38, 49] for the second. Either approaches, as computational experiments and actual physical tests show, turn out to be acceptable for solving various nonlinear problems as well. However, these results are not as rigorous at the time of writing.

At the time a revaluation of old approaches took place with respect to a solution for the problems considered, for which a principle of *natural step regularization* had been accurately formulated (see Chaps. 4 and 5), there appeared new formulations of inverse problems and new areas were revealed for the application of direct and regularized methods of solution, due to the numerous practical investigations having been carried out. However, a certain distrust of the methods used for studying heat transfer processes based on a solution of inverse problems remained for a long while. Unfortunately, some publications aggravated this situation: either the principal property of the problems under consideration (ill-posed!) was disregarded completely or not taken into account very accurately. As a result of which little use or even false procedures for solving inverse problems have been suggested. But at present the possibility of a successful solution and effective application of methods for inverse heat conduction problems under various scientific and engineering investigation is considered to have been proven in principle.

This chapter is devoted to general questions of formulation of inverse problems of heat transfer and classification. Its purpose is to show that inverse problems are methodical bases of the modern approach to diagnostics and the identification of heat transfer processes. Namely,

---

[9] Tikhonov, A.N.: Glasko, V.B.: Methods of determining the surface temperature of a body. Zh. Vychisl. Mat. i Mat. Phys., 1967, Vol. 7, N. 4, pp. 910–914 (in Russian).

- the major concepts are given in Sect. 1.2 for simulating heat transfer processes including identification as a basic section of simulation;
- Section 1.3 treats questions on general formulation of the inverse problem concept, the mathematical statement of heat transfer inverse problems of various types and classification by signs (for the main forms of heat transfer and types of unknown characteristics);
- Section 1.4 presents a number of examples to analyse the most important areas of the practical application of methods based on a solution of inverse heat transfer problems, and one class of problems—that of inverse heat conduction, are investigated further.

Finally, the contents and structure of the book are outlined in Sect. 1.5.

1.2 Simulation of Heat Transfer Processes

Both direct and inverse heat transfer problems presuppose, a preliminary schematization (simulation) of a real heat transfer process in the form of some mathematical representation. This question is now considered at some length.

By *simulation* in general we take a substitution of a model M for a studied object O (original) with the view to investigating the properties of O with the help of M. Simulation is a universal method of scientific cognition; it plays an important part in thermophysics and heat engineering.

Later, by thermal simulation it is understood to mean the compilation and substantiation of mathematical models (MM) of heat transfer processes. These models are used for various researches in thermophysics and heat engineering, including heat conditions computation and designing and testing thermally loaded engineering units.

Mathematical thermal models can be expressed by various means, from functional analysis and differential equations to computational algorithms and computer programs. Each of the stages is important for the final results of simulation, and it also incorporates errors and inaccuracies of the previous stage of the mathematical description. There exists a natural co-subordination of these models and their development. Problem programming is conducted after the computational algorithm has been compiled. Algorithmization, in its turn, becomes possible after the differential form of the problem statement has been completely determined. Thus the program can be regarded as the MM of the algorithm, while the computational algorithm is the model for the differential equation system.

Differential models are widely used as initial heat MMs. We distinguish between *models with distributed parameters*, the basis of which is constituted by equations with partial derivatives, and *models with lumped parameters* described by systems of ordinary differential equations.

Heat models can be stationary and non-stationary (dynamic), linear and non-linear, one-dimensional and multidimensional.

In practice the scientist often possesses a heat model of a process or an engineering system. This model can be complete and accurate, but it can also be complex and unwieldy which necessitates creating a simpler model and simpler computational method. In such cases the first and more accurate model serves as the original and the feasibility of the simpler model is found from the more complex one.

Thus we distinguish between an experimental (physical) and a purely mathematical heat simulation. In the former, the study of thermal processes or heat conditions are conducted as a result of actual experimental investigations on mock-ups, models and specimens. So, the above situation is a purely mathematical simulation. Calculation of the thermal state of a real object can also be interpreted as mathematical simulation which usually forms part of a wider class of investigation, i.e. experimental simulation.

Generally, MM development for a heat transfer process can be divided into two stages, called the structure and parametric identification.

The structural identification consists of determining the general structure of the mathematical description of a process which is in the form of certain equations. *The parametric identification* presupposes necessary quantitative information having been determined for the selected model, i.e. determining unknown characteristics of the model.

Both identification processes involve the correlating of a phenomenon model for the physical [19].

1.3 Inverse Heat Transfer Problems (IHTP)

1.3.1 The Division of Problems into Direct and Inverse

When formulating general statements and choosing the main classes of inverse problems, the statements of direct problems, i.e. classical problems of heat conduction theory, are assumed to be known [93, 114, 172]. Each direct problem within the framework of an accepted MM can be compared with a certain set of inverse problems.

Below, all problem statements of heat transfer between a solid body or a system and the environment are considered from the "cause–effect" relationship. Here, according to the accepted model, by *causal characteristics* of heat transfer in the body or in the system we mean boundary conditions and their parameters, initial conditions, thermophysical properties, internal sources of heat and conductivities as well as geometric characteristics of the body or the system. Then the *effect* is a heat state which is determined by the temperature field of an object studied.

The purpose of *direct heat transfer problems* is to specify the cause-and-effect relationships. On the other hand, if it is required to recover causal characteristics from definite information about the temperature field we have a statement of the *inverse heat transfer problem.*

The statements of inverse problems, unlike those of direct ones, cannot be reproduced in actual experiment: it is not possible to reverse the cause-and-effect relationship physically instead of mathematically. For example, it is impossible to reverse the course of a heat transfer process or to change the course of time. Therefore, in mathematical formalization this property manifests itself as "incorrect" mathematical conditioning and inverse problems present a typical example of ill-posed problems in the heat transfer theory.

The violation of cause-and-effect relations that is presupposed in the initial inverse problem statement creates serious difficulties of their solution. In the first place, difficulties are met in the development of methods and algorithms which would give true results. Nevertheless, as will be shown further, it is quite possible to overcome the difficulties mentioned here.

Proceeding from their general purpose, all inverse problems can be divided into three classes:

- inverse problems that arise in the diagnostics and identification of physical processes,
- inverse problems that arise in the design of engineering products,
- inverse problems that arise in the control of processes and systems.

Inverse problems of the first class are usually experimental research. In these cases it is necessary to reconstruct causal characteristics on the basis of certain measured "output" effect characteristics. These problems are primary, both with respect to direct problems and with respect to the other classes of inverse problems since they are connected with the construction of MMs and determination of different characteristics of the models.

Inverse problems of design type consist in determining design characteristics of an engineering unit on the basis of given quality indices within certain limits. Required characteristics are causal with respect to these indices and limits.

In the case of the control, the role of causal characteristics is played by controlling influences the change in which creates the control action expressed by the system state, i.e. the effect.

It should be noted that there exists a fundamental difference between the two types of problems, i.e. between the problems of diagnostics and identification and the problems of design and control. In the case of design and control problems, the widening of the class of acceptable solutions usually simplifies things, since it is then necessary to find any practically feasible solution that would ensure the extremum of quality criterion with the given accuracy. At the same time, for identification and diagnostics problems the wider the class of possible solutions, the worse the situation. Specifically, the errors of causal characteristics determined can increase which will make obligatory the use of regular methods of solution.

The theory and methodology of solution of inverse problems (that appear with diagnostics and identification of physical processes) are less developed than those for the other two classes of problems. Hence the aim of the present book is to fill this void for inverse problems of heat transfer.

Let us consider the types of inverse problems within the framework of each of the above-mentioned classes. In correspondence with three main forms of heat transfer introduce three groups of inverse problems: inverse problems of heat conduction, inverse problems of convective heat transfer, and inverse problems of radiative heat transfer. If combined (complex) heat transfer is considered, corresponding statements of inverse problems will appear.

According to causal characteristics required it is possible to divide inverse problems of each group into various kinds. Most often, MMs of heat transfer processes are based on equations with partial derivatives. In the general case, four kinds of inverse problems are introduced for them, viz. boundary, coefficient, retrospective problems, and geometric ones. *Boundary* problems consist in finding functions and parameters that form boundary conditions; *coefficient problems* involve determining functions and parameters that form part of equation coefficients; *retrospective* problems (i.e. time reversed ones) consist in finding initial conditions; *geometric* problems presuppose reconstructing geometric characteristics of a domain or some characteristic points, lines, or surfaces within a domain (e.g. determining the co-ordinates of a phase transfer boundary or of a contact line of mediums with different physical properties).

Let us now consider a more concrete formulation of various problems.

1.3.2 Inverse Problems of Heat Conduction

Problems of this kind are the most investigated and widely used.

As an example, let us consider a one-dimensional problem of heat conduction in a two-layer plate (Fig. 1.1) assuming that the layer materials have different thermophysical properties and that in one of them there occurs a phase transition, e.g. melting. Layer boundaries $b_1(\tau)$, $b_2(\tau)$, $b_3(\tau)$ can move with time as a result of some physical process (ablation, thermal expansion or shrinking, mechanical deformation). The inner front of phase transfer $\eta(\tau)$ is also moveable.

We will assume that the temperature field $T(x, \tau)$ in the plate in the time interval $\tau \in (0, \tau_m]$ is described by the equation system for generalized heat conduction

$$C_j \frac{\partial T_j}{\partial \tau} = \frac{\partial}{\partial x}\left(\lambda_j \frac{\partial T_j}{\partial x} \right) + K_j \frac{\partial T_j}{\partial x} + S_j, \quad j = 1, 2, 3 \tag{1.1}$$

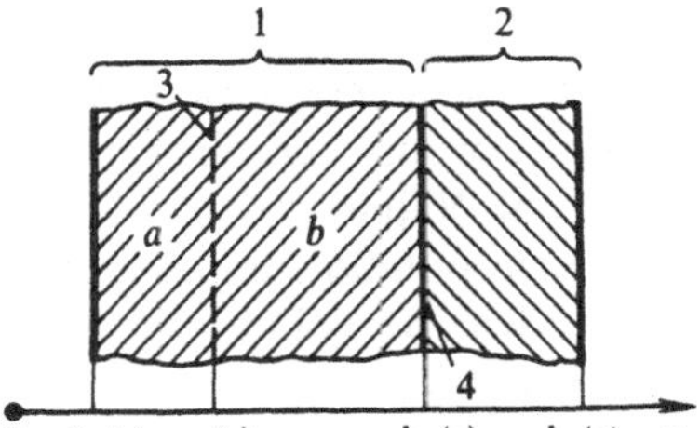

Fig. 1.1. A two-layer plate with phase transition: 1, 2 – the first and the second layers; 3 – the boundary of phase transition; 4 – the line of contact between layers; a, b – two phase states of material in the first layer

in domains $b_1(\tau) < x < \eta(\tau)$, $\eta(\tau) < x < b_2(\tau)$, $b_2(\tau) < x < b_3(\tau)$ respectively. Conjunction conditions on lines $\eta(\tau)$ and $b_2(\tau)$ are

$$T_1(\eta(\tau) - 0, \tau) = T_2(\eta(\tau) + 0, \tau),$$

$$\lambda_1 \frac{\partial T_1}{\partial x}\bigg|_{x=\eta(\tau)-0} - \lambda_2 \frac{\partial T_2}{\partial x}\bigg|_{x=\eta(\tau)+0} = r\frac{\partial \eta}{\partial \tau},$$

$$T_2(b_2(\tau) - 0, \tau) = T_3(b_2(\tau) + 0, \tau) - R\lambda_2 \frac{\partial T_2}{\partial x}\bigg|_{x=b_2(\tau)-0}$$

$$\lambda_2 \frac{\partial T_2}{\partial x}\bigg|_{x=b_2(\tau)-0} = \lambda_3 \frac{\partial T_3}{\partial x}\bigg|_{x=b_2(\tau)+0}.$$

To the system (1.1) let us also add initial temperature distributions

$$T_j(x, 0) = \xi_j(x), \quad j = 1, 2, 3.$$

at $b_1(0) \leqslant x \leqslant \eta(0)$, $\eta(0) \leqslant x \leqslant b_2(0)$, $b_2(0) \leqslant x \leqslant b_3(0)$ respectively, and conditions on the plate boundaries. As the boundary conditions we can regard temperatures

$$T_j(b_j(\tau), \tau) = t_j(\tau), \quad j = 1, 3;$$

or heat fluxes

$$-\lambda_j \frac{\partial T_j}{\partial x}\bigg|_{x=b_j(\tau)} = q_j(\tau), \quad j = 1, 3;$$

or Newton conditions of convective heat transfer

$$-\lambda_j \frac{\partial T_j}{\partial x}\bigg|_{x=b_j(\tau)} = \alpha_j[T_j(b_j(\tau), \tau) - T_j^*(\tau)], \quad j = 1, 3;$$

or conditions counting for body heat transfer with the surroundings by means of convection and radiation as well as surface heat sources (or sinks) caused by other processes (melting, sublimation, atom recombination, etc.)

$$-\lambda_j \frac{\partial T_j}{\partial x}\bigg|_{x=b_j(\tau)} = \alpha_j[T_j(b_j(\tau), \tau) - T_j^*(\tau)] + A_j q_{rj} - \varepsilon_j \sigma T_j^4(b_j(\tau), \tau) + g_j,$$
$$j = 1, 3,$$

where q_r is the incident radiant flow; σ – Stephan–Boltzmann constant. Various combinations of the above-mentioned boundary conditions on lines $b_1(\tau)$ and $b_3(\tau)$ are also possible.

Coefficients C_j, λ_j, K_j and the source term S_j in the general case can be functions of co-ordinate x, time τ, and temperature T_j, or any combination of these variables; in the simplest case they will be constant. Values r, R, α_j, A_j, ε_j, g_j can be considered as functions of the time and the corresponding temperature.

In the given problem, the causal characteristics will be volumetric heat capacities C_j, thermal conductivities λ_j, convection coefficients K_j, sources S_j, movements of boundaries b_1, b_2, b_3, and phase transition front η, volumetric heat of phase transition r, contact thermal resistance R, boundary temperatures t_j, heat fluxes q_j, ambient temperatures T_j^*, absorption factors A_j, emissivities ε_j, and surface heat sources g_j. The inverse problem of any kind consists in determining certain values of the sum total of causal characteristics adduced above. Certain additional conditions should be given: in most cases they are temperature measurements $T(d_i, \tau) = f_i(\tau)$, $i = \overline{1, N}$ in N stationary or moving points d_i of a body; it is seldom that spatially continuous temperatures are considered.

According to the above-introduced causal characteristics of heat transfer process, the following kinds of inverse problems can be introduced:

1) *a retrospective heat conduction problem*, or the problem with reverse time – the finding of temperature distributions in previous moments (in other words – the determining of a prehistory of the given heat state);

2) *a boundary inverse problem* – the reconstruction of thermal conditions at the boundary of the body. A problem connected with the continuation of heat conduction equation solution from boundary $x = M(\tau)$, where temperature $T(M(\tau), \tau)$ and heat flux $q(M(\tau), \tau)$ being assigned simultaneously, belongs to this type of problems;

3) *a coefficient inverse problem of heat conduction* – the specification of coefficients of heat conduction equation (the identification of heat conduction operator).

4) finally, it is possible to introduce one more kind of inverse problem, *a geometric* one that consists in finding some geometric characteristics of a heated body, e.g. in reconstructing the movement of heat transfer boundary of a body on the basis of the results of temperature measurements within the body.

Some additional comments connected with the formulation of the inverse heat conduction problem are required.

1. The determination of functions and parameters in boundary conditions (heat transfer coefficient α in the boundary conditions of the third kind, contact thermal resistance R in the boundary conditions of the fourth kind, heat absorption factor and emissivity A, ε can be usually reduced to a boundary problem. For example, coefficient $\alpha(\tau)$ is calculated from convective heat flux $q_k(\tau)$, body surface temperature $T_w(\tau)$ and characteristic temperature of gas (liquid) flow $T^*(\tau)$:

$$\alpha(\tau) = q_k(\tau)/[T^*(\tau) - T_w(\tau)].$$

Value $q_k(\tau)$ is found from the heat balance equation on the surface of the body by the known conductive heat flux assuming heat radiation of gas as well as ablation and injection into the boundary layer are not available

$$q_k(\tau) = q(\tau) + \varepsilon\sigma T_w^4(\tau).$$

In such a statement, function $T^*(\tau)$ and the coefficient ε must also be given. Functions $q(\tau)$ and $T_W(\tau)$ are calculated from solution of the boundary inverse problem.

1. The problem of defining certain causal characteristics could be overspecified, i.e. several additional conditions instead of only one could be given.

2. *Combined* statements are possible when causal characteristics of different types are sought simultaneously. For example, we can simultaneously estimate boundary conditions and the temperature field in the past moments of time in the problem without initial conditions: this is a combination of a boundary problem and a retrospective one. There can exist natural combinations of a boundary problem and a coefficient one as well as combinations of a boundary problem and a geometric inverse problem of heat conduction.

Generally, inverse heat conduction problems are divided into one-dimensional and multidimensional, linear and non-linear, with fixed or moving boundaries, or single and multiple-connected, depending on the process model used and the form of the domain of independent varible change.

1.3.3 Inverse Problems of Convective Heat Transfer

These problems are still at the statement stage and under investigation which is why we will not consider their mathematical formulation and will only note that these types of inverse problems should have important practical applications. Thus, of immediate practical interest are the problems of recovering temperature distribution across the boundary layer, evaluating viscosity and heat conductivity, and determining catalytic activity of a solid wall along which there exists a non-equilibrium flow of dissociated gas.

1.3.4 Inverse Problems of Radiation Heat Transfer

Such problems occur in determining the proper law of radiation intensity distribution for infrared heat sources which serve to simulate given heat flows on a heated body. A system of discretely radiating modules (Fig. 1.2) is then considered: this system is divided into groups with similar intensity of module radiators within each group. The number of modules in such a group equals n. Assuming that the number of surface elements d_i of an object tested is equal to n and using the zone method, it is possible to make heat balance equations which would characterize the equality of heat flows received by each element under such radiation conditions of this particular heating system and given heat

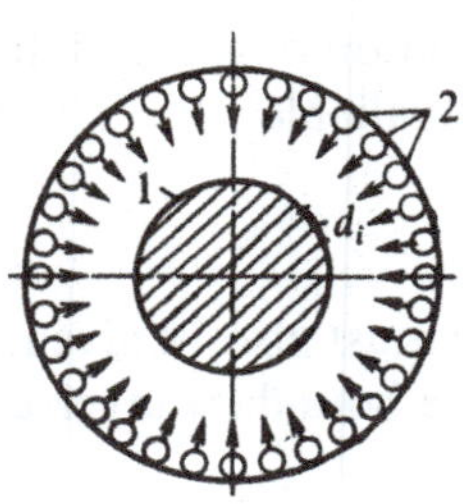

Fig. 1.2. An infra-red simulator: 1 – test object; 2 – radiators

flows q_i. For bodies with convex form, these equations are

$$\pi \sum_{j=1}^{m} J_j \sum_{k=1}^{n} \varphi_{di-jk} A_{i-j} = q_i, \quad i = \overline{1, m}\,, \tag{1.2}$$

where φ_{di-jk} is the local angular coefficient of the i-th heat-receiving element and k-th module of j-th group; A_{i-j} is the absorptivity of the i-th element with respect to the radiation of the j-th group.

It is necessary to find radiation intensities of modules J_j, i.e. causal characteristics which induce heat flows on the object. Thus this problem in its initial statement is *discrete* and can be directly reduced to an algebraic solution of system (1.2).

Another example of inverse problems of radiation heat transfer will be considered with respect to the mathematical model which is given in a *continuous* form by means of an integral operator. Let there be some arbitrary closed system of diffusively radiating grey bodies separated by a transparent medium (Fig. 1.3). In this case one of the inverse problem statements consists in finding temperature distribution on body surfaces $T(M)$ on the basis of known field of half-spherical density of incident radiation $E_i(M)$. From the law of energy conservation, this problem for steady-state conditions is reduced to a solving of the integral equation of the first kind

$$\int_F E_{ef}(N) \frac{\cos\theta_M \cos\theta_N}{\pi r_{MN}^2} \mathrm{d}F_n = E_i(M)$$

for half-spherical density of effective radiation E_{ef} with a consequent computation of temperature field $T(M)$ by equation

$$T(M) = \left\{ \frac{1}{\sigma\varepsilon} [E_{ef} - (1 - A)E_i] \right\}^{1/4}.$$

Here, the incident radiation E_i is a sum of radiation absorbed and reflected by the surface. Effective radiation coming from the grey body surface is a sum of radiations, being its own and reflected.

Problems of this kind arise in determining the working conditions of radiation heating appliances on the test rigs and in various technological processes.

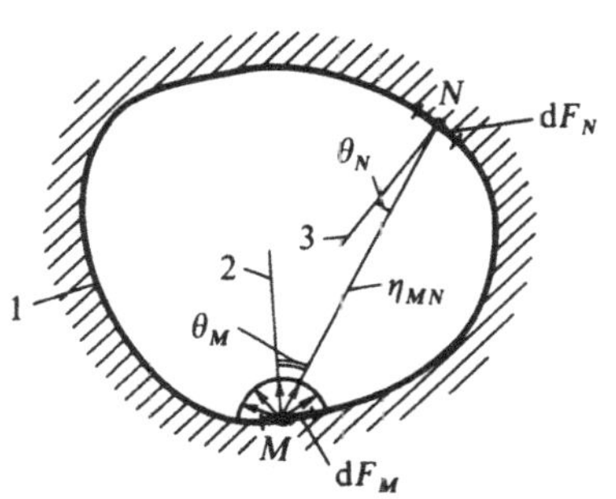

Fig. 1.3. A closed set of bodies: 1 – diffusively radiating surface F; 2, 3 – normals towards to the surface at points M and N, respectively

Inverse problems of radiation heat transfer thus formulated are stationary and linear, i.e. radiation-exchanging bodies are considered to be non-transparent. There are problem statements of great practical interest which count for partial transparency of bodies, e.g. temperature distribution recovery in partially transparent layers by solving the non-linear inverse problem for the radiation transfer equation. In this case the input data given are the measurement results for the monochromatic radiation intensity of this layer for various wavelengths and directions.

1.3.5 Inverse Problems of Combined Heat Transfer

This is the most general and varied group of IHTPs, the solution of which is increasingly used in engineering. However, because of a number of specific methodological difficulties, solutions to these problems are not yet sufficiently well developed. Let us consider three typical inverse problems of combined heat transfer.

In the first case, complex heat transfer is considered for a system of bodies (Fig. 1.4). The bodies in the system exchange heat energy with the environment and between themselves. In the case given the basic heat transfer equation is obtained from the analysis of the heat balance under the assumption that the engineering system can be divided into a finite number L of isothermic elements:

$$C_l \frac{\mathrm{d}T_l}{\mathrm{d}\tau} = \sum_{j=1}^{L+r} \lambda_{lj}(T_j - T_l) + \sum_{j=1}^{L+S} K_{lj}\sigma(T_j^4 - T_l^4)F_l + Q_{sl} + Q_l, \quad l = \overline{1, L},\ s \geqslant 0,\ r \geqslant 0\,, \tag{1.3}$$

where C_l is the thermal capacity of the l-th element, λ_{lj} the heat exchange coefficients (conductive coupling) between elements l and j, ε_{lj} the angular coefficients, σ the Stephan–Boltzmann constant, Q_{sl} the quantity of heat supplied to the element l from the environment, Q_l the quantity of heat released in the l-th element, r and s the number of elements interacting with the environment via convection and radiation, respectively and F_l is the element surface area.

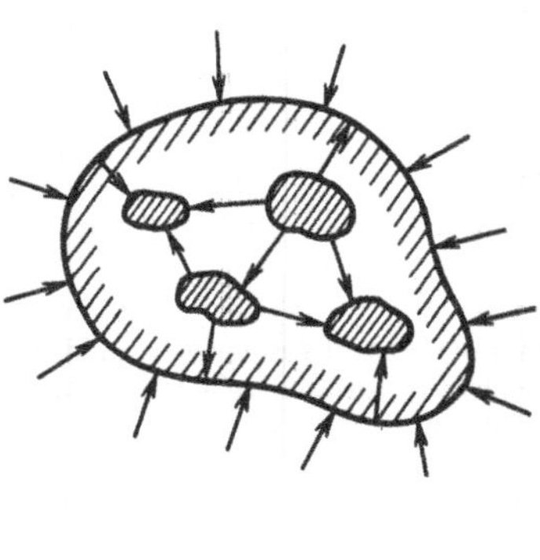

Fig. 1.4. A system of bodies (the heat fluxes are arbitrary shown by arrows)

The coefficients λ_{lj} are calculated by different relations depending on the type of heat conduction between the elements l and j, i.e. heat conduction, convection, or heat conduction plus convection.

The heat transfer process described by Eq. (1.3) is determined by the parameters of the boundary conditions and the heat balance equations by the relations of conductive, convective and radiative types, by effective thermophysical properties and heat sources, by geometry and relative position of bodies, as well as by the thermal initial state of the system.

If we are to calculate a heat state (temperature conditions) of a system by the given causal characteristics, such a calculation will be the objective of the *direct problem of heat transfer in a system.*

In the case when individual causal characteristics are not known and it is required to define them by known information about the system heat state (true for simulation and admissible for design) the *inverse heat transfer problems* are being solved then. The model as in Eq. (1.3) is often used in description and optimization of heat transfer processes in various engineering systems (electroengineering and radioelectronic appliances, flight vehicle modules, various heat exchangers, metallurgic furnaces, etc.). Hence, the inverse problems for Eq. (1.3) were called *inverse problems of heat transfer in engineering systems.* One should note that a simultaneous determination of all parameters C_l, λ_{lj}, K_{lj}, Q_{sl}, Q_l is possible only with the accuracy up to a constant multiplier.

In conducting heat tests it often becomes necessary to obtain more details of temperature field in separate elements of an engineering system rather than by means of the combined model given by Eq. (1.3). Applying the method of combined model division into simple models it is possible to get statements of local inverse heat transfer problems (usually, inverse heat conduction problems).

Another inverse problem of combined heat transfer which is of great practical interest is connected with determining of thermophysical properties of partially transparent scattering materials. It is formulated as the problem of determining coefficients $C(T)$ and $\lambda(T)$ of heat conduction equation in the following form:

$$C\frac{\partial T}{\partial \tau} = \operatorname{div}(\lambda \operatorname{grad} T) - \operatorname{div} \boldsymbol{q}_r \, . \tag{1.4}$$

The radiation energy flow vector $\boldsymbol{q}_r$ at some point of the body P is calculated from radiation spectral intensity I_ν which is characterized by frequency ν

$$\boldsymbol{q}_r = \int_0^\infty d\nu \int \boldsymbol{l} I_\nu dl \, ,$$

where $\boldsymbol{l}$ is a unit-equal vector of direction of radiation propagation at point P.

The magnitude of I_ν is found from the radiation transport equation in the absorbing, emitting, and scattering medium:

$$\boldsymbol{l} \operatorname{grad} I_\nu = (K_\nu + \beta_\nu)[-I_\nu(\boldsymbol{l}, P) + (1 - Sc_\nu)B_\nu(P) + \frac{Sc_\nu}{4\pi} \int_{4\pi} I_\nu(\boldsymbol{l}', P)\theta_\nu(\boldsymbol{l}, \boldsymbol{l}')d\omega, \tag{1.5}$$

where K_ν, β_ν are the absorption and scattering coefficients, respectively for radiation with the frequency ν, $Sc_\nu = \beta_\nu/(K_\nu + \beta_\nu)$ the Schuster criterion, B_ν the spectral radiation intensity of an absolutely black body, θ_ν the indicatrisse of scatter, and ω is the body angle.

This problem is solved with proper boundary conditions for Eqs. (1.4) and (1.5). Additional data are obtained from the results of temperature measurements in the body studied.

Such *inverse problems of conductive-radiative heat transfer* appear when thermal shielding, thermal insulating and optic materials are being developed and studied.

One more typical problem is connected with the development and testing of porous cooling systems of various structures. In these cases it is necessary to have information on the following characteristics: heat fluxes on blown surfaces; thermal conductivity λ_s and internal heat transfer coefficient α_V of a porous body, heat transfer coefficient α_0 at a coolant inlet into a porous material. The determination of these parameters from transient temperature measurements in a porous structure is reduced to a solution of an *inverse problem of conductive-convective heat transfer*. In the one-dimensional case for a flat layer of a porous material with gaseous coolant (Fig. 1.5), MM of heat and mass transfer has the form

$$C_s \frac{\partial T_s}{\partial \tau} = \frac{\partial}{\partial x}\left(\lambda_s \frac{\partial T_s}{\partial x}\right) - \frac{\alpha_V}{1-\Pi}(T_s - T_g), \ x \in (0, b), \ \tau \in (0, \tau_m] ; \tag{1.6}$$

$$(\rho C_p)_g \frac{\partial T_g}{\partial \tau} = \frac{\partial}{\partial x}\left(\lambda_g \frac{\partial T_g}{\partial x}\right) - (\rho v C_p)_g \frac{\partial T_g}{\partial x} + \frac{\alpha_V}{\Pi}(T_s - T_g) ,$$
$$x \in (0, b), \ \tau \in (0, \tau_m] ; \tag{1.7}$$

$$T_s(x, 0) = \xi_s(x), \ T_g(x, 0) = \xi_g(x) , \tag{1.8}$$

$$-\lambda_s \frac{\partial T_s(b, \tau)}{\partial x} = \alpha_0 [T_s(b, \tau) - T_{g0}] ; \tag{1.9}$$

$$(\rho v C_p)_g T_g(b, \tau) = (\rho v C_p)_g T_{g0} + \alpha_0 [T_s(b, \tau) - T_{g0}] ; \tag{1.10}$$

$$-\lambda_s \frac{\partial T_s(0, \tau)}{\partial x} = q(\tau) ; \tag{1.11}$$

$$\frac{\partial^2 T_g(b, \tau)}{\partial x^2} = 0 ; \tag{1.12}$$

$$-\frac{dP_g}{dx} = \alpha(\mu v)_g + \beta(\rho v)_g^2 ; \tag{1.13}$$

$$\rho_g = \frac{P_g M_g}{8314 T_g} , \tag{1.14}$$

where indices s and g mean solid and gaseous phases respectively, c_p is the specific heat capacity at constant pressure, ρ the density, v the velocity, Π the

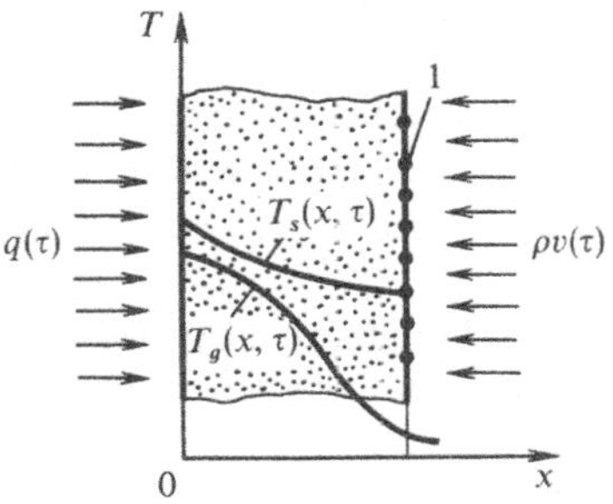

Fig. 1.5. The porous cooling: 1 – temperature-sensitive elements

body porosity, P the pressure, μ the viscosity, M the molecular weight, α, β the hydraulic coefficients and T_{g0} the initial temperature of the insufflated gas.

The model contains energy Eqs. (1.6) and (1.7) for solid and liquid phases with proper initial and boundary conditions (1.8)–(1.12), Darcy's modified law (1.13) and equation of state for gas (1.14). Condition (1.12) is one of the variants of the natural boundary condition which provides the uniqueness of the direct statement solution and yields results that are in good agreement with those corresponding to actual boundary conditions of the first and the second kind.

Values g, λ_s, α_v, α_0 are among the causal characteristics required. Measurement results are given by conditions

$$T_s(d_i, \tau) = f_i(\tau), \quad \tau \in [0, \tau_m] ,$$
$$i = \overline{1, N}, \; N \geqslant 1, \; 0 \leqslant d_1 < d_2 < \cdots < d_N \leqslant b .$$

Inverse problems enumerated in 1.3 give an idea of the diversity and scope of practical uses of the corresponding techniques.

1.4 Practical Applications and the Role of Inverse Problems in Thermal Investigations

The need for a statement and solution for inverse heat transfer problems arises in various thermophysical investigations and also in designing and operating thermally loaded engineering units, [1, 14, 17, 81, 87, 95, 100, 116, 127, 160–162, 184]. Below we will consider a number of possible applications of inverse problem techniques which could have been even greater. However, the practical applications discussed give one an idea of their importance and untapped potential.

Non-stationary heat measurement and thermal loading control. In heat testing of various engineering systems, or in studying heat transfer processes on experimental stands in thermal probing of hot gas flows, and in other cases there appears the problem of determining temperatures, heat fluxes and coefficients of heat transfer on the surfaces of bodies (various structural elements, shielding and insulating coatings, samples, and heat flux sensors, etc.). Since the intensity of heat transfer to a body usually changes with time because of the changes in heating (cooling) rates, and the non-stationariness of experimental installations parameters, etc. it is especially important to be able to determine non-stationary heat loadings.

As a rule, it is impossible to actually measure time-changing heat fluxes and heat transfer coefficients. The surface temperature of the objects studied often remains unaccessible for direct measurements. At the same time, there exists the possibility to measure temperatures at separate points within a body or on some surface part. Thus it becomes necessary to solve boundary inverse problems of heat transfer, i.e. to determine thermal boundary conditions by calculation on the basis of temperature measurements.

Such problems of heat measuring are often encountered in the simulation of thermal conditions on test gas-dynamic facilities [22, 28, 29, 123, 134, 167, 173] in thermal vaccum chambers [101, 181], in studying heat transfer processes in pipelines and ducts [109, 139], in testing various engine installations [65, 127, 128, 186], in the course of flight simulation [174, 177, 199, 204, 206], in full-scale tests of flying vehicles [155, 156, 194, 195], and so on.

Heat measuring techniques based on the solution of boundary inverse problems make practically inertialess heat diagnostics possible of both slowly developing and transient heat transfer processes. Investigations have demonstrated that it is possible to ensure the accuracy of the recovery of heat fluxes and heat transfer coefficients which could be commensurable with the accuracy of temperature measurements in a solid body.

General Principles of One-dimensional Heat Measuring. Among the most widely used are the sensors of heat fluxes with "one-dimensional" sensing elements (SE) composed of the materials from thermophysical properties which have been much studied. The sensors are designed so that the temperature field in the SE may be identified (with acceptable accuracy) with a one-dimensional temperature field in a flat-parallel infinite plate. In particular, the SE has the form of a uniform cylinder with a small enough diameter and with a thermally isolated side surface (Fig. 1.6). One of its ends is acted upon by a changing heat flux $q(\tau)$

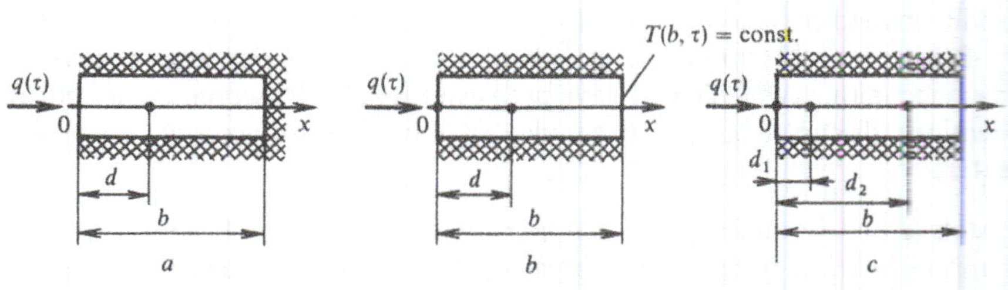

Fig. 1.6. One-dimensional sensors of heat fluxes (the heat-insulated boundaries are arbitrary outlined by shading)

while the other end is either thermally isolated (Fig. 1.6a) or at a constant temperature (Fig. 1.6b). This latter condition is provided by intensive heat removal, e.g. by a cooling liquid that flows along the surface at a considerable rate. Then, for an inverse problem solution in determining the value $q(\tau)$, it is necessary to conduct heat measurement at some point of the SE with coordinate d ($0 \leqslant d \leqslant b$ for the thermally isolated end, $0 \leqslant d < b$ for the thermostated end). If it is difficult to ensure or monitor the above-mentioned boundary conditions at $x = b$ it may be necessary to measure the temperatures at least at the two points d_1 and d_2, $0 \leqslant d_1 \leqslant b$, $0 \leqslant d_2 \leqslant b$, $d_1 \neq d_2$ (Fig. 1.6c). In this case solving the inverse problem of heat transfer makes it possible to find the heat fluxes both at the boundary $x = 0$ and at the boundary $x = b$.

Comparative testing technique. Inverse problems of heat transfer make it possible to create techniques for calculation-cum-experiment determination of various functions and parameters included in the equations of heat balance on the body surface. These techniques are based on *conducting comparative tests.* This approach can be explained in the following example. First find a local coefficient of convection heat transfer on a solid body surface as a time function $\alpha(\tau)$ and an integral coefficient of the surface radiation depending on its temperature $\varepsilon(T_w)$ at a known environment temperature $T^*(\tau)$. To solve this problem one should employ two sensing elements, 1 and 2, placed within a body (Fig. 1.7). The SE are made of materials with differing thermophysical properties. The heat exchanging ends of these elements have a coating that is being studied while the side surfaces of the SE are thermally isolated. Assuming conditions

$$\alpha_1(\tau) = \alpha_2(\tau) = \alpha(\tau), \quad \tau \in [0, \tau_m],$$
$$\varepsilon_1(\tau) = \varepsilon_2(\tau) = \varepsilon(\tau), \quad \tau \in [0, \tau_m],$$

it is possible to write a thermal balance equation on the element surface

$$\begin{aligned} \alpha(T^* - T_{w_1}) &= \varepsilon\sigma T_{w_1}^4 + q_1, \\ \alpha(T^* - T_{w2}) &= \varepsilon\sigma T_{w2}^4 + q_2, \end{aligned} \tag{1.15}$$

where q is the conduction heat flux entering the SE.

These equations hold true for any moment of time $\tau \in [0, \tau_m]$. Function $T_{w_1}(\tau)$, $T_{w_2}(\tau)$, $q_1(\tau)$, $q_2(\tau)$ contained in the equations are found by solving an

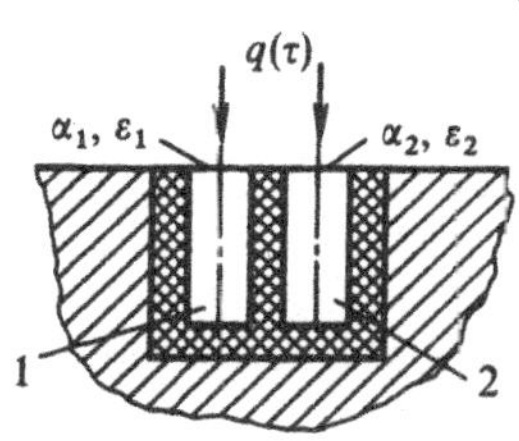

Fig. 1.7. A sensor with two sensitive elements 1 and 2

inverse heat conduction problem for each element by the temperature measurements in them. By solving the system (1.15) at each moment τ_n on a chosen grid, the dependencies $\alpha(\tau)$ and $\varepsilon(\tau)$ are obtained. By finding the mean temperatures T_{w_1} and T_{w_2} for each τ_n, it is possible to make a transition from the dependencies $\alpha(\tau)$, $\varepsilon(\tau)$ to the dependencies $\alpha(T_w/T^*)$, $\varepsilon(T_w)$.

Likewise, by adding the third SE it is possible to find the vector function $\{\alpha(\tau), \varepsilon(\tau), T^*(\tau)\}$, if the temperature $T^*(\tau)$ is unknown.

The diagnostics of space-time conditions of heat transfer. One-dimensional sensors make it possible to measure non-stationary heat fluxes and local heat transfer coefficients. To determine the discrete fields of these values it is necessary to introduce a sufficient number of sensors into various points of space, in particular, into various points of the streamlined surface of a solid body (Fig. 1.8). However, in solving two- and three-dimensional IHCPs it is possible to recover time and dimension dependencies $q(s, \tau)$, $\alpha(s, \tau)$ that characterize heat transfer on a body surface (here s is a surface point). In this case temperature measurements are usually conducted on a thermally isolated boundary part of a body (on a line for a two-dimensional case, on a surface for a three-dimensional one). Sensors with such sensing elements can be mounted (Fig. 1.9) on a model or a mock-up of an object studied, as well as on a full-scale object whose thermal conditions are being determined in a test or full-time operation. Sometimes temperature measurements can be conducted within a solid body.

This methodology to determine heat loadings and heat transfer coefficients is based upon boundary inverse problems for a uniform heat conduction equation. In many cases, in particular in diagnostics of heat release to wall structures, a different approach is more convenient where the unknown value of heat loading is determined as *a source* in a heat conduction equation. Let one of the surfaces of an isothropic plate of thickness b be acted upon by a heat flux $q(x, y, \tau)$ (Fig. 1.10). Assuming, for simplicity, that the reverse surface is thermally isolated and that across the plate thickness the temperature gradient equals

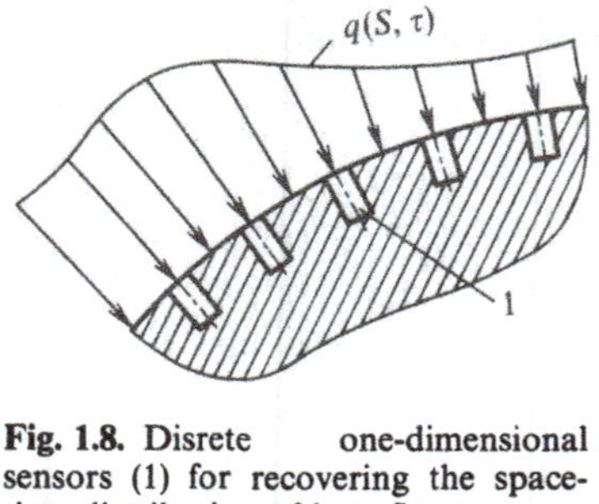

Fig. 1.8. Disrete one-dimensional sensors (1) for recovering the space-time distribution of heat fluxes

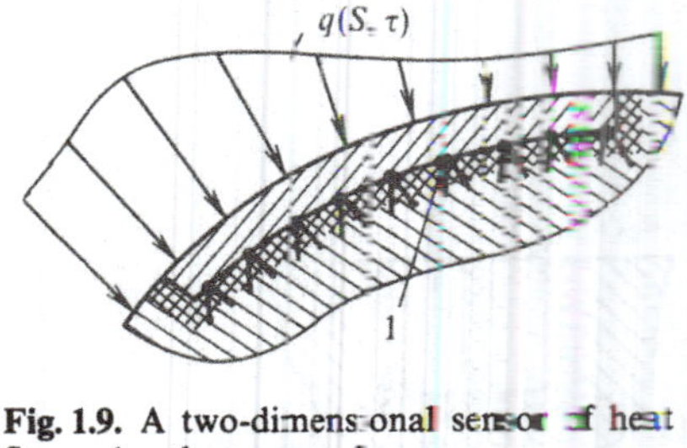

Fig. 1.9. A two-dimensional sensor of heat fluxes: 1 – thermocouples

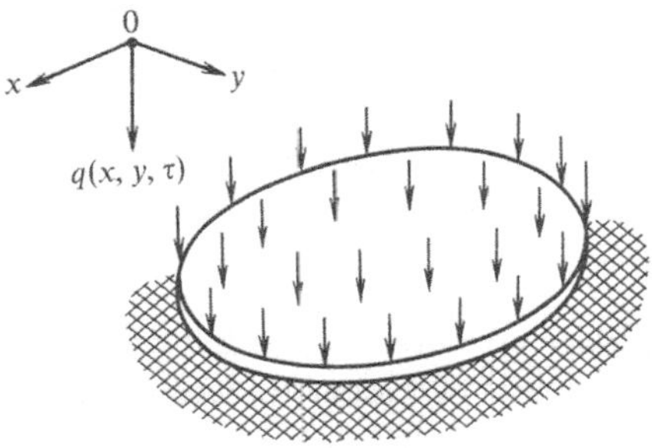

Fig. 1.10. A heat-loaded thin plate

zero, it is possible to write the heat conduction equation

$$C\frac{\partial T}{\partial \tau} = \frac{\partial}{\partial x}\left(\lambda \frac{\partial T}{\partial x}\right) + \frac{\partial}{\partial y}\left(\lambda \frac{\partial T}{\partial y}\right) + \frac{q(x, y, \tau)}{b},$$

where $q(x, y, \tau)$ is an unknown value.

In accordance with the uniqueness of this inverse problem solution it is necessary to know the temperature field of a plate $T(x, y, \tau)$ which can be measured by means of thermovisor devices.

IHTPs can be of great benefit when combined with other methods (heat setting ink, different techniques of flow visualization) for studying thermal conditions on the surface of a body with complex geometry. To such problems heat transfer studies are carried out where there exist separation zones, gap effects, changes of flow conditions in the boundary layer, surface catalytic effects, etc.

In the course of the thermal experiment it may become necessary to control heat loadings in accordance with a certain time program within the limits of the assigned accuracy of its representation. The creation of a monitoring system for heat loading control presupposes obtaining information from monitor sensors of non-stationary heat flows. A solution algorithm of the boundary IHTP (as a condition of its working on a real-time basis) can act as the main algorithm of information processing in the system of automatic-control system software. Specifically, such a control is required when conducting spacecraft thermo-vacuum tests to compensate for various errors (non-uniformity of heat flow field from solar and planetary simulators, nonparallelism of beams, influence of a studied object on temperature conditions of simulators, background heat sources, etc.). In such cases the effective adjustment of heat sources for the required mode of operation is performed by a computer as the source information is being processed. The information is gained by heat flow sensors whose SE are fitted with temperature detectors.

The possibility of using IHTP methods to monitor the heat transfer rate and temperature on the heat-loaded surface areas of reentry vehicles and spacecrafts during the flight has not been excluded. In this case, the algorithm of the inverse problem solution is realized through an onboard computer.

One more field of practical application of boundary IHPT techniques comprises the *technology processes* that are connected with the heating and cooling of processed prefabs and parts, in particular in tempering and hot punching. For such technologies, it is necessary to identify heat exchange conditions on a body surface in order to determine the thermal rates of processing that are required.

The study of porous cooling, heating and failure of materials. In the course of experimental studies of porous cooling systems it is necessary to determine the non-stationary thermal boundary conditions on the surface of a porous body and to identify the heat effect of gas insufflation onto a boundary layer. The actual measuring of values included into the boundary conditions of a heated surface is either very difficult or downright impossible, but the temperature on the reverse side of a solid matrix can be measured. In this case we are faced with the necessity of solving a boundary inverse problem for an equation system for heat and mass transfer (see, for instance, (1.6)–(1.14)) in a porous structure when there is a flow of gaseous component through it [26].

Of great practical importance is the problem of studying the heating and destruction of thermal shielding materials [3, 113, 116, 117, 121, 125, 152, 165] including the investigation of the lessening of convective heat flow with insufflation of gaseous products from the ablated surface. The main types of measurements in experimental investigation of thermal shielding materials are temperature measurements within the bodies (by means of thermocouples) and on the outer surfaces (by optic methods) and measurements of the ablation rate. The processing and interpretation of measurement results can be performed by methods based on the solution of inverse heat conduction problems.

The recovery of temperature fields and heat flows. The problem of reconstruction (continuation) of the temperature field and the heat flow field in a solid body from the results of interior temperature measurements is often reduced to a solution of an inverse problem.

The study of heat resistance characteristics of materials and structures and hence, the reproduction of temperature gradients required for specimens and models under test conditions is closely connected with the above problem. The fitting of models and specimens with numerous thermosensors is, as a rule, impossible due to technological, structural and purely methodical reasons (because of its violating the integrity and strength properties of materials, introducing distortions into the temperature field and into the field of thermal stresses, and also due to the difficulties in providing good thermal contacts of thermosensors within the material, etc.). Therefore, the initial information from such investigations is generally small and often conditioned only by indirect temperature measurements at a large enough distance from the outer surfaces that are being heated.

The study of non-stationary heat transfer. The problem of nonstationary heat transfer is of great importance when studying different heat and mass transfer processes in the system composed of solid-gas (liquid) [86, 109, 129, 134, 148]. In many important cases, the experiment still remains the major technique of such studies. Therefore it is required not only to correctly conduct and successfully carry out experimental research, but also (and this is very important) to find effective ways of processing the obtained data. It was found [28, 29, 53, 65, 81, 151] that inverse problems form an effective means to obtain the necessary results in experimental information processing.

Identification of heat transfer processes in engineering devices. With the help of inverse problems we can identify and correct the mathematical models of heat transfer needed in engineering systems. For example, if a spacecraft module is under consideration, its heat condition depends on the integral factors of absorption A and radiation ε of the outer surface, on the coefficient of heat transfer between the selected elements of the module and on their volumetric heat capacity. In the process of mathematical modelling a problem appears that consists of specifying effective (that is the most suitable for the heat model selected) values of parameters from the data of non-stationary temperature tests, [17, 101, 191]. As has already been mentioned in Sect. 1.3.5, this problem can be solved by using the inverse problem of heat transfer in the system of N bodies and also with the help of inverse heat conduction problems when dividing a combined model into simple elements.

Determination of thermophysical parameters of mediums. Recently methods to determine the thermophysical properties of materials became widespread: they are based on solving coefficient inverse problems. This fact is accounted for by the superiority of this approach over the traditional methods of measuring thermophysical properties, especially in studying properties over wide temperature ranges, at large velocities and temperature change gradients, and in the presence of phase transitions within the material.

For example, among the problems that are of great practical importance are those to determine the temperature dependence of thermal conductivity and temperature field of a cooled ingot during steel tempering. Large temperature drops and the wide range of this change often makes the use of conventional methods for thermophysical measurements impossible. At the same time coefficient inverse problems prove to be an effective means of solving this.

Another aspect is that of the determination of thermophysical properties and kinetic characteristics of thermal shielding materials which are ablated when interacting with a high temperature gas flow. However, thermophysical measurements based on classical approaches for the majority of materials can be conducted at temperatures which are considerably less than those realized under operating conditions. This is especially so when dealing with composite materials containing components which decompose on heating and give off gaseous products. In fact studying thermophysical properties in special furnaces does not correspond to the actual conditions of heating and ablation of materials for a number of other reasons (coke by-products disposal, scale effects, time factors, and heating rates). But it is possible by using the concept of "effective thermophysical properties" and by heating specimens on special stands that produce high enthalpy gaseous flows. This is then followed by processing the data of temperature measurements by methods using inverse heat conduction problems [25, 32, 40–47, 68, 80–82, 110, 112, 113, 115, 122, 124, 152].

Thermophysical properties thus obtained correspond to heating conditions approximating natural ones and allow calculation of more precise temperature fields in thermal shielding coatings and the selection of thickness of ablated layers with lesser margins (closer to optimum).

Coefficient inverse problems provide an effective means of determining thermophysical properties of freezing-and-melting soils which is extremely important for various engineering projects in a permafrost regions.

Contact heat transfer studies. The determination of contact resistances which characterize the heat transfer between the connected parts of structures as well as the prediction of their change in the course of time is a problem of great significance especially for structures where there is a great number of bolted and rivetted joints, connectors, hinges, etc.

For multi-layer thermal shielding coatings it is necessary to know the resistance of adhesive films, and this problem can be interpreted as one of contact resistance specification. The boundary inverse problem method can be successfully applied to processing the results of specially conducted experiments in solving the problem of contact heat transfer under non-stationary conditions.

Evaluation and monitoring of optic radiation characteristics. During lengthy space flights it is important to be accurate at monitoring and forecasting the optic radiation properties of the spacecraft outer surfaces, the radiators of thermoregulation systems, and the solar batteries. Modification of these properties takes place as a result of sublimation of the surface layers of materials and coatings and due to the action of ionizing radiation and meteoric matter.

The monitoring of properties of reflecting surfaces of heaters and cryogenic panels must be carried out during tests in thermal vacuum chambers.

The impossibility of installing temperature sensors on the surface with unknown properties necessitates solving boundary inverse problems through the results of temperature measurements at inner points of the body. To set up a calculation scheme it is necessary to prescribe a reflection model, to estimate angular coordinates of the surface element studied and to check for shading or mutual irradiance.

Determination of heat loadings during friction. One more application of inverse heat conduction problems consists of the use of corresponding methods to determine heat release during the friction of two solid surfaces.

Such a problem has its own engineering importance, it makes the studying of heat transfer influence in the friction zone on friction mechanism possible, and also the experimental determination of one of the most important operating characteristics of bearings, e.g. the friction moment which is determined by the heat release intensity in the contact zone.

Finally it should be noted that methods using heat transfer inverse problems can be used in other research fields whenever there is a necessity to identify processes described by parabolic type equations. In particular, are those to determine filtration parameters in studying mass transfer processes in underground hydrodynamics (oil-and-gas dynamics, hydrotechnics, hydrogeology, land improving) and the problems of determining coefficients of quasilinear parabolic equations that describe the hydrodynamics of a liquid mirror in reservoirs.

Optimization of design decisions with regard to heat limitations. The importance of this part of design work is evident: it is at this stage that engineering decisions are made within the framework of a chosen concept of the heat conditions support system, and that the design parameters are selected and certain characteristics of the whole unit are specified.

The fundamental part of optimum thermal design is contained in the mathematical heat model (models) of a system created and in the special-purpose function that is being extremized. The model connects unknown design characteristics (i.e. causal when applied to the inverse problem statement) with parameters that describe the thermal state of a unit (effect characteristics) and with loading influences, e.g. outer and inner heat flows. Definite heat transfer and geometric characteristics of an engineering system may correspond with variants in design solutions.

Thus, if one observes the concept of cause and effect characteristics, the problem of optimum heat designing in itself can be regarded as an inverse problem of heat transfer in an extremum statement. On the basis of known conditions which define the permissible thermal state of a unit (i.e. the given domain of the change in effect characteristics) one should find causal characteristics which satisfy both this state and the optimum criterion of heat designing.

In particular, regarding the optimization of multilayer thermally insulated coatings such a statement of the inverse problem has already been formulated [see 14] and the algorithm for its solution is given in [98].

Let us note in conclusion that the diagnostics and identification of processes of heat- and mass-transfer can be related to solutions of different types of inverse problems. But, later on, in setting out the material, special attention will be given to a study of boundary IHCPs.

First, boundary problems belong to one of the most important and largest classes of problems in thermal model testing. In fact, in any heat experiment, testing involves the need to define heat loads (or the values to define them) on the specimens under investigation. In many cases the only means is by using boundary IHCPs.

Up till now other inverse problems have not been so widespread. However, this does not diminish their significance in all important practical applications. First, it concerns coefficient inverse problems, which present an effective instrument in investigating thermophysical and thermokinetic properties of materials under non-stationary conditions of heating. Nevertheless, there are other methods as well for investigating thermophysical properties. They are based on the laws of stationary and quasi-stationary heat processes (see, [84, 97, 120, 149, 153]).

Second, boundary IHCPs from the standpoint of obtaining stable solutions are of methodical interest. The experience of solving various inverse problems proves that boundary IHCPs compared with coefficient and geometry problems show a greater tendency to distortion of results. It should be added here that it is difficult to predict the behaviour of a desired solution, since the a priori information about it is limited. From this point of view a better situation is usually observed using coefficient and geometry IHCPs. Therefore, the development of methods to solve non-stationary inverse problems of heat conduction is better performed using boundary IHCPs. Many of the methodical techniques developed for boundary problems are generally used with other types of inverse heat conduction problems. Iterative regularization is known as such a universal method which, when applied to boundary IHCPs, is considered in Chap. 6 the extension of this approach onto other inverse problems is given in Chap. 8.

One last note in connection with the dimensionality of inverse heat conduction problems: In many cases, at the expense of special design solutions when developing sensors and models and by selecting the conditions necessary to conduct the experiments, it becomes possible to investigate the heat exchange processes in the design elements on the base of a one-dimensional heat conduction equation. This simplifies the data processing considerably by using inverse problem methods, and under definite conditions makes it even more precise.

A one-dimensional model of a planar slab is the most widespread. More rare are the models used for a cylinder wall and spherical layer, it being underlined that if a curvilinear wall has the ratio of thickness to a small radius of curvature $b/r \leqslant 0, 1$, then it is usually possible to use the model of a planar slab. A model of semi-infinite body (thermally thick wall) is used in the case when the Fourier number $Fo_b = a\tau/b^2 \leqslant 0, 1 - 0, 2$.

Of course, there are situations when one-dimensional models of heat conduction cannot give satisfactory results and more complex problem statements are required. In such cases, as a rule, we manage to confine ourselves to two-dimensional models.

The true definition of causal characteristics by IHCP methods can to a large extent depend on the quality of initial data. That is why it is important that the initial heat measurements should be made with accuracy [63, 153].

1.5 The Contents and Structure of the Book

The aim of the book is to state the main questions of applied theory and methodology of inverse problems of heat conduction, including the various aspects of a general statement and classification and an analysis of conditions of inverse problem correctness, the formulation of principles for their solution and to develop on this base a complex of methods and algorithms for heat conduction process parametric identification and experimental data processing.

As will be shown later, the solution for ill-posed inverse problems is related to the use of any information known in advance about the quantities sought and the initial data. Depending on the type and amount of this information, the methods of solution can be arbitrary divided into *universal* and *problem-oriented*. In the first case we consider a priori data of a general character and see how these methods cope with the wide classes of problems. In the second case we include in our consideration more specific data on the problem and, hence, the application of problem-oriented methods is usually more narrow. Both of these are considered and analyzed here. The Tikhonov *regularization method* is known as the universal method, on the base of variational form in which the universal algorithms of a boundary IHCP solution are constructed later on. The author suggests *an iterative regularization* based on non-linear gradient algorithms; it has achieved a greater degree of universality and is the approach most used here.

So-called *direct methods* can be referred to problem-oriented problems, which make stricter demands of the experiment conditions, but in their field of application can *exceed universal algorithms through their operation and accuracy*. That is why to understand the specifications of IHCP solution more clearly the author gives a description and investigation of various *direct methods*.

Great importance is attached to *computational experiments*, which are an integral component of investigations carried out using inverse problem methods. Numerical simulation enables the real capacities of any algorithm to be found out and exposes the influence of uncertainties in the mathematical model of the heat transfer process and in the parameters, and to investigate the characteristics of the solution depending on the type of dependences defined, temperature sensor location, and the type and value of the initial data errors, etc. As a result, it appears possible to select the most suitable method and the IHCP solution algorithm (in every case) and to establish the necessary parameters of this algorithm.

Proceeding from this analysis of the results of an extensive computational experiment the author gives some practical recommendations regarding the most rational areas of application for the methods considered here.

A brief description of Chaps. 2 to 8 follows.

Chapter 2 gives information about the conditions necessary for IHTP solution *existence, uniqueness and stability*. The inverse heat conduction problems and an inverse problem of heat transfer in the engineering systems are considered. The analysis is based on the studies of various analytical and numerical forms of solution showing that all of these problems were potentially unstable or poorly conditioned. Physical and mathematical causes of this phenomenon are also discussed. Efforts are made to find possible ways to obtain stable results, which use the three principles of regularization: *step, iterative and regularization according to Tikhonov's variational principle*. Some main results are given on the conditions pertaining to the uniqueness of the IHCP solution.

Problems are considered in **Chapter 3** connected with the possible *analytic formulations* of boundary inverse problems for the heat conduction equation with constant coefficients. The appropriate integral forms of non-stationary IHCP are obtained. *One-dimensional statements* are considered separately for bodies with fixed and moving boundaries. *Multidimensional problems* are considered as well. It is established that in all of the above cases the inverse problems in the boundary-value formulation are reduced to a solution for ill-posed integral equations of the first kind with respect to either unknown boundary functions or thermal potential densities. Types of kernel and right-hand sides of these integral equations are given. Integral forms are deduced for analytical continuation of a temperature field (IHCP in a Cauchy statement). In this case an integral representation of the inverse problem is based on the system of two linear integral equations. It is shown that the IHCP integral statements, obtained by thermal potentials, are most suitable in practical realization compared with the formulations obtained from the method using Green's functions (provided sufficiently complicated curvilinear forms of the domains of the inverse problem solution are considered).

A fictitious boundary method is described, which is designed to reduce an inverse problem with moving boundaries of the body to a problem with fixed boundaries.

Based on the principle of *natural step regularization*, an effort is made in **Chapter 4** to construct economic (in terms of machine time expense) recurrent algorithms of computing temperatures and heat fluxes at the boundary of the body. On the basis of these algorithms lie the approximate solutions of IHCP integral forms in linear formulation for one-dimensional and two-dimensional cases. Numerical stability of this direct method is investigated in detail and conditions of step regularization are established for a one-dimensional case. Physical considerations are given explaining the results obtained and recommendations are made for practical utilization of the algorithms constructed.

Chapter 5 is devoted to a study of the application of various *explicit and implicit finite-difference schemes* of numerical continuation of the temperature field when using a step regularization process of computations. The appropriate difference algorithms of non-linear IHCP solution are constructed and a study is made on their regularized and exact characteristics. A property of *scheme viscosity* is exposed, which can be used to increase the algorithm computing stability. Physical and mathematical aspects of this phenomenon are considered. At the end of the chapter a question is studied on the possible *artificial hyperbolization* of a heat conduction equation in solving boundary IHCP. Such a technique allows increased stability of a numerical solution of the problem.

Direct (approximate-analytical and numerical) methods, described in Chaps. 4 and 5, are checked on solutions of methodical examples. Expediency is established here for use of these methods in combination with a preliminary *smoothing* of temperature measurements. A correct option for smoothing algorithms is of great importance here.

It appears that the use for this purpose of algorithms of least-squares type, in particular for the smoothing procedure with the help of a sliding five-point picture (using the method of fourth differences), does not generally guarantee fine recovery quality of heat loads. At the same time, the smoothing of temperature data with the use of regularizing algorithms of the second order of smoothness can provide acceptable results of IHCP solution with the help of the direct methods considered.

Chapter 6 is an introduction to *the iterative regularization method*, when the stability of the inverse problem solution is achieved by matching the number of iterations in the gradient methods of the first order with the error of initial data. At a heuristic level of rigour the concepts are analyzed, which lie on the basis of iterative regularization. With the frames of this approach the algorithms of parametric and functional optimization are constructed for solving boundary IHCP in one-dimensional formulation and a detailed numerical simulation is carried out of characteristics with respect to stability and accuracy of any unknown results. Based upon computational experiment a legitimacy is established for a discontinuation of iterations using the *residual criterion* for disturbed input data. A possibility is found to utilize standard techniques (elaborated for solving well-posed problems) to halt iterative processes and restart the iterations, provided that the results of temperature measurements have been smoothed in advance. At the end of the chapter a gradient algorithm is constructed to solve the boundary IHCP in the case of a two-dimensional heat propagation.

The methods in Chaps. 4 to 6 do not assume the possibility of introducing any special constraints for smoothness on the unknown functions under reconstruction. However, in many cases it is noted that an unknown causal characteristic of the heat transfer process has a certain degree of smoothness. This important a priori information should be taken into account in problem solving, in particular, when studying high-dynamic and short-term heat actions on structural elements. As a result, one can noticeably increase the accuracy of computations. Methods given in Chaps. 7 and 8 enable this question to be solved.

Chapter 7 is devoted to the development of *regularized procedures of solving* boundary IHCP, in which a condition of the assumed order of smoothness of restored heat loads is taken into account. The study of this question begins with a general formulation of the problem on solution regularization of integral equations of the first kind with a variable upper limit, many one-dimensional inverse problems for a heat conduction equation with constant coefficients being reduced to these equations. Based on the method using Lagrangian multipliers, a smoothing functional is constructed and in order to find its extreme the Euler equation is derived. A regularized computational procedure is obtained from a solution of the problem on minimization of a finite-dimensional analogue of the smoothing functional with stabilizers of the first and second order. Algebraic forms of IHCP solution obtained in Chap. 4 are utilized here. To choose a parameter of regularization the *residual principle* and *quasioptimal parameter technique* is recommended. The workability and efficiency of the approach discussed is supported by an extensive computational experiment. Regularization of numerical algorithms for a solution of non-linear IHCP is proposed using two methods. The first utilizes a sequential linearization of the problem at several time intervals and then a finite-dimensional regularization procedure of the first order is constructed for each interval. This method is known as *by-interval regularization.* The second considers regularization of a numerical solution of the inverse problem using a Cauchy statement, which is performed by introducing *a special system of regularized residuals* corresponding to various spatial layers of the difference grid. The chapter ends with a description of solution regularization of a two-dimensional IHCP, represented in the form of an integral equation of the first order.

Further development of *an iterative regularization* is outlined in Chapter 8. At the beginning rigorous mathematical results are stated in brief, substantiating this approach with the most widely spread gradient algorithms – the *steepest descent and the conjugate gradients.* Namely, information is presented on the regularization of pointed methods by the number of iterations and legitimate use of a residual criterion to choose the number of the last iteration or restart the iterative process. The following three questions are also considered

- the computation of a gradient residual for a generalized IHCP statement, including all the main types of heat conduction inverse problems;
- the consideration of qualitative and quantitative a priori information about the solution;
- the search for other methods for a regularizing halt of iterative processes, which do not demand immediate knowledge of the value of the input data error. Regularized gradient methods are considered in the chapter for solving both linear and non-linear, one-dimensional and two-dimensional inverse problems. Examples are presented for constructing concrete algorithms with respect to boundary and coefficient IHCP as well as the results of computational experiments.

In conclusion there is a comparative qualitative analysis of the main methods of IHCP solution considered in the book and a decision made of the best universal application of the iterative regularization method.

1.6 Summary

1. Based on the character of causality-consequence relationships, general formulations of heat transfer inverse problems are given and a classification is suggested. Typical inverse problems of heat conduction, radiation and combined heat transfer are considered.
2. An analysis is carried out of some areas of application of the methods based on the solution of inverse problems for the investigation of heat transfer processes and the design and development of heat-loaded structures. It is shown that these methods are an effective means of diagnostics and identification of heat transfer processes, thus increasing information validity and verification of the thermophysical investigations.

Chapter 2

Analysis of Statements and Solution Methods for Inverse Heat Transfer Problems

Introduction

The effective application of methods based on solving the inverse problems in thermal simulation and in processing the results of thermal tests is determined by the depth of the mathematics required connected with the statement and algorithmic presentation of the problems, by clarifying the specific difficulties in their solution. Misunderstanding of the nature of problems that are poorly based can lead to errors in problem solving. Even in cases when a proper method is used its effective application and specific features of the problem may not be fully realized. Such negligence of the formulation of problems of a given class as well as the methods of their solving can lead to doubt as to the suitability of the very concept of inverse problems for practical research. Experience shows that it is only through understanding the physical, technical and mathematical principles of a given problem that one finds it possible to use inverse problem methods effectively and creatively in thermal research.

This chapter deals with the problems of existence, uniqueness and stability of solutions for inverse heat transfer problems. General concepts of well-posed and ill-posed problems are introduced in Sect. 2.1, based upon the Hadamard definition. To analyze the stability conditions of boundary IHCP and then solve them, three possible formulations of these problems are outlined: *a boundary-value formulation, the Cauchy formulation and a variational formulation.* Such a systematization will help us to come to a selection of either form of the inverse problem shown depending on the conditions of its possible practical use and solution specifications. The given classification covers almost all problem formulations known at present on restoration of boundary conditions. Following the proposed classification, refer to the first solutions of boundary IHCP made by Shumakov [150], Mirsepassi [192] and Stolz [203] for a boundary-value formulation. Methods of IHCP solution in variational formulation were first developed by Frank[1] and Beck [158]. The solution of Tiomkin [132] and Burggraf [169] corresponded to a Cauchy formulation. However none of these three formulations were stable. Mathematical causes and the physical nature of

[1] I. Frank: An application of least squares method to the solution of the inverse problem of heat condition. *J. Heat Transfer*, 1963, **85c**.

the instability of a boundary IHCP are now discussed. Similarly, solution instability occurs, as follows from the given analysis, for other IHCP – coefficient, retrospective and geometric and for an inverse problem of heat transfer in the engineering system. Two examples of IHCP formulation are given in Sect. 2.2, demonstrating possible violation of *the condition of existence* of a solution for no accurate input data.

Section 2.3 contains an analysis on *the conditions of the uniqueness* of the boundary IHCP solutions and references to the works which treat the problem of uniqueness of unknown functions for other types of IHCP. For a boundary IHCP with constant coefficients there is a proof of uniqueness of solution performed without any complicated mathematical constructions. Where coefficients depend on the coordinate and time, this property is shown to be rigorous. Works are also mentioned where the reader can find exact mathematical results on this question. As for coefficient IHCP, most typical situations that occur are studied for uniqueness of solution seen in the works of Muzylev [106, 107], Klibanov [77, 78], Alifanov and Klibanov [32], Alifanov, Balakovskii and Klibanov.[2] It is shown in Sect. 2.4 that in the case of linear boundary IHCP, the degree of solution instability (in other words, the problem conditionality) depends on the kind of kernel of a corresponding integral equation which for bodies with fixed boundaries is defined by Green's function. A so-called *pseudoinverse problem* (when the temperature measurements are made at the boundary of the body) has the best conditionality. It is shown that such a formulation of IHCP is well-posed at certain, not too onerous for practice conditions.

To solve unstable inverse problems we can use two approaches: reduction to *a conditionally well-posed formulation* and *problem regularization*. The first approach assumes contraction (considerable, at times) of a set of acceptable solutions to differing well-based levels, as explained in Sect. 2.5, where for illustration we offer simple examples showing apparent separation of these levels based upon information concerning the problem known beforehand. In most cases in practice this method is difficult to realize due to lack of such a priori information. Hence, later on in the book, we study various principles and regularization methods for solving ill-posed problems which can be divided into *natural regularization* (*self-regularization*) and *regularization by Tikhonov* (Sect. 2.6.). In the first case these methods used for IHCP solution were known as "*direct*". *Step regularization* is most often used here, when a value of the approximate step in time is chosen according to the type and level of errors in the initial data. Sources of natural regularization are described; they can be conditioned by stabilization of the body heating conditions and/or by viscous properties of computational procedures. The main directions and specifications for practical use of direct methods are summed up. Section 2.6 ends with

[2] Alifanov, O.M.; Balakovskii, S.L.; Klibanov, M.V.; Reconstruction of causal characteristics of the thermal conductivity process from the solution of the combined inverse problem. J. Eng. Phys., 1987, Vol. 52, No. 5, pp. 617–622.

a description of the general regularization method suggested by Tikhonov which can be applied to a solution of a wide range of problems. One of its major varieties – *a variational method* – is specially outlined. It is also pointed out that lately a new regularization method – *an iterative regularization* – has been considerably developed and found practical application as elaborated upon by the author and his colleagues.

2.1 Inverse Problems Formulation and Stability of Their Solution

2.1.1 General Mathematical Formulation

The general statement of inverse heat transfer problems has been introduced in Sect. 1.3. It assumes the following formal mathematical representation.

Let u and f denote the desired and observable characteristics of a heat model and be treated as elements of metric spaces U and F, respectively: u and f can be some complex formations.

Operator is assigned as A acting from U into F setting causality between unknown characteristics of a model and input data. The domain of operator definition $D_A \subseteq U$ and the range of operator $Q_A = A(D_A) \subseteq F$. The following operator equation of the first kind can be then written

$$Au = f. \tag{2.1}$$

Problem (2.1) is called *well-posed* if its solution meets the following requirements (the Hadamard conditions [180, 88, 140]):

1 it exists for any $f \in Q_A = F$ (*solvability condition*);

2 it is the only one in U (*uniqueness condition*);

3 and it is continuously dependent of f (*stability condition*).

In other words, the problem of solving Eq. (2.1) is well-posed when there exists a unique and continuous inverse mapping A^{-1}, the domain of which is $D_{A^{-1}} = Q_A = F$.

If one of the above requirements is violated problem (2.1) becomes *ill-posed*. This is the very situation which is observed in solving inverse heat conduction problems.

In spite of the seemingly obvious nature of problems to define causal characteristics of heat conduction processes it is nevertheless easy to arrive at a wrong mathematical statement if one follows the formal process described above.

Let us first study this singularity for a boundary IHCP. In order to avoid too many technical details a simple problem is considered which is widespread in practice in thermal research.

The process of heat conduction on a slab is described by the equation

$$\frac{\partial T}{\partial \tau} = a\frac{\partial^2 T}{\partial x^2}, \quad 0 < x < b,\ 0 < \tau \leqslant \tau_m, \tag{2.2}$$

where $a = \text{const} > 0$.

The initial temperature distribution is known as

$$T(x, 0) = \varphi(x), \quad 0 \leqslant x \leqslant b. \tag{2.3}$$

At the point $x = d$ $(0 < d \leqslant b)$ the temperature time history is preset

$$T(d, \tau) = T^*(\tau), \quad 0 \leqslant \tau \leqslant \tau_m. \tag{2.4}$$

On the right boundary $x = b$ is the condition of the second kind known

$$-\lambda \frac{\partial T(b, \tau)}{\partial x} = q^*(\tau), \quad 0 \leqslant \tau \leqslant \tau_m, \tag{2.5}$$

where $\lambda = \text{const} > 0$.

The determination of the heat flux on the left boundary $q(\tau) = -\lambda \partial T(0, \tau)/\partial x$ is required.

In the first place, let us analyse the third condition of correctness, i.e. the stability of the given problem, assuming that the existence and uniqueness of solution have taken place. As a preliminary, let us specify the computational statements of the boundary IHCP. For this, take three possible cases: *a boundary-value statement*, *a statement in the Cauchy form*, and *a variational statement of the problem* [12]. They will be the subject of discussion in Sects. 2.1.2, 2.1.3, and 2.1.4, respectively.

2.1.2 Boundary-Value Statement of a Boundary Inverse Heat Conduction Problem

Assume an algorithm is available for solving a corresponding direct problem $f = Au$; a method for its "inversion" has been found to establish an unknown relation: input data → boundary conditions $(u = R(f))$.

A solution of the direct problem for Eq. (2.2), according to superposition principles and Duhamel integral [143, 172], can be presented in the form

$$T(x, \tau) = \int_0^\tau q(\xi) \frac{\partial \vartheta(x, \tau - \xi)}{\partial \tau} d\xi + \int_0^\tau q^*(\xi) \frac{\partial \vartheta(b - x, \tau - \xi)}{\partial \tau} d\xi, \tag{2.6}$$

where $\vartheta(x, \tau)$ is the temperature response of the body to a unit-equal heat flux at one of the boundaries $(x = 0)$, provided that the other one $(x = b)$ is thermally insulated.

Value $\vartheta(x, \tau)$ can be calculated by the formula

$$\vartheta(x, \tau) = \frac{1}{\lambda} \left\{ \frac{a\tau}{b} + \frac{3(b - x)^2 - b^2}{6b} + \frac{2b}{\pi^2} \sum_{k=1}^{\infty} \frac{(-1)^{k+1}}{k^2} \exp\left[-k^2 \pi^2 \frac{a\tau}{b^2} \right] \cos\left(k\pi \frac{b - x}{b} \right) \right\}. \tag{2.7}$$

Thus, the inverse problem in this case is reduced to the next integral form

$$Au \equiv \int_0^\tau u(\xi)K(\tau-\xi)\mathrm{d}\xi = f(\tau), \quad 0 < \tau \leqslant \tau_m, \tag{2.8}$$

where

$$u(\tau) = q(\tau); \quad K(\tau-\xi) = \frac{\partial \vartheta(d, \tau-\xi)}{\partial \tau};$$

$$f(\tau) = T^*(\tau) - \int_0^\tau q^*(\xi)\frac{\partial \vartheta(b-d, \tau-\xi)}{\partial \tau}\mathrm{d}\xi .$$

The integral Eq. (2.8) has a varying upper limit τ and, by definition, is the Volterra equation with respect to an unknown function $u(\tau)$. Since $u(\tau)$ is only kept under the symbol of an integral such an equation is called an equation of the first kind. A more widespread and simple method of its solution may be reduced to an equation of the second kind (with an unknown function outside the integral) by performing a differentiation operation

$$u(\tau) + \frac{1}{K(0)}\int_0^\tau u(\xi)\frac{\partial K(\tau-\xi)}{\partial \tau}\mathrm{d}\xi = \frac{f'(\tau)}{K(0)} .$$

Such a representation, however, is only possible when the kernel of the original equation $K(\tau)$ is not equal to zero anywhere in the segment $[0, \tau_m]$. In Eq. (2.8), $K(\tau-\xi) \to 0$, $\xi \to \tau$, and therefore the method cannot be used. Note, repeated differentiation does not yield the result desired and may incur great loss of accuracy.

Therefore, we must primarily solve the original integral equation of the first kind. The solution will be a typically unstable problem. This fact clearly demands illustration. In the example that follows let $u(\tau)$ be a unique solution of an integral equation. Take the function

$$u_\delta(\tau) = u(\tau) + \Delta u(\tau) \tag{2.9}$$

where $\Delta u(\tau) = c\sin\omega\tau$. C, ω are any fixed numbers here.

Assume $u(\tau) \in U$ and $u_\delta(\tau) \in U$. With large frequency ω for integral values we have

$$\int_0^\tau u_\delta(\xi)K(\tau-\xi)\mathrm{d}\xi \to \int_0^\tau u(\xi)K(\tau-\xi)\mathrm{d}\xi .$$

At the same time the values of functions $u_\delta(\tau)$ and $u(\tau)$ can differ greatly, when C is large enough, i.e. when, however small, the deviations of the right part $f(\tau)$ are able to correspond to however large the values of an unknown function. Or, in other words, the operator A does not have any limited inverse, and the third condition of correctness is not observed. It is possible to show that a similar situation appears when solving some other inverse problems in the boundary-value statement.

2.1.3 Boundary Inverse Heat Conduction Problem in the Cauchy Statement

Provided the number of spatial variables is more than one, let us search for a continuation of the heat conduction equation solution from the $\partial\Omega_1$ boundary of a certain small domain with the given temperature and heat flux (under fourth kind conditions – the Cauchy data) up to the boundary $\partial\Omega_2$ (with an unknown condition) of a large domain (Fig. 2.1). When applied to a one-dimensional problem (2.2)–(2.5) the Cauchy statement corresponds to continuation of Eq. (2.2) solution up to the $x = 0$ boundary with known conditions at a point $x = d_1\,(d \leqslant d_1 \leqslant b)$:

$$T(d_1, \tau) = T_1^*(\tau); \qquad -\lambda\frac{\partial T(d_1,\tau)}{\partial x} = q_1^*(\tau). \tag{2.10}$$

In this case we have to calculate in advance the initial data at a point d_1 as a result of direct problem solving in the domain $d \leqslant x \leqslant b$, $0 \leqslant \tau \leqslant \tau_m$.

Suppose that functions $T_1^*(\tau)$ and $q_1^*(\tau)$ are differentiable for an infinite number of times

$$\left(T_1^{*(n)} \equiv \frac{\mathrm{d}^n T_1^*}{\mathrm{d}\tau^n},\quad q_1^{*(n)} \equiv \frac{\mathrm{d}^n q_1^*}{\mathrm{d}\tau^n},\ n \to \infty\right)$$

Then, we can draw an exact solution for temperature field determination. The solution was obtained for the first time by Stephan [202, 188]

$$T(x,\tau) = \sum_{n=0}^{\infty}\left[\frac{T_1^{*(n)}(\tau)(x-d_1)^{2n}}{(2n)!\,a^n} - \frac{q_1^{*(n)}(\tau)(x-d_1)^{2n+1}}{(2n+1)!\,a^n\lambda}\right]. \tag{2.11}$$

Differentiating expression (2.11) with respect to x and confining ourselves with the first N numbers, we may write a finite expression for the heat flow rate at the body boundary with $x = 0$:

$$q(\tau) = \sum_{n=0}^{N-1}\frac{q_1^{*(n)}(\tau)d_1^{2n}}{(2n)!\,a^n} + \lambda\sum_{n=1}^{N}\frac{T_1^{*(n)}(\tau)d_1^{2n-1}}{(2n-1)!\,a^n}. \tag{2.12}$$

For the body surface temperature $T_w(\tau) = T(0,\tau)$:

$$T_W(\tau) = \sum_{n=0}^{N-1}\left[\frac{T_1^{*(n)}(\tau)d_1^{2n}}{(2n)!\,a^n} + \frac{q_1^{*(n)}(\tau)d_1^{2n+1}}{(2n+1)!\,a^n\lambda}\right]. \tag{2.13}$$

When applied to the IHCP solution refer to [132, 133, 169] in which a similar procedure has been suggested.

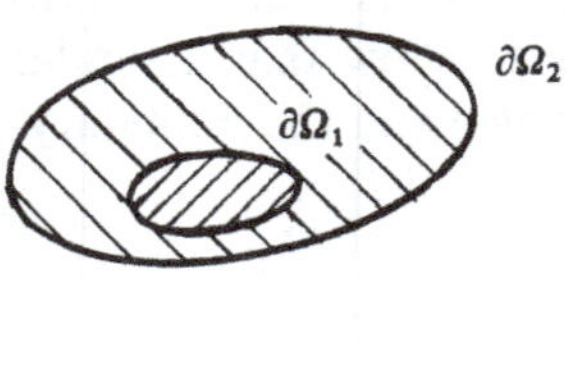

Fig. 2.1. The domain of an IHCP solution

Thus, in determining the boundary conditions in accordance with the given form of IHCP solution, it becomes necessary to compute values of derivatives up to the N-th order. At the same time, the $d/d\tau$ differentiation operator is a typical example of an unbounded operator whose computation in itself is an ill-posed problem. Let us explain the above by the following example. Assume that an input function (temperature or heat flux $f(\tau)$), apart from a useful (true) signal $\bar{f}(\tau)$ with a $\bar{f}'$ derivative contains an additional error component $\delta(\tau)$

$$f(\tau) = \bar{f}(\tau) + \delta(\tau),$$

where $\delta(\tau) = \frac{1}{\sqrt{\omega}} \sin \omega\tau, \omega > 0$, in this case with $\omega \to \infty$ $f(\tau) \to \bar{f}(\tau)$ but $|f'| \to \infty$.

For practical calculations, the loss in accuracy by numerical differentiation will become greater with the increase of the order of any derivative being computed.

Thus, any solution for an inverse problem using the Cauchy statement may be aligned with instable results.

2.1.4 Variational Statement of an Inverse Heat Conduction Problem

Let us define the objective functional as a residual $\rho_F(Au, f)$ corresponding to a deviation of the left side of Eq. (2.1) from the right side using the metric of the F space. At first sight, it may seem that in order to obtain unknown functions one should resort to ρ_F minimization with no additional conditions. With numerical realization, however, such an approach can cause serious difficulties. On the one hand, it can be explained by the fact that from a functional convergence no convergence of approximated solutions necessarily follows and, on the other hand, one usually has to solve a problem with inaccurate data. A "solution" thus obtained may differ considerably from a real one.

Such an event is a result of the fact that $Au = f$ problem is ill-posed. The conditionality[3] of a variational problem in many cases turns out to be worse than that of an initial problem. For instance, when minimizing a quadratic functional one can take advantage of the Euler equation

$$A^* Au - A^* f = 0, \tag{2.14}$$

formally an equivalent to the initial problem.

It is known, however, that Eq. (2.14) has much worse conditionality than $Au - f = 0$ because the A^*A operator provides stronger smoothing properties than operator A.

For Eq. (2.8) and normalization corresponding to the space L_2 (the Hilbert space of real-valued square integrable functions) a residual functional has the

[3] Property characterizing the degree of the effect of errors in the operator and in the right side of the equation on the solution.

form

$$\rho_{L_2}^2 = \int_0^{\tau_m} \left[\int_0^{\tau} u(\xi) K(\tau - \xi) \mathrm{d}\xi - f(\tau) \right]^2 \mathrm{d}\tau .$$

It is possible to show (see Sect. 7.2) that its minimization is equivalent to the solution of a Fredholm equation of the first kind (the limits of integration are unchangeable, the $u(\tau)$ unknown function only stands when using the symbol of integral)

$$\int_0^{\tau_m} u(\zeta) \bar{K}(\xi, \zeta) \mathrm{d}\zeta = \bar{b}(\xi), \quad 0 \leqslant \xi \leqslant \tau_m ,$$

where

$$\bar{K}(\xi, \zeta) = \begin{cases} \int_{\xi}^{\tau_m} K(\tau - \xi) K(\tau - \zeta) \mathrm{d}\tau, & 0 \leqslant \zeta < \xi ; \\ \int_{\zeta}^{\tau_m} K(\tau - \xi) K(\tau - \zeta) \mathrm{d}\tau, & \xi \leqslant \zeta \leqslant \tau_m ; \end{cases}$$

$$\bar{b}(\xi) = \int_{\xi}^{\tau_m} f(\tau) K(\tau - \xi) \mathrm{d}\tau .$$

Deliberate incorrectness of such a problem has been demonstrated as earlier for the Volterra equation of the first kind.

2.1.5 Other Examples of Unstable Inverse Problems

Despite the above considerations of IHCP for one particular case of one-dimensional body heating, a conclusion about the violation of the third condition of correctness can be extended to other problems.

Let us consider a two-dimensional boundary problem for a homogeneous heat conduction equation with constant coefficients assigned in the rectangular domain of the spatial variables. Assume three domain boundaries ($x = 0$, $x = b$, $y = d$) as thermally insulated and on the fourth ($y = 0$) there operates an unknown heat flux $q(x, \tau)$. Assigned additionally is temperature distribution $T^*(x, \tau) = T(x, y^*, \tau)$ along the line $y = y^* = \text{const}$, $0 < y^* \leqslant d$.

The integral representation of the boundary-value statement of this IHCP is in the form of

$$\int_0^{\tau} \mathrm{d}\xi \int_0^{d} q(x', \xi) K(x, y^*; x', 0; \tau - \xi) \mathrm{d}x' = T^*(x, \tau) - T_0 , \qquad (2.15)$$

where $T_0 = \text{const}$ is the initial temperature of the body and $(\lambda/a) K(x, y; x', y'; \tau - \xi)$ is the Green's function of the given boundary-value problem.

Equation (2.15) is a two-dimensional integral equation of the first kind with a Volterra behaviour in τ and Fredholm in x. Its solution is unstable due to errors of input data and approximation.

Let us prove next that transition to a stationary inverse problem does not generally make it well-posed. It is required to find a solution for the stationary heat conduction equation

$$\frac{\partial^2 T}{\partial x^2} + \frac{\partial^2 T}{\partial y^2} = 0 \tag{2.16}$$

in the semi-band $y > 0$, $-\frac{\pi}{2} \leqslant x \leqslant \frac{\pi}{2}$ satisfying the conditions

$$T\left(-\frac{\pi}{2}, y\right) = T\left(\frac{\pi}{2}, y\right) = T(x, 0) = 0; \quad \frac{\partial T(x, 0)}{\partial y} = \varphi(x),$$
$$\varphi\left(-\frac{\pi}{2}\right) = \varphi\left(\frac{\pi}{2}\right) = 0, \tag{2.17}$$

where $T = T(x, y)$ is a twice continuously differentiable function of the x and y variables. Physically, such a statement means that three boundary planes of a solid body $x = \pm\pi/2$, $y = 0$ are being kept at zero temperature and a steady-state heat flux of $-\lambda\varphi(x)$ density acts on $y = 0$ boundary. The initial temperature of the body is zero. With these data it is necessary to reconstruct the temperature field inside the domain under study.

Setting $\varphi(x) = 0$, the only solution of the given problem will be the identity $T(x, y) = 0$. Suppose now that $\varphi(x) = \exp[-\sqrt{2n+1}] \times \cos[(2n+1)x]$. It is possible to show that in this case the solution is

$$T(x, y) = \frac{1}{2n+1}\exp[-\sqrt{2n+1}]\cos[(2n+1)x]\,\mathrm{sh}[(2n+1)y].$$

Hence, with $y \neq 0$ $\lim_{n\to\infty} |T| = \infty$. But $\lim_{n\to\infty} \varphi = 0$. Therefore, a solution of the problem in (2.16)–(2.17) can differ greatly from a zero solution.

The same result is obtained if the problem is treated in a somewhat different way using the following Cauchy conditions on the $y = 0$ boundary: the heat flow is equal to zero (the isolated wall, $\partial T(x, 0)/\partial y = 0$), the temperature $T(x, 0) = p(x)$. Note that the superposition of the solutions to these two problems allows a solution of the inverse problem for the Laplace equation (2.16) to be written in the case of non-zero conditions: $T(x, 0) = p(x)$ and $\partial T(x, 0)/\partial y = \varphi(x)$. Any of these problems can be reduced to the Fredholm integral equation of the first kind. So, for a second problem, we have (a solution $T(x, y^*)$ corresponding to $y = y^*$ being sought)

$$Au \equiv \int_{-\frac{\pi}{2}}^{\frac{\pi}{2}} u(x')K(x, x', y^*)\mathrm{d}x' = f(x), \quad -\frac{\pi}{2} \leqslant x \leqslant \frac{\pi}{2}, \tag{2.18}$$

where $K(x, x', y^*) = \sum_{j=0}^{\infty} ch^{-1}(jy^*)\sin\left[j\left(x + \frac{\pi}{2}\right)\right]\sin\left[x' + \frac{\pi}{2}\right]$;

$$u(x') = T(x', y^*); \quad f(x) = p(x).$$

The solution of Eq. (2.18) with respect to u does not continuously depend on the right side f.

Let us sum up the results of the analysis performed. *If a boundary inverse problem of heat conduction is formulated similarly to a direct one, with no special constraints, then we could not guarantee a continuous dependence of results on input data.* In its initial mathematical statement, the inverse problem is ill-posed by Hadamard (i.e. it is unstable).

2.1.6 Physical Explanation of Boundary Inverse Problem Instability

In its background the physical nature of inverse problem instability contains a property of heat conduction process which is characterized by the strong smoothing and time lagging of inherent features of boundary functions while the point studied moves inside the body from the heat exchange surface. For example, if the heat exchange surface of a thermally thick wall (a semi-infinite body model) has a temperature varying in time according to the harmonic law. Then, while penetrating inside the body considerable temperature equalization occurs as in the decrease of amplitude oscillations and the time shift (lag) of temperature waves. Let the temperature on the boundary be $T_w = \bar{T}_w - T_m \cos \omega\tau$, where $\bar{T}_w$ is the mean value of surface temperature, T_m are the temperature oscillation amplitude and ω is the frequency. Then, the temperature field in a thermally thick wall is determined by an expression [93, 172]

$$T(x, \tau) = \bar{T}_w + T_m \exp[-\sqrt{\omega/2a}\,x] \cos(\omega\tau - \sqrt{\omega/2a}\,x),$$

from which it is evident that temperature oscillation amplitude exponentially decrease while moving deeper inside the body (Fig. 2.2). At some distance from the heat exchange surface a complete temperature oscillation damping takes place. Two domains can be seen along the coordinate: I and II (see Fig. 2.2). If a temperature meter is set in domain I, then a fundamental possibility to reconstruct the boundary condition exists. There is no such a possibility if

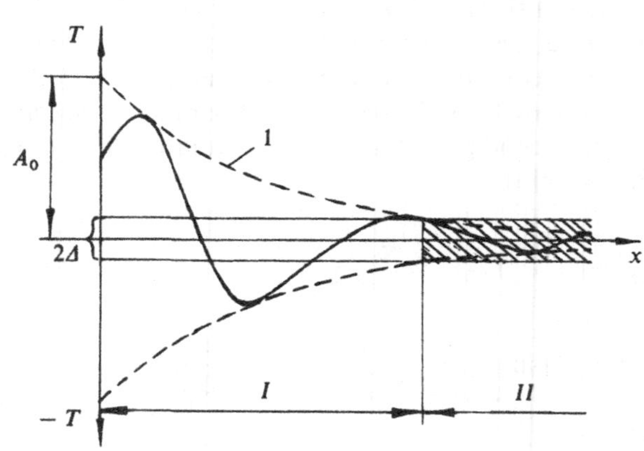

Fig. 2.2. Temperature curve change in depth of a thermally thick wall at a fixed moment of time: 1 – an envelope of temperature limiting values $A \exp[-x\sqrt{\omega/2a}]$

a temperature-sensitive element is shifted into the shaded domain II. Location of the boundary between these domains depends on the thermal diffusivity of the body material and sensitivity threshold Δ of a measuring system, and hence it is impossible to distinguish the signal value from zero if this value differs from zero less than the threshold itself.

If variations inherent in boundary conditions are less apparent and they do smooth while moving from the body surface, then, on the contrary, occurrence of even small oscillations in the temperature of distant points must correspond to considerable time variations of the boundary condition.

Such physics of heat propagation lead to a known "pathological" singularity of inverse problems, i.e. to considerably greater errors in solution compared with those from input data. As a result, small f deviations will lead to large deviations of the desired solution u.

As measuring and registering of input data may be performed using inaccurate technical equipment, so the data obtained may be approximate, and an IHCP solution corresponding to input data that has already been measured can differ greatly from a true one and may carry a pronounced oscillating nature.

The representation of initial operator A through an approximating operator A_n (with h approximation accuracy) as well as the errors of rounding-off in calculation play quite a definite role in distorting results. If an inverse problem is solved by an analog computer additional errors from operational units and different noise may be introduced.

Thus, methods for solving boundary inverse problems must take into account the high sensitivity of results to various kind of errors. Otherwise, it is easy to have a solution which will be very far from true.

2.1.7 On Stability of Coefficient Inverse Heat Conduction Problems

Consider as an example the following problem. Let the process of heat transfer in region $0 < x < b$, $0 < \tau \leqslant \tau_m$ be described by a heat conduction equation

$$C(T)\frac{\partial T}{\partial \tau} = \frac{\partial}{\partial x}\left(\lambda(T)\frac{\partial T}{\partial x}\right). \tag{2.19}$$

and it is known that thermal diffusivity $a = \dfrac{\lambda}{C}$ does not change with temperature.[4] It is required, by knowing the conditions of the second boundary-value problem $T(x, 0) = 0$, $-\lambda\,\partial T(0,\tau)/\partial x = q_1(\tau)$, $-\lambda\,\partial T(b,\tau)/\partial x = q_2(\tau)$ and measuring the temperature $T(d,\tau)$ at a certain point $x = d$, $0 \leqslant d \leqslant b$, to find the dependence of the thermal conductivity on temperature $\lambda(T)$ provided that the a magnitude is considered given.

[4] This condition with an accuracy that may be adjusted in practice is seen in many metallic and non-metallic materials (vanadium, molybdenum, nickel, niobium, rhenium, tantalum, certain brands of graphite and fused quartz, etc.) within sufficiently wide temperature intervals.

To solve this problem, one can set the following procedure. Use the Kirchoff substitution

$$\theta(x, \tau) = \frac{1}{\lambda_0} \int_0^{T(x, \tau)} \lambda(T) \mathrm{d}T , \tag{2.20}$$

where λ_0 is a certain constant value; re-write Eq. (2.19) with regard for $a = \text{const}$:

$$\frac{\partial \theta}{\partial \tau} = a \frac{\partial^2 \theta}{\partial x^2} . \tag{2.21}$$

The boundary conditions in terms of a new dependent variable will acquire the form

$$\theta(x, 0) = 0; \quad -\lambda_0 \frac{\partial \theta(0, \tau)}{\partial x} = q_1(\tau); \quad -\lambda_0 \frac{\partial \theta(b, \tau)}{\partial x} = q_2(\tau) . \tag{2.22}$$

Thus, by the known functions $q_1(\tau)$ and $q_2(\tau)$, we can calculate the time change of the "model" temperature θ at the point $x = d$. Since the coefficients are constant in the problem given by Eqs. (2.21) and (2.22) we can make use of the solution (2.6)

$$\theta(d, \tau) = \int_0^{\tau} q_1(\xi) \frac{\partial \vartheta(d, \tau - \xi)}{\partial \tau} \mathrm{d}\xi + \int_0^{\tau} q_2(\xi) \frac{\partial \vartheta(b - d, \tau - \xi)}{\partial \tau} \mathrm{d}\xi .$$

Referring now to expression (2.20), to define $\lambda(T)$ we get the integral Volterra equation of the first kind with a unit-equal kernel

$$\int_0^T \lambda(t) \mathrm{d}t = f(T), \quad 0 \leqslant T \leqslant T_{\max}(d, \tau) , \tag{2.23}$$

where $f(T) = \lambda_0 \theta(T)$. Function $\theta(T)$ is obtained by excluding a variable from the dependences $T(d, \tau)$ and $\theta(d, \tau)$.

Thus, the $\lambda(T)$ determination from Eq. (2.23) is analogous to ill-posed differentiation problem (see Sect. 2.1.3)

$$\lambda(T) = \lambda_0 \frac{\partial \theta}{\partial T} + \lambda(0) .$$

This example has shown that even with such a relatively simple formulation the stability condition can be violated. Nevertheless, the degree of the instability condition for thermophysical properties (TPP) usually turns out to be less than that in the case with a boundary IHCP, and the transition to a well-posed statement can be performed by less restrictive conditions for the problem statement. It is shown in [71] that with a proper selection of functional spaces the coefficient problems appear to be correct and can be solved with the help of known numerical methods.

The identification of the source in a heat conduction equation from the stability viewpoint can present rather a complicated problem.

Example. Function $g(x)$ is being sought from the equation

$$\frac{\partial T}{\partial \tau} = a\frac{\partial^2 T}{\partial x^2} + s(\tau)g(x),$$

prescribed in the area $x > 0$, $0 < \tau < \tau_m$ under the following conditions:

$$T(x, 0) = T_0 = \text{const}, \quad x \geqslant 0; \qquad T(0, \tau) = \varphi(\tau), \quad 0 \leqslant \tau \leqslant \tau_m;$$

$$T(d, \tau) = \psi(\tau), \quad 0 \leqslant \tau \leqslant \tau_m; \qquad \frac{\partial T(\infty, \tau)}{\partial x} = 0, \quad 0 \leqslant \tau \leqslant \tau_m,$$

where $s(\tau)$, $\varphi(\tau)$ and $\psi(\tau)$ are known functions.

Using the Green's function, let us write a solution of the corresponding direct problem

$$T(x, \tau) = T_0 + \frac{x}{2}\sqrt{a\pi}\int_0^\tau \varphi(\xi)\frac{\exp[-x^2/4a(\tau-\xi)]}{(\tau-\xi)^{3/2}}\mathrm{d}\xi$$

$$+ \frac{1}{2\sqrt{a\pi}}\int_0^\infty g(x')\mathrm{d}x'$$

$$\times \int_0^\tau s(\xi)\frac{\exp[-(x-x')^2/4a(\tau-\xi)] - \exp[-(x+x')^2/4a(\tau-\xi)]}{\sqrt{\tau-\xi}}\mathrm{d}\xi .$$

Supposing that $x = d$ in the last formula obtain an integral equation of the first kind with respect to an unknown function $g(x)$

$$\int_0^\infty g(x')H(d, x'; \tau)\mathrm{d}x' = f(\tau), \quad 0 \leqslant \tau \leqslant \tau_m, \tag{2.24}$$

where

$$H(d, x'; \tau) = \frac{1}{2\sqrt{a\pi}}\int_0^\tau s(\xi)\frac{\exp\left[-\dfrac{(d-x')^2}{4a(\tau-\xi)}\right] - \exp\left[\dfrac{(d+x')^2}{4a(\tau-\xi)}\right]}{\sqrt{\tau-\xi}}\mathrm{d}\xi,$$

$$f(\tau) = \psi(\tau) - T_0 - \frac{d}{2\sqrt{a\pi}}\int_0^\tau \varphi(\xi)\frac{\exp\left[-\dfrac{d^2}{4a(\tau-\xi)}\right]}{(\tau-\xi)^{3/2}}\mathrm{d}\xi .$$

Generally the solution of Eq. (2.24) will be unstable regarding the errors in the assignment of the functions $\psi(\tau)$ and $\varphi(\tau)$. A similar result can also be obtained for other formulations of coefficient inverse problems as, for example, in a two-dimensional case.

2.1.8 Retrospective Heat Conduction Problem

Temperature field determination also presents ill-posed problem. Let us consider a solution of the heat conduction equation $\partial T/\partial \tau = \partial^2 T/\partial x^2$ in a rectangular area with negative time $\tau < 0$, $0 < x < \pi$ satisfying the conditions $T(0, \tau) = T(\pi, \tau) = 0$; $T(x, 0) = \psi(x)$.

For $\psi(x) = 0$, there is a solution $T(x, \tau) = 0$. If we suppose $\psi(x) = \exp[-\sqrt{n}]\sin nx$, then we can obtain a solution

$$T(x, \tau) = \exp[-\sqrt{n}]\exp[-n^2\tau]\sin nx\,.$$

It is evident that for any fixed $x = x^*$ and sufficiently large n the absolute values of $T(x^*, \tau)$, $\tau < 0$ function can be very large in spite of the input function $\psi(x)$ tending to zero. Thus, to infinitely small changes of input data $\psi(x)$ there are sufficiently large deviations in the solution. From such a sufficiently abstract example of a retrospective problem there follows a conclusion as to the irreversibility of a heat conduction problem with respect to time.

Consider now a problem that is more strict. Suppose heating of the body with finite thickness is described by a one-dimensional equation of heat conduction with TPP constants $\partial T/\partial\tau = a\partial^2 T/\partial x^2, 0 < x < b, \tau > 0$, the body boundaries being thermally insulated $\partial T(0, \tau)/\partial x = \partial T(b, \tau)/\partial x = 0$.

Thermal sensors are installed along the body thickness enabling necessary information to be acquired about the temperature distribution $T_\delta(x, \tau^*)$ at a certain moment of time $\tau = \tau^*$ with a definite degree of accuracy δ. Up to this moment, in the time period $0 \leqslant \tau < \tau^*$, the signals from the sensors for some reasons were not obtainable.

A problem arises as to continue the solution of a heat conduction equation towards the previous time moments $0 \leqslant \tau < \tau^*$. We shall seek temperature distribution at an initial moment of time $T(0, x) = \psi(x)$. If $\psi(x)$ function is obtained, then, by solving a direct problem of heat conduction, we may find a temperature field in the area $0 \leqslant x \leqslant b, 0 < \tau < \tau^*$.

We write the solution of the second boundary-value problem with zero boundary conditions and a non-zero initial distribution [93, 172]

$$T(x, \tau) = \int_0^b \psi(\eta) G(x, \eta, \tau)\,d\eta\,,$$

where

$$G(x, \eta, \tau) = \frac{1}{b}\left\{1 + 2\sum_{j=1}^{\infty} \cos\frac{j\pi\eta}{b}\cos\frac{j\pi x}{b}\exp\left[-\frac{a\pi^2 j^2}{b^2}\tau\right]\right\}.$$

From it we obtain an integral form of a retrospective inverse problem

$$\int_0^b \psi(\eta) G(x, \eta, \tau^*)\,d\eta = T_\delta(x, \tau^*), \quad 0 \leqslant x \leqslant \delta\,.$$

Thus, we have an integral equation of the first kind, i.e. an unstable problem.

2.1.9 Geometric Inverse Heat Conduction Problem

Let us follow a stability condition violation of a geometric problem in the case of a thermally thick wall heating, with a moving boundary acted upon by a heat flux $q(\tau)$:

$$\frac{\partial T}{\partial \tau} = a\frac{\partial^2 T}{\partial x^2}, \quad X(\tau) < x, \quad 0 < \tau \leqslant \tau_m;$$

$$T(x, 0) = 0, \quad X(0) \leqslant x;$$

$$-\lambda\frac{\partial T(X(\tau), \tau)}{\partial x} = q(\tau); \quad \frac{\partial T(\infty, \tau)}{\partial x} = 0, \quad 0 \leqslant \tau \leqslant \tau_m.$$

The law of boundary movement $X(\tau)$ is unknown. Only its initial position $X(0) = 0$ is given. It is known that $X(\tau)$ is a continuous differentiable function. Time dependence $T^*(\tau) = T(d, \tau)$ at a certain point $x = d$ is assigned as additional information.

Solution of a direct problem can be presented with the help of simple layer thermal potentials [113]:

$$T(x, \tau) = \frac{a}{2\sqrt{\pi}}\int_0^\tau v(\xi)\frac{\exp\left[-\frac{(x - X(\xi))^2}{4a(\tau - \xi)}\right]}{\sqrt{a(\tau - \xi)}}\mathrm{d}\xi,$$

where $v(\tau)$ is thermal potential density.

Function $v(\tau)$ is found from a solution of the Volterra integral equation of the second kind (see Sect. 3.1.2):

$$v(\tau) + \int_0^\tau v(\xi)\frac{X(\tau) - X(\xi)}{2va\pi(\tau - \xi)^3}\exp\left[-\frac{(X(\tau) - X(\xi))^2}{4a(\tau - \xi)}\right]\mathrm{d}\xi = \frac{2q(\tau)}{\lambda},$$

$$0 \leqslant \tau \leqslant \tau_m. \tag{2.25}$$

The problem (2.25) with a known function $X(\tau)$ is well-posed, its solution $v(\tau)$ exists and is a continuous function provided that $q(\tau)$ is continuous.

Determination of $X(\tau)$ through the results of internal temperature measurements $T^*(\tau)$ and by the known boundary condition $q(\tau)$ is reduced to a solution of the following system of integral equations with respect to $v(\tau)$ and $X(\tau)$:

$$\frac{a}{2\sqrt{\pi}}\int_0^\tau v(\xi)\frac{\exp[-(d - X(\xi))^2/4a(\tau - \xi)]}{\sqrt{a(\tau - \xi)}}\mathrm{d}\xi = T^*(\tau);$$

$$v(\tau) + \int_0^\tau v(\xi)\frac{X(\tau) - X(\xi)}{2\sqrt{a\pi(\tau - \xi)^3}}\exp\left[-\frac{(X(\tau) - X(\xi))^2}{4a(\tau - \xi)}\right]\mathrm{d}\xi = \frac{2q(\tau)}{\lambda},$$

$$0 \leqslant \tau \leqslant \tau_m.$$

These integral equations with respect to $X(\tau)$ are nonlinear equations of the first kind of Volterra type, i.e. the problem thus formulated is ill-posed.

2.1.10 Inverse Heat Transfer Problems in Engineering Systems

We consider now inverse heat transfer problems in technical systems (see Sect. 1.3.5) which, generally, lack the property of continuous dependence of their solutions relative to input data.

Consider a heat model consisting of L-elements and write Eq. (1.3) in the following form:

$$d\bar{T}(\tau)/d\tau = A[\bar{\alpha}, \bar{\beta}, \bar{T}(\tau)], \tag{2.26}$$

where $\bar{T}(\tau)$ is a vector-function of L-dimensionality, the components of which $T_l(\tau)$ represent temperature time dependences in the system's elements. A is an operator, non-linear in temperature, connecting the vectors of the known causal $\bar{\alpha}$ and desired $\bar{\beta}$ characteristics with the $\bar{T}(\tau)$ vector-function.

The components of the $\bar{\alpha}$ and $\bar{\beta}$ vectors are the parameters of the heat model

$$\lambda_{lj}/C_l, \quad K_{lj}/C_l, \quad Q_{sj}/C_l, \quad Q_l/C_l.$$

The results of temperature measurements $T_l^*(\tau)$ in K-elements ($K \leqslant L$) of model known to have some error are used as initial information.

To determine unknown causal characteristics it is natural to proceed to an extreme problem statement when a solution is sought from a condition of the minimum temperature residual. If a mean-square criterion is taken in the capacity of an incompatibility measure of Eq. (2.26), then we obtain the following problem:

$$\sum_{l=1}^{K} \int_0^{\tau_m} [T_l(\beta, \tau) - T_l^*(\tau)]^T [T_l(\beta, \tau) - T_l^*(\tau)] d\tau - \inf_{\bar{\beta} \in B \subseteq E_p},$$

where B is a domain of admissible values of the desired parameters in the p-dimensional Euclidean space E_p (with no constraints for desired values of parameters $B = E_p$), $T_l^*(\tau)$ is the temperature measured in the l-th element, τ_m is the duration of experiment and T is the transposition symbol.

The problem under study would be stable if its $\bar{\beta}$ solution converges to an exact β_T with an error of measurements tending to zero. There are, however, examples of practical calculations (see in particular, [101, 181]) performed using a method of least squares and the Kalmann recurrent linear estimation, where the results obtained are to the contrary, namely that small errors in measurements, approximation and round-off errors all give large deviations in a solution wanted.

Thus, in summary of the analysis performed in Sect. 2.1 we can make the following conclusion: *if classes of admissible solutions of inverse problems are not limited in some specific way, then their common bond is the absense of continuous dependence of desired results on input data.*

2.2 Existence of Inverse Problem Solutions

In the above, while analysing stability conditions, we have postulated the validity of the other two classical conditions of well-posed statements. Now, let us dwell on the question of the legitimacy of such a statement by analyzing a boundary inverse problem taken as an example.

From the standpoint of physical representation, it would be natural to believe that if a consequential response has taken place (in the given case, it is the temperature at a certain point of the body), then there exists an inducing cause (in the given case, we speak about a boundary-value condition). Assume that by mathematical description one has managed to find a differential model adequate to the actual heat transfer process, i.e. an operator A is specified, thus putting the function $f \in F$ in correspondence with any solution $u \in U$, where U and F are functional spaces. The question of existence of an inverse problem solution given in the form of an operator Eq. (2.1) includes a mathematical study of f belonging to a set of functions which is a domain of operator A values. That is why U and F spaces must be in agreement with each other. For instance, if a solution is sought in a certain given class of functions, then selection of f cannot be arbitrary.

Consider a heat conduction problem corresponding to semi-infinite body heating:

$$\partial T/\partial \tau = a\,\partial^2 T/\partial x^2, \quad x > 0,\ \tau > 0;\ T(x, 0) = 0,\ x \geqslant 0;$$

$$T(0, \tau) = T_w(\tau), \quad T(\infty, \tau) = 0,\ \tau \geqslant 0.$$

Let an experimental function $T^*(\tau) = T(d, \tau)$ be known at point $x = d > 0$; it is required to define the $T_w(\tau)$ function. Suppose that the form of a desired boundary condition is postulated as a polynomial of half-degree $T_w(\tau) = \sum_{k=0}^{n} a_k \tau^{k/2}$ and it is necessary to find a_k coefficients from an exact solution of the inverse problem. Such a solution, however, will be possible only in the case when the assigned temperature takes the form

$$T^*(\tau) = \sum_{k=0}^{n} a_k \tau^{k/2} \frac{i^k \operatorname{erfc}\left[\dfrac{d}{2\sqrt{a\tau}}\right]}{i^k \operatorname{erfc}[0]},$$

which follows from a direct problem solution given in Ref. [111].

Yet, if $T^*(\tau)$ is considered simply as a certain continuous function on the interval $[0, \tau_m]$, then an inverse problem thus formulated will not have any solution. As another and more general example, consider the integral equation (2.8) under the assumption that $q^* = 0$. Then,

$$Au \equiv \int_0^{\tau} u(\xi) K(\tau - \xi)\,d\xi = f(\tau), \quad 0 < \tau \leqslant \tau_m, \tag{2.27}$$

where $f(\tau) = T^*(\tau)$. $u(\tau) = q(\tau)$. We shall seek the solution of Eq. (2.27) in the class $L_2[0, \tau_m]$ of square integrable functions. The right-hand side is also considered as belonging to the $L_2[0, \tau_m]$ space.

It is known that any solution of a homogeneous heat conduction equation with constant coefficients and which is twice continuously differentiable for spatial coordinates and continuously differentiable with respect to time would have derivatives for time of up to any order. Therefore, the domain of values of the A operator in L_2 presents a set of all continuous and infinitely differentiable functions; any $q(\tau)$ function would be transformed by the kernel of Eq. (2.27) into one of the functions of this set. That is why it is not for every right-hand side $T^*(\tau) \in L_2$ that a solution of Eq. (2.27) can exist in L_2 space. Usually, instead of the exact function $T^*(\tau)$, only its approximation without a required smoothness is available and then an "exact" solution of Eq. (2.27) does not exist ($f \notin AU$).

A solution of an inverse problem can exist for one pair of spaces U and F and may not exist for another.

2.3 Uniqueness of Solution of Inverse Heat Conduction Problems

Consider an inverse problem that is solvable; it is required to study the aspect of uniqueness for its solution. The uniqueness of a solution for an inverse problem does not follow from the fact that a solution of the corresponding direct problem is unique. The study is in itself an independent and complex problem if applied to different inverse heat conduction problems. There are now sufficient results available in the mathematical literature; see [32, 52, 59, 60, 64, 71, 77, 78, 89, 106, 107, 146, 166, 170, 171, 190, 198].

Take as an example the procedure for obtaining uniqueness conditions for boundary IHCP. The question will be put separately for other types of inverse problems.

2.3.1 Heuristic Approach to the Question of Uniqueness of a Boundary Inverse Problem Solution

Let us first demonstrate the uniqueness of a boundary IHCP solution on a physical level. The problem will be taken as being linear but general, regarding, in particular, the mobility of the body boundaries, which is often the case in practice for thermal model testing.

Consider that a heat conduction process is described by an equation:

$$C(x,\tau)\frac{\partial T}{\partial \tau} = \frac{\partial}{\partial x}\left(\lambda(x,\tau)\frac{\partial T}{\partial x}\right) + S(x,\tau), \quad X_1(\tau) < x < X_3(\tau),$$

$$0 < \tau \leqslant \tau_m . \tag{2.28}$$

The initial and boundary conditions are

$$T(x,0) = \varphi(x), \quad X_1(0) \leqslant x \leqslant X_3(0),$$

$$-\lambda\frac{\partial T(X_1(\tau),\tau)}{\partial x} = q(\tau), \quad T(X_3(\tau),\tau) = f_3(\tau), \quad 0 \leqslant \tau \leqslant \tau_m . \tag{2.29}$$

Thermophysical properties $C(x,\tau)$ and $\lambda(x,\tau)$ are positive, continuous and bounded functions. Initial and boundary conditions are in accordance with each other.

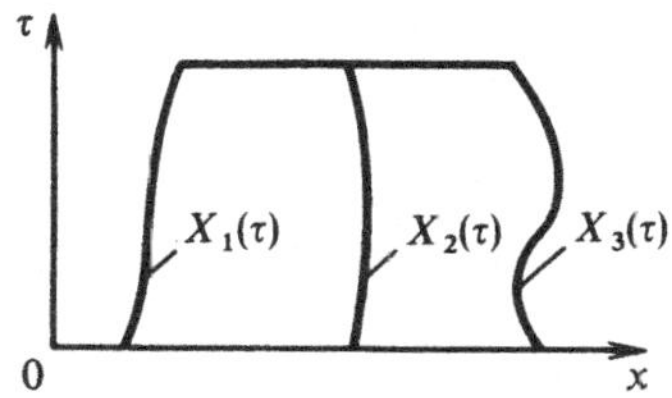

Fig. 2.3. A domain with mobile boundaries

Assume that $q(\tau)$ value is the one desired and assigned as an additional condition

$$T(X_2(\tau), \tau) = f_2(\tau), \quad X_1(\tau) < X_2(\tau) < X_3(\tau). \tag{2.30}$$

Let there be two solutions of the given problem: $q'(\tau)$ and $q''(\tau)$. Introduce the functions $\Delta q(\tau) = q'(\tau) - q''(\tau)$; $\Delta T(x, \tau) = T'(x, \tau) - T''(x, \tau)$, where T' and T'' are solutions of the problem given by equations (2.28)–(2.29) for $q = q'$ and q'', respectively,

Value $\Delta T(x, \tau)$ as a consequence of the problem linearity must satisfy a homogeneous heat conduction equation with the homogeneous initial condition

$$C(x, \tau)\frac{\partial \Delta T}{\partial \tau} = \frac{\partial}{\partial x}\left(\lambda(x, \tau)\frac{\partial \Delta T}{\partial x}\right), \quad X_1(\tau) < x < X_3(\tau), 0 < \tau \leqslant \tau_m,$$

$$\Delta T(x, 0) = 0, \quad X_1(0) \leqslant x \leqslant X_3(0); \tag{2.31}$$

$$-\lambda\frac{\partial \Delta T(X_1(\tau), \tau)}{\partial x} = \Delta q(\tau), \quad 0 \leqslant \tau \leqslant \tau_m;$$

$$\Delta T(X_3(\tau), \tau) = 0, \quad 0 \leqslant \tau \leqslant \tau_m. \tag{2.32}$$

Also on the line $X_2(\tau)$ we have

$$\Delta T(X_2(\tau), \tau) = 0, \quad 0 \leqslant \tau \leqslant \tau_m. \tag{2.33}$$

It follows from conditions (2.31), (2.32) and (2.33) that the $\Delta T(x, \tau)$ function everywhere in the area $X_2(\tau) \leqslant x \leqslant X_3(\tau)$, $0 \leqslant \tau \leqslant \tau_m$ is equal to zero, i.e. apart from the condition $\Delta T(X_2(\tau), \tau) = 0$ an additional condition $\partial \Delta T(X_2(\tau), \tau)/\partial x = 0$, $0 \leqslant \tau \leqslant \tau_m$ is known.

Since the accepted model of heat conduction corresponds to the infinite speed of heat propagation, so, for $\Delta q(\tau) \neq 0$, the obtained condition of thermal insulation of the $X_2(\tau)$ boundary would have led to a change of $\Delta T(X_2(\tau), \tau)$ with respect to a zero value for arbitrarily small periods of time $\tau > 0$. Hence, $\Delta q(\tau) = 0$ and the initial assumption about the availability of two solutions will be accepted as untrue, i.e. the problem has only one solution.

Similarly, at a physical level that is not absolutely rigorous, it is possible to demonstrate the uniqueness of a boundary IHCP solution in two-dimensional and three-dimensional statements.

2.3.2 Uniqueness of Solution of a Boundary Inverse Problem with Constant Coefficients

Consider a more rigorous proof of uniqueness of a boundary IHCP solution as stated in Eqs. (2.28)–(2.30) providing the invariability of the physical parameters $C = \text{const}$ and $\lambda = \text{const}$.

Let us first turn to a boundary-value problem statement. Divide the domain $D = \{(x, \tau): X_1(\tau) \leqslant x \leqslant X_3(\tau), 0 \leqslant \tau \leqslant \tau_m\}$ into two sub-domains (Fig. 2.4) with a common boundary parallel to the time axis: $D_1 = \{X_1(\tau) \leqslant x < b, 0 \leqslant \tau \leqslant \tau_m\}$ and $D_2 = \{b \leqslant x \leqslant X_3(\tau), 0 \leqslant \tau \leqslant \tau_m\}$.

Choose the value b arbitrarily in the interval $X_{2\,\max} \leqslant b \leqslant X_{3\,\min}$ ($X_{2\,\max}$ and $X_{3\,\min}$ are absolute extrema of functions $X_2(\tau)$ and $X_3(\tau)$: it is assumed that $X_{3\,\min} > X_{2\,\max}$).

Introduce then a new domain D'_1, broadening the domain D_1 by inserting a fictitious boundary $x = d$, $d = X_{1\,\min}$: $D'_1 = D_1 + S$, where $S = \{(x, \tau)$ $d \leqslant x < X_1(\tau), 0 \leqslant \tau \leqslant \tau_m\}$ is a fictitious domain.

Suppose that Eq. (2.28), a heat conduction equation, is true in the domain $D' = D'_1 + D_2$. Divide the procedure for solving the original IHCP into two stages:

(1) Reconstruction of the "fictitious" boundary function $q_\varphi(\tau) = -\lambda\,\partial T(d, \tau)/\partial x$ using the conditions $T(b, \tau)$ and $-\lambda\,\partial T(b, \tau)/\partial x$, i.e. an inverse heat conduction problem. Note that it is a problem as to how to continue the solution of Eq. (2.28) up to the fictitious boundary, which is possible for functions that are sufficiently smooth and if set at $x = b$ boundary.

(2) Determination of the desired boundary function $q(\tau)$ on the $X_1(\tau)$ line under conditions $q_\varphi(\tau)$ and $-\lambda\,\partial T(b, \tau)/\partial x$, i.e. a direct heat conduction problem. It is accepted here that the functions $T(b, \tau)$ and $-\lambda\,\partial T(b, \tau)/\partial x$ are unambiguously determined by solving a heat conduction equation with the conditions set on the $X_2(\tau)$ and $X_3(\tau)$ lines.

For a direct problem with corresponding constraints, e.g. the sufficiently smooth curves $X_1(\tau)$ and $q_\varphi(\tau)$, the uniqueness of the solution is considered to be an established fact. Then, the uniqueness of an inverse problem solution would appear from the identity $\Delta q_\varphi(\tau) = 0$, since in this case $\Delta q(\tau) = 0$: all boundary conditions in the direct problem of heat conduction are zero.

By such a method (referred to as *a method of fictitious boundaries*), the inverse problem in the domain with moving boundaries is reduced to an inverse

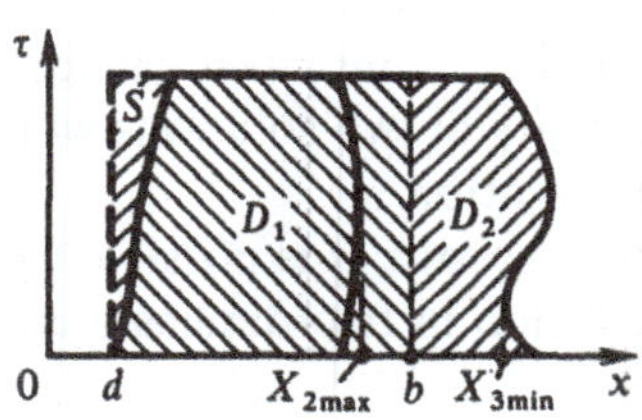

Fig. 2.4. Domains obtained by fictitious boundaries

problem in a rectangular domain without changing a notation of the heat conduction equation.

Let $q'_\varphi(\tau)$ and $q''_\varphi(\tau)$ be two solutions of an inverse problem. Then, function $\Delta T(x,\tau) = T(x,\tau)|_{q'_\varphi} - T(x,\tau)|_{q''_\varphi}$ must satisfy a boundary-value problem

$$\frac{\partial \Delta T}{\partial \tau} = a\frac{\partial^2 \Delta T}{\partial x^2}, \quad d < x < b, \quad 0 < \tau \leqslant \tau_m; \tag{2.34}$$

$$\Delta T(x, 0) = 0, \quad d \leqslant x \leqslant b; \tag{2.35}$$

$$-\lambda\frac{\partial \Delta T(d,\tau)}{\partial x} = \Delta q_\varphi(\tau); \quad -\lambda\frac{\partial \Delta T(b,\tau)}{\partial x} = 0, \quad 0 \leqslant \tau \leqslant \tau_m. \tag{2.36}$$

At the same time, at point $x = b$ satisfaction of the following condition is necessary

$$\Delta T(b,\tau) = 0, \quad 0 \leqslant \tau \leqslant \tau_m. \tag{2.37}$$

From Eq. (2.6) one can obtain an integral representation of the problem shown in Eqs. (2.34) and (2.36) in the Duhamel's form

$$\Delta T(x,\tau) = \int_0^\tau \Delta q_\varphi(\xi)\frac{\partial \vartheta(x,\tau-\xi)}{\partial \tau}\mathrm{d}\xi.$$

Subject to equation (2.37), it leads to an equality

$$\int_0^\tau \Delta q_\varphi(\xi)\frac{\partial \vartheta(b,\tau-\xi)}{\partial \tau}\mathrm{d}\xi = 0. \tag{2.38}$$

According to the Titchmarch's convolution theorem [69], if functions $\Delta q_\varphi(\tau)$ and $\partial\vartheta(b,\tau)/\partial\tau$ are continuous in the domain $0 \leqslant \tau \leqslant \tau_m$, $\tau_m < \infty$, then it follows from equality (2.38) that at least one of these functions in the domain $0 \leqslant \tau \leqslant \tau_m$ equals zero everywhere. It is known from the mathematical theory about heat conduction that

$$\frac{\partial \vartheta(b,\tau-\xi)}{\partial \tau} > 0, \quad \tau \neq \xi.$$

Then follows the uniqueness of this inverse problem solution.

Note that the result obtained on the uniqueness of boundary IHCP for a single-layer plate can be extended onto other one-dimensional inverse problems such as for a semi-infinite body, solid and hollow cylinders, solid and hollow spheres and for composite bodies as well.

A question on the uniqueness of some boundary IHCPs in linear statements has been already analysed. A rigorous proof of this property for an inverse problem under more general conditions should be the subject of a pure mathematical study.

Note in conclusion that the existence and uniqueness of an inverse problem solution in the Cauchy statement for a general heat conduction equation with coefficients dependent upon spatial coordinates and time can be determined from the Kowalewski theorem [119]. It is therefore required that:

1) the equation coefficients should be analytical functions for their arguments;
2) the Cauchy conditions should be set on an analytical surface;
3) the input data and a solution should belong to a class of analytical functions.

With fewer limitations the uniqueness of non-characteristic Cauchy problem solving for a linear parabolic equation of N-order and with varying coefficients has been proved (see [89]). For a linear parabolic equation of the second order the result had been established earlier (see [198 .]). In particular, consider a one-dimensional heat conduction problem of the form

$$\frac{\partial T}{\partial \tau} = a(x, \tau)\frac{\partial^2 T}{\partial x^2} + b(x, \tau)\frac{\partial T}{\partial x}, \quad (x, \tau) \in \bar{\Omega} = [0, d] \times [0, \tau_m],$$

where $a(x, \tau) = \lambda(x, \tau)/C(x, \tau) > 0$, $b(x, \tau) = 1/C(x, \tau)\ \partial\lambda(x, \tau)/\partial x$.

The uniqueness of a Cauchy problem solution based on the determination of a temperature field $T(x, \tau)$ in Ω domain by known boundary conditions of the fourth kind at point $x = d$

$$T(d, \tau) = f(\tau), \qquad -\lambda\frac{\partial T(d, \tau)}{\partial x} = q(\tau)$$

will take place provided that

$$T(x, \tau) \in C^2; \quad f(\tau), q(\tau) \in C^1; \quad a(x, \tau) \in C^2 \quad \text{and} \quad b(x, \tau)$$

is a bounded function.

2.4 Degree of Instability of a Boundary Inverse Heat Conduction Problem

In practical calculations the instability of an inverse problem manifests itself as an oscillating solution. In this sense, one can speak about the degree of instability of IHCP algorithms as implying greater or lesser oscillation swings and frequencies of a reconstructed function, other conditions being equal. The degree of instability can noticeably change depending on a heat transfer model studied, on the body thickness and depth of sensors setting, and on thermophysical properties, i.e. upon those factors which impose an effect on smoothing properties of a heat conduction operator.

It is natural to expect that the greater the coefficient of thermal diffusivity and the closer the sensor setting to the body boundary with the desired condition, the weaker would the IHCP ill-posed statement affect the results.

Let us turn back to the integral Eq. (2.8). Its smoothing properties are disclosed by the form of kernel $K(\tau - \xi) = \frac{a}{\lambda}G(x, 0; \tau - \xi)$. In the given case, the kernel corresponds to the influence function (Green's function) $G(x, x'; \tau - \xi)$ of the second boundary-value problem for a one-dimensional equation of heat conduction in the domain $0 < x < b$. Recall the physical meaning of the influence function: generally it is the temperature occurring at point s of the body at moment τ of time due to the effect of a unit-equal instantaneous heat source at point s' at moment ξ provided that a zero boundary condition is valid on the boundary of the spatial area at all successive moments of time $\tau > \xi$ and at moment ξ the temperature of the points of spatial area equals zero.

In our case, the current value of the x coordinate corresponds to point s and value $x = 0$ – to s'.

Consider a case with a semi-infinite body $\left(b \to \infty, \dfrac{\partial T(\infty, \tau)}{\partial x} = 0\right)$, for which the corresponding influence function has the simplest form [93, 172]

$$G(x, 0; \tau - \xi) = K(x, \tau - \xi)\frac{\lambda}{a} = \frac{1}{\sqrt{a\pi}} \frac{\exp[-x^2/4a(\tau - \xi)]}{\sqrt{\tau - \xi}}.$$

Assume that temperature $T^*(\tau)$ is known at point $x = d$. Then, an IHCP upon determining a heat flux $q(\tau)$ on boundary $x = 0$ is reduced to an integral equation

$$\int_0^\tau q(\xi)K(d, \tau - \xi)\mathrm{d}\xi = T^*(\tau), \quad 0 < \tau \leqslant \tau_m. \tag{2.39}$$

Value $z = \lambda\sqrt{\dfrac{\pi}{a}}K(d, \tau - \xi)$, depending on the difference in the arguments $\delta\tau = \tau - \xi$ for different quantities of d^2/a, is shown on Fig. 2.5. As the relation $(d^2/a) > 0$ increases, the function $z(\delta\tau)$ becomes more smooth and its maximum

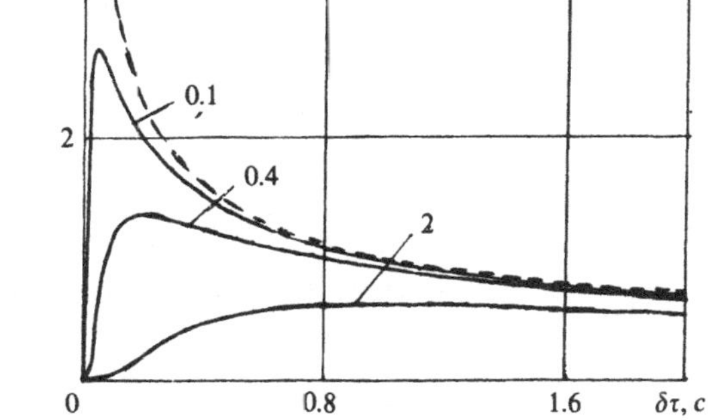

Fig. 2.5. The kernel of an integral equation for a semi-infinite body with a boundary condition of the second kind ($z = \lambda\sqrt{\dfrac{\pi}{a}}K(d, \tau - \xi)$; $\delta\tau = \tau - \xi$)

decreases and is shifted towards larger values of $\delta\tau$. It is inevitable that with direct problem solving the smoothing effect of the singularities of the $q(\tau)$ function become stronger. But with inverse heat conduction problem solving, any errors of input data would distort the results of calculations still further as value d^2/a increases.

Such a conclusion arises from physical considerations. It is readily confirmed by a mathematical example shown in Eqs. (2.8) and (2.9), with the help of which the ill-posed statement of an IHCP integral form has been illustrated. We may assume that the rate of convergence $\int_0^\tau u_\delta(\xi)K(\tau-\xi)\mathrm{d}\xi \rightarrow \int_0^\tau u(\xi)K(\tau-\xi)\mathrm{d}\xi$ with frequency ω increasing would be greater for smoother kernels of an integral equation, i.e. the harmonic component of the right-hand side of Eq. (2.8), $\Delta f(\tau) = C\int_0^\tau \sin\omega\xi K(\tau-\xi)\mathrm{d}\xi$, would become infinitely small with lesser quantities of ω. It follows that in reconstructing the boundary conditions oscillations of the function $u(\tau)$ will appear to be larger for the same fluctuating errors in the input data $f(\tau)$, provided that the kernel has a smoother ("spreading") form.

Of special interest is the analysis of the stability condition when recalculating a boundary condition of the first kind into a boundary condition of the second kind, i.e. the heat flux is determined on the very boundary where temperature measurements are taken. For a semi-infinite body this case will correspond to $d = 0$.

The problem studied is in effect not an inverse heat conduction problem, since in determining $q(\tau)$ continuation of the heat conduction equation solution is not required. In fact, it is a nearly direct heat conduction problem which should be solved by the conditions set on the two boundaries of the domain with a recalculation of the boundary conditions. Let us call this problem a *pseudo-inverse* one. It is possible to demonstrate that a pseudo-inverse heat conduction problem is stable using general assumptions for the input function [5].

Let us prove it using the example of an integral equation for an inverse heat conduction problem with a semi-infinite body.

For proof use the Holmgren method. Multiply both sides of Eq. (2.39) by $(z-\tau)^{-1/2}$ and integrate with respect to τ from 0 to z:

$$\int_0^z \mathrm{d}\tau \int_0^\tau (z-\tau)^{-1/2} K(d,\tau-\xi)q(\xi)\mathrm{d}\xi = \int_0^z (z-\tau)^{-1/2} T^*(\tau)\mathrm{d}\tau .$$

Interchange the order of integration by a triangular area

$$\int_0^z q(\xi)V(\xi,z)\mathrm{d}\xi = \int_0^Z (z-\tau)^{-1/2} T^*(\tau)\mathrm{d}\tau,$$

where $V(\xi,z) = \int_\xi^z K(d,\tau-\xi)(z-\tau)^{-1/2}\mathrm{d}\tau$.

Differentiate this expression with respect to z. The right-hand derivative in it makes sense if the $T^*(\tau)$ function satisfies the Hölder condition

$$|T^*(\tau') - T^*(\tau'')| \leqslant \kappa|\tau'-\tau''|^\alpha, \quad (\tfrac{1}{2} < \alpha \leqslant 1),$$

where κ is a positive constant. In this case it can be specified by expression [73]

$$\frac{d}{dz}\int_0^z (z-\tau)^{-\frac{1}{2}} T^*(\tau)d\tau = z^{-\frac{1}{2}} T^*(z) - \tfrac{1}{2}\int_0^z (z-\xi)^{-\frac{3}{2}}(T^*(\xi) - T^*(z))d\xi.$$

In the result we obtain

$$q(z)\,V(z,z) + \int_0^z q(\xi)\frac{\partial V(\xi,z)}{\partial z}d\xi = z^{-\frac{1}{2}} T^*(z)$$

$$-\frac{1}{2}\int_0^z (z-\xi)^{-\frac{3}{2}}(T^*(\xi) - T^*(z))d\xi\,.$$

It is not difficult to show that for $x = 0$ value $V(\xi, z) = \frac{1}{\lambda_0}\sqrt{a\pi}$. Substituting z with τ, we finally obtain

$$q(\tau) = \frac{\lambda}{\sqrt{a\pi}}\left[\frac{T_w(\tau)}{\sqrt{\tau}} - \frac{1}{2}\int_0^\tau \frac{T_w(\xi) - T_w(\tau)}{(\tau-\xi)^{3/2}}d\xi\right].$$

If the Hölder condition is satisfied which, as a rule, is not a constraining one, then the sub-integral function in the last expression has weak singularity and, therefore, this integral does exist. Its calculation can be performed using the known rules of approximate integration of functions which have discontinuity points.

Thus, the problem of solving an integral equation of the first kind (2.39) to specify a heat flux in the semi-infinite body by known temperatures at the boundary has been reduced to a well-posed problem of calculation.

2.5 Conditionally-Well-Posed Statement of Inverse Problems

While analysing the stability of inverse heat transfer problems, it has been established that their solution can contain no property of stability. With this singularity of inverse problems are connected the main difficulties of constructing effective computational algorithms. If the initial statement of an unstable problem is not changed, then the methods worked out for solving well-posed problems can be unsuitable when applied to an inverse problem. An attempt to obtain very accurate or, if possible, an exact causal characteristic u answering the fine structural distinctions of its consequent image f assigned with some error would lead to an unstable result.

For example, in applying a variational IHCP form, the specific nature of an ill-posed problem would develop into an ambiguous solution, i.e. the initial unstable IHCP statement is, in effect, somewhat undefined and one cannot be sure of the smallness of error in the inverse problem solution however small the errors in the input data.

A major conclusion made in the analysis is that another, i.e. a well-posed, statement of the problem should be formulated for an inverse problem. Mathematical theory suggests two possible approaches to resolve this situation. The first is based on the change of the classical concept of well-posed problem and on the transition to its conditionally well-posed statement (well-posed by Tikhonov [38]). The second approach is concerned with approximation of the inverse operator A^{-1} which is not continuous by a family of continuous operators (see Sect. 2.6.4).

Consider the first approach. *A conditionally well-posed* problem statement assumes a compulsory satisfaction of the following conditions:

1. it is postulated that a solution of the $Au = f$ equation exists and belongs to a certain set of feasible solutions $M \subseteq D_A$ (D_A is the domain of the operator A);
2. a solution is unique on the set M;
3. the infinitesimal variations of the solution correspond to infinitesimal variations of the right-hand side f, which do not derive a solution from the M class.

These conditions represent a generalization of the concept of stability by Hadamard for many problems which are ill-posed in the classical sense. The most essential part in the Tikhonov definition of well-posed statement deals with the narrowing of a set of feasible solutions in such a way that the problem could become stable. That is why the main question of practical application of this approach is concerned with a substantiation of the method for selecting a class of feasible solutions M, i.e. a *class of correctness*.

A compact set[5] is often taken as the above class. In this case, the following result is true [136]: if the compact set $M \subseteq D_A$ satisfies the conditions of 1 and 2 items of Tikhonov's stability and A is a continuous operator, then an inverse mapping $A^{-1} f = u$ would also be continuous. A compact set in every particular case is assigned individually on the basis of the physical aspect of the problem. Note that the condition of compactness is not necessary but is sufficient enough to obtain a stable solution for an ill-posed problem.

The approach presented in the theory of ill-posed problems has been considerably developed by M.M. Lavrentiev, V.K. Ivanov and other Soviet mathematicians (a survey of these studies is given, for example, in [103]).

Let us now turn to a question of how to apply the above approach to the solution for ill-posed inverse heat conduction problems. It should be noted here that the mathematical result already stated has not yet found practical application in thermophysical investigations. At the same time, some of the known methods of heat studies, based on different, usually pure physical, concepts, can be referred to a conditionally well-posed IHCP statement and, thus, mathematically substantiated.

There are several examples. Usually, it is known in advance that a causal characteristic sought (it can be a heat boundary condition, or thermophysical

---

[5] A set of M-metric space U is called compact when any sequence of elements of this set contains a convergent sub-sequence.

properties, etc.) is regular in kind, i.e. it is a sufficiently smooth function. It is evident that we should regard this moment for an inverse problem statement as a certain regularity condition for singling out an admissible class of functions.

The strongest regularity condition incorporates the assumption of constancy of an unknown quantity within the whole interval of argument variation. In the result we obtain the most simple case for our analysis.

Consider the one-sided heating of a flat slab (one of its boundaries is thermally insulated) using the flux of constant density q. In practice the problem given is of great importance for measuring stationary heat fluxes as, for example, in the jets of high-temperature gas. For homogeneous initial temperature distribution and invariable thermo-physical properties we obtain a classical result

$$T(d, \tau) - T_0 = q\vartheta(d, \tau), \tag{2.40}$$

from which follows

$$q = \frac{1}{\vartheta(d, \tau)}[T(d, \tau) - T_0], \tag{2.41}$$

where T_0 is the initial temperature and $T(d, \tau)$ is the temperature at d distance from the heating boundary.

Function $\vartheta(d, \tau)$ is determined by formula (2.7). Dependence (2.41) with the exact data $T^*(\tau) = T(d, \tau)$ allows an exact solution of the inverse problem to be obtained at any moment of time. The temperature values measured are different from the true values by a magnitude of error $\delta(\tau)$: $T^*_\delta(\tau) = T^*(\tau) + \delta(\tau)$. Thus value $q^*_\delta(\tau)$ calculated directly using formula (2.41) can greatly differ from the exact constant magnitude q. Such a situation has occurred when a constancy condition for heat flux has been virtually taken into account when solving not an inverse but a direct problem of heat conduction. Despite the fact that for q we have obtained an explicit expression, its direct application is hardly possible. The point is that function $T^*_\delta(\tau)$ does not belong to the AM set, i.e. an inverse problem has no solution in the compact set prescribed (functions which are constant in the $[0, \tau_m]$ segment represent the simplest example of the M compact set in space $C\,[0, \tau_m]$ of the continuous functions in the said segment).

A way out from this situation is in the application of such methods which can define function $\tilde{q}(\tau) \in M$, the latter minimizing the $\rho\,(Au, f_\delta)$ functional, that is, giving minimum deviation of the left side of Eq. (2.8) from the right side on the M compact set.

As a discrepancy functional we can take a mean-square deviation

$$\rho^2(Au, f_\delta) = \int_0^{\tau_m} [Au(\tau) - f(\tau)]^2\, d\tau\,.$$

In this case, $Au(\tau) = T(\mathrm{d}, \tau), f(\tau) = T^*(\tau)$ and $u(\tau) = q = \text{const}$. Calculating the partial derivative $\partial\rho^2/\partial q$ and equating it to zero, we obtain an expression for determining an optimal value $q = \tilde{q}$

$$\tilde{q} = \int_0^{\tau_m} \vartheta(d, \tau)[f(\tau) - T_0]\,d\tau \Big/ \int_0^{\tau_m} \vartheta^2(d, \tau)\,d\tau\,.$$

A natural method of simplifying the data processing procedure consists in using the heat regularization property. After a definite time lapse, the function $\vartheta(d, \tau)$ can be presented by the expression [93]

$$\vartheta(d, \tau) = \frac{a}{b\lambda}\tau + \frac{3(b-d)^2 - b^2}{6b\lambda}, \tag{2.42}$$

i.e. the time dependence of temperature at point $x = d$ becomes linear. The start-up time of a regular heating regime depends on the location of the point inside the slab and on thermal diffusivity quantity, and lies within $0,35b^2/a \leqslant \tau_\rho \leqslant 0.5\, b^2/a$. From the moment of regularization of a temperature curve to determine the heat flux we can use a formula

$$q = \frac{b\lambda}{a}\frac{dT^*}{d\tau} \simeq \frac{b\lambda}{a}\frac{\Delta T^*}{\Delta\tau},$$

obtained by differentiating (2.40) in terms of formula (2.42). Here, ΔT is an increment in temperature during $\Delta\tau = \tau - \tau_\rho$. Since the experimental data carry different errors in themselves, the function $T_\delta^*(\tau)$ measured in the regularization section would not be a direct line and, in order to match the input data with an IHCP statement, it is necessary to approximate them in advance by a linear dependence following the method of least squares. To increase the accuracy of results one needs a sufficient observation interval.

Thus, for the simplest conditionally well-posed statement of boundary IHCP (providing a condition of heat-flux constancy) and after expiry of a certain period of time, the temperature $T(d, \tau)$ becomes regularized, considerably simplifying calculation of the q value.

Using the classical method of regular regime [84], one can make up easy and convenient procedures to determine the boundary conditions (constant coefficients of convective heat transfer and heat fluxes) for bodies of different shapes and dimensions. The procedures are based upon the universal property of temperature field regularization. Starting from a certain period of time, the temperature of the body at any point varies according to the exponential law $(T_e - T)/(T_e - T_0) \sim \exp[-m\tau]$, where T_e is the ambient temperature, T_0 the initial temperature of the body and $m = d/d\tau \ln(T_e - T)$ the rate of heating $(0 \leqslant m \leqslant m_1 < \infty)$.

Note that the exponents with negative powers present another example of a compact set in the space $C\,[0, \tau_m]$ of continuous functions in the segment. A change in the initial statement of the inverse heat conduction problem at the expense of the constancy condition of the heat flux q or convective heat transfer coefficient α, and also the condition of the exponential temperature change T_w of the body surface leads to the most simple IHCP statements noted as well-posed by Tikhonov.

It is possible to simulate more general cases of stable IHCP solutions under the conditionally well-posed statement. For instance, such a solution would take place if q or T_w were being sought among the continuous monotonic functions.

Thus, one of the principles enabling us to obtain stable approximations in inverse problems consists of narrowing the domain of admissible solutions as a result of accounting for a priori information for an unknown function. The information should be physically substantiated.

With the sources of a priori information on IHCP solution we can use properties assumed for an unknown function, namely: continuity and definite signs at certain intervals of argument change; the most and least guaranteed values; time intervals within which the characteristic points must lie (extremal, zero and inflection points), etc.

In the domain thus isolated, an approximate solution is being sought as most suitable to experimental data. It is natural that, with such an approach, an inverse problem is stated as being extremal and consists of minimizing deviations in calculated and prescribed temperatures in the selected class of unknown functions.

It should be said, however, that, in many cases when solving a boundary IHCP, one cannot set conditions in which the plot deviations of an unknown function conditioned by the instability of the initial problem would not be too large. Moreover, in the statement mentioned we have to solve a problem for a conditional extremum leading to familiar computational difficulties.

A somewhat better situation is observed for coefficient inverse problems. First of all, it follows from physical considerations that a solution (which includes the coefficients of heat conduction, heat capacity and thermal diffusivity) should be sought in the class of positive functions. The upper bounds of the solutions can be well-grounded in many cases. This is much better in comparison with a boundary IHCP as one can predict the behaviour of unknown functions and then preset sufficiently narrow classes of feasible solutions that provide stable results.

Let us sum up the above considerations. A conditionally well-posed statement of the inverse problem, being one of the possible ways of obtaining stable results, requires corresponding modification of the initial problem statement. This is at times the most difficult aspect in such an approach to the solution. From a practical point of view, a conditionally well-posed statement of a IHCP often has no universal application.

2.6 Regularization Principles of Ill-Posed Inverse Problem Solutions

The material in this section can be considered as an introduction to the problem of inverse heat conduction problem regularization. Basic statements of the given analysis formulated in the general form will be studied in detail in the following chapters.

Let us consider possible principles for obtaining stable solutions of inverse problems, which are not associated with an a priori assignment of a correctness class. We shall divide them into two groups:

1. *self-regularization* of inverse problems;
2. *regularization* according to the Tikhonov method.

2.6.1 Direct Methods and Some Techniques of Natural Regularization

The term introduced as natural regularization (self-regularization) means that an inverse problem solution is obtained in its initial statement but with the help of such approximate methods which allow control over the measure of proximity of a solution being obtained as compared to an exact one by changing the usual parameters of computational algorithms (such as steps in arguments, number of terms of a series in the function expansion, etc.). We refer to them as *direct methods*.

Such an approach to IHCP solution is more popular due to the simplicity of the computational algorithms used for their realization. But in practical application of direct methods we may encounter difficulties due to methods employed.

Let us first consider the problem of temperature field continuation in the case of a flat body and for which we have an explicit solution [expression (2.11)]. It is clear that by truncating a series with the ever decreasing number of terms N we would decrease the degree of problem instability (i.e. to regularize a solution) at the expense of eliminating the highest derivatives of exponential functions. With $N = 1$, we obtain a formula for a heat flux:

$$q(\tau) = -q^*(\tau) + \lambda \frac{d}{a} \frac{\mathrm{d}T^*(\tau)}{\mathrm{d}\tau},$$

which expresses the heat balance in a slab of small thickness and is used in the difference method for numerical solution of a heat conduction equation. But even this constrained case is not entirely free from instability, one has to calculate the $\mathrm{d}T^*(\tau)/\mathrm{d}\tau$ derivative. For this purpose we shall make use of another form of natural regularization.

An approximated computation of the derivative can be performed by the formula

$$\frac{\mathrm{d}T}{\mathrm{d}\tau} \simeq \frac{T(\tau + \Delta\tau) - T(\tau)}{\Delta\tau}, \tag{2.43}$$

where $\Delta\tau$ is a numerical differentiation step.

If T are disturbed by a fluctuating error, then, with $\Delta\tau \to 0$ we have an unstable computational problem. In the actual calculations when applying formula (2.43) we select the step by intuition, i.e. in such a way that, first, eliminates unwanted oscillations in the numerical solution and, second, does not lose specific features of the $T'(\tau)$ curve. This is the way in which an operator acts when engaged in "manual" treatment of temperature oscillograms. The given approach – *a step regularization* of numerical solution – can be referred to as natural regularization of an ill-posed problem. Its concept consists in matching the computational step $\Delta\tau$ with an error of input data.

Note that the processing of experimental data $T_\delta(\tau)$, so widespread in practice and included in a certain smoothing procedure by using the least-square technique may give no applicable results for a successive numerical differentiation. It can be explained by the fact that such an approach enables us to obtain a convergence only on the average for the function but it does not guarantee good approximation to its derivative.

So, it was shown by the above example on continuation of the heat conduction equation solution that we can achieve stable results by using natural regularizing properties set within the algorithms of approximate computations. But we can immediately notice two dangers in applying these direct methods of inverse problem regularization. Since no constructive condition has been formulated in choosing the regularizing parameters of computational algorithms (in the example given N and $\Delta\tau$), we can in practical calculations either "undersmooth" a solution or over-rough the model. In either case, any solution obtained may appear far from true and therefore inaccurate.

A natural step-type regularization method can be applied for approximate calculation of the first derivative but it is of little use for successive differentiation.

A contradiction arises: on the one hand, it is required to increase solution accuracy by expanding the number of series terms and, on the other hand, vice versa, to reduce this number since errors do increase sharply because of instability of the differentiation operation. *A regularization by series truncation* would confine us to several first numbers (usually $N = 1 \div 3$). This can lead to a noticeable distortion of results. It should be noted that with an approximate calculation of derivatives the series convergence slows down as compared with their exact values. That is why an IHCP solution expressed in the form of series in derivatives of experimental functions (such an approach is described in [132, 133, 169] and other ones close to it – in [79, 185]) should not be recommended for processing the results of intensive and dynamic heat tests. It can be used more successfully for slow developing heat exchange processes, provided that temperature-sensitive elements are positioned close enough to the body surface, and when very precise measurements are made.

The aim of the above was, firstly, to illustrate some principles of IHCP natural regularization and, secondly, to warn against arbitrary usage of direct methods. One should not, however, draw too a hasty conclusion that direct methods are not applicable at all for solving the inverse problems. In many cases they are a sufficiently accurate and simple means of processing experimental information.

2.6.2 Sources of Natural Regularization

The sources of natural regularization of inverse problem solution can be connected with:

1. a regularization effect of heating conditions of the body point where a temperature-sensitive element is mounted;

2. "viscous" properties of computational algorithms, which are a natural filtration of high-frequency components in the causal characteristic required.

The first principle is usually realized in the form of a step IHCP regularization. The desired causal characteristic, being a continuous curve $u(z)$, it can be changed by a piecewise homogeneous approximate dependence, i.e. the whole portion of z argument change is broken into intervals with $\Delta z_n = z_n - z_{n-1}$, $n = 1, 2, \ldots, N$ steps, the curve being described at each interval by dependences of one and the same type. It should be noted that these dependences are expressed by the more regular functions which have a minimum number of non-zero derivatives. Usually, a piecewise-constant approximation is applied and, seldom a piecewise-linear one. The Δz_n step quantity is selected on condition that a temperature change at the point of measurement caused by the action of an approximate causal factor at the given interval could be distinguishable against the background of various errors (noise, measurement errors, data recording and interpretation, approximation and roundoff errors).

For example, it is required to determine a heat flux change $q(\tau)$ on the body boundary by internal temperature measurements. The curve $q(\tau)$ is replaced by a stepwise line, the duration of each step being selected so that an increment in temperature at a preset point of the body would be large enough for the accompanying errors.

With regard to regularization of body heating conditions, we do not mean that one should strictly maintain a regular thermal mode in the classical sense of the problem [84, 93]. But in certain cases it is possible to trace close relations between the principle of the step regularization and selection of the time interval from a condition of regular regime occurrence. This is an approach to the boundary IHCP solution which has been suggested for the first time in [150].

For heat measurements, a situation is widespread when the approximation and roundoff errors in solving a heat conduction problem are small compared with temperature data errors and, hence, they can be neglected. In this case, selection of N number of intervals with z homogeneous digitization can be performed based on matching the temperature mean-square residual with the magnitude of these errors:

$$\frac{1}{K}\sum_{k=1}^{K}\int_{0}^{\tau_m}[T_k(N,\tau) - T_k^*(\tau)]^2\,d\tau \simeq \delta^2 ,$$

where K is the number of temperature-sensitive elements in the body, $T_k(N, \tau)$ is a temperature depending on time at the k-th point of the body obtained through computation for N intervals, $T_k^*(\tau)$ is the temperature at the k-th point of the body, $\delta^2 = \int_0^{\tau_m} \sigma^2(\tau)\,d\tau$ is an estimate of the input data integral error, and $\sigma(\tau)$ is a mean-root-square error of measurements.

As a result of such step selection in the $u(N, z)$ approximate solution, there will be smoothing over the harmonics connected with a pulsating change of errors in the registered curves $T_k^*(\tau)$, i.e. the components with frequencies higher than a certain limited frequency will not be extracted.

The second principle of natural regularization based on viscous properties of computational algorithms is also aimed at suppressing the unwanted harmonics in the inverse problem solution by the choice of some parameters of the algorithms that control the smoothness of approximate solutions. This principle can be realized by:

1. a corresponding selection of approximation steps of the initial problem through different computational methods of IHCP solving;
2. a corresponding selection of the degree of the approximate polynomial for an unknown quantity;

In all of these cases, an IHCP solution is "roughened" on purpose in comparison with an "exact" solution which corresponds to the disturbed initial data.

Direct methods based (explicitly or inexplicitly) on principles of natural regularization are suggested for solving boundary and coefficient IHCP as, for example, in [3, 6, 23, 24, 40, 53, 65, 70, 72, 96, 110, 111, 112, 122, 127, 128, 150, 151, 157–160, 162, 163, 165, 168, 182–185, 200, 203].

The criterion for selection of control parameters in an algorithm is the matching of temperature discrepancy with the input information error.

2.6.3 Capabilities of Direct Methods

These methods were subject to investigation for a solution of boundary inverse heat conduction problems. Let us briefly consider the main results of this analysis.

1. A step regularization being a simple means of obtaining stable approximations can be successfully used for a boundary IHCP solution in linear and non-linear boundary-value statements and in the Cauchy statements in areas with moving and fixed boundaries [1, 2, 6, 8, 12, 22, 24, 70, 72, 110, 150, 151, 157–159, 162, 165, 200]. For sufficiently smooth input information, it becomes possible to establish limiting values of non-dimensional time steps in the form of the Fourier number increment $\Delta Fo_l = a\Delta\tau/l^2$ which sets applicability limits to a step-type principle of natural regularization in order to solve different inverse problems. In the case of boundary IHCP with constant coefficients, we have managed to formulate [12] a common principle for sampling a computational step quantity from the stability of results, without resorting to an experimental technique applied earlier for revealing the instability boundary in solving methodical examples. Chapters 4 and 5 are devoted to detailed description of all these questions.

It is important to note that a step regularization can also be used to solve other types of inverse problems.

2. A step regularization combined with some finite difference schemes for the numerical solution of boundary IHCP using the Cauchy statement broadens the applicability constraints of direct methods compared with an analytical approach (see Chap. 5).

3. Regularization of an IHCP solution by choosing a degree of algebraic polynomial to approximate an unknown function in the whole interval of argument change is of little use as a method of determining the boundary conditions save for simple forms of curves reconstructed which are close to the class of polynomials. This conclusion seems obvious since in approximating a certain curve $y(x)$ by a polynomial of n-th degree it is generally difficult to obtain a suitable result. On the one hand, with small values n an approximating dependence appears to be far from a true solution but, at the same time, with greater n, an approximation contains unnecessary oscillating components.

Nevertheless, we can use this this method in solving other types of inverse problems such as coefficient ones, provided the unknown functions can be described by polynomials with a degree not higher than a third one.

Applying the principle of step regularization of solutions combined with a recurrent parameter evaluation of the approximate dependence using least squares, we can diminish the admissible values $\Delta\tau$ compared to those chosen using a pure step-type principle [70, 158, 159–162, 165].

To specify the causal characteristics and heat state of an object, we can use the Kalmann linear estimation theory [23, 65, 96, 127, 128, 181]. The Kalmann method is recurrent and, from a theoretical point of view, it gives estimates whose accuracy increases with new measurements. When applied to ill-posed problems, however the Kalmann method is in itself not regular and it must be realized in combination with the principle of matching the results with errors of the input information or with the smoothing of the input temperature curves and natural (usually a step-type) regularization of the solution (for details see [23]). A trial-and-error method has been successfully used in [80, 82] for solving IHCPs of different types.

Thus, with the proper use of natural regularizing properties conditioned by the very process of heat propagation and also using properties incorporated in computational schemes, direct methods can be successfully applied for solving different inverse problems.

2.6.4 Regularization of Unstable Problems Using the Tikhonov Method

Despite the arguments given in favour of the natural principles of IHCP regularization, there are many situations when direct methods are not satisfactory.

First, they do not always enable us to reconstruct the most detailed structures of the causal characteristics desired. First, this is related to transient boundary conditions with varying dynamic character. At the same time, these are the very problems which are widely met in the process of thermal model testing and mathematical data processing of thermal tests of a space vehicle and its parts. Efforts have been made under these conditions to gain better results at the expense of raising the degree of approximation of a desired function or by increasing the number of iterations and selecting a minimized discrepancy following the Sobolev space W_2^k-generalized differentiable functions (instead of L_2); they have not yet yielded the anticipated effect and, inevitably, with some errors in the input information, lead to oscillations in the solution.

Second, many experimental thermal investigations are of so short duration that critical values of time steps with temperature field continuation and boundary condition reconstruction turn out to be comparable to a full length of experiment or they can even exceed it.

In all of the above cases, we could use a general method of regularization devised by A.N. Tikhonov and designed to solve unstable problems [137, 138, 140].

Consider an equation

$$Au = f, \quad u \in U, \, f \in F , \tag{2.44}$$

where U and F are metric spaces, A is the assigned operator ($A: U \to F$) and u, f the desired and observable elements, respectively (u and f can be vectors, functions, vector-functions). Later, we shall not discuss the conditions of solvability and uniqueness, assuming that problem (2.44) in this sense to be well-posed. Nevertheless, we consider that a stability condition for Eq. (2.44) can be violated.

The general principle for solving unstable problems suggested by A.N. Tikhonov uses the construction of regularizing operators.

Diverging temporarily from the effect of errors in the initial data, let us take *a regularizing operator* for problem (2.44) to be the operator R_α dependent upon the scalar parameter α to answer the following conditions:

1. R_α is determined in the whole space F for any $\alpha > 0$;
2. R_α is continuous in F;
3. for any $u \in U$: $R_\alpha Au \underset{\alpha \to 0}{\to} u$. (2.45)

The parameter α is called a *regularization parameter*. Hence, the introduction of the R_α regularizing operator is connected to an approximation of the A^{-1} inverse operator (which is not necessarily continuous) in the family of continuous operators.

The condition (2.45) provides convergence to an exact solution of Eq. (2.44). Considering now the problem with an error in the right side and in the operator, one can easily find that such a condition will yield an unstable result. That is why it is an important addition to the determination of a regularizing operator which assumes a condition of matching the regularization parameter α with an error of the initial data. Let us consider this question further.

The degree of approximation of the known element f_δ to an exact (but unknown) right side is characterized by the error δ:

$$\rho_F(f_\delta, f) \leqslant \delta ,$$

where ρ_F is a metric (a distance between elements) in space F.

Inaccuracy in the A operator assignment is usually combined with its approximation by a certain operator A_h (from a family of approximating operators) and it occurs when proceeding to a computational form of the problem.

As a deviation measure of the operators A and A_h specified on the U space, we introduce a metric

$$d_U(A_h, A) = \sup_{u \in U} \rho(A_h u, Au)$$

and consider the following condition as being satisfied

$$d_U(A_h, A) \leqslant h .$$

Thus, instead of exact initial data $\{A, f\}$ in problem (2.44) we introduce the approximated data $\{A_h, f_\delta\}$, the error of such substitution being characterized by a two-component vector $\sigma = \{\delta, h\}$, where δ and h are non-negative numbers.

Let us now give a full definition of a regularizing operator. The operator $R_\alpha = R[f_\delta, A_h, \alpha]$ is called a regularizing one, provided it has the following properties:

1. there is such a pair of numbers $\sigma_0 = \{\delta_0, h_0\}$, $\delta_0 > 0$, $h_0 > 0$, that the R_α operator is specified for every $\alpha > 0$ and any A_h, f_δ for which

$$\rho_F(f_\delta, f) \leqslant \delta \leqslant \delta_0, \quad d_U(A_h, A) \leqslant h \leqslant h_0 ;$$

2. there are such functions $\alpha = \alpha(\sigma)$ and $\sigma = \sigma(\varepsilon)$ that if for any $\varepsilon > 0$ the conditions $\delta \leqslant \delta(\varepsilon)$ and $h \leqslant h(\varepsilon)$ are satisfied, the solution being obtained $u_{\alpha(\sigma)} = R[f_\delta, A_h, \alpha(\sigma)]$ would deviate from an exact solution u_T (on the exact data A and f) not more than for ε:

$$\rho_U(u_{\alpha(\sigma)}, u_T) \leqslant \varepsilon .$$

The selection of the α parameter is performed on the basis of certain information about the problem being solved. With a properly chosen value of α the solution $u_\alpha(\sigma)$ deviates little from u_T, the accuracy of approximation being specified by accuracy of the assignment of f and A.

A regularizing operator together with the method for regularization parameter selection which answer the above formulated condition 2, is called a *regularizing algorithm* for problem (2.44).

An ill-posed problem can be called a *regularized one*, provided it has at least one regularizing algorithm.

There are different methods in use now for constructing the regularizing operators. *A variational method* [137, 138, 140–142] based on the minimization of a smoothing functional is the most used:

$$\Phi_\alpha[u, A, f] = \rho_F^2(Au, f) + \alpha\Omega[u], \quad u \in D(\Omega),$$

where $\Omega[u]$ is a stabilizing functional, $D(\Omega)$ is the domain of $\Omega[u]$.

By $\Omega[u]$ we assume such a functional that a set M_c assigned by the condition

$M_c = \{u: \Omega[u] \leqslant C\}$, $c \geqslant 0$ is compact in space U. As Ω we can take the deviation $\rho_V(u, u^*)$ of a desired solution u from some fixed element u^* in the metric of space V. Here, $V \subset U$ represents a subspace with a stronger metric, i.e. for any elements $u_1, u_2 \in V$ an equality is satisfied

$$\rho_U(u_1, u_2) \leqslant \rho_V(u_1, u_2).$$

It is proved that if the operator A is continuous, single-valued and additive (generally A can be non-linear), the functional $\Omega[u]$ is rigorously convex, then for any $f \in F$ there is a single element $u_\alpha \in D(\Omega)$, the latter minimizing the functional Φ_α, $\alpha > 0$.

With a proper selection of the α parameter matched with an error of input data, the $u_\alpha(\sigma)$ solution of the problem of minimizing the smoothing functional would converge (with the errors δ and h tending to zero) to a u_T, i.e. to an exact solution of the initial problem with exact data.

The problem of constructing algorithms to solve inverse heat conduction problems regularized according to the above considered variational method has been studied by various authors [1, 4, 7, 10, 16, 21, 22, 162, 164]. Some items on the theory of regularization methods applicable to IHCP solution and the corresponding results are given in Chap. 7.

Another effective method for constructing regularizing algorithms is treated in Chaps. 6 and 8. It is *the method of iterative regularization* for solving problems of thermal identification. The method has been suggested by the author [11–13, 15] and developed and substantiated further (see [18, 20, 25, 33–38, 48, 49, 125]).

2.7 Summary

1. General computational statements for boundary inverse heat transfer problems have been formulated to reveal the main specific features of these problems and to find suitable approaches to solve them.

   The IHCP formulations, being defined concretely and specified, will be referred to later. The integral forms of boundary and Cauchy statements, including multidimensional cases, will be fully represented in Chap. 3. Variational statements will be studied in greater detail in Chaps 6–8.
2. The question is considered on the correctness of initial statements of heat transfer inverse problems; their common property is set which is in violation of stability condition.
3. Major approaches are outlined for regularization of inverse problems, with the possibility of finding stable approximate solutions.

   These approaches will be fully realized in the form of appropriate computational algorithms in Chaps. 4–8.

Chapter 3

Analytical Forms of Boundary Inverse Heat Conduction Problems

Introduction

This chapter treats the problem statements on specification of transient heat loads, the concept of which involves *integral representations* of inverse heat conduction problems. To stabilize solutions obtained later in integral forms, one can employ different principles. A method of step regularization is suggested for this purpose in Chap. 4. Some other methods of stability damping are outlined in Chaps. 6 and 7, the integral statements described herein being suitable for them as well.

Section 3.1 treats *one-dimensional IHCP formulations* for a semi-infinite body and a plate with fixed and moving boundaries. The surface heat fluxes are sought while measurements are made inside or at the boundary of the body. In these cases, as is shown, the instability of the problem is related to the need to solve one or a set of Volterra equations of the first kind, which are obtained by Green's functions or the heat potential theory. One of the varieties of boundary IHCP formulation in which the input data are the results of *differential measurements* (i.e. when temperature differences are taken between some points of the body – with the help of differential thermocouples or thermopiles) is analyzed in Sect. 3.2.

Section 3.3 is devoted to different formulations of *three-dimensional* boundary IHCP in the multiply-connected domain. They are also obtained based on the methods of Green's functions and heat potentials, including ill-posed integral equations of the Volterra-Fredholm type.

Section 3.4 treats the integral form of a boundary IHCP statement in a *two-dimensional* case. The results presented here can be considered as a special case of the results in Sect. 3.3. A method is proposed in Sect. 3.5 on transformation of the original IHCP statement, which in some cases makes the algorithms of solving these problems much more simple. It is called *a method of fictitious boundaries* and is based on special transitions from domains of complex form, in particular, for bodies with moving boundaries, to domains of simple forms for which sufficiently simple Green's functions are already known or obtained.

3.1 Determination of Transient Boundary Conditions in a One-Dimensional Case

3.1.1 Bodies with Fixed Boundaries

It was shown in Sect. 2.1.2 that for a slab with constant thermophysical properties a boundary inverse problem statement consisting of the determination of heat flux at one of the boundaries of the slab is formulated as an integral equation of the first kind. The known solution of a direct heat conduction problem is used as reference material. Generalizing this approach for one-dimensional bodies of different geometry, flat and cylinder-spherical, boundary conditions of the first and second kind sought and assigned, one can write a common integral form of the boundary inverse problem for a one-dimensional equation of heat conduction with constant coefficients

$$\int_0^\tau u(\xi)\, K(\tau - \xi)\, \mathrm{d}\xi = f(\tau), \quad 0 < \tau \leqslant \tau_m\,, \tag{3.1}$$

where $u(\xi)$ is a solution sought (e.g. a heat flux or the body surface temperature).

For uniform problem statements, when a heat source in the body is not available, and the initial distribution and second (known) boundary conditions are zero, the function $f(\tau)$ represents a temperature change at a certain point of the body. In some other cases, the right-hand side includes integral members taking into account the inhomogeneity of a problem.

Actually, the $f(\tau)$ function and the integral equation kernel form $K(\tau - \xi)$ depend on the statement of a corresponding direct problem and they can be found by known methods of mathematical physics and heat conduction theory, such as Green's functions method, a superposition principle, Duhamel's integral, operational method and so on. A long list of literature concerning these methods and their applications [51, 69, 93, 94, 118, 143, 172 and other] gives tensive information about the techniques for specifying the functions $f(\tau)$ and $K(\tau - \xi)$ and helps to find corresponding expressions in every particular case.

Consider now the principle of constructing integral forms of boundary IHCP for domains with moving boundaries. We can illustrate this by the examples with a thermally thick wall and a slab of finite thickness provided that the points of temperature measurements do not move in the process of heating. Note that inverse problems in domains with fixed boundaries can be interpreted as a special case for the results obtained below.

3.1.2 A Thermally Thick Wall with a Moving Boundary

The second boundary value heat conduction problem for a semi-infinite solid body with a moving boundary written generally in terms of Kirchhoff's model temperature $\Theta = \dfrac{1}{\lambda_0} \int_0^T \lambda(T)\, \mathrm{d}T$ providing $a = \text{const}$ is

$$\frac{\partial \Theta}{\partial \tau} = a\frac{\partial^2 \Theta}{\partial x^2}, \quad X(\tau) < x, \quad 0 < \tau \leqslant \tau_m; \tag{3.2}$$

$$\Theta(x,0) = \varphi(x), \quad X(0) \leqslant x,$$

$$\lambda_0 \frac{\partial \Theta(X(\tau),\tau)}{\partial x} + q(\tau) = 0, \quad 0 \leqslant \tau \leqslant \tau_m, \tag{3.3}$$

$$\frac{\partial \Theta(\infty,\tau)}{\partial x} = 0, \quad 0 \leqslant \tau \leqslant \tau_m,$$

where $X(\tau)$ is the known law of boundary movement (a continuous function) and $\varphi(x)$ is the assigned continuous differentiable function, matched with boundary conditions.

From the physical point of view, the boundary movement can be conditioned by ablation, thermal shrinkage or linear expansion of the body material.

Assume that $q(\tau)$ is unknown and we have to find it using additional data about the temperature at some point $x = d$, $d > X(\tau)$: $\Theta(d,\tau) = f(\tau)$, $0 \leqslant \tau \leqslant \tau_m$. Reduce this problem to a problem with zero initial condition. For this, consider a continuous along with its derivative function $\psi(x)$, $x \geqslant C_1$, $C_1 < X(\tau)$ is such that $\psi(x) = \varphi(x)$ for $x \geqslant X(0)$. It is known that the expression

$$z(x,\tau) = \int_{C_1}^{\infty} \psi(\eta)\frac{\exp[-(x-\eta)^2/4a\tau]}{\sqrt{4\pi a\tau}}\,d\eta$$

is a solution of Eq. (3.2), satisfying the initial condition. The problem is then transformed into the following statement:

$$\frac{\partial \omega}{\partial \tau} = a\frac{\partial^2 \omega}{\partial x^2}, \quad X(\tau) < x, \quad 0 < \tau \leqslant \tau_m;$$

$$\omega(x,0) = 0, \quad X(0) \leqslant x;$$

$$\lambda_0 \frac{\partial \omega(X(\tau),\tau)}{\partial x} + q(\tau) + \lambda_0\frac{\partial z(X(\tau),\tau)}{\partial x} = 0, \quad 0 \leqslant \tau \leqslant \tau_m; \tag{3.4}$$

$$\frac{\partial \omega(\infty,\tau)}{\partial x} = 0, \quad 0 \leqslant \tau \leqslant \tau_m,$$

where $\omega(x,\tau) = \Theta(x,\tau) - z(x,\tau)$. A supplementary condition will acquire the form $\omega(d,\tau) = f(\tau) - z(d,\tau)$, $0 \leqslant \tau \leqslant \tau_m$.

Write the $\omega(x,\tau)$ as a thermal potential of simple layer

$$\omega(x,\tau) = \frac{a}{2\sqrt{\pi}}\int_0^{\tau} v(\xi)\frac{\exp[-(x-X(\xi))^2/4a(\tau-\xi)]}{\sqrt{a(\tau-\xi)}}\,d\xi, \tag{3.5}$$

where $v(\tau)$ is the potential density.

The representation (3.5) is a twice continuously differentiable function for the x coordinate and is continuously differentiable with respect to time τ; it satisfies a heat conduction homogeneous equation and zero initial condition. When passing through the $x = X(\tau)$ boundary, the thermal potential remains continuous, while the potential derivative for x undergoes a final discontinuity.

The derivative limiting values approaching the domain boundary from within or without are determined by the known discontinuity formula [143]

$$\frac{\partial\omega(X(\tau)\pm 0,\tau)}{\partial x} = \frac{\partial\omega^{\circ}(X(\tau),\tau)}{\partial x} \mp \frac{v(\tau)}{2}, \tag{3.6}$$

where

$$\frac{\partial\omega^{\circ}(X(\tau),\tau)}{\partial x} = -\frac{1}{4\sqrt{a\pi}}\int_0^{\tau} v(\xi)\frac{X(\tau)-X(\xi)}{(\tau-\xi)^{3/2}}\exp\left[-\frac{(X(\tau)-X(\xi))^2}{4a(\tau-\xi)}\right]\mathrm{d}\xi .$$

Taking into account that spatial derivatives in the boundary condition (3.4) are one-sided, we have

$$q(\tau) = -\lambda_0\frac{\partial z(X(\tau),\tau)}{\partial x} + \frac{\lambda_0}{2}\left\{v(\tau) + \int_0^{\tau} v(\xi)\frac{X(\tau)-X(\xi)}{2\sqrt{a\pi(\tau-\xi)^3}}\exp\left[-\frac{(X(\tau)-X(\xi))^2}{4a(\tau-\xi)}\right]\mathrm{d}\xi\right\}.$$

Note that for $X(\tau) = C = \mathrm{const}$ $q(\tau) = -\lambda_0\dfrac{\partial z(C,\tau)}{\partial x} + \dfrac{\lambda_0}{2}v(\tau)$.

The potential density $v(\tau)$ is to be obtained by the known function $f(\tau)$ from the solution of the Volterra integral equation of the first kind which follows from the representation (3.5) with $x = d$:

$$\int_0^{\tau} v(\xi)\frac{\exp[-(d-X(\xi))^2/4a(\tau-\xi)]}{\sqrt{a(\tau-\xi)}}\mathrm{d}\xi = \frac{2\sqrt{\pi}}{a}f^*(\tau), \quad 0<\tau\leqslant\tau_m ,$$

where

$$f^*(\tau) = f(\tau) - z(d,\tau) .$$

The instability of the IHCP under consideration is concentrated in this integral equation; its solution is the most important step in determining $q(\tau)$ and it must be regularized by some method.

Note that if besides $q(\tau)$ we have to specify the temperature $T(s,\tau)$ at some point of the body $x = s$, the value $\Theta(s,\tau)$ has to be found first:

$$\Theta(s,\tau) = \omega(s,\tau) + z(s,\tau) ,$$

and then turn to $T(s,\tau)$ using the relation $T(\Theta)$ following from the Kirchoff transformation $T = \lambda_0\int_0^{\Theta}\mathrm{d}\Theta/\lambda(\Theta)$. The $\lambda(\Theta)$ dependence must be calculated in advance with the help of the Kirchoff direct formula by the known function $\lambda(T)$.

If $\lambda(T) = \lambda_0 = \mathrm{const}$, we obtain an identity $T = \Theta$.

3.1.3 A Slab with a Moving Boundary

Let a slab with b original thickness be heated (cooled) from two sides by varying in time the heat fluxes $q_1(\tau)$ and $q_2(\tau)$. One of its boundaries is moveable and the displacement law $X(\tau)$ is known. Temperature distribution at the initial moment of time equals zero, or the reduction to a zero condition has been performed in a way similar to that described in Sect. 3.1.2. A model temperature in this case is given by the system of equations

$$\frac{\partial \Theta}{\partial \tau} = a\frac{\partial^2 \Theta}{\partial x^2}, \quad X(\tau) < x < b, \quad 0 < \tau \leqslant \tau_m;$$

$$\Theta(x, 0) = 0, \quad X(0) \leqslant x \leqslant b;$$

$$\lambda_0 \frac{\partial \Theta(X(\tau), \tau)}{\partial x} + q_1(\tau) = 0, \quad 0 \leqslant \tau \leqslant \tau_m;$$

$$\lambda_0 \frac{\partial \Theta(b, \tau)}{\partial x} + q_2(\tau) = 0, \quad 0 \leqslant \tau \leqslant \tau_m.$$

Search a solution of the direct problem as a sum of the simple layer thermal potentials

$$\Theta(x, \tau) = \int_0^\tau v_1(\xi) K(x, X(\xi); \tau, \xi)\, \mathrm{d}\xi + \int_0^\tau v_2(\xi) K(x, b; \tau, \xi)\, \mathrm{d}\xi, \qquad (3.7)$$

where v_1, v_2 are the potential densities to be estimated;

$$K(x, s; \tau, \xi) = \frac{1}{2}\sqrt{\frac{a}{\pi}}\frac{1}{\sqrt{\tau - \xi}} \exp\left[-\frac{(x - s)^2}{4a(\tau - \xi)}\right].$$

Suppose now that the temperature at two points of the slab $T(d_1, \tau)$ and $T(d_2, \tau)$, $d_1 < d_2$, $X(\tau) \leqslant d_1, d_2 \leqslant b$ is known, while the heat fluxes $q_1(\tau)$ and $q_2(\tau)$ are unknown quantities. In this case, the expression (3.7) enables us to write a system of two integral equations in order to find the unknown functions $v_1(\xi)$ and $v_2(\xi)$. The system can be represented as a Volterra equation of the first kind:

$$\int_0^\tau \boldsymbol{K}(\tau, \xi)\, \boldsymbol{v}(\xi)\, \mathrm{d}\xi = \boldsymbol{f}(\tau), \quad 0 < \tau \leqslant \tau_m. \qquad (3.8)$$

The bi-component vector functions are introduced here

$$\boldsymbol{v}(\xi) = [v_1(\xi), v_2(\xi)]^T, \quad \boldsymbol{f}(\tau) = [\Theta(d_1, \tau), \Theta(d_2, \tau)]^T,$$

where

$$\Theta(d_k, \tau) = \frac{1}{\lambda_0} \int_0^{T(d_k, \tau)} \lambda(T)\, \mathrm{d}T,$$

and a matrix of kernels

$$\boldsymbol{K}(\tau, \xi) = \begin{bmatrix} K(d_1, X(\xi); \tau, \xi) & K(d_1, b; \tau, \xi) \\ K(d_2, X(\xi); \tau, \xi) & K(d_2, b; \tau, \xi) \end{bmatrix}.$$

The given problem is unstable for small disturbances of the operator and those of the right-hand side. That is why its solution must be regularized.

A transition to unknown values of heat flux $q_1(\tau)$ and $q_2(\tau)$ can be performed after having calculated the $\partial\Theta/\partial x$ gradient at the slab boundaries found through the thermal potential derivative discontinuity property shown in equality (3.6). As a result, we receive the following expressions

$$q_1(\tau) = \frac{\lambda_0}{2}\left\{v_1(\tau) + \int_0^\tau v_1(\xi)\, V(X(\tau), X(\xi); \tau, \xi)\,d\xi + \int_0^\tau v_2(\xi)\, V(X(\tau), b; \tau, \xi)\,d\xi\right\};$$

$$q_2(\tau) = \frac{\lambda_0}{2}\left\{-v_2(\tau) + \int_0^\tau v_1(\xi)\, V(b, X(\xi); \tau, \xi)\,d\xi\right\},$$

where $V(z, s; \tau, \xi) = \dfrac{z - s}{2\sqrt{a\pi(\tau - \xi)^3}} \exp\left[-\dfrac{(z - s)^2}{4a(\tau - \xi)}\right]$.

Thus, the most complicated stage in solving the assigned inverse problem is to determine densities of the thermal potentials v_1 and v_2 from a solution of system (3.8).

3.2 Recovery of Boundary Conditions with a Differential Method of Measurement

In carrying out different thermophysical experiments as well as full-scale and model thermal tests of an aircraft and its elements, we observe a situation when it is difficult to provide thermostatting of "cold" junctions in the thermocouples. As a result, it becomes impossible to measure accurately enough the temperatures in the object under study. A procedure for recovering the absolute temperature meanings is suggested below. A solution of the boundary IHCP according to the results of differential tests (by differential thermocouples or thermopiles) is also considered.

3.2.1 Determination of Absolute Temperature Meanings

Let the temperature differences ΔT_1 and ΔT_2 be measured in the one-dimensional body between its points x_1, x_3 and x_2, x_3 respectively (Fig. 3.1). It is required to determine temperature $T_3(\tau)$ at point x_3. We assume that a heat transfer in the body is described by equation $\partial T/\partial\tau = a(\partial^2 T/\partial x^2)$ and it is

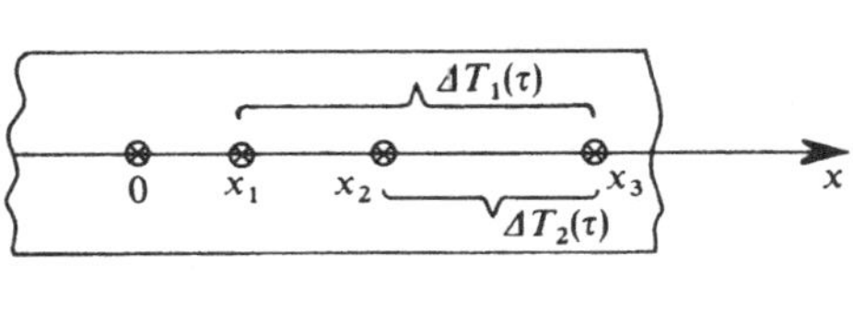

Fig. 3.1. Spatial temperature difference

known that the body heat condition at $\tau = 0$ corresponds to the homogeneous temperature distribution $T(x,0) = T_0$.

Introduce the substitution $v(x,\tau) = T(x,\tau) - T_3(\tau)$ and, assigning $x_1 = 0$ for a new dependent variable, write the problem with overdefined conditions:

$$\frac{\partial v}{\partial \tau} = a\frac{\partial^2 v}{\partial x^2} - Q(\tau); \tag{3.9}$$

$$v(x,0) = 0; \; v(0,\tau) = T_1(\tau) - T_3(\tau) \equiv \Delta T_1(\tau);$$

$$v(x_2,\tau) = T_2(\tau) - T_3(\tau) \equiv \Delta T_2(\tau); \quad v(x_3,\tau) = 0,$$

where $Q(\tau) = \mathrm{d}T_3(\tau)/\mathrm{d}\tau$.

Now, let us state an inverse problem: find the source $Q(\tau)$ in Eq. (3.9). According to the superposition principle, we write a solution of the direct problem for value $v(x_2,\tau)$:

$$v(x_2,\tau) = v_I(x_2,\tau) + v_{II}(x_2,\tau).$$

Here, v_I and v_{II} are the solutions corresponding to a homogeneous equation ($Q(\tau) = 0$) with a known condition ΔT_1 and to a nonhomogeneous Eq. (3.9) with a zero value ΔT_1:

$$v_I(x_2,\tau) = \int_0^\tau \Delta T_1(\xi)\, S(\tau - \xi)\,\mathrm{d}\xi; \quad v_{II}(x_2,\tau) = \int_0^\tau Q(\xi)\, K(\tau - \xi)\,\mathrm{d}\xi,$$

where [93, 172]

$$S(\tau - \xi) = \frac{2\pi a}{x_3^2}\sum_{k=1}^{\infty} k\exp\left[-\frac{k^2\pi^2 a}{x_3^2}(\tau - \xi)\right]\sin\frac{k\pi x_2}{x_3};$$

$$K(\tau - \xi) = \frac{4a}{\pi\lambda}\sum_{k=0}^{\infty}\frac{(-1)^k}{2k+1}\exp\left[-\frac{4a(2k+1)^2\pi^2(\tau - \xi)}{x_3^2}\right]$$

$$\times\cos\left[\frac{(2k+1)\pi\left(x_2 - \dfrac{x_3}{2}\right)}{x_3}\right].$$

Thus, for determining $T'_3(\tau)$, we have an integral equation

$$\int_0^\tau Q(\xi)\, K(\tau - \xi)\,\mathrm{d}\xi = f(\tau), \tag{3.10}$$

where $f(\tau) = \Delta T_2(\tau) - \int_0^\tau \Delta T_1(\xi)\, S(\tau - \xi)\,\mathrm{d}\xi$.

To obtain $T_3(\tau)$, we can integrate $T'_3(\tau)$ beginning from the known value $T_3(0) = T_0$. If the function is not of special interest now we can act otherwise and make use of the following property of integral convolution with two functions:

$$\int_0^\tau Q(\xi)\, K(\tau - \xi)\,\mathrm{d}\xi = \int_0^\tau\left(\int_0^\xi Q(\eta)\,\mathrm{d}\eta\right)\frac{\partial K(\tau - \xi)}{\partial \xi}\,\mathrm{d}\xi.$$

Then, in order to obtain $T_3(\tau)$ it is necessary to solve an integral equation

$$\int_0^\tau (T_3(\xi) - T_0)\frac{\partial K(\tau - \xi)}{\partial \xi} d\xi = f(\tau) . \tag{3.11}$$

Let us consider the right-hand side $f(\tau)$ of Eqs. (3.10) and (3.11) to be a continuous differentiable function and also note that $K(0) \neq 0$ and $\partial K(0)/\partial \xi \neq 0$. Then, having differentiated Eqs. (3.10) and (3.11) with respect to τ and dividing the obtained equalities into $K(0)$ and $\partial K(0)/\partial \xi$ respectively, we receive

$$Q(\tau) + \int_0^\tau Q(\xi)\frac{K_\tau(\tau - \xi)}{K(0)} d\xi = \frac{f'(\tau)}{K(0)} ; \tag{3.12}$$

$$(T_3(\tau) - T_0) + \int_0^\tau (T_3(\xi) - T_0)\frac{K_{\xi\tau}(\tau - \xi)}{K_\xi(0)} d\xi = \frac{f'(\tau)}{K_\xi(0)} . \tag{3.13}$$

Equations (3.12) and (3.13) are the Volterra integral equations of the second kind; if $f'(\tau)$ is a continuous function, their solutions $Q(\tau)$ and $T_3(\tau)$ are also continuous. Hence, if we can reconstruct the continuous first derivative by the experimentally assigned function $\Delta T_2(\tau)$, the equations obtained then allow us to calculate "good" approximations to the functions being sought. For this purpose we can use the algebraic recurrent method outlined in Chap. 4.

3.2.2 The Solution of a Boundary Inverse Problem by the Results from Temperature Difference Measurements

Consider the problem of recovering the heat flux at the slab boundary by the known condition on the other boundary and by the temperature drop prescribed between some inside points of the body x_1 and x_2. For definiteness, assume the following statement:

$$\partial T/\partial \tau = a\partial^2 T/\partial x^2 , \quad 0 < x < b , \quad 0 < \tau \leqslant \tau_m ;$$

$$T(x,0) = T_0 ; \quad T(x_1,\tau) - T(x_2,\tau) = \Delta T(\tau); \quad \partial T(b,\tau)/\partial x = 0 .$$

The task is to find value $q = -\lambda\, \partial T(0,\tau)/\partial x$.

Write the solution of the second boundary-value problem with a zero condition at one of the boundaries

$$T(x,\tau) = T_0 + \int_0^\tau q(\xi) K(x,\tau - \xi) d\xi , \tag{3.14}$$

where $K(x,\tau - \xi)$ is specified by a corresponding influence function (see formula (4.8)).

By substituting $x = x_1$ and $x = x_2$ into Eq. (3.14) and subtracting one expression from the other, we have an integral equation for defining $q(\tau)$ by the known temperature difference:

$$Aq \equiv \int_0^\tau q(\xi) [K(x_1,\tau - \xi) - K(x_2,\tau - \xi)] d\xi = \Delta T(\tau), \quad 0 < \tau \leqslant \tau_m . \tag{3.15}$$

Considering the physical sense of the influence function, it becomes clear that the conditionality of this problem will worsen as the points of setting the thermocouples come together. With values $\Delta x = x_1 - x_2$ less than some definite value, Eq. (3.15) can be considered as degenerate: $A \to 0$, $\Delta T(\tau) \to \omega$, where 0 is a zero operator and ω is the noise of measurements.

Another limiting case is of interest when measuring the temperature difference between two boundaries of the body $(x_1 = 0, x_2 = b)$. In this case, the IHCP conditionality in the given statement is the best.

Taking advantage of this technique, it is not difficult to write the IHCP integral forms for a whole number of other cases including some for domains with the moveable boundaries. We can also elaborate upon the corresponding numerical procedures of the IHCP solution in the non-linear statement, when thermophysical properties of the body are dependent on temperature.

3.3 Analytical Forms of Multidimensional Inverse Problems

In simulating the heat conditions for structural elements of an aircraft and other machines and equipment, a situation can arise when we have to solve the IHCPs in two- or even three-dimensional statements. Let us formulate some approaches to a solution for the multidimensional boundary IHCPs.

3.3.1 The Statement of a Three-Dimensional Inverse Problem

Let a solid body be given definite geometry and dimensions with heat sources inside; the body is in heat interaction with the environment. Introduce in our consideration the dimensionless coordinates $\bar{x} = x/d$, $\bar{y} = y/d$, $\bar{z} = z/d$, where d is any characteristic dimension of the body; associate the body with a certain domain Ω of three-dimensional space $E_3(\bar{x}0\bar{y}\bar{z})$ bounded by a surface $\partial\Omega$ (Fig. 3.2). Let $\bar{\tau}$ denote the nondimensional time $\bar{\tau} = a\tau/d^2$.

If thermal diffusivity a is thought of as constant, we can write an equation defining the body temperature field within the points of the Ω domain (for simplification, all dashes over independent variables are conventionally suppressed)

$$\partial T(P,\tau)/\partial\tau = \partial^2 T(P,\tau)/\partial x^2 + \partial^2 T(P,\tau)/\partial y^2 + \partial^2 T(P,\tau)/\partial z^2 + Q(P,\tau)\,, \tag{3.16}$$

where $Q(P,\tau)$ is the density of inside heat sources (known function), $P \in \Omega$.

The solution of Eq. (3.16) is sought in the cylindrical area $R = \Omega \times [0, \tau_m]$ with a lateral surface $\Sigma_\Omega = \partial\Omega \times (0, \tau_m]$ having a formation element parallel to the axis τ.

Suppose that a temperature distribution is set at zero time

$$T(P,0) = \psi(P)\,; \quad P \in \bar{\Omega} = \Omega + \partial\Omega, \tag{3.17}$$

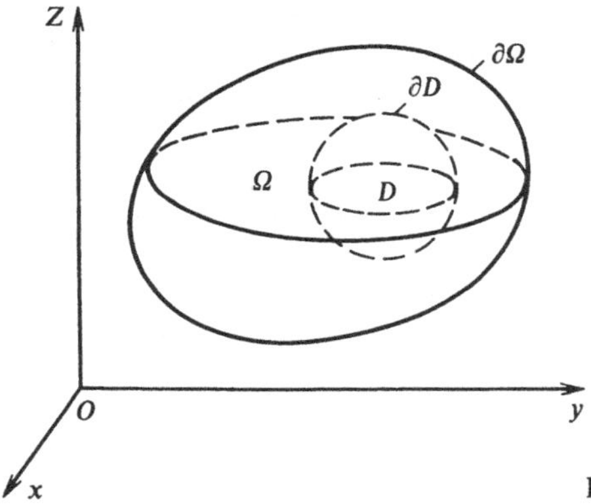

Fig. 3.2. The domain in three-dimensional space

the boundary condition of the second kind is known

$$\left.\frac{\partial T}{\partial n_\omega}\right|_{\partial\Omega} = -\frac{q(\omega,\tau)}{\lambda} \equiv \rho(\omega,\tau), \quad \omega \in \partial\Omega, \quad \tau \geqslant 0 \tag{3.18}$$

and matched with the initial distribution

$$\left[\left.\frac{\partial\psi(P)}{\partial n_\omega}\right|_{P\in\partial\Omega} = \rho(\omega,0)\right].$$

Here, n_ω is the direction of the inner normal to the surface $\partial\Omega$ and λ is the body thermal conductivity.

The direct problem is stated as follows: find the temperature field $T(P,\tau)$ in the domain $\bar{\Omega}$, which is twice continuously differentiable in the x, y, z coordinates and continuously differentiable in τ time, and which satisfies the heat conduction equation (3.17) and initial and boundary conditions (3.17) and (3.18).

Now, let the domain D with the boundary ∂D be located inside $\bar{\Omega}$, the surface $\partial\Omega$ enveloping the surface ∂D. At the boundary ∂D a temperature is given

$$T|_{\partial D} = f(l,\tau), \quad l \in \partial D, \quad 0 \leqslant \tau \leqslant \tau_m, \tag{3.19}$$

matched with the initial condition, i.e. with $\tau = 0$ $\psi(P)|_{P\in\partial D} = f(l,0)$. The boundary conditions on the surface $\partial\Omega$ are unknown. It is required to solve an inverse problem, i.e. knowing the condition (3.19) and the initial distribution (3.17) we are to find a heat flux density $q(\omega,\tau)$ or, which is the same, a normal derivative $\left.\dfrac{\partial T}{\partial n_\omega}\right|_{\partial\Omega} = \rho(\omega,\tau)$ at all points $\omega \in \partial\Omega$.

After this the temperature field definition in the domain $\bar{\Omega}$ will correspond to a direct heat conduction problem.

3.3.2 The Inverse Problem Integral Form Constructed via the Influence Function Method

At first, consider one of the possible approaches to a solution of the inverse problem, which rests on the basis of the method of influence functions of point heat sources (Green's functions). If the Green's function G_{II} of the second boundary-value heat conduction problem is known, the definition of an unknown function $\rho(\omega, \tau)$ is reduced to a solution of the following integral equation of the first kind:

$$-\int_0^\tau \int_{\partial\Omega} p(\omega, \xi)\, G_{II}(l, \omega; \tau, \xi)\, \mathrm{d}S_\omega\, \mathrm{d}\xi = \gamma(l, \tau), \quad l \in \partial D, \quad 0 < \tau \leqslant \tau_m;$$

where

$$\gamma(l, \tau) = f(l, \tau) - \int_\Omega \psi(P')\, G_{II}(l, P'; \tau, 0)\, \mathrm{d}P'$$

$$- \int_0^\tau \int_\Omega Q(P', \xi)\, G_{II}(l, P'; \tau, \xi)\, \mathrm{d}P'\, \mathrm{d}\xi;$$

$\int_{\partial\Omega} \ldots \mathrm{d}S_\omega$ is a double integral over the surface $\partial\Omega$;

$\int_\Omega \ldots \mathrm{d}P'$ is a triple integral over the three-dimensional domain Ω;

$\mathrm{d}S_\omega$ is an element of the surface $\partial\Omega$; $\mathrm{d}P' = \mathrm{d}x'\, \mathrm{d}y'\, \mathrm{d}z'$.

With general assumptions concerning the surface bounding the domain Ω ($\partial\Omega$ must be a Lyapunov surface, see Sect. 3.3.3), the Green's function G_{II} is known (see [135]). It is like

$$G_{II}(P, \omega^*; \tau, \xi) = U(P, \omega^*; \tau, \xi)$$
$$+ \int_\xi^\tau \mathrm{d}\xi' \int_{\partial\Omega} \sum_{k=0}^{\infty} 2^k v_k(P, \omega; \tau, \xi')\, U(\omega, \omega^*; \xi', \xi)\, \mathrm{d}S_\omega, \tag{3.20}$$

where $U(P, \omega^*, \tau, \xi) = \dfrac{1}{8[\pi(\tau - \xi)]^{3/2}} \exp\left[-\dfrac{r^2_{P\omega^*}}{4(\tau - \xi)}\right]$;

$r_{P\omega^*}$ is the distance between some fixed points $P \in \Omega$ and $\omega^* \in \partial\Omega$;

$$U(P, \omega^*; \tau, \xi) \equiv v_0(P, \omega^*, \tau, \xi),$$

$$v_k(P, \omega^*; \tau, \xi) = \int_\xi^\tau \mathrm{d}\xi' \int_{\partial\Omega} v_{k-1}(\omega, \omega^*; \xi', \xi) \frac{\partial U(P, \omega; \tau, \xi')}{\partial n_\omega}\, \mathrm{d}S_\omega;$$

n_ω is direction of the inside normal at the point $\omega \in \partial\Omega$. A series in Eq. (3.20) converges absolutely and uniformly.

Thus, we have obtained a formal notation of an integral equation for solving a boundary IHCP under study, consisting in the specification of heat flux on the surface of the body from temperature measurements at some inside surface ∂D. As the inside points l move away from the boundary ω, i.e. with the increase of

$r_{l\omega}$ distance and as the surface ∂D contracts to a point, an inverse problem would become even more difficult to find a solution.

For the whole number of really interesting domains Ω, the influence function can be obtained (see [172]) as an explicit analytical expression (the half-space $x > 0$; the infinite slab $0 < x < b$; the rectangular parallelepiped $0 < x < b$, $0 < y < c$; $0 < z < d$; the dihedral angle of π/m, m magnitude is a natural number; the domains bounded by outside and inside spherical or cylindrical surfaces; cones, etc.). Under these conditions, an IHCP solution by the above method is noticeably simplified. But generally such an approach is highly labour-consuming since we have to solve a three-dimensional integral equation of the first kind where the kernel and the right-hand side are functions whose calculating involves multiple numerical integration and infinite series computation. That is why another analytical approach is suggested in the next chapter for solution of the given inverse problem in the boundary-value statement.

3.3.3 Thermal Potentials in a Three-Dimensional IHCP Integral Representation

Introduce a new dependent variable

$$u(P,\tau) = T(P,\tau) - T_0(P,\tau) - T_Q(P,\tau)\,. \tag{3.21}$$

Assume that function $T_0(P,\tau)$ is continuously differentiable in the direction of the normal n_ω to $\partial\Omega$ everywhere in $\bar{\Omega}$; it satisfies a homogeneous heat conduction equation ($Q(P,\tau) = 0$) in R and with $\tau = 0$ it becomes equal to the initial temperature distribution $T_0(P,0) = \psi(P), P \in \bar{\Omega}$. Function $T_Q(P,\tau)$ must have a continuous derivative in n_ω everywhere in $\bar{\Omega}$ and satisfy Eq. (3.16) as well as be set equal to zero at initial time $T_Q(P,0) = 0$, $P \in \bar{\Omega}$. Then, for the $u(P,\tau)$ determination, we can present a simpler boundary-value problem

$$\frac{\partial u(P,\tau)}{\partial \tau} = \frac{\partial^2 u(P,\tau)}{\partial x^2} + \frac{\partial^2 u(P,\tau)}{\partial y^2} + \frac{\partial^2 u(P,\tau)}{\partial z^2}, \quad (P,\tau) \in R\,; \tag{3.22}$$

$$u(P,0) = 0\,, \quad P \in \bar{\Omega}\,; \tag{3.23}$$

$$\left.\frac{\partial u}{\partial n_\omega}\right|_{\partial\Omega} = \bar{p}(\omega,\tau) \equiv p(\omega,\tau) - \left.\frac{\partial T_0}{\partial n_\omega}\right|_{\partial\Omega} - \left.\frac{\partial T_Q}{\partial n_\omega}\right|_{\partial\Omega}, \quad (\omega,\tau) \in \textstyle\sum_\Omega. \tag{3.24}$$

Reduction to a homogeneous statement given by Eqs. (3.22)–(3.24) can be performed with the following selection of auxillary functions:

$$T_0(P,\tau) = \int_\Omega \psi(P')\,G(r_{P'P},\tau)\,\mathrm{d}P'\,; \tag{3.25}$$

$$T_Q(P,\tau) = \int_0^\tau \int_\Omega Q(P',\xi)\,G(r_{P'P},\tau-\xi)\,\mathrm{d}P'\,\mathrm{d}\xi\,, \tag{3.26}$$

where $G(r_{P'P},\tau-\xi)$ is a fundamental solution of a three-dimensional heat conduction equation:

$$G(r_{P'P}, t) = \begin{cases} \dfrac{1}{8(\pi t)^{3/2}} \exp\left[-\dfrac{r^2_{P'P}}{4t}\right] & \text{with } t > 0\,, \\ 0 \text{ with } t \leqslant 0\,, \\ (\text{except for } t = 0,\ r_{P'P} = 0)\,; \end{cases}$$

$r_{P'P}$ is the distance between the points $P'(x', y', z')$ and $P(x, y, z) \in \Omega$;

$$r_{P'P} = |P' - P| = \sqrt{(x' - x)^2 + (y' - y)^2 + (z' - z)^2}\,.$$

Let function $\psi(P)$ be continuous in the domain $\bar{\Omega}$ and continuously differentiable at some vicinity of the boundary $\partial\Omega$. Assume also that function $Q(P, \tau)$ holds a Hölder continuity with respect to P in $\bar{\Omega}$ uniformly with respect to τ, i.e. for all P and $P' \in \bar{\Omega}$ an inequality is satisfied

$$|Q(P, \tau) - Q(P', \tau)| \leqslant A\, r^{\beta}_{PP'}\,, \tag{3.27}$$

where A and β are the constants, $0 < \beta \leqslant 1$.

The uniformity condition with respect to τ additionally requires the existence of a certain minimum value A, for which condition (3.27) is satisfied at all values of τ.

Note that to satisfy condition (3.27), it is not sufficient to have only the continuity of the function. At the same time, if a scalar field $Q(P, \tau)$ for any τ is differentiable in the domain $\bar{\Omega}$, i.e. if it is differentiable at every point of this domain, the function $Q(P, \tau)$ will satisfy the Hölder condition in advance for an arbitrary index $0 < \beta \leqslant 1$.

The above requirements set for functions $\psi(P)$ and $Q(P, \tau)$ are in no way too limiting in practical applications, enabling use of the results proven in reference [179] and to substantiate the validity of the statement in Eqs. (3.22)–(3.24) using functions (3.25) and (3.26). Specifically, the solution $T(P, \tau)$ of a direct heat conduction problem, satisfying the conditions set out in Eqs. (3.21)–(3.24) with the given selection of functions $T_0(P, \tau)$, $T_Q(P, \tau)$ and continuous function $\rho(\omega, \tau)$, $(\omega, \tau) \in \Sigma_{\Omega}$, does exist and is unique.

Consider now a problem inverse to the statement in Eqs. (3.22)–(3.24), and namely, with respect to a known function

$$u|_{\partial\Omega} = \varphi(l, \tau)\,, \quad l \in \partial D\,, \quad 0 \leqslant \tau \leqslant \tau_m$$

define a normal derivative $\left.\dfrac{\partial u}{\partial n_{\omega}}\right|_{\partial\Omega}$ at point ω of the boundary $\partial\Omega$. Having solved this simpler problem, we may find an unknown function $\rho(\omega, \tau)$ from condition (3.24).

To solve this inverse problem we can make use of the temperature field representation by means of surface thermal potentials. Let us introduce into our consideration a thermal potential of the simple layer, induced by some simple sources distributed along the boundary surface $\partial\Omega$:

$$U(P,\tau)=\int_0^\tau\int_{\partial\Omega} v(\omega,\xi)\,G(r_{\omega P},\tau-\xi)\,\mathrm{d}S_\omega\,\mathrm{d}\xi\,,\quad P(x,y,z)\in\Omega\,,$$

$$\omega(\eta,\zeta,\gamma)\in\partial\Omega\,, \tag{3.28}$$

where function $v(\omega,\xi)$ characterizes the density of sources.

Next we assume that $v(\omega,\xi)$ is continuous in the Σ_Ω function and $\partial\Omega$ belongs to the $C^{1+\beta}$ class. Let us elucidate the last condition. First, let the surface $\partial\Omega$ at each of its points ω^* have a normal. Then introduce a local Cartesian system of coordinates η^*,ζ^* and γ^* with the origin at point ω^*; the axis ζ^* is pointed by the norm to $\partial\Omega$ (Fig. 3.3). It is known that $\partial\Omega\in C^k$ (k is a natural number), provided that the number $d=\text{const}>0$ does exist and it has the following property: a sphere of radius d with the centre at some arbitrary point $\omega^*\in\partial\Omega$ cuts out a section from the $\partial\Omega$ surface, which in the local coordinate system η^*,ζ^*,γ^* can be assigned by an equation $\zeta^*=f(\eta^*,\gamma^*)$, where all partial derivatives of the f function up to k-power inclusive do exist and they are continuous functions. If, in addition, all derivatives of k-order are continuous as according to Hölder with β index [see inequality (3.27)] and neither the index β nor the Hölder constant A are dependent of ω^*, then we can say that $\partial\Omega\in C^{k+\beta}$ and is a Lyapunov surface.

Note that if $\partial\Omega$ is a twice continuously differentiable surface, it belongs in advance to the $C^{1+\beta}$ class with any index β, i.e. a more powerful condition $\partial\Omega\in C^2$ is sufficient for further consideration.

Under these assumptions the thermal potential [Eq. (3.28)] satisfies a heat conduction equation everywhere in E_3 and, with $\tau\to 0$, tends to 0. A derivative of the potential $\partial U/\partial n_\omega$ in the direction of normal to $\partial\Omega$ surface has a discontinuity of the first kind in going via Σ. A limiting value of the normal derivative at some point $\omega^*\in\partial\Omega$ (Fig. 3.4), gained from within the domain, i.e. when $P\to\omega^*$, $P\in\Omega$, is determined by a discontinuity formula [108]

$$\left.\frac{\partial U(\omega^*,\tau)}{\partial n_{\omega^*}}\right|_i=-\frac{v(\omega^*,\tau)}{2}+\int_0^\tau\int_{\partial\Omega} v(\omega,\xi)\frac{\partial}{\partial n_{\omega^*}}G(r_{\omega\omega^*},\tau-\xi)\,\mathrm{d}S_\omega\,\mathrm{d}\xi\,. \tag{3.29}$$

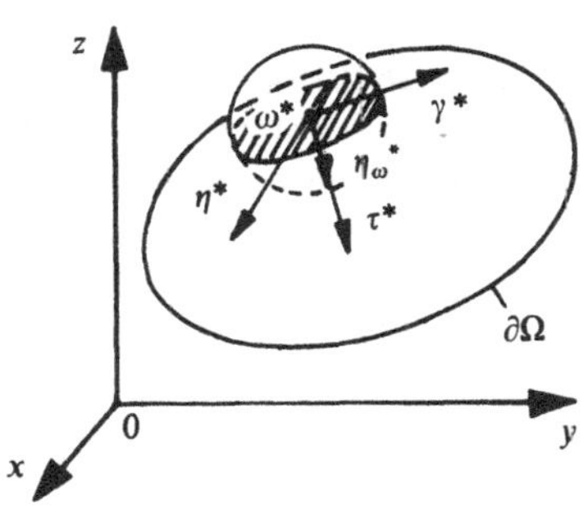

Fig. 3.3. Local coordinate systems

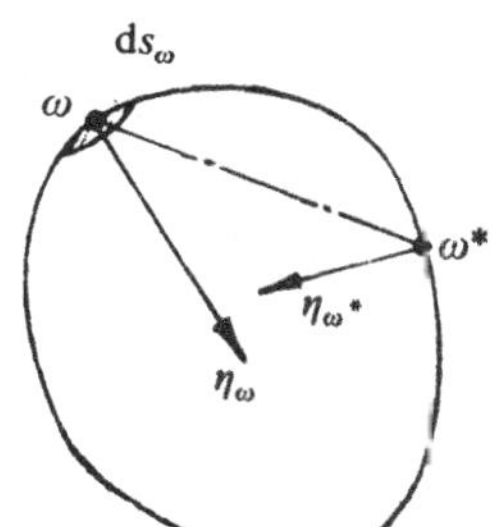

Fig. 3.4. Points ω, ω^* at the boundary $\partial\Omega$

Now let function $v(\omega, \xi)$ satisfy the input conditions of an inverse problem, namely, an equality must be satisfied for $P = l \in \partial D$

$$U(l,\tau) \equiv \int_0^\tau \int_{\partial\Omega} v(\omega,\xi) G(r_{\omega l}, \tau - \xi) \, dS_\omega \, d\xi = \varphi(l,\tau), \quad 0 < \tau \leqslant \tau_m, \tag{3.30}$$

where $\varphi(l,\tau) = f(l,\tau) - T_0(l,\tau) - T_Q(l,\tau)$. In this case, a solution of the boundary-value problem in Eqs. (3.22)–(3.24) will be defined by the potential $u(P,\tau) = U(P,\tau), (P,\tau) \in R$ and the desired function $\rho(\omega^*,\tau)$ will be calculated in accordance with condition (3.24) and formula (3.29):

$$\rho(\omega^*.\tau) = \bar{\rho}(\omega^*,\tau) + \left.\frac{\partial T_0}{\partial n_{\omega^*}}\right|_{\partial\Omega} + \left.\frac{\partial T_Q}{\partial n_{\omega^*}}\right|_{\partial\Omega},$$

where $\bar{\rho}(\omega^*,\tau) = -\dfrac{v(\omega^*,\tau)}{2} + \int_0^\tau \int_{\partial\Omega} v(\omega,\tau) \dfrac{\partial}{\partial n_{\omega^*}} G(r_{\omega\omega^*}, \tau - \xi) \, dS_\omega \, d\xi$.

The derivative of $G(r_{\omega\omega^*}, \tau - \xi)$ in the direction of the normal is

$$\frac{\partial}{\partial n_{\omega^*}} G(r_{\omega\omega^*}, \tau - \xi) = \frac{r_{\omega\omega^*}}{16\pi^{3/2}(\tau - \xi)^{5/2}} \exp\left[-\frac{r_{\omega\omega^*}}{4(\tau - \xi)}\right] \cos(r_{\omega\omega^*}, n_{\omega^*}). \tag{3.31}$$

With the result (3.31) we have a formula

$$\bar{\rho}(\omega^*,\tau) = -\frac{v(\omega^*,\tau)}{2} + \frac{1}{16\pi^{3/2}} \int_{\partial\Omega} dS_{\omega^*} \int_0^\tau v(\omega,\xi) \frac{r_{\omega\omega^*}}{(\tau - \xi)^{5/2}}$$
$$\times \exp\left[-\frac{r^2_{\omega\omega^*}}{4(\tau - \xi)}\right] \cos(r_{\omega\omega^*}, n_{\omega^*}) \, d\tau .$$

Thus, the major part of IHCP solution includes defining the $v(\omega, \xi)$ function from a solution of the integral Eq. (3.30) which is an equation of the first kind, a Volterra-type with respect to τ and a Fredholm-type with respect to P. It follows that a solution as shown in Eq. (3.30) must be regularized in some way.

3.3.4 A Three-Dimensional Inverse Problem in the Cauchy Statement

Let us consider now another form of the boundary inverse problem solving in a three-dimensional space, which will correspond to a Cauchy statement. Suppose that a temperature and the temperature normal derivative are given at the inside boundary ∂D (the Cauchy conditions):

$$T\Big|_{\partial D} = f_1(l,\tau), \quad \left.\frac{\partial T}{\partial n_l}\right|_{\partial D} = f_2(l,\tau), \quad l \in \partial D, \quad 0 \leqslant \tau \leqslant \tau_m,$$

matched with the initial temperature distribution, where n_l is a direction of the normal towards surface ∂D at point l; f_1 and f_2 are known functions.

The task is to define function $\rho(\omega, \tau) = \left.\frac{\partial T}{\partial n_\omega}\right|_{\partial\Omega}$ at all points $\omega \in \partial\Omega$. In other respects, we shall regard an IHCP statement as identical to that considered above. The problem thus formulated is called an "outside" IHCP in the Cauchy statement since the boundary with a desired function encloses the boundary with initial data.

As previously, let the initial inverse problem be reduced to a homogeneous statement assigned by the following conditions

$$\frac{\partial u(P,\tau)}{\partial \tau} = \frac{\partial^2 u(P,\tau)}{\partial x^2} + \frac{\partial^2 u(P,\tau)}{\partial y^2} + \frac{\partial^2 u(P,\tau)}{\partial z^2}, \quad (P,\tau) \in R; \tag{3.32}$$

$$u(P,0) = 0, \quad P \in \Omega; \tag{3.33}$$

$$\left.\frac{\partial u}{\partial n_l}\right|_{\partial D} = \bar{f}_2(l,\tau), \quad \left. u\right|_{\partial D} = \bar{f}_1(l,\tau), \quad l \in \partial D, \quad 0 \leqslant \tau \leqslant \tau_m, \tag{3.34}$$

where

$$\bar{f}_1 = f_1(l,\tau) - T_0(l,\tau) - T_Q(l,\tau);$$

$$\bar{f}_2 = f_2(l,\tau) - \left.\frac{\partial T_0}{\partial n_l}\right|_{\partial D} - \left.\frac{\partial T_Q}{\partial n_l}\right|_{\partial D}$$

Continue the solution of the heat conduction equation with the Cauchy data in the direction of the $\partial\Omega$ surface, presenting $u(P,\tau)$ as a sum of thermal potentials of the simple and double layer $u(P,\tau) = U(P,\tau) + W(P,\tau)$, $P \in (\bar{\Omega} - \bar{D})$.

The $U(P,\tau)$ potential is determined by the action of simple sources of the v strength (density) attached to the ∂D surface:

$$U(P,\tau) = \int_0^\tau \int_{\partial D} v(l,\xi)\, G(r_{lP}, \tau - \xi)\, \mathrm{d}S_l\, \mathrm{d}\xi, \quad 0 < \tau \leqslant \tau_m, \quad P \in (\bar{\Omega} - \bar{D}),$$
$$l \in \partial D.$$

The potential $W(P,\tau)$ is created by dipolar sources with a dipole axis directed along the outside normal to ∂D when the positive heat sources of $\mu(l,\xi)/h$ intensity are distributed on the one side of the ∂D surface at a small distance $h/2$ from it and the negative heat sources (run-offs)-$\mu(l,\xi)/h$ are located on the other

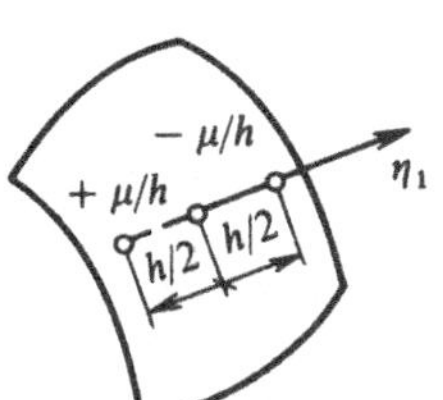

Fig. 3.5. The dipolar sources

side of the surface and at the same distance from it:

$$W(P,\tau)=\int_0^\tau\int_{\partial D}\mu(l,\xi)\frac{\partial}{\partial n_l}G(r_{lP},\tau-\xi)\,dS_l\,d\xi,$$
$$=\frac{1}{16\pi^{3/2}}\int_0^\tau\int_{\partial D}\mu(l,\xi)\frac{r_{lP}}{(\tau-\xi)^{5/2}}\exp\left[-\frac{r_{lP}^2}{4(\tau-\xi)}\right]\cos(r_{lP},n_l)\,dS_l\,d\xi\,.$$

From a combined action of simple and double sources within the $(\bar{\Omega}-\bar{D})$ domain a certain temperature field is realized $u(P,\tau)$ satisfying the heat conduction Eq. (3.32) and the initial condition (3.33); following from the properties of thermal potentials. Thus functions $u(P,\tau)$ and $\partial u(P,\tau)/\partial n_{l*}$ pass into some limiting values u_e and $\partial u_e/\partial n_{l*}$ on the ∂D boundary (the symbol e stands for approach along the normal to boundary ∂D of domain D from outside $P\to l^*$, $P\in\bar{\Omega}-\bar{D}$) in accordance with the transition formulae for the potentials [108]:

$$W_e(l^*,\tau)=-\frac{\mu(l^*,\tau)}{2}+\int_0^\tau\int_{\partial D}\mu(l,\xi)\frac{\partial}{\partial n_l}G(r_{ll^*},\tau-\xi)\,dS_l\,d\xi\,;$$

$$\left.\frac{\partial U(l^*,\tau)}{\partial n_{l^*}}\right|_e=\frac{v(l^*,\tau)}{2}+\int_0^\tau\int_{\partial D}v(l,\xi)\frac{\partial}{\partial n_{l^*}}G(r_{ll^*},\tau-\xi)\,dS_l\,d\xi\,.$$

The normal derivative of potential W and potential U itself are continuous passing through the ∂D surface and for these functions such a transition is performed in a purely formal way.

Thus, to determine the unknown densities v and μ of potentials with simple and double layers we can apply the following system of equations:

$$-\frac{\mu(l^*,\tau)}{2}+\int_0^\tau\int_{\partial D}\mu(l,\tau)\frac{\partial}{\partial n_l}G(r_{ll^*},\tau-\xi)\,dS_l\,d\xi$$
$$+\int_0^\tau\int_{\partial D}v(l,\xi)\,G(r_{ll^*},\tau-\xi)\,dS_l\,d\xi=\bar{f}_1(l^*,\tau)\,;$$
$$\int_0^\tau\int_{\partial D}\mu(l,\tau)\frac{\partial^2}{\partial n_l^2}G(r_{ll^*},\tau-\xi)\,dS_l\,d\xi+\frac{v(l^*,\tau)}{2} \tag{3.35}$$
$$+\int_0^\tau\int_{\partial D}^{\tau}v(l,\xi)\frac{\partial}{\partial n_{l^*}}G(r_{ll^*},\tau-\xi)\,dS_l\,d\xi=\bar{f}_2(l^*,\tau)\,,$$

where $(l^*,\tau)\in\sum_D=\partial D\times(0,\tau_m)\,.$

Note that Eqs. (3.35) is a system of Volterra linear integral equations of the second kind (with respect to τ) and for its solution it is possible to apply a successive approximation method.

Having determined the desired functions $\mu(l^*,\tau)$ and $v(l^*,\tau)$, we can calculate a temperature field continuation $u(P,\tau)$ in the domains $0<\tau\leqslant\tau_m$, $P\in(\bar{\Omega}-\bar{D})$. This process, with large distances of P points from the l boundary will entail a considerable loss of accuracy.

Note that an IHCP statement described above satisfies the case when there is an inner cavity in the solid body on which we can measure (or assign) temperature and heat flux.

3.3.5 The Interior Problem in the Cauchy Statement

Suppose that the Cauchy conditions $f_1(\omega,\tau)$ and $f_2(\omega,\tau)$ are known at the exterior boundary $\partial\Omega$ of the Ω domain and we have to solve an inverse problem by temperature field extension into the interior of the Ω domain. The domain Ω, therefore, can be multiply connected with the closed surface $\partial\Omega$, embracing all inside spaces with boundaries $\partial D_1, \partial D_2, \ldots, \partial D_n$ (Fig. 3.6). The conditions are not assigned for interior boundaries, but their action is reflected in either dependences $f_1(\omega,\tau)$ and $f_2(\omega,\tau)$, characterizing the exterior heat conditions.

Consider a homogeneous IHCP statement

$$\partial u(P,\tau)/\partial\tau = \partial^2 u(P,\tau)/\partial x^2 + \partial^2 u(P,\tau)/\partial y^2 + \partial^2 u(P,\tau)/\partial z^2 ,$$

$$P \in \Omega , \; 0 < \tau \leqslant \tau_m ,$$

$$u(P,0) = 0, \quad P \in \bar{\Omega} ;$$

$$u\Big|_{\partial\Omega} = \bar{f}_1(\omega,\tau); \quad \frac{\partial u}{\partial n_\omega}\Big|_{\partial\Omega} = \bar{f}_2(\omega,\tau) , \quad \omega \in \partial\Omega , \quad 0 \leqslant \tau \leqslant \tau_m ,$$

As before, write a solution $u(P,\tau)$ via the sum of two potentials

$$u(P,\tau) = U(P,\tau) + W(P,\tau) , \quad P \in \Omega ,$$

$$\text{where } U(P,\tau) = \int_0^\tau \int_{\partial\Omega} v(\omega,\xi) G(r_{\omega P}, \tau - \xi) \, \mathrm{d}S_\omega \, \mathrm{d}\xi ;$$

$$W(P,\tau) = \int_0^\tau \int_{\partial\Omega} \mu(\omega,\xi) \frac{\partial}{\partial n_\omega} G(r_{\omega P}, \tau - \xi) \, \mathrm{d}S_\omega \, \mathrm{d}\xi ,$$

$$P \in \Omega , \; \omega \in \partial\Omega .$$

Passing onto the $\partial\Omega$ boundary, from within the domain we can obtain a system of linear integral equations for determining the potential densities

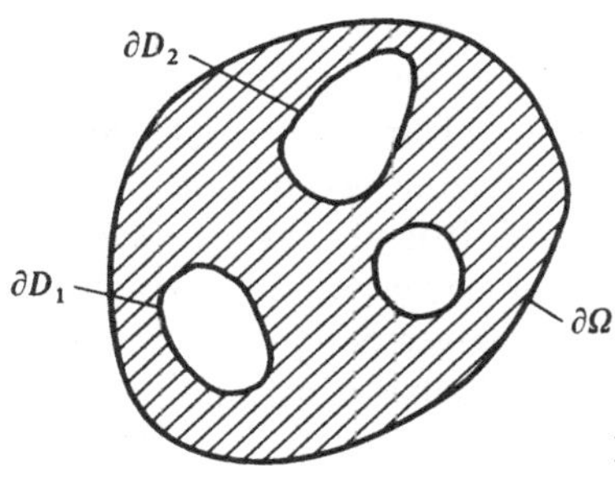

Fig. 3.6. A multiply-connected domain of IHCP solution

$\nu(\omega,\xi)$ and $\mu(\omega,\xi)$:

$$\int_0^\tau \int_{\partial\Omega} \nu(\omega,\xi)\, G(r_{\omega\omega^*},\tau-\xi)\,\mathrm{d}S_\omega\,\mathrm{d}\xi + \frac{\mu(\omega^*,\tau)}{2}$$

$$+\int_0^\tau \int_{\partial\Omega} \mu(\omega,\xi)\frac{\partial}{\partial n_\omega} G(r_{\omega\omega^*},\tau-\xi)\,\mathrm{d}S_\omega\,\mathrm{d}\xi = \bar{f}_1(\omega^*,\tau)\,;$$

$$-\frac{\nu(\omega^*,\tau)}{2} + \int_0^\tau \int_{\partial\Omega} \nu(\omega,\xi)\frac{\partial}{\partial n_{\omega^*}} G(r_{\omega\omega^*},\tau-\xi)\,\mathrm{d}S_\omega\,\mathrm{d}\xi$$

$$+\int_0^\tau \int_{\partial\Omega} \mu(\omega,\xi)\frac{\partial^2}{\partial n_\omega^2} G(r_{\omega\omega^*},\tau-\xi)\,\mathrm{d}S_\omega\,\mathrm{d}\xi = \bar{f}_2(\omega^*,\tau)\,,$$

where

$$(\omega^*,\tau)\in \sum\nolimits_\Omega = \partial\Omega\times(0,\tau_m]\,.$$

3.3.6 Conclusions

Several boundary inverse problem statements have been formulated for a three-dimensional heat conduction equation. Other statements are possible too; for example, in the case when temperature measurements are taken at some non-closed surface within a solid body, arbitrarily separating it in parts, or when temperature data and a heat flux on some part of the body's exterior surface (overdefined boundary) are known, while elsewhere the boundary conditions are not assigned. Section 4.5 deals with one of these problems for a rectangular two-dimensional domain.

It must be especially mentioned that the question on properties and realization of three-dimensional IHCP solutions remains for the present open and requires investigation both from a purely mathematical aspect and that of creating computational algorithms for practical application.

It is obvious that a transition to two- and three-dimensional statements of inverse problems incurs a noticeable increase in machine time. At the same time there is the growing influence of smoothing and time-lag factors of heat conduction and approximation errors on the solution. It may even turn out that the anticipated gain in accuracy in the estimates of causal characteristics of heat conduction processes connected with the use of a truer model of heat propagation, may not, in effect, take place and would in effect reduce the accuracy of results. Hence, in every case, expediency for a transition from a one-dimensional inverse problem statement to two-dimensional and from a two-dimensional to three-dimensional must be thoroughly substantiated. It is always necessary to reduce problem dimensionality at the expense of corresponding structural decisions in developing the models, specimens and gauges designed for experimental study and introduction of a number of admissible constraints on the testing program.

3.4 Statement of a Two-Dimensional Inverse Problem

In practical applications, we can very often regard a heat conduction process as being independent on one spatial coordinate, say, z, i.e. the process proceeds almost uniformly in all planes, parallel to a plane under study. That is why the analysis of the plane case acquires an independent meaning in spite of the fact that the general approach to the construction of integral forms of a two-dimensional IHCP is, on the whole, identical to a three-dimensional one. Later on we shall consider the inverse problem in the boundary-value statement.

Let a heat conduction equation be given in the Ω domain of a two-dimensional space

$$\partial T/\partial \tau = a[\partial^2 T/\partial x^2 + \partial^2 T/\partial y^2]\,, \quad (x, y) \in \Omega\,,\ 0 < \tau \leqslant \tau_m\,.$$

The initial condition $T(x, y, 0) = \psi(x, y)$, $(x, y) \in \bar{\Omega} = \Omega + \partial\Omega$ and temperature on the ∂D closed line, located within the domain Ω, $T(\tau)|_{x, y \equiv \partial D} = f(x, y, \tau)$ are known. The heat flux at the $\partial\Omega$ boundary is unknown.

Following the method formulated in Sect. 3.3.3 we introduce into our consideration a thermal potential of the simple layer, which is caused by the action of some simple sources distributed in the line $\partial\Omega$,

$$U(x, y, \tau) = \int_0^\tau \int_{\partial\Omega} v(\omega, \xi)\, G(r_{\omega P}, \tau - \xi)\, d\omega\, d\xi\,, \tag{3.36}$$

where

$$G(r_{\omega P}, \tau - \xi) = \frac{1}{4\pi(\tau - \xi)} \exp\left[- \frac{r_{\omega P}^2}{4a(\tau - \xi)} \right]$$

is a fundamental solution of a two-dimensional heat conduction equation, $r_{\omega P}$ is the distance between points $\omega(\eta, \gamma) \in \partial\Omega$ and $P(x, y) \in \Omega$ (vector $r_{\omega P}$ is believed to be directed from ω to P) and $v(\omega, \xi)$ is the source density dependent upon the ω location in the line and the action time ξ.

The potential (3.36) outside the $\partial\Omega$ line with $\tau > 0$ is a continuously differentiable in its arguments function satisfying a heat conduction equation. The

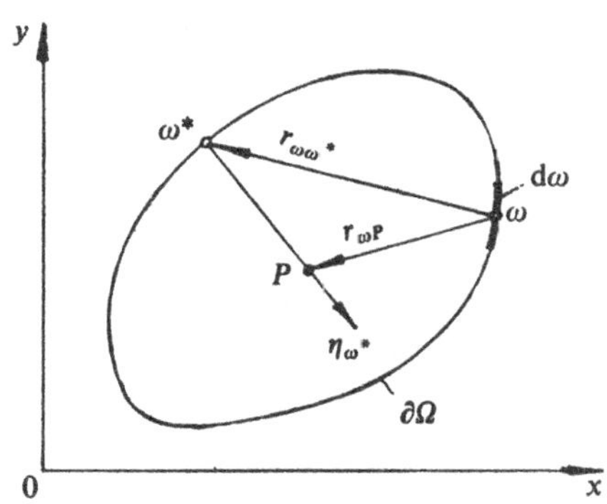

Fig. 3.7. Points ω, ω^* at the boundary $\partial\Omega$

potential is continuous in the line and at the same time its normal derivative suffers a finite discontinuity. In passing of the normal derivative from the inside and outside points P to some point ω^* situated in the $\partial\Omega$ line (see Fig. 3.7), we have respectively

$$\left.\frac{\partial U(\omega^*,\tau)}{\partial n_{\omega^*}}\right|_i = -\frac{v(\omega^*,\tau)}{2} + \Phi_* ;$$

$$\left.\frac{\partial U(\omega^*,\tau)}{\partial n_{\omega^*}}\right|_e = \frac{v(\omega^*,\tau)}{2} + \Phi_* ,$$

where n_{ω^*} is the direction of the interior normal to the line $\partial\Omega$ at ω^* point;

$$\Phi_* = \int\limits_0^\tau \int\limits_{\partial\Omega} v(\omega,\xi)\frac{\partial}{\partial n_{\omega^*}} G(r_{\omega\omega^*}, \tau-\xi)\,\mathrm{d}\omega\,\mathrm{d}\xi = \int\limits_0^\tau \int\limits_{\partial\Omega} v(\omega,\xi)\frac{r_{\omega\omega^*}}{8\pi a(\tau-\xi)^2}$$
$$\times \exp\left[-\frac{r^2_{\omega\omega^*}}{4a(\tau-\xi)}\right]\cos(r_{\omega\omega^*}, n_{\omega^*})\,\mathrm{d}\omega\,\mathrm{d}\xi .$$

Introduce a notation

$$\mathrm{d}\beta = \frac{\cos(r_{\omega\omega^*}, n_{\omega^*})}{r_{\omega\omega^*}}\mathrm{d}\omega$$

where β is any angle.

Then

$$\Phi_* = \int\limits_0^\tau \int\limits_{\partial\Omega} v(\omega,\xi)\frac{r^2_{\omega\omega^*}}{8a\pi(\tau-\xi)^2}\exp\left[\frac{r^2_{\omega\omega^*}}{4a(\tau-\xi)}\right]\mathrm{d}\beta\,\mathrm{d}\xi .$$

Thus, a relation of the unknown heat flux at the $\partial\Omega$ boundary with the potential density follows from the expression for the temperature normal derivative

$$\left.\frac{\partial T}{\partial n_{\omega^*}}\right|_{(x^*,y^*)\in\partial\Omega} = -\frac{v(\omega^*,\tau)}{2} + \Phi_*(v(\omega,\xi)) .$$

Specification of $v(\omega,\tau)$ for IHCP in the boundary-value statement is obtained from the solution of a two-dimensional integral equation of the first kind of mixed type (for the $\partial\Omega$ line this is a Fredholm type equation, for time – a Volterra type):

$$\int\limits_0^\tau \int\limits_{\partial\Omega} v(\omega,\xi)\,G(r_{\omega\sigma}, \tau-\xi)\,\mathrm{d}\omega\,\mathrm{d}\xi = T(\tau)|_{\sigma\in\partial D},$$

where $T(\tau)|_{\sigma\in\partial\Omega}$ is the temperature set at some interior line ∂D, bounding the domain $D \subset \Omega$ and $\sigma = \sigma(x,y)$ are the points of ∂D line.

3.5 Fictitious Boundary Method for Solving Inverse Boundary Problems

3.5.1 General Scheme of the Method

The solution of a boundary IHCP for domains of complex shape (two- and three-dimensional problems) and those with moveable boundaries (one- and multi-dimensional) has even greater theoretical and computational difficulties. Also it is not always appropriate to make a solving procedure sufficiently universal since any change in domain geometry may cause changes in the algorithms and in the program. A method for solving a one-dimensional IHCP with moving boundaries has been suggested in [9], which simplifies the algorithmic part of the problem. There are prerequisites for further useful generalizations of this approach including the case with a multidimensional statement. Let us consider the main content of the method.

Let domain Ω with a $\partial\Omega$ boundary (a surface, line, points depending on the problem dimensionality) be assigned at some space of the body coordinate change. Another domain D is located within Ω and temperature $T(P, \tau)$, $P \in \partial D$ is known at the ∂D boundary. Here, some sections $\partial\Omega$ and ∂D may be common. Let $\bar{\Omega}$ and $\bar{D}$ denote the closures of these domains ($\bar{\Omega} = \Omega + \partial\Omega$, $\bar{D} = D + \partial D$). It is required to find a temperature field in the whole domain $\bar{\Omega}$.

For IHCP in the boundary-value statement it is suggested that a solution is obtained in four stages.

1. Change the geometry of domains $\bar{\Omega}$ and $\bar{D}$ into new domains $\bar{\Omega}'$ and $\bar{D}'$: this would meet the embedding conditions $\bar{\Omega} \subseteq \bar{\Omega}'$, $\bar{D}' \subseteq \bar{D}$, i.e. the domain $\bar{D}'$ compared with D can be narrowed while $\bar{\Omega}'$ is expanded; this would enable a suitable and, possibly, universal algorithm of the inverse problem solving for a heat conduction equation in the domain $\bar{\Omega}' - \bar{D}'$ to be formulated; this would ensure that the $\partial\Omega'$ and $\partial D'$ boundaries were located extremely close to the initial boundaries $\partial\Omega$ and ∂D; this condition should considerably reduce the degree of increased instability of the problem at the expense of $\partial\Omega'$ moving away from the boundary with initial data; usually, the boundaries $\partial\Omega'$ and $\partial\Omega$, as well as $\partial D'$ and ∂D, have common sections or points.
2. Using a certain algorithm of inverse problem solving, find a temperature field in the $\bar{D}$ domain and, hence, determine the temperature $T^{*\prime}(\tau)$ and the heat flux $q^{*\prime}(\tau)$ at the $\partial D'$ boundary.
3. According to $T^{*\prime}$ or $q^{*\prime}$ find temperature T_φ at $\partial\Omega'$ boundary (see Sects. 3.3 and 3.4).
4. Using the obtained values $T_\varphi(\tau)$ and $T^{*\prime}(\tau)$ solve a direct problem in the $\bar{\Omega}' - \bar{D}'$ domain and define a boundary condition desired at $\partial\Omega$.

For IHCP in the Cauchy statement, a change of boundaries in the domain of the inverse problem solving ($\bar{\Omega} \to \bar{\Omega}'$) is not obligatory. At the same time, the transition $\bar{D} \to \bar{D}'$ may be useful.

Thus, the essence of *the fictitious boundary method* is in the initial inverse problem transformation to a new and simpler statement without initial differential equation modification. For example, in the case of a one-dimensional problem with moving boundaries of the body and with travelling points of temperature sensors, such an approach includes a substitution of initial domains (x, τ) with moving boundaries for rectangular domains, for which the development of corresponding computational algorithms is simpler and is a problem that has been much studied. In particular, the selection of step $\Delta\tau$ for natural regularization of a solution (see Sect. 2.6 and Chap. 4) and the construction of algorithms regularizing the solution using the Tikhonov method are performed in a much simpler way than in the initial case.

A drawback of such an approach is that we can identify actual deterioration of IHCP conditionality owing to the increased maximum distance away from the boundary with a desired condition to boundaries with known data. Smoothing and time lag in the heat signal transfer can lead to a situation when a change in the desired boundary condition for $\partial\Omega'$ does not in effect cause any temperature change at $\partial D'$ for a long time. In such a situation, it is evident that if admissible values of time steps appear too large and the accuracy of boundary condition recovery is not satisfactory, we would not manage to apply a step regularization principle. However, it is possible to point out a large number of practical problems when the fictitious boundary method does not enter into contradiction with a step regularization principle.

In all other cases, application of the fictitious boundary method is possible in combination with other regularization techniques. It is helpful here to use a *special procedure* consisting of expansion of the domain for a time coordinate towards higher time values to compensate for the loss of information about the temperature registered at the beginning of the process. A time interval, at which the temperature is being registered, must be longer than that of the boundary condition recovery. The investigations specially conducted for these purposes have shown that such a procedure enables boundary conditions to be reconstructed precisely by means of the regularization method in the case of considerable loss of information about the temperature measured.

Note that the use of the fictitious boundary method is most helpful in those cases when a heat conduction equation is linear and a transition to a new domain of solution $(\bar{\Omega}' - \bar{D}')$ enables us to construct, or to use the already known, Green's function in its explicit form.

3.5.2 Some Applications of a Fictitious Boundary Method

Let us consider two practical problems, whose objective is to define conditions at the moving heat transfer boundaries of the body according to temperature measurements inside the body with constant thermophysical properties.

In the *first problem* a heat experiment is made from observing the conditions of one-dimensionality of a temperature field in the body. It is required to define boundary conditions at the two moving boundaries $X_1(\tau)$ and $X_4(\tau)$ (Fig. 3.8) by the known time dependences of temperature at two interior points of the body. The coordinates of these points can vary with time. The laws of their movement $X_2(\tau)$ and $X_3(\tau)$, as well as $X_1(\tau)$ and $X_4(\tau)$, are known. Following this method, let us perform two transformations for transition to a problem with stationary boundaries with initial data and with stationary boundaries having the requisite functions (see Sect. 2.3.2).

The first transformation is a transition to the new input data $T_2(\tau)$ and $T_3(\tau)$ at some straight lines $x_2 = \text{const}$ and $x_3 = \text{const}$, which are gained from solution of a direct heat conduction problem in the domain $D_2 = \{X_2(\tau) \leqslant x \leqslant X_3(\tau), 0 \leqslant \tau \leqslant \tau_m\}$ (see Figs. 2.3 and 2.4). To be precise, assume that $x_2 = X_{2\,\max}$ and $x_3 = X_{3\,\min}$ (it is believed that $X_{3\,\min} > X_{2\,\max}$). If the input functions are assigned at stationary points, it is evident that we do not need such a transformation. The second transformation is a final transition to rectangular domains

$$D_1' = \{0 \leqslant x \leqslant x_2, 0 \leqslant \tau \leqslant \tau_m\}, \quad D_3' = \{x_3 \leqslant x \leqslant X_{4\,\max}, 0 \leqslant \tau \leqslant \tau_m\},$$

incorporating the domains D_1 and D_3 ($D_1 \subseteq D_1', D_3 \subseteq D_3'$).

Now solve inverse problems for defining fictitious temperatures $T_{\varphi 1}(\tau)$ and $T_{\varphi 4}(\tau)$, corresponding to the boundaries $x = 0$ and $x = X_{4\,\max}$. This can be performed via the Duhamel integral using a superposition principle of solutions. For example, for $T_{\varphi 1}$ we have the following integral equation:

$$\int_0^\tau T_{\varphi 1}(\xi) \frac{\partial \vartheta(x_2, \tau - \xi)}{\partial \tau} \mathrm{d}\xi = g(\tau), \quad 0 < \tau \leqslant \tau_m,$$

where

$$g(\tau) = T_2(\tau) - \int_0^\tau T_3(\tau) \frac{\partial \vartheta(x_3 - x_2, \tau - \xi)}{\partial \tau} \mathrm{d}\xi;$$

$\vartheta(x, \tau - \xi)$ is a solution of the first boundary-value problem in the domain $0 \leqslant x \leqslant x_3$ providing zero temperature at the $x = 0$ boundary and a unit-equal temperature at the other boundary.

The last step in defining the desired conditions at the boundaries $X_1(\tau)$ and $X_4(\tau)$ is solution of the direct heat conduction problems in the domains D_1' and D_3', respectively. These problems present no particular difficulties. A technique of their solution can be based on the Duhamel principle or on the theory of thermal potentials.

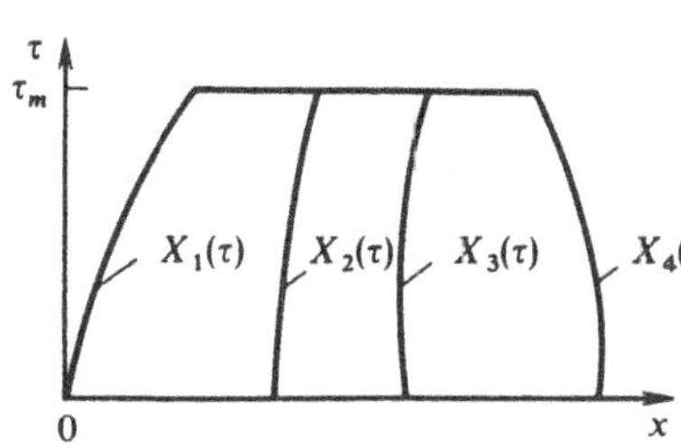

Fig. 3.8. The domain with moving boundaries

In the *second problem*, the conditions of an experiment correspond to a two-dimensional heat conduction process in the domain formed by three rectangular segments, two of which are parallel to one another, the third being perpendicular to them, and by one moving boundary (Fig. 3.9a). A moving boundary of the body in a thermal sense is active and can move inside the body at the expense of ablation according to the known law $S(x, \tau)$. Temperature measurements are taken at that side of the domain which is opposite to a moveable boundary. The task is to determine temperature of the ablating boundary.

Introduce a rectangular coordinate system (x, y) so that the axis x at zero time $\tau = 0$ would touch the point of a moving boundary ranging at a greater distance from the boundary with known temperature and the axis y would coincide with a lateral side of the domain (Fig. 3.9b).

Formalize the problem in the following way:

$$\partial T(x, y, \tau)/\partial\tau = a[\partial^2 T(x, y, \tau)/\partial x^2 + \partial^2 T(x, y, \tau)/\partial y^2],$$
$$(x, y) \in \Omega, \quad 0 < \tau \leqslant \tau_m,$$
$$T(x, y, 0) = T_0, \quad (x, y) \in \bar{\Omega};$$
$$\partial T(0, y, \tau)/\partial x = \partial T(b, y, \tau)/\partial x = \partial T(x, c, \tau)/\partial y = 0,$$
$$0 \leqslant \tau \leqslant \tau_m;$$
$$T(x, c, \tau) = T^*(x, \tau), \quad 0 \leqslant x \leqslant \tau_m;$$
$$T(S(x, \tau), \tau) = ?, \quad 0 < \tau \leqslant \tau_m,$$

where

$$\Omega = \{0 < x < b, \quad S(x, \tau) < y < c\}.$$

Since at one of the fixed boundaries the input data are known, there is no need to construct the domain $\bar{D}'$. As a domain $\bar{\Omega}'$ we can take a rectangular $0 \leqslant x \leqslant b$, $0 \leqslant y \leqslant c$, for which an explicit form of the G influence function corresponding to the second boundary-value problem is known (see Sect. 4.5). Then, an IHCP solution in the domain $\bar{\Omega}'$ will be equivalent to the solution of a two-dimensional

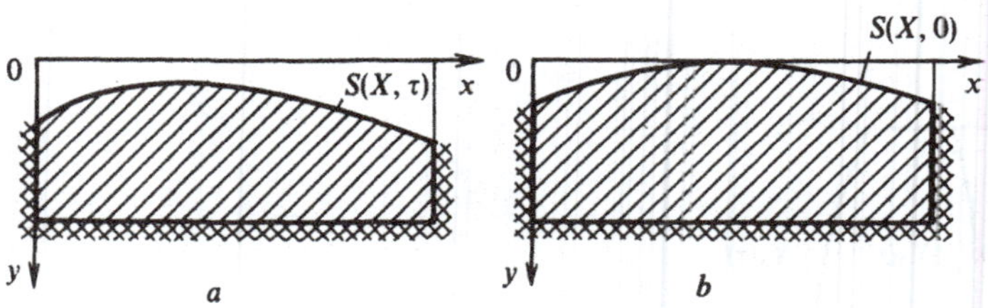

Fig. 3.9. A two-dimensional domain of spatial variables with a moving boundary: *a* is the boundary current position and *b* is the boundary initial position.

integral equation relative to the heat flux $q_\varphi(x', \xi)$ at the fictitious boundary $y = 0$:

$$\frac{a}{\lambda}\int_0^\tau\int_0^b q_\varphi(x', \xi)\, G(x, c; x', 0; \tau - \xi)\, dx'\, d\xi = f(x, c, \tau)\,, \tag{3.37}$$

where

$$f(x, c, \tau) = T^*(x, \tau) - T_0\,.$$

Equation (3.37) is a linear integral equation of the first kind, it is of Volterra type for time and Fredholm-type for space coordinates. Its solution must be regularized by one of the methods pointed out in Chap. 2.

Having determined $q_\varphi(x', \xi)$ we can solve a direct heat conduction problem in the domain $\bar{\Omega}'$ and find the temperature at the boundary $S(x, \tau)$.

We can list many other problem statements of interest for practical use which are still pertinent for the approach discussed above. For example, inverse problems are of great significance in axially symmetric statements. Similar heat conduction models are widespread in studying heat conditions of different structural elements. In these cases, solid and hollow cylinders, cones and spheres are taken as expanded domains $\bar{\Omega}'$.

3.6 Summary

1. Based on methods of mathematical physics and heat conduction theory (Green's functions method, Duhamel principle, thermal potentials method), analytical statements of boundary inverse problems are obtained for one-dimensional and multi-dimensional equations of heat conduction. In all cases the common mathematical notation of IHCP is reduced to corresponding integral equations of the first and second kind.
2. To solve inverse problems with moveable boundaries of the body and in domains of complex geometrical shapes a fictitious boundary method is suggested, enabling us to simplify the algorithms of computation of heat boundary conditions. The integral forms of IHCP presented above should be treated as "blank", upon which it is possible to study the methods further and develop different regularized algorithms of inverse problem solving. This will be the subject of succeeding chapters.

Chapter 4

Direct Algebraic Method of Determining Transient Heat Loads

Introduction

Data processing methods for transient heat experiments, based on solution of integral forms of IHCPs in the boundary-value statement, are the most widely used techniques at present [1, 2, 3, 6, 12, 70, 72, 150, 151, 158, 159, 162, 175, 192, 200, 203]. A number of algorithms for solving virtually one and the same inverse problems have been developed. In this connection, it is necessary to formulate some general criteria for a comparative analysis of algorithms of the given type, especially, as in most cases they are created on approximately equal principles of numerical solution using *a step regularization* principle.

Among the most effective direct methods, which use a step regularization, there is that of *function specification* suggested by J.V. Beck [158]. It utilizes a by-interval (usually, steptype) approximation of unknown functions by means of the least-squares technique and, according to accepted classification, is referred to as the variational method. This is in widespread use in the USA for solving various practical problems due to the algorithmic simplicity of its realization and fine stability characteristics. This last can be explained by the fact that the problem in finding a value of unknown functions at every time interval is specially overdefined, i.e. several values of input temperature per interval are used, and, certainly, by the averaging action of the least-squares technique. Thus the effect of step regularization increases because of these factors. As a result it is possible to use relatively small time steps. Accordingly, the accuracy of computations is sufficiently high. However, like most other direct methods, that of Beck has no rigorous mathematical substantiation. An exhaustive description of such an approach with corresponding algorithmic constructions and numerous computational examples can be found in [162], which is why it is not considered here apart from a small comparison in Sect. 8.7.1 of computational results, obtained with the results corresponding to iterative regularization.

A purely *steptype technique of natural regularization* of IHCP solutions is investigated in this chapter, the heuristic interpretation of which was illustrated in Sect. 2.6. For linear problem statements considered in domains with fixed geometry, the principles of constructing computational algorithms are stated as well as the conditions for their practical use being analysed. Following the method of fictitious boundaries, the results obtained below are directly applicable to a problem of specifying heat loads for bodies whose boundaries move in the process of heating. This general approach to approximation would not have to change for inverse problems with moving boundaries, the integral statements of which have been given in Sects. 3.1.2 and 3.1.3.

A piecewise-constant approximation of unknown function is mainly used in the algorithms considered below. They are *recurrent* and can be utilized to define heat boundary conditions in two-dimensional cases with constant thermo-physical properties. The material of the chapter is outlined as follows. First, in Sect. 4.1 a general approach will be described for the construction of algorithms for the solution of one-dimensional boundary inverse problems, represented in integral form, and the concrete results will be given for cases of a semi-infinite body and a slab with boundary conditions of the first and second kind. The question of unknown function restorability depending on the value of the approximation step will be discussed in Sect. 4.2.

The conditionality of computational algorithms and constructive ways of approximation step selection, which provide stability in the IHCP solution, are studied in Sect. 4.3. In particular, a simple rule will be established for defining the minimum possible steps providing the input data are not disturbed. Then, in Sect. 4.4, it will be shown that a transition to a higher order of approximation (e.g. to piecewise-linear) will necessitate larger values of time steps to provide stability of the computations compared with a piecewise-constant approximation of the unknown dependence. Two approaches are shown here to use this method for solving the IHCPs which have the input data including fluctuating errors. Finally, in Sect. 4.5, it is expanded into a solution of IHCP in a two-dimensional formulation.

4.1 The Recurrent Algorithm Construction

Consider a computational aspect of an algebraic solution method, which goes back to the works of T.J. Mirsepassi [192] and G. Stolz [203]. It has been shown above that for a one-dimensional heat conduction equation with constant coefficients the boundary inverse problems have a common integral representation (see Eq. (3.1)). Replace Eq. (3.1) by an approximate discrete analog. For this, divide a complete time section into a finite number of intervals with variable steps:

$\Delta\tau_n = \tau_n - \tau_{n-1}$, $n = 1, 2, \ldots, m$ and re-write Eq. (3.1) as

$$\sum_{i=1}^{n} \int_{\tau_{i-1}}^{\tau_i} u(\xi) K(\tau_n - \xi)\,\mathrm{d}\xi = f(\tau_n), \quad n = 1, 2, \ldots, m\,.$$

Introduce a step approximation of the unknown function $u(\tau)$ assuming it as equal to a constant value $\hat{u}_n$ at each interval. Then,

$$\sum_{i=1}^{n} \hat{u}_i \Delta\vartheta_{in} = f_n, \quad n = 1, 2, \ldots, m\,, \tag{4.1}$$

where

$$f_n = f(\tau_n), \quad \Delta\vartheta_{in} = \int\limits_{\tau_{i-1}}^{\tau_i} K(\tau_n - \xi)\,d\xi .$$

To calculate coefficients $\Delta\vartheta_{in}$ we could have used a finite-difference approximation, but this should not be done for the following reason. According to the Duhamel's principle [143, 172] a solution of the linear heat conduction equation for $u(\tau)$, $\tau > 0$ boundary condition, given at one of the body boundaries ($x = 0$) with the zero supplementary condition at the other boundary ($x = b$), can be written as

$$T(x, \tau) = \int\limits_0^\tau u(\xi) \frac{\partial\vartheta(x, \tau - \xi)}{\partial\tau}\,d\xi ,$$

where $\vartheta(x, \tau)$ is a solution of the given boundary-value problem with $u(\tau) = 1$.

Thus, the kernel of the integral Eq. (3.1) is a derivative with respect to time of the unit-equal solution of the corresponding boundary-value heat conduction problem: $K(\tau - \xi) = \partial\vartheta(d, \tau - \xi)/\partial\tau$. That is why, knowing the exact solution ϑ, we can write an expression for coefficients of the (4.1) algebraic system

$$\Delta\vartheta_{in} = \vartheta(d, \tau_n - \tau_{i-1}) - \vartheta(d, \tau_n - \tau_i) .$$

Such a technique allows errors connected with the transition to a computational form of the problem to be reduced.

After approximation we have a system of linear algebraic equations $A_\Delta u = f$, $u = \{\hat{u}_i\}_1^m$, $f = \{f_n\}_1^m$ with the lower triangular matrix A_Δ

$$A_\Delta = \begin{bmatrix} \Delta\vartheta_{11} & & & \\ \Delta\vartheta_{12} & \Delta\vartheta_{22} & & \\ \vdots & & \ddots & \\ \Delta\vartheta_{1m} & \Delta\vartheta_{2m} & \dots & \Delta\vartheta_{mm} \end{bmatrix} .$$

The determinant of the A_Δ matrix is equal to the product of its diagonal elements: $\det A_\Delta = \Delta\vartheta_{11}\Delta\vartheta_{22}\dots\Delta\vartheta_{mm}$. From the physical sense of the $\vartheta(d, \tau)$ value it follows that all $\Delta\vartheta_{in} > 0$. From here $\det A_\Delta \neq 0$. Hence, the matrix A_Δ is theoretically non-degenerate and there exists the inverse matrix A_Δ^{-1}. So, at first glance, a solution of system (4.1) presents no difficulty. It is easy to get the following recurrent formula, which allows us, step by step, to find all components of m-dimensional vector u:

$$\hat{u}_n = \frac{1}{\Delta\vartheta_{nn}}\left(f_n - \sum_{i=1}^{n-1} \hat{u}_i \Delta\vartheta_{in}\right), \quad n = 1, 2, \dots, m . \tag{4.2}$$

To complete the construction of a given algorithm, it is necessary to point out to what points of a time interval would correspond to the obtained values of $\hat{u}_n$. The solution procedure stated can be considered as a multiple use of the mean-value theorem where such values of $\hat{u}_n$ are sought which give equal areas for the true and stepwise curves at each interval. Had points τ_n' corresponding to every mean value of $\hat{u}_n$ been known, the approximation errors would have been reduced to the minimum. However it is not possible to find the points of averaging, which is why we shall consider that the $\hat{u}_n$ values, calculated using the formula (4.2) are referred to as the midpoints of every time interval $\hat{u}_n = u(\tau_n - (\tau_n - \tau_{n-1})/2)$.

Consider now the more interesting case when time steps are recognized as constant and equalling $\Delta\tau = \tau_m/m$. Allowing for this condition, we can simplify the computational algorithm sufficiently. The point is that with the constant $\Delta\tau$ the matrix A_Δ has a property of by-diagonal equality of elements $\Delta\vartheta_{1n} = \Delta\vartheta_{2,n+1} = \ldots = \Delta\vartheta_{m-n+1,m}$, i.e. the coefficients are equal with the same difference of indexes. Thus we can introduce a new designation

$$\Delta\vartheta_{n-i} \equiv \Delta\vartheta_{in} = \vartheta[d, \Delta\tau(n-i+1)] - \vartheta(d, \Delta\tau(n-i)]\,,$$
$$i, n = 1, 2, \ldots, m; \quad n \geqslant i$$

and re-write the formula in Eq. (4.2):

$$\hat{u}_n = \frac{1}{\Delta\vartheta_0}\left(f_n - \sum_{i=1}^{n-1} \hat{u}_i \Delta\vartheta_{n-i}\right), \quad n = 1, 2, \ldots, m\,. \tag{4.3}$$

Therefore, instead of the $m(m+1)/2$ coefficients of the matrix it is necessary to calculate and store in the computer memory only the m coefficients.

For calulation convenience the coefficients $\Delta\vartheta_{n-i}$ are better presented in dimensionless form. Present the expressions for $\Delta\vartheta_{n-i}$ and the right sides f_n, corresponding to some statements of inverse problems for a heat conduction equation with uniform initial distribution of temperature $T(x, 0) = T_0$ (T_n^* is a temperature value at the point $x = d$) that are more usable in practice.

Case A: a semi-infinite body with boundary condition of the first kind (Fig. 4.1a) –

$$f_n = T_n^* - T_0; \quad \Delta\vartheta_{n-i} = -\operatorname{erf}\left[\frac{1}{2\sqrt{\Delta Fo_d(n-p)}}\right]_{p=i}^{p=i-1}.$$

Case B: a semi-infinite body with boundary condition of the second kind (Fig. 4.1b) –

$$f_n = \frac{\lambda}{d}(T_n^* - T_0); \quad \Delta\vartheta_{n-i} = 2\sqrt{\Delta Fo_d(n-p)}\operatorname{ierfc}\left[\frac{1}{2\sqrt{\Delta Fo_d(n-p)}}\right]_{p=i}^{p=i-1}.$$

Case C: a slab with boundary conditions of the first kind (Fig. 4.1c) –

$$f_n = T_n^* - T_0;$$

$$\Delta\vartheta_{n-i} = \sum_{k=0}^{\infty}\left\{\operatorname{erfc}\left[\frac{2k + d/b}{2\sqrt{\Delta Fo_b(n-p)}}\right] - \operatorname{erfc}\left[\frac{2(k+1) - d/b}{2\sqrt{\Delta Fo_b(n-p)}}\right]\right\}_{p=i}^{p=i-1}$$

for small times as $(n-p)\Delta Fo_b < 0.1 - 0.4$;

$$\Delta\vartheta_{n-i} = \frac{2}{\pi}\sum_{k=1}^{\infty}\frac{(-1)^k}{k}\exp\left[-k^2\pi^2\Delta Fo_b(n-p)\right]\sin\left[k\pi\frac{b-d}{b}\right]_{p=i}^{p=i-1}$$

for large times as $(n-p)\Delta Fo_b > 0.1 - 0.4$.

Case D: a slab with boundary conditions of the second kind (Fig. 4.1d) –

$$f_n = \frac{\lambda}{b}(T_n^* - T_0);$$

$$\Delta\vartheta_{n-i} = 2\sqrt{Fo_b(n-p)}\sum_{k=0}^{\infty}\left\{\operatorname{ierfc}\left[\frac{2k + d/b}{2\sqrt{\Delta Fo_b(n-p)}}\right]\right.$$

$$\left. + \operatorname{ierfc}\left[\frac{2(k+1) - d/b}{2\sqrt{\Delta Fo_b(n-p)}}\right]\right\}_{p=i}^{p=i-1}$$

for small times;

$$\Delta\vartheta_{n-i} = \left\{\Delta Fo_b(n-p) + \frac{2}{\pi^2}\sum_{k=1}^{\infty}\frac{(-1)^{k+1}}{k^2}\exp\left[-k^2\pi^2\Delta Fo_b(n-p)\right]\right.$$

$$\left.\times\cos\left[k\pi(1-d/b)\right]\right\}_{p=i}^{p=i-1}$$

for large times.

Case E: a slab with unknown boundary condition of the first kind and known one of the second kind (Fig. 4.1e) –

$$f_n = T_n^* - T_0;$$

$$\Delta\vartheta_{n-1} = \sum_{k=1}^{\infty}(-1)^{k+1}\left\{\operatorname{erfc}\left[\frac{2k - d/b}{2\sqrt{\Delta Fo_b(n-p)}}\right]\right.$$

$$\left. + \operatorname{erfc}\left[\frac{2(k-1) + d/b}{2\sqrt{\Delta Fo_b(n-p)}}\right]\right\}_{p=i}^{p=i-1}$$

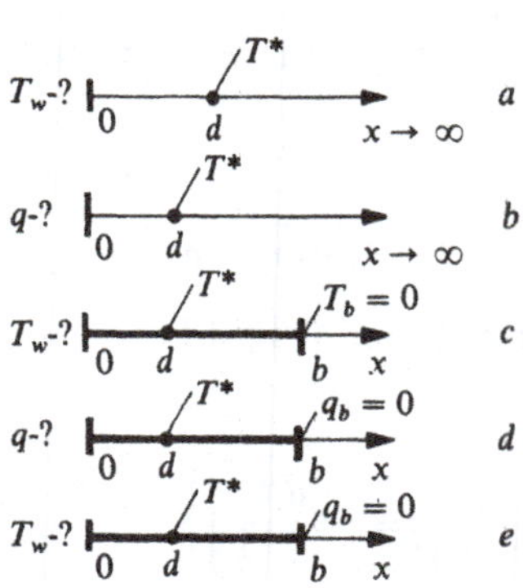

Fig. 4.1a–e. The models of one-dimensional heat conduction. **a, b** – a semi-infinite body; **c, d, e** – a slab

for small times;

$$\Delta\vartheta_{n-i} = -\frac{4}{\pi}\sum_{k=1}^{\infty}\frac{(-1)^{k+1}}{2k-1}\exp\left[-\frac{\pi^2}{4}(2k-1)^2\,\Delta Fo_b(n-p)\right]$$

$$\times\cos\left[\frac{\pi}{2}(2k-1)(1-d/b)\right]_{p=i}^{p=i-1}$$

for large times.

The coefficients $\Delta\vartheta_{n-i}$ for some other statements of IHCPs are given in [7, 70].

4.2 The Boundary Condition Recoverability

Having completed the formal part of the computational algorithm, let us analyse its consistency. The initial problem is unstable, i.e. small deviations of the right side can lead to a large error in the solution (see Sect. 2.1.2). It is natural to suppose that problem (3.1) becomes even more unstable, when the degree of the $u(\tau)$ unknown function effect on temperature at point $x = d$ decreases. This property is closely related with the principal possibility of the approximate recovery of $u(\tau)$ for the prescribed function $f(\tau)$. We now consider this question in detail.

Any coefficient $\Delta\vartheta_{n-i} = \vartheta_{n-i+1} - \vartheta_{n-i}$ in the expression (4.3) can be interpreted as the temperature value at point $x = d$ upon the $\tau = \Delta\tau(n-i)$ moment of time, corresponding to the impulse action of the unit value boundary condition (from the side of the heat exchange surface with a desired condition) during $[\tau = -\Delta\tau, \tau = 0]$ (Fig. 4.2a).

Meanwhile, on the other surface of the body at all moments of time a zero boundary condition is kept. Such interpretation follows from linearity of the problem, allowing use of the principle of solutions superposition. In this case

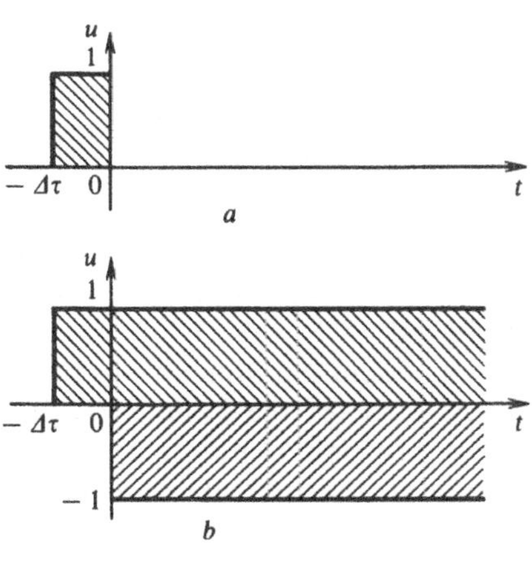

Fig. 4.2. The superposition of step functions: a is a unit-equal impulse of the boundary function; b are positive and negative boundary effects, the superposition of which is the boundary function impulse

a "summary" solution $\Delta\vartheta_{n-i}$ is given by the temperature difference: $\vartheta_{n-i+1}(\tau)$, which corresponds to a boundary condition in the form of a positive "step" of the unit value in the whole time axis from the moments $\tau = -\Delta\tau$ and $\vartheta_{n-i}(\tau)$, corresponding to a negative unit-equal step beginning at the moment $\tau = 0$ (Fig. 4.2b). From here, a differential analog of the point source is a unit-equal impulse, acting over the period of one time step. It is obvious that as $\Delta\tau$ decreases, the temperature responses $\Delta\vartheta_{n-i}$ decrease in value, tending to zero. We can imagine a situation when errors related with the truncation of infinite series while calculating the coefficients of A_Δ matrix and f right side, as well as errors of rounding-off in calculations by computer, would become comparable with a value of some elements of the A_Δ matrix. It is evident that use of the direct method in this case would be useless. For points set deep in the body, the coefficients on the main diagonal $\Delta\vartheta_0$ would tend to zero more quickly with decreasing $\Delta\tau$. As the determinant of the A_Δ matrix for $\Delta\tau = \text{const}$ equals $\det A_\Delta = (\Delta\vartheta_0)^m$ so for a certain step $\Delta\tau$ larger than that at which an effect of actual nullification of elements $\Delta\vartheta_0$ could develop, the A_Δ matrix would appear in the vicinity of the singular matrix, and this procedure for its inversion would make no sense. It happens as soon as the lesser is the thermal diffusivity of the body a and greater is the distance of point d from a heat transfer boundary.

In relation to this topic we can introduce a notion of *recoverability* of boundary condition relative to solution $T(d, \tau)$ of the direct heat conduction problem, assuming the property of the initial statement of IHCP which calls for the uniqueness and accuracy of its approximate solution in the time interval under consideration for the prescribed input function. For example, if point $x = d$ is located so deep that it does not "feel" any changes in the boundary condition during the given time interval, we shall find the problem in Eq. (3.1) in homogeneous statement with the following conditions: $f(\tau) \sim 0$, $K(\tau - \xi) \sim 0$. Such a problem when realized has an infinite set of solutions, i.e. a so-called ambiguity of solutions develops, stemming from different errors.

The recoverability of the boundary condition will be determined not only by the quality of the input information and "thermal" depth of point $x = d$ ($\delta_T = d/\sqrt{a}$), but also by the algorithms for inverse problem solving. For example, the direct method of recurrent type (when values of boundary function are found successively at every moment of time) cannot provide the necessary quality of the boundary condition recoverability, if the initial period of invariance of temperature measured with respect to the boundary condition exists and if the approximation step in time be taken less than this period. In this case the matrix elements and the right side of the problem are small (almost disappear) and the uncertainty of solution at the initial interval does not allow calculations by a recurrent formula to be continued.

It is obvious that a deviation of the approximate solution from an exact one for the given method will also depend on the error of the initial problem approximation. In direct methods which use the principle of natural regularization by approximation steps it is possible to indicate the optimal step at which the accuracy of solution is maximum. With small steps $\Delta\tau$ the approximate solutions of IHCPs differ considerably from the unknown solution due to the instability effect, and at the same time with greater $\Delta\tau$ the approximation errors become quite noticeable. That is why it is always useful to have a step for calculations as close as possible to the minimum value, which can guarantee the required smoothness of results.

Thus, the main question for the practical use of the direct algebraic IHCP solution method is in the proper selection of the design step value, i.e. in the formulation of the *step regularization principle* of inverse problem solving.

4.3 Step Regularization Principle and Limits of Method Applicability

4.3.1 Conditionality of an Approximating Set of Algebraic Equations

It has been stated above that with sufficiently small time steps $\Delta\tau$ the direct method will not give a satisfactory IHCP solution, since the algebraic system (4.1) appears to be unstable (the A_Δ matrix is *ill-conditioned*)

Further on, it will be of interest to consider two interrelated problems. The first problem includes a comparative analysis of the stability characteristics of different statements and schemes of IHCP solution within the frames of the described method. The second problem deals with the study of criteria needed for natural step regularization.

Let us consider the first problem. Introduce into our consideration such concepts as an inner product $(\boldsymbol{x}, \boldsymbol{y})$ of some $\boldsymbol{x}$ and $\boldsymbol{y}$ vectors, a Euclidean norm $\|\boldsymbol{x}\|$ of some $\boldsymbol{x}$ vector and a spectral norm $\|\boldsymbol{A}\|$ of some $\boldsymbol{A}$ matrix:

$$(\boldsymbol{x}, \boldsymbol{y}) = \sum_{n=1}^{m} x_n y_n; \quad \|\boldsymbol{x}\| = (\boldsymbol{x}, \boldsymbol{x}) = \sqrt{\boldsymbol{x}^T \boldsymbol{x}};$$

$$\|\boldsymbol{A}\| = \max_{x \neq \boldsymbol{\theta}} \frac{\|\boldsymbol{A}\boldsymbol{x}\|}{\|\boldsymbol{x}\|},$$

where T is a transposition index and $\boldsymbol{\theta}$ is a null vector.

A *spectral condition number of matrix* $\boldsymbol{A}_\Delta$ is the product of the $\boldsymbol{A}_\Delta$ spectral norm and the spectral norm of the inverse matrix $\boldsymbol{A}_\Delta^{-1}$: $C(\boldsymbol{A}_\Delta) = \|\boldsymbol{A}_\Delta\| \, \|\boldsymbol{A}_\Delta^{-1}\|$. The norms of the $\boldsymbol{A}_\Delta$ and $\boldsymbol{A}_\Delta^{-1}$ matrices are determined by the largest and the least eigenvalues of a corresponding normal matrix $\boldsymbol{B} = \boldsymbol{A}_\Delta^T \boldsymbol{A}_\Delta$:

$$\|\boldsymbol{A}_\Delta\| = \sqrt{\lambda_{\boldsymbol{B}\max}}; \quad \|\boldsymbol{A}_\Delta^{-1}\| = \sqrt{\lambda_{\boldsymbol{B}\min}^{-1}}; \quad \lambda_{\boldsymbol{B}\max} > 0, \quad \lambda_{\boldsymbol{B}\min} > 0.$$

The condition number $C(\boldsymbol{A}_\Delta)$ puts an upper bound on the ratio of the relative root-mean-square error of the system solution to the relative root-mean-square error of the right side vector [178]:

$$\frac{\|\delta \boldsymbol{u}\|}{\|\boldsymbol{u}_T\|} \leqslant C(\boldsymbol{A}_\Delta) \frac{\|\delta \boldsymbol{f}\|}{\|\boldsymbol{f}_T\|},$$

where u_T and f_T are "exact" vectors; δf are "perturbations" of the right side; δu: $Af(u + \delta u) = f + \delta f$.

Considering the case when the input information f is prescribed exactly, but the matrix elements are disturbed $(A_\Delta + \delta A_\Delta)$, we can write

$$\frac{\|\delta u\|}{\|u_T + \delta u\|} \leqslant C(A_\Delta)\frac{\|\delta A_\Delta\|}{\|A_\Delta\|}.$$

From here, the deviation $\|\delta u\|$, referred to $\|u_T + \delta u\|$, is bounded by a relative "error" of matrix A_Δ, multiplied by the condition number.

Thus, the $C(A_\Delta)$ value can be considered as the A_Δ matrix quality for a comparative analysis of different algorithms with various $\Delta\tau$ values. The higher $C(A_\Delta)$, the more unstable is the inverse matrix A_Δ^{-1}.

To specify the condition number we have to find the eigenvalues of the B matrix which is a complicated problem for the ill-conditioned matrix A_Δ. In this case, when the main conclusion of the analysis must be in the selection of any algorithm, the product of Euclidean norms $N = \left(\sum_{i,n} a_{in}^2\right)^{1/2}$ of A_Δ and A_Δ^{-1} matrices can be taken as an evaluating criterion of comparison, since the condition number $C(A_\Delta)$ satisfies the following inequality [145]

$$\frac{1}{m}\gamma(A_\Delta) \leqslant C(A_\Delta) \leqslant \gamma(A_\Delta),$$

where

$$\gamma(A_\Delta) = N(A_\Delta)N(A_\Delta^{-1}).$$

The $\Delta\vartheta_{n-i}$ coefficients of matrix A_Δ can be given as functions of dimensionless variables: $\Delta Fo_d = a\Delta\tau/d^2$ for a semi-infinite body $(x \geqslant 0)$; $\Delta Fo_b = a\Delta\tau/b^2$ and d/b for a slab $(0 \leqslant x \leqslant b)$. Since the degree of instability of a computational process depends on the form of a kernel of the integral equation and on its time approximation, an increment in the Fourier number ΔFo_d or ΔFo_b for the given inverse problem is the best considered parameter for natural regularization. Such an approach helps an analysis to be made in general criterion form.

The results of computations in the form of dependences $\gamma(A_\Delta, \Delta Fo)$ are illustrated in Fig. 4.3. They are given for different IHCP statements. It was assumed in this analysis that values $\gamma(A_\Delta)$, which majorize the corresponding condition numbers $C(A_\Delta)$, are arranged in the proper order to each other i.e. if $C_1 > C_2$, then $\gamma_1 > \gamma_2$.

In addition to these computations performed experiments have been made for the solution of different model inverse problems. The bounds of stability–instability of the results have been established for the statements with input information that has been exactly prescribed. The analysis of the data has been in agreement with dependences $\gamma(A_\Delta, \Delta Fo)$: if the numerical stability of IHCP solution for any case has been achieved with greater ΔFo in comparison another problem statement, the first curve $\gamma(\Delta Fo)$ would then be placed above the

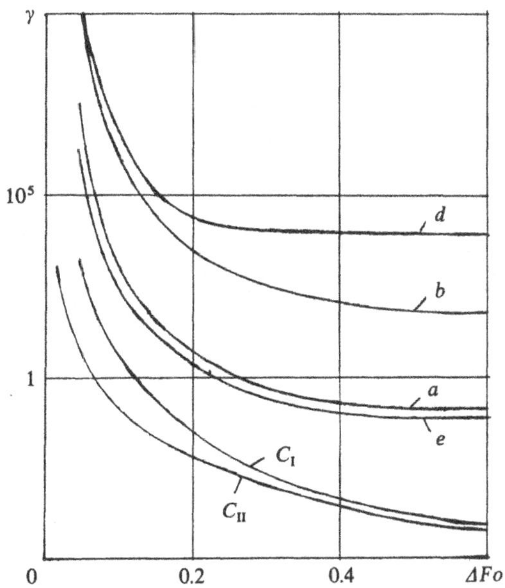

Fig. 4.3. The majorant relative values of the condition numbers for various models of IHCP with a step approximation of the unknown function ($m = 50$) (symbols of the curves correspond to models on Fig. 4.1): d, $e - d = b$; $c_I - d = 0.95b$; $c_{II} - d = 0.5b$

second one. From this point of view the technique of qualitative comparison of stability characteristics of different algorithms for inverse problem solving can be considered as justified.

4.3.2 The Admissible Approximation Steps of Boundary Conditions. The Condition Sufficient for Step Regularization

Here we introduce the concept of *critical step* of time approximation for the boundary-value inverse problem ΔFo_{cr}. By this we shall differentiate the domain of stable results, satisfying the condition $\Delta Fo \geqslant \Delta Fo_{cr}$, and unstable results ($\Delta Fo < \Delta Fo_{cr}$). In general, it is more correct to talk about the admissible value (or the range of values) ΔFo_{adm} at which an approximate solution has a reasonable (physically substantiated) smoothness, i.e. if with $\Delta Fo < \Delta Fo_{adm}$ the solution contains unwanted oscillations, then on the contrary with $\Delta Fo > \Delta Fo_{adm}$ the unwanted smoothness of the solution fine structure takes place at the expense of approximation filtering of useful information, corresponding to frequencies in solution which are physically explainable. Also, the approximation systematic error develops more strongly as ΔFo increases, thus resulting in the "departure" of the recoverable curve from true values.

As calculations above show, by using the direct method a "break-down" of the computational process into the instability domain usually occurs in a comparatively narrow range of numbers ΔFo which is why we can use the concept ΔFo_{cr} in this case.

To solve the second problem stated in Sect. 4.3.1, let us first make use of the fact known in linear algebra that an increase in the absolute value of the A_Δ matrix coefficients positioned at the main diagonal (compared with other coefficients) results in improving the matrix conditionality. To achieve this is only possible at the expense of a time step increase because kernels of the corresponding integral equations $K(\tau - \xi)$ have the maxima $K_{max} = K(\tau - \xi^*)$ (with $\xi \to \tau$ $K(\tau - \xi) \to 0$). Physically it means that on exposure to a unit-equal boundary condition at one surface and a zero boundary condition at the other surface the rate of temperature increase at the point studied begins to diminish after a passage of time $\xi = \xi^*$. This is true except in the case with a slab thermally isolated on the one side when the boundary condition is specified on the other side of the slab by the known temperature on the thermally isolated side (or somewhere in its vicinity). In this case the kernel of the integral equation is a monotonous increasing function, being asymptotic to a constant value.

For other inverse problems there is such a number ΔFo that with $\Delta Fo > \Delta \bar{F}o$ the elements standing on the A_Δ diagonal will appear the largest. In particular, if we use a step approximation of the boundary condition of the second kind with equal intervals, such a situation for a semi-infinite body rises with values $\Delta Fo > \kappa$, $\kappa \approx 1$. It is possible to show that the following equality of coefficients will correspond to a sufficient condition of stability of the results with respect to disturbance of the input data $f(\tau)$:

$$\Delta \vartheta_0 = \Delta \vartheta_1 \, . \tag{4.4}$$

Since function $\Delta \vartheta(\tau)$, representing a temperature change at point $x = d$ by action of the unit-value impulse, is a unimodal function and by performing the condition shown in Eq. (4.4) its maximum lies between points $\tau = 0$ and $\tau = \Delta\tau$, the rest of the coefficients will satisfy the inequalities (Fig. 4.4)

$$\Delta \vartheta_1 > \Delta \vartheta_2 > \Delta \vartheta_3 > \dots .$$

Consider the following example, characterizing a method of point error. Let value f_1 have a disturbance δf_1. This disturbance will lead to an error in computing values $\hat{u}_n$, $n = 1, 2, \dots$. Whether the action of δf_1 on $\hat{u}_n$ with time will grow or not depends on the algorithm instability or stability relative to errors in the input information. Follow a behaviour of the error in the solution

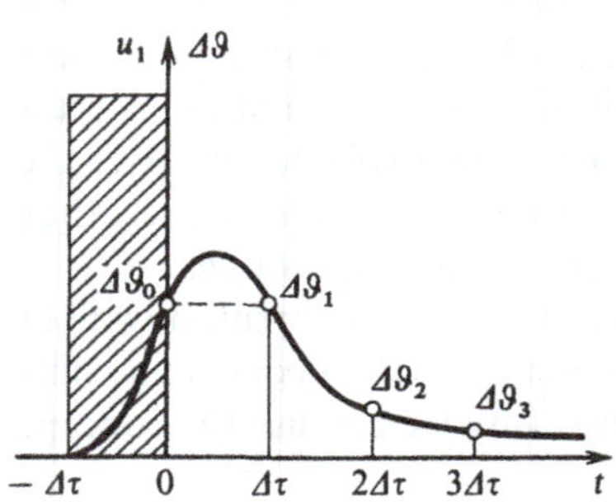

Fig. 4.4. Value $\Delta\vartheta(\tau)$ change at step value $\Delta\tau$ corresponding to a sufficient stability condition $\Delta\vartheta_0 = \Delta\vartheta_1$

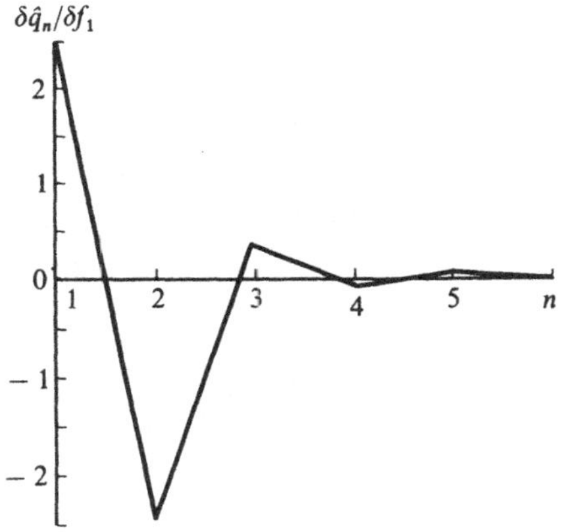

Fig. 4.5. Change of relative error in the heat flux at the boundary of a thermally thick wall by action of local disturbance in the input data ($\Delta Fo_d = 1$)

for several first time steps. Denote this error $\delta \hat{u}_n = \hat{u}_n - \tilde{u}_n$. Then we get from the (4.3) formula

$$\frac{\delta \hat{u}_1}{\delta f_1} = \frac{1}{\Delta \vartheta_0}; \quad \frac{\delta \hat{u}_2}{\delta f_1} = -\frac{1}{\Delta \vartheta_0} \frac{\Delta \vartheta_1}{\Delta \vartheta_0}; \quad \frac{\delta \hat{u}_3}{\delta f_1} = -\frac{1}{\Delta \vartheta_0}\left[\frac{\Delta \vartheta_2}{\Delta \vartheta_0} - \left(\frac{\Delta \vartheta_1}{\Delta \vartheta_0}\right)^2\right];$$

$$\frac{\delta \hat{u}_4}{\delta f_1} = -\frac{1}{\Delta \vartheta_0}\left[\frac{\Delta \vartheta_3}{\Delta \vartheta_0} - 2\frac{\Delta \vartheta_1 \Delta \vartheta_2}{(\Delta \vartheta_0)^2} + \left(\frac{\Delta \vartheta_1}{\Delta \vartheta_0}\right)^3\right]; \tag{4.5}$$

$$\frac{\delta \hat{u}_5}{\delta f_1} = -\frac{1}{\Delta \vartheta_0}\left[\frac{\Delta \vartheta_4}{\Delta \vartheta_0} - 2\frac{\Delta \vartheta_1 \Delta \vartheta_3}{(\Delta \vartheta_0)^2} + 3\frac{(\Delta \vartheta_1)^2 \Delta \vartheta_2}{(\Delta \vartheta_0)^3} - \left(\frac{\Delta \vartheta_2}{\Delta \vartheta_0}\right)^2 - \left(\frac{\Delta \vartheta_1}{\Delta \vartheta_0}\right)^4\right] \ldots$$

For stability of the results, i.e. when $\Delta \hat{u}_2 \geqslant \Delta \hat{u}_3 \geqslant \Delta \hat{u}_4 \geqslant \ldots$, it is necessary that positive and negative members in every one of these expressions could in the end compensate each other. We can expect that this condition to be observed at such value of ΔFo which yields equality (4.4). The results of computation for a thermally thick body are shown in Fig. 4.5.

4.3.3 The Method of Defining Critical Values of Calculation Intervals

The principle outlined above for selecting the approximation step in the direct algebraic method of IHCP solving corresponds to a sufficient condition, of stability, i.e. the recommended increment value of the Fourier number appears to be larger than the critical value $\Delta \bar{F}o > \Delta Fo_{cr}$.

The change of the boundary heat condition being dynamic in nature, the increase of the time step may result in a definite distortion of results because of a large approximation error in the unknown function. As computational experiments showed, there is a possibility to considerably reduce values ΔFo in comparison with estimates obtained before an irregular solution emerges providing the input data have sufficient smoothness. The method suggested below enables critical values of the step ΔFo_{cr} to be found without experimental "grouping" for an instability boundary as the result of solving many methodical examples.

The degree of proximity of the discrete form of the inverse problem (4.1) to the initial integral statement (3.1) is defined by value of the time step ΔFo, which cannot be done asymptotically small. For a stable problem solving it is necessary to select such a time interval beginning from the moment of the next step change of function $\hat{u}(\tau)$ by which the temperature reaction of the body at point $x = d$ to this change may be distinguished. Obviously, the time for "expecting" a desired response (a step ΔFo) would be entirely specified by the kernel of the integral equation (see Sect. 2.4 and Fig. 2.5). If the given function acquires a monotonically decreasing form, it means that the highest temperature change appeared due to a boundary condition disturbance at the moment $\tau = \tau^*$ corresponds to the moment τ^*, i.e. there is no time lag. It is natural to expect that such IHCP will be sufficient "good", since a heat signal has been passed in a moment. This is the case for a limiting inverse problem statement, when by using measurements of the temperature on the surface of the body it is required to recover the heat flux penetrating into the body, i.e. it is a pseudo-inverse problem.

In some other case when a thermosensor is positioned inside the body at distance d from the heat exchange surface, the information on structural changes in the boundary condition will reach the point studied in a less clearly defined form. Attenuation of the causal phenomenon for deep points will be considerable, and a section of the curve $T(d, \tau)$ corresponding to the highest increment of temperature (which intuitively is the one most saturated with information on any change in the boundary condition) will develop in a definite time $\Delta\tau_m$ after the signal is passed at the surface of the body. It is possible therefore, to assume that an attempt to recover the initial signal with only using part of the temperature information, obtained during $\tau^* + \Delta\tau$, where $\Delta\tau$ is noticeably less than $\Delta\tau_m$, would hardly yield suitable results, even if the temperature data were known to be accurate. Numerical experiments prove this assumption. Inverse problems have been solved by algebraic method according to the results of calculations of the corresponding direct problems. Approximation of the boundary condition in both cases has been one and the same with equal steps in time. Actually, the errors of approximation have been excluded, the round-off errors remaining only in the computations, which are connected with the finite word length. Computations have been performed with numbers, with an accuracy corresponding to decimal numbers with 10–11 digits. Results of computations showed that with $\Delta\tau < \Delta\tau_m$ negligible errors in round-off even so could lead to oscillating solutions of IHCP. At the same time for all $\Delta\tau \geqslant \Delta\tau_m$ the solutions were stable.

The stability of pseudo-inverse heat conduction problem (see Sect. 2.4) as well as the stated physical aspects along with the computational analysis gave rise to the following hypothesis: to obtain the stability of the results when using step regularization for algebraic solution of the boundary IHCP, an admissible value of time interval in step approximation of the boundary condition can be taken as approximately equal (but not less) to the time between a step-wise change of the curve $\hat{u}(\tau)$ and maximum manifestation of this instantaneous impulse in the temperature of the body point under study. Otherwise, one may demand that by piecewise-constant approximation of the unknown function, ΔFo values should satisfy the condition $\Delta Fo_l > \Delta Fo_{l_{cr}} = Fo_l^*$, $\Delta Fo_l \simeq \Delta Fo_{l_{cr}}$ where $Fo_l^* = a(\tau - \xi)^*/l^2$ is a Fourier number corresponding to the point of maximum of $K(\tau - \xi)$; $l = d$ or b depending on the problem to be solved.

Then, when using exact (or close to exact) input information the algebraic solution given by formula (4.3) should become sufficiently smooth.

4.3.4 Examples for Calculating Critical Steps

According to the method suggested for evaluating critical values of approximation steps let us analyse some statements of boundary IHCP using known expressions for $K(\tau - \xi)$ [51, 172]. Consider first the problems of *recovering the boundary condition of the second kind.* In this case kernels of integral equations of the type shown in Eq. (3.1) with accuracy up to a constant factor are equal to corresponding influence functions providing a transmission of the heat signal at the boundary of the body with an unknown condition.

For a **semi-infinite body**

$$K(Fo_d) = \frac{a}{\lambda} G(Fo_d), \tag{4.6}$$

where

$$G(Fo_d) = \frac{1}{d\sqrt{\pi}} \frac{\exp\left[-\frac{1}{4Fo_d}\right]}{\sqrt{Fo_d}}$$

is the influence function at point $x = d$ (Fig. 4.6).

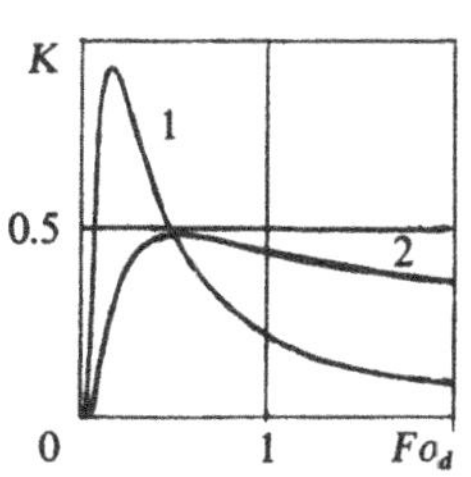

Fig. 4.6. The kernels of integral equations for a semi-infinite body: 1 – for the boundary condition of the first kind [formula (4.10), $\bar{K} = Kd^2/a$]; 2 – for the boundary condition of the second kind $\left[\text{formula (4.6)}, \bar{K} = K\frac{d\lambda}{a}\right]$

Putting a derivative $dK(Fo_d)/dFo_d$ equal to zero, we get value $Fo_d^* = \frac{1}{2}$, from which

$$\Delta Fo_{d_{cr}} \simeq \tfrac{1}{2}. \tag{4.7}$$

This is the very stability condition or one close to it which has been formulated as the result of solving different methodical problems in [70, 158].

Consider now a **flat-and-parallel slab**. Let one of its boundaries $x = b$ be thermally isolated, while a heat flux of unknown magnitude acts on the other boundary. The kernel of the integral Eq. (3.1) is

$$K(Fo_b) = \frac{a}{\lambda} G(d/b, Fo_b), \tag{4.8}$$

where $$G(d/b, Fo_b) = \frac{1}{b} + \frac{2}{b} \sum_{k=1}^{\infty} \exp[-k^2\pi^2 Fo_b] \cos\frac{k\pi d}{b}.$$

The plots $G(d/b, Fo_b)$ constructed in Fig. 4.7 show that the behaviour of the curves at $d < 0.5b$ and $d \geqslant 0.5b$ differs qualitatively. While the influence function in domain $d < 0.5b$ reaches its maximum, a maximum value for distance $d \geqslant 0.5b$ cannot be reached, i.e. function $G(Fo_b)$ is monotonously increasing and become asymptotic to value $1/b$. This value corresponds to the temperature uniformly distributed in the body section, which reaches this steady state with $Fo_b \to \infty$ after a pulsed supply to boundary $x = 0$ of the heat flow qdt equals one. That is why this method of finding $\Delta Fo_{b_{cr}}$ is only appropriate with $x < 0.5b$. Defining a critical value of step $\Delta\tau$ for $x < 0.45b$ can be made from the condition

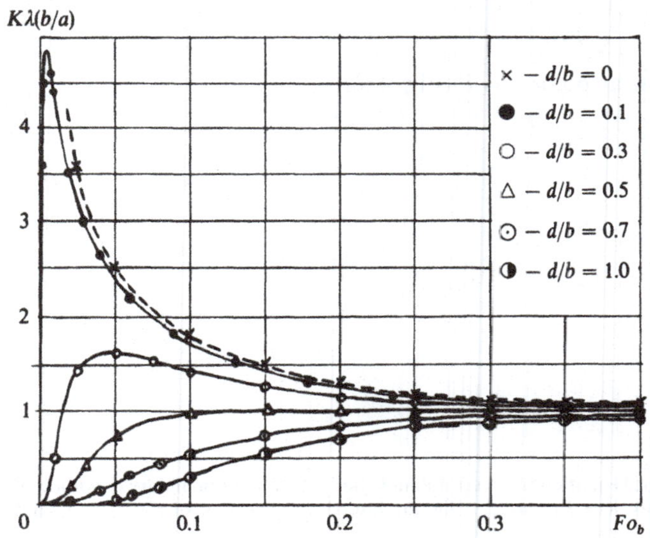

Fig. 4.7. The kernel of integral Eq. (4.8)

(4.7), i.e. $\Delta\tau_{cr} \cong d^2/2a$ since in this range of x for values of Fourier number $Fo_d = \frac{a\tau}{d^2} \leqslant Fo_d^*$ heat propagation is practically identical to a semi-infinite body.

The dependence $Fo_b^*(d/b)$ for a thermally isolated slab with an unknown boundary condition of the second kind is constructed in Fig. 4.10.

Another IHCP statement corresponds to a prescribed temperature at one boundary of the slab and to an unknown heat flux at the other boundary. In this case the pulsed body heating from side $x = 0$ with a zero temperature at boundary $x = b$ yields the following influence function (Fig. 4.8):

$$G(d/b, Fo_b) = \frac{2}{b}\sum_{k=0}^{\infty} \exp\left[-\frac{(2k+1)^2\pi^2 Fo_b}{4}\right]\cos\left[\frac{(2k+1)\pi d}{2b}\right]. \tag{4.9}$$

Values Fo_b^*, corresponding to small values of temperature G at some point $x = d$, can be specified by the plot in Fig. 4.10, it being known that at $d \cong b$ $Fo_b^* = 0.167$. Within the range of $0 < d \leqslant 0.4b$ the character of heat conduction coincides with the model of a semi-infinite body and therefore we can use the formula (4.7).

Now consider statements of inverse problems when the *body surface temperature* is taken as an *unknown function*. In these cases the kernel of the integral equation is proportional to the temperature at point d which emerges as a result of body heating by a temperature impulse from the side of the boundary with an unknown condition.

If the domain of coordinate x corresponds to a semi-bounded straight line, the kernel is (see Fig. 4.6)

$$K(Fo_d) = \frac{a}{2\sqrt{\pi}d^2}\,\frac{\exp[-\frac{1}{4}Fo_d]}{Fo_d^{3/2}}. \tag{4.10}$$

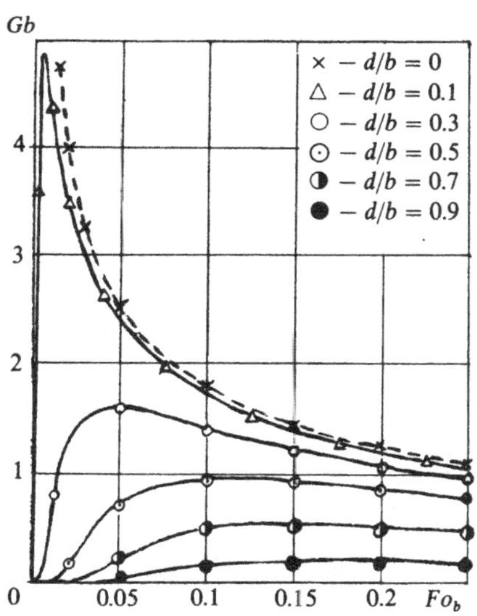

Fig. 4.8. The influence function (4.9)

It is easy to calculate that the maximum transmission time of this function after the impulse supply will be equal to $Fo_d^* = 1/6$, from which $\Delta Fo_{cr} \cong 1/6$.

For a slab of finite thickness consider two cases, when the boundary conditions of the first and second kind are prescribed correspondingly at the boundary $x = b$.

In the first case the kernel of integral Eq. (3.1) is defined by the expression

$$K(d/b, Fo_b) = \frac{a}{2b^2\sqrt{\pi Fo_b^{3/2}}} \sum_{k=1}^{\infty} (-1)^{k+1} \left\{ (2k - d/b) \exp\left[-\frac{(2k - d/b)^2}{4Fo_b} \right] + [2(k+1) + d/b] \exp\left[\frac{(2(k-1) + d/b)^2}{4Fo_b} \right] \right\}, \quad (4.11)$$

in the second case,

$$K(d/b, Fo_b) = \frac{a}{2b^2\sqrt{\pi Fo_b^{3/2}}} \sum_{k=-\infty}^{+\infty} \left\{ (2k + d/b) \exp\left[-\frac{(2k - d/b)^2}{4Fo_b} \right] \right\} \quad (4.12)$$

The corresponding plots are shown in Figs. 4.9 and 4.10.

Figure 4.11 illustrates the solution errors found by formula (4.5) for the case of recovering the heat flux into a semi-infinite body. It is seen that as the number of steps increases, the errors are of non-increasing character, provided $\Delta\tau \cong \Delta\tau_{cr}$. At the same time a small decrease of the design step compared with the value found from the kernel's maximum led to the oscillations in solution. Here, if the kernel carries a clearly marked maximum, the data obtained, ΔFo_{cr}, will clearly define the boundary of the IHCP solution stability. In this case, for steps ΔFo, a bit less than ΔFo_{cr}, the solution begins to oscillate at high amplitude (Fig. 4.12a, b). For $K(\tau - \xi)$ functions, which carry no such sharp

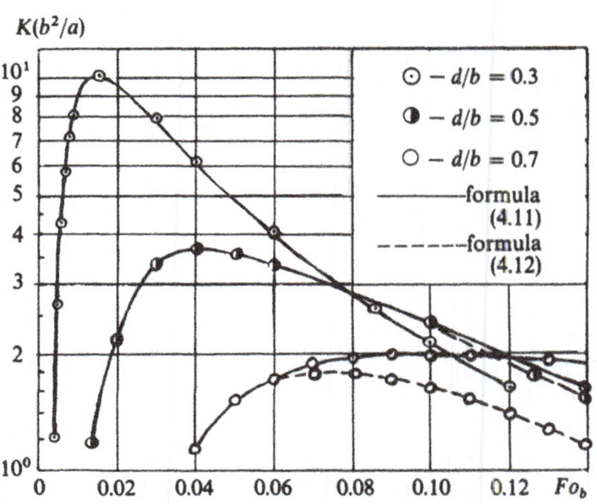

Fig. 4.9. The kernels of Eqs. (4.11) and (4.12)

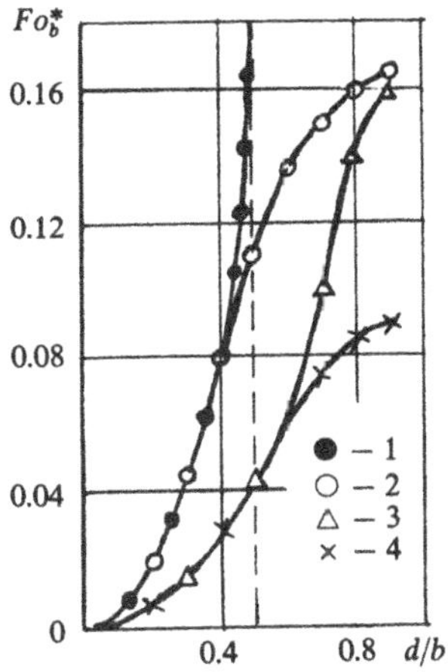

Fig. 4.10. Fourier numbers, corresponding to points of kernels maximum: *1* – (4.8); *2* – (4.9); *3* – (4.11); *4* – (4.12)

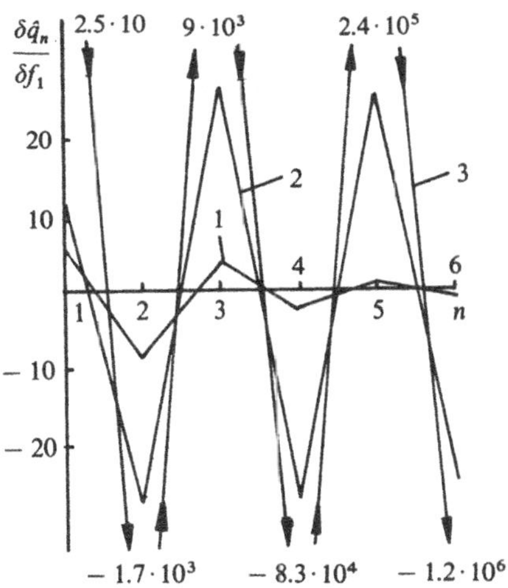

Fig. 4.11. The relative error change in the heat flux at the boundary of a thermally thick wall under the action of local disturbance in the input data (vector of disturbance $\delta f = [\delta f_1, 0, 0, \ldots, 0]^T$): 1 – $\Delta Fo_d = 0.5$; 2 – $\Delta Fo_d = 0.3$; 3 – $\Delta Fo_d = 0.1$

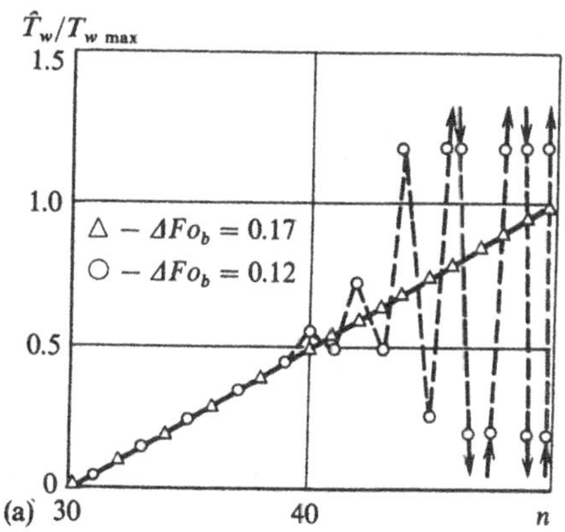

Fig. 4.12a continued

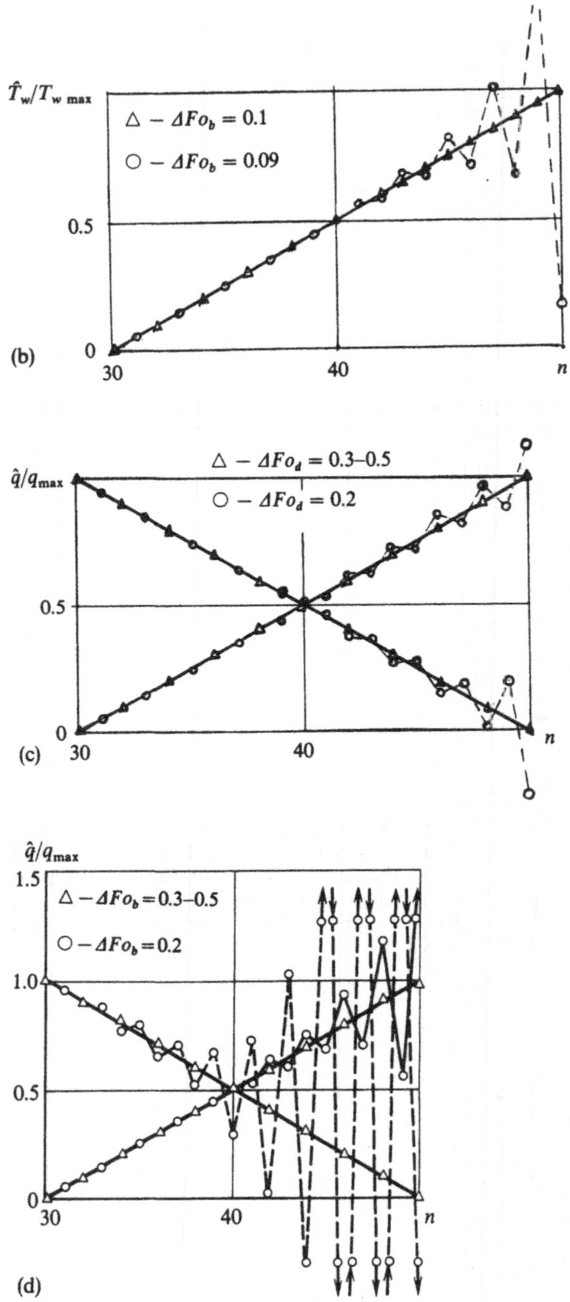

Fig. 4.12a–d. Design examples of boundary conditions using the exact data (symbols for heat conduction models a, b, c, d correspond to Fig. 4.1, full lines are exact linear dependences): **a** is a slab (e), $d = b$, $\Delta Fo_{b_{cr}} \simeq 0.17$; **b** is a slab (c), $d = 0.95b$, $\Delta Fo_{b_{cr}} \simeq 0.1$; **c** is a thermally thick body (b), $\Delta Fo_{d_{cr}} \simeq 0.5$; **d** is a slab (d), $d = b$, $\Delta Fo_{b_{cr}} \simeq 0.5$

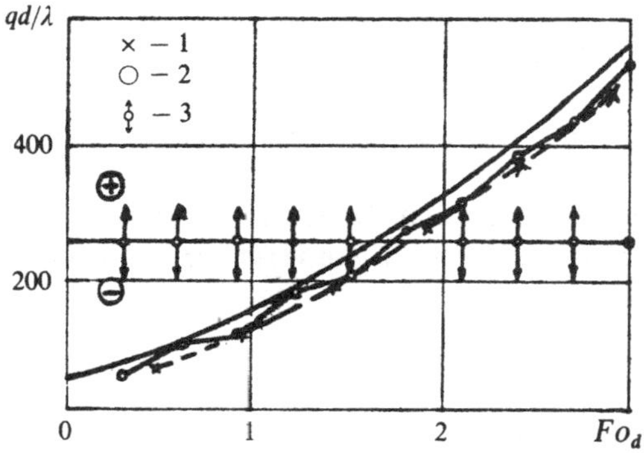

Fig. 4.13. The results of 1HCP solving for a thermally thick body using the undistributed data: *1*, $\Delta Fo_d = 0.5$; *2*, $\Delta Fo_d = 0.32$; *3*, $\Delta Fo_d = 0.07$, full curve is an exact function: $\bar{q}_T \equiv \frac{qd}{\lambda} = 48 + 88Fo_d + 27Fo_d^2$

a peak, the "shaking" of the solution is less abrupt (Figs. 4.12c and 4.13). However, it is also here that the emergence of even negligible errors in the results of measurements makes us use in our calculation values ΔFo, which are closer to ΔFo_{cr}.

Thus, this numerical analysis confirms this hypothesis. The described process of finding ΔFo_{cr} is a convenient and universal procedure in the technique of direct methods for solving the integral forms of boundary inverse heat conduction problems.

As for those IHCP statements in which kernels of integral equations do not reach their maximum at the finite time intervals, it is possible to find a desired value of the step from the condition $\Delta\tau_{cr} = \frac{l^2 Fo_l^{**}}{a}$, where Fo_l^{**} is a Fourier number, at which the curve $K(Fo_l)$ with a sufficient degree of accuracy becomes asymptotic (see Fig. 4.7). For a flat slab with a temperature sensor positioned on a thermally isolated side opposite to the boundary with the unknown heat flux, $\Delta Fo_{b_{cr}} \cong 0.5$ (see Fig. 4.12d).

4.3.5 Stability Comparative Analysis of Inverse Heat Conduction Problems in the Boundary-Value Statements of Different Type

Let us deal with the analysis of properties of the direct method stability by solving the inverse problems in the boundary-value statements. Two factors (smoothing and time lag) have been mentioned in Sect. 2.1.6, as being the causes of mathematical instability of IHCPs. If the heat condition at the boundary of the body experiences disturbance, then, while moving from the body boundary, this disturbance would, apparently, "smear" in space and time. The largest deviation of temperature related to the disturbance in the boundary condition would shift to the side with greater time. These effects develop more strongly as the point moves into the depths of the body. It is obvious that the inverse problem here, from stability of the results, proceeds from bad to worse, i.e. the unstable operation of counting occurs through greater ΔFo steps.

Consider one of the aspects of this phenomenon. A stepwise change of the boundary function $\hat{u}(\tau)$ at a definite moment of time $\tau = \tau^*$, connected with the approximation of the real curve $u(\tau)$, causes an instantaneous temperature change over the whole points of the body. The heat signal amplitude, however, decays sharply both by x, and by τ. If we apply a Fourier method to solve a direct heat conduction problem, these effects will be counted by corresponding exponential members and a reduction of the rate of convergence in the sign-changing series of solution.

So, for a boundary-value problem

$$\frac{\partial T}{\partial \tau} = a\frac{\partial^2 T}{\partial x^2}, \quad 0 < x < b, \ \tau > \tau^*; \qquad T(x, \tau^*) = f(x), \ 0 \leqslant x \leqslant b;$$

$$-\lambda\frac{\partial T(0, \tau)}{\partial x} = q = \text{const}; \quad \frac{\partial T(b, \tau)}{\partial x} = 0, \qquad \tau \geqslant \tau^*,$$

according to a superposition principle, we receive $T(x, \tau) = T_{i.c.}(x, \tau) + T_{b.c.}(x, \tau)$ where $T_{i.c.}$ and $T_{b.c.}$ are the components of a temperature field from the action of initial and boundary conditions, respectively.

These components are expressed by formulae [93, 172]

$$T_{i.c.}(x, \tau) = \sum_{k=0}^{\infty} c_k \cos(\mu_k x) \exp[-\mu_k^2 a\tau];$$

$$T_{b.c.}(x, \tau) = \frac{q}{\lambda}\left\{\frac{a\tau}{b} - \frac{b^2 - 3(b-x)^2}{6b} + 2b\sum_{k=1}^{\infty}\frac{(-1)^k}{\mu_k^2}\exp[-\mu_k^2 a\tau]\cos[\mu_k(b-x)]\right\},$$

where

$$c_k = \frac{2}{b}\int_0^b f(\eta)\cos(\mu_k\eta)\,d\eta, \ k = 1, 2, \ldots; \ c_0 = \frac{1}{b}\int_0^b f(\eta)\,d\eta;$$

μ_k are the eigenvalues of the problem.

Note that the series convergence and the form of exponential factors are interrelated, namely, the decrease of component values in $\sum_k$ depends on arguments in the exponential factor, which are defined by the eigenvalues of a corresponding problem of Sturm–Liouville. We can conclude that the less the eigenvalues and the less the order of their increase, the worst is a conditionality of inverse problem for small times τ. The eigenvalues are defined by formulating a heat conduction problem. It is known [117] that if we consider the problems with boundary conditions of the first, second and third kind, the largest eigenvalues would rest with the first boundary-value problem, while the least numbers with the second. The eigenvalues μ_k^2 for a slab with boundary conditions of the first and second kind are equal to $\mu_k^I = \pi k/b$, $k = 1, 2, \ldots$, $\mu_k^{II} = \pi k/b$,

$k = 0, 1, \ldots, (\mu_1^I > \mu_0^{II}, \mu_2^I > \mu_1^{II}, \ldots)$, respectively. That is why we can make the following assumption: other things being equal the main characteristic of stability ΔFo_{bcr}, will have the largest value for the inverse problem in the second boundary-value statement and the smallest value for an IHCP in the statement corresponding to the first boundary-value problem. This conclusion is in full agreement with the results obtained in the previous section

4.4 The Solution of an Inverse Heat Conduction Problem Using Some Other Methods of Approximation and with Disturbed Data

We could now set the task to improve the quality of problem approximation, applying a piecewise-linear or piecewise-quadratic description for the unknown function. However, upgrading the order of the approximation formulae will lead to an increase of the minimum admissible step ΔFo_{cr} and which is why it can but give better results. Let us explain this using an example of piecewise-linear approximation.

Let a direct heat conduction problem be the first boundary-value problem for a semi-infinite body

$$\frac{\partial T}{\partial \tau} = a\frac{\partial^2 T}{\partial x^2}, \quad 0 < x,\ 0 < \tau \leqslant \tau_m; \tag{4.13}$$

$$T(x, 0) = \psi(x), \quad 0 \leqslant x; \qquad T(0, \tau) = f(\tau), \quad 0 \leqslant \tau \leqslant \tau_m. \tag{4.14}$$

Differentiate Eq. (4.13) in time and change the uniqueness conditions in a proper way:

$$\frac{\partial T'}{\partial \tau} = a\frac{\partial^2 T'}{\partial x^2}, \quad 0 < x,\ 0 < \tau \leqslant \tau_m; \tag{4.15}$$

$$T'(x, 0) = \varphi(x), \quad 0 \leqslant x; \qquad T'(0, \tau) = f'(\tau), \quad 0 \leqslant \tau \leqslant \tau_m, \tag{4.16}$$

where $T' = \partial T/\partial \tau$.

Solutions of the problems (4.13)–(4.16) are presented with the help of integral operators of identical kind

$$T(x, \tau) = \int_0^\infty \psi(\eta) G(x, \eta, \tau)\,\mathrm{d}\eta + a\int_0^\tau f(\xi)\frac{\partial G(x, 0, \tau - \xi)}{\partial \eta}\mathrm{d}\xi, \tag{4.17}$$

$$T'(x, \tau) = \int_0^\infty \varphi(\eta) G(x, \eta, \tau)\,\mathrm{d}\eta + a\int_0^\tau f'(\zeta)\frac{\partial G(x, 0, \tau - \xi)}{\partial \eta}\mathrm{d}\xi, \tag{4.18}$$

where $G(x, \eta, \tau - \xi)$ is the influence function of the first boundary-value heat conduction problem for a half-line.

Assume now that function $f(\tau)$ is unknown and it is to be found by known values of temperature $T(d, \tau)$ at a definite point of the body $x = d > 0$. A corresponding integral equation follows from expression (4.17)

$$\int_0^\tau f(\xi)K(\tau-\xi)\,d\xi = F(\tau), \quad 0<\tau\leqslant\tau_m\,, \tag{4.19}$$

where $K(\tau-\xi) = a\dfrac{\partial G(d,0,\tau-\xi)}{\partial\eta} = \dfrac{\partial\vartheta(d,\tau-\xi)}{\partial\tau}$ [$\vartheta(x,\tau)$ is the solution of the problem in Eqs. (4.13) and (4.14) with $\psi(x)=0$ and $f(\tau)=1$]; $F(\tau) = T(d,\tau) - \int_0^\infty \psi(\eta)G(x,\eta,\tau)\,d\eta$.

Integrating expression (4.18) in time between 0 and t, changing the order of integration in the second term of the right side and passing to an integral equation with respect to function $f'(\xi)$ we receive

$$\int_0^t f'(\xi)S(t-\xi)\,d\xi = H(\tau), \quad 0<t\leqslant\tau_m\,, \tag{4.20}$$

$$\text{where } S(t-\xi) = a\int_\xi^t \frac{\partial G(d,0,\tau-\xi)}{\partial\eta}\,d\tau = \vartheta(d;\tau-\xi)\,; \tag{4.21}$$

$$H(t) = T(d,\tau) - \psi(d) - \int_0^t d\tau \int_0^\infty \varphi(\eta)G(x,\eta,\tau)\,d\eta\,.$$

By step approximation of $f(\tau)$ the least admissible step $\Delta Fo_{d_{cr}}$ will be defined by the time interval $[0, (\tau-\xi)^*]$, where $(\tau-\xi)^*$ is the moment, corresponding to the point of maximum of the kernel in Eq. (4.19). If for $f(\tau)$ approximation we use a piecewise-linear dependence, the solution of problem (4.19) is equivalent to the solution of the integral Eq. (4.20) with a piecewise-constant approximation of $f'(\tau)$. The step of stable solution will be defined by the form of the kernel. In this case the kernel is obtained by integration according to the formula (4.21) and is a monotonously increasing function $\vartheta(t-\xi)$, becoming asymptotic to a steady-state value $\vartheta=1$. That is why a piecewise-linear approximation of the inverse problem should have larger step values providing stability of solutions compared to the case of approximation of the boundary function by the piecewise-constant function. Plots of functions

$$\bar{K}(Fo_d) = \frac{d^2}{a}K(Fo_d) = \frac{\exp[-1/(4Fo_d)]}{2\sqrt{\pi}Fo_d^{3/2}} \quad\text{and}\quad \vartheta(Fo_d) = \operatorname{erfc}\left[\frac{1}{2\sqrt{Fo_d}}\right]$$

are shown in Figs. 4.6 and 4.14, respectively.

Lastly, let us consider natural regularization of the boundary inverse problem with *disturbed input data.* Two approaches can be suggested here for using the direct algebraic method of IHCP solving.

The first approach assumes preliminary smoothing of the initial temperature information. The question of selecting a smoothing algorithm is of prime significance here. As the results of numerical simulation showed, if the method of regularization by Tikhonov with the second derivative in the regularizator [24] or the method of smoothing cubic splines [130, 199] be used for smoothing the

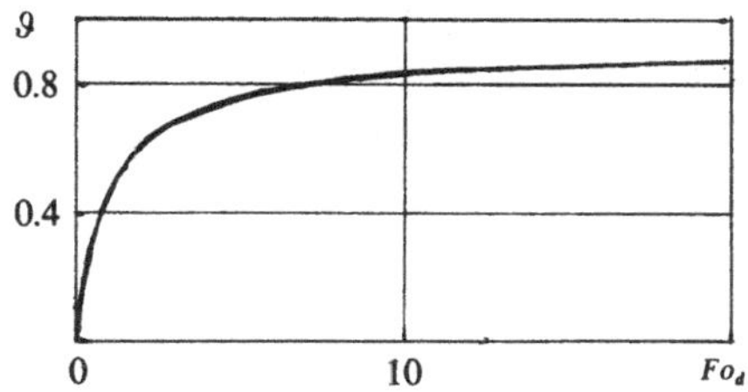

Fig. 4.14 The plot of the kernel of (4.20) integral equation

experimental functions, the selection of ΔFo_{cr} can be done following the procedure suggested above.

The second approach assumes no preliminary smoothing. This excludes selection of a time step according to the above procedure, since it does not filtrate random errors in the input data. In the given case matching of temperature discrepancy with the value of measurements error can serve as the criterion for choosing the step ΔFo_{adm}. Should an approximation error in the definite range of steps ΔFo appear considerably lower than errors in the assignment of the right side of Eq. (3.1), the selection of the number of intervals m ($m = \tau_m/\Delta\tau$) in defining the boundary function could be done, for example, from the following condition for a "weighted" mean-square residual:

$$\sum_{n=1}^{m} [(T_n(m) - T_n^*)/\bar{\sigma}_n]^2 \simeq m\,,$$

where $T_n(m)$ is the temperature at the nth moment of time, calculated for the m-dimensional vector $\boldsymbol{u}$, found by formula (4.3); T_n^* is an experimental value of temperature at the nth moment of time.

The weights $\bar{\sigma}_n$ are estimates of the root mean-square error of experimental data for the nth moment of time.

The analysis of these two approaches for accuracy of the results favours the first method.

4.5 Algorithmic Presentation of a Two-Dimensional Inverse Heat Conduction Problem

Let us formulate the main principles of algebraic solution of IHCP integral forms in a two-dimensional statement, when the unknown function $u(x, \tau)$ (the body surface temperature, heat flux, thermal potential density) satisfies the integral equation

$$Au \equiv \int_0^{\tau} d\xi \int_0^{b} u(x', \xi) K(x, x'; \tau - \xi)\, dx' = f^*(x, \tau)\,,$$

$$0 < \tau \leqslant \tau_m, \quad 0 < x < b\,, \tag{4.22}$$

where $f^*(x, \tau)$ is the prescribed right side.

Example. Consider an inverse problem in the case of a one-sided heating of a flat slab (Fig. 4.15). Suppose that the problem can be formulated as two-dimensional, i.e. the process of heating is the same at all points on the axis z. Then we have

$$\frac{\partial T}{\partial \tau} = a\left(\frac{\partial^2 T}{\partial x^2} + \frac{\partial^2 T}{\partial y^2}\right), \quad x \in (0, b), \ y \in (0, d), \ \tau \in [0, \tau_m],$$

$$T(x, y, 0) = \psi(x, y), \quad x \in [0, b], \quad y \in [0, d];$$

$$-\lambda \frac{\partial T(x, 0, \tau)}{\partial y} = q(x, \tau), \quad \tau \in [0, \tau_m];$$

$$\frac{\partial T(x, d, \tau)}{\partial y} = \frac{\partial T(0, y, \tau)}{\partial x} = \frac{\partial T(b, y, \tau)}{\partial x} = 0, \quad \tau \in [0, \tau_m],$$

where function $q(x, \tau)$ is unknown and needs to be found using a complementary condition – the temperature is given at some line $y = y^* = \text{const}$, $0 < y^* \leqslant d$

$$T(x, y^*, \tau) = f(x, \tau), \quad \tau \in [0, \tau_m].$$

Consider the boundary-value statement of this inverse problem, which is obtained with the help of Green's function $G(x, y; x', y'; \tau - \xi)$:

$$\frac{a}{\lambda} \int_0^\tau \mathrm{d}\xi \int_0^b q(x', \xi) G(x, y^*; x', 0; \tau - \xi)\, \mathrm{d}x' = \bar{f}(x, y^*, \tau), \tag{4.23}$$

$$\tau \in [0, \tau_m], \ x \in (0, b),$$

where $\bar{f}(x, y^*, \tau) = f(x, \tau) - \int_0^b \mathrm{d}x' \int_0^d \psi(x', y') G(x, y^*; x', y'; \tau - \xi)\, \mathrm{d}y'$;

$$G(x, y; x', y'; \tau - \xi) = \frac{1}{bd}\left\{1 + 2\sum_{r=1}^{\infty} \cos\frac{r\pi x}{b} \cos\frac{r\pi x'}{b} \times \exp\left[-\frac{ar^2\pi^2(\tau - \xi)}{b^2}\right]\right\}\left\{1 + 2\sum_{r=1}^{\infty} \cos\frac{r\pi y}{d} \cos\frac{r\pi y'}{d} \times \exp\left[-\frac{ar^2\pi^2(\tau - \xi)}{d^2}\right]\right\}.$$

Equation (4.23) is identical in its form to Eq. (4.22).

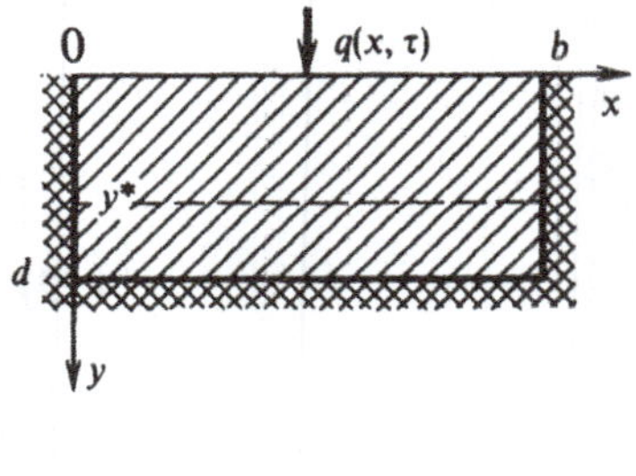

Fig. 4.15. The rectangular domain of spatial variables

Introduce a uniform grid with nodes $x_l = l\Delta x$, $l = 0, 1, \ldots, L$; $\tau_n = n\Delta\tau$, $n = 0, 1, \ldots, m$ and replace function $u(x, \tau)$ with vector $\boldsymbol{u}$ using step approximation with respect to time and space, e.g. in the form $\boldsymbol{u} = [\tilde{u}_{11}, \tilde{u}_{21}, \ldots, \tilde{u}_{L1}; \tilde{u}_{12}, \tilde{u}_{22}, \ldots, \tilde{u}_{L2}; \ldots; \tilde{u}_{1m}, \tilde{u}_{2m}, \ldots, \tilde{u}_{Lm}]^T$,

where

$$\tilde{u}_{ln} = \frac{\hat{u}_{l,n-1} - \hat{u}_{ln}}{2}; \; \hat{u}_{ln} = \frac{u_{l-1}(\tau_n) + u_l(\tau_n)}{2};$$

$$u_l(\tau_n) = u(x_l, \tau_n).$$

Such a representation means averaging along the spatial and time intervals as a result of which a continuous surface $u(x, \tau)$ is approximated by a "tile-type" surface (Fig. 4.16). Instead of the continuous right side $f^*(x, \tau)$ take the corresponding vector $\boldsymbol{f}^*: \boldsymbol{f}^* = [f^*_{11}, f^*_{21}, \ldots, f^*_{L1};\; f^*_{12}, f^*_{22}, \ldots, f^*_{L2}; \ldots; f^*_{1m}, f^*_{2m}, \ldots, f^*_{Lm}]^T$.

Using quadrature formulae, we shall receive a finite-dimensional (matrix) approximation for the integral Eq. (4.22)

$$A_\Delta \boldsymbol{u} = \boldsymbol{f}^* . \tag{4.24}$$

Note that if the kernel $K(x, x'; \tau - \xi)$ admits analytical integration by x' and ξ [e.g. Eq. (4.23)], we should make use of such a possibility when defining the elements of the A_Δ matrix in order to increase the approximation accuracy of the problem.

Matrix A_Δ has a triangular-block structure

$$A_\Delta = \begin{bmatrix} A_{11} & & & \\ A_{21} & A_{22} & & \\ A_{31} & A_{32} & A_{33} & \\ \vdots & & & \ddots \\ A_{m1} & A_{m2} & \ldots & A_{mm} \end{bmatrix} . \tag{4.25}$$

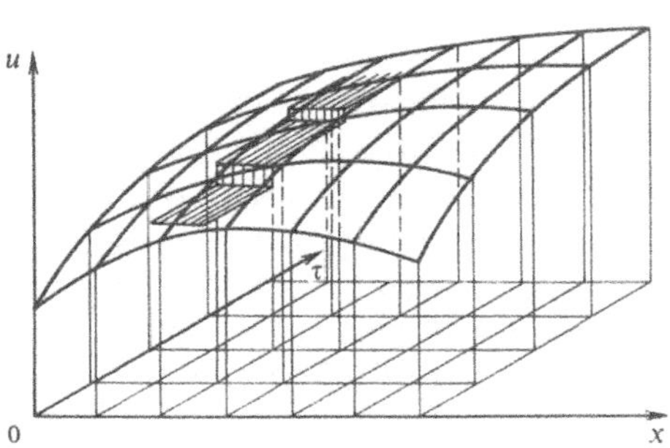

Fig. 4.16. Step approximation of the surface $u(x, \tau)$

Matrices-blocks A_{in} have the property of by-diagonal equality

$$A_{11} = A_{22} = \ldots = A_{mm} \equiv A_0\,;$$

$$A_{21} = A_{32} = \ldots = A_{m,m-1} \equiv A_1\,;$$

$$\ldots\ldots\ldots\ldots\ldots\ldots\ldots$$

$$A_{m-1,1} = A_{m2} \equiv A_{m-2}\,.$$

Each matrix A_s is a square matrix of L dimensionality:

$$A_s = \begin{bmatrix} a_{11} & a_{12} & \ldots & a_{1L} \\ a_{21} & a_{22} & \ldots & a_{2L} \\ \vdots & \vdots & & \vdots \\ a_{L1} & a_{L2} & \ldots & a_{LL} \end{bmatrix}_{s,\ s=0,1,\ldots,m-1}$$

having the elements defined by the form of the kernel in Eq. (4.22). The computation of vector $\boldsymbol{u}$ components can be based on two variants of the direct method.

Variant I is based on the matrix A_0 one-fold inversion and recurrent calculation of vectors $\boldsymbol{u}_n = [u_{1n}, u_{2n}, \ldots, u_{Ln}]^T$ for fixed times (beginning from $n = 1$ up to $n = m$) multiplying A_0^{-1} by the new right side [see the formula in Eq. (4.3) for comparison]:

$$\boldsymbol{u}_n = A_0^{-1}\left(\boldsymbol{f}_n^* - \sum_{i=1}^{n-1} A_{n-i}\boldsymbol{u}_i\right),$$

where $\boldsymbol{f}_n^* = [f_{1n}^*, f_{2n}^*, \ldots, f_{Ln}^*]^T$.

Note that the elements of matrices A_s, $s = 0, 1, \ldots, m-1$ and A_0^{-1} can be calculated once for given values of Δx and $\Delta\tau$ and stored in the machine for further use.

Since the procedure of matrix inversion can lead to a great loss of accuracy, it is more reasonable (see [178]) to make use of the second variant of solution. Variant II assumes a successive solution of the system $A_0\boldsymbol{u}_n = \boldsymbol{b}_n$, $n = 1, 2, \ldots, m$ in which the right side for every time n is given by a recurrent expression $\boldsymbol{b}_n = \boldsymbol{f}_n^* - \sum_{i=1}^{n-1} A_{n-i}\boldsymbol{u}_i$.

This algorithm for solving a two-dimensional non-stationary inverse problem calls for some regularizing conditions, which define the admissible level of approximation to an "exact" (unstable) solution. Otherwise, unsatisfactory results can be obtained because of the sensitivity of IHCP solution to errors of approximation and in the input data. This main property of the inverse problem becomes stronger for a two-dimensional statement than in a one-dimensional case. Lag and smoothing effects of the heat conduction process become stronger combined with approximation errors in variable x. For this reason the direct method of solution based on algorithm step regularization will not be so effective as for one-dimensional IHCPs and can only be used for comparatively low dynamic processes of heating with heat fluxes weakly changing along the heat exchange surface of the body. It is in this case when the constraints concerning stability of results and which are imposed on the approximation steps do not interfere with accuracy, i.e. they do not lead to a considerable distortion of the form of the unknown functions. Less constraining, from this point of view, are some other principles of regularization of the given algorithm. They are outlined in Chaps. 6 and 7.

4.6 Summary

1. An approximate-analytical method is developed for solving boundary IHCP; it is an outgrowth of the approach initiated even at the end of 1950s and the beginning of 1960s by T. Mirsepassi, G. Stolz and N. Shumakov. The integral statements of inverse problems and the step regularization principle form the basis of this method. Conditions are formulated, guaranteeing a step regularization of the solution as well as suggestions made for constructive techniques in determining critical values of time steps.
2. The algorithms are described for solving linear one-dimensional and two-dimensional inverse problems of heat conduction. Recommendations are given for practical applications. In studying an approximate solution of the integral form of one-dimensional IHCP it is possible, as described in this chapter, to show a "pure" regularization effect, conditioned by a selection of the time step of approximation. It is of interest to analyse how this effect will vary if in addition to a time approximation of the problem, a need appears for approximation with another (spatial) variable. The answer to this question can be found in the next chapter.

Chapter 5

Solution of Boundary Inverse Heat Conduction Problems by Direct Numerical Methods

Introduction

In studying high-temperature processes of special interest are boundary IHCP in non-linear statements, when thermophysical properties of the body are dependent upon temperature. In particular, for many metals the effects of temperature-dependent thermal conductivity and specific heat must be allowed for at $T \geqslant 600 \div 800°C$. The general approach to the solution of non-linear inverse problems uses numerical methods.

One of the first in this direction is the work of Beck and Wolf [165], in which a method of function specification was generalized for a non-linear case. In Beck, Blackwell and Saint-Clare [162] this technique was developed further, a bibliography of works written by American scholars is also given where numerical methods have been used to construct algorithms for IHCP solution.

This chapter, based on the studies by the author and on those of his colleagues, demonstrates the development of finite-difference algorithms for heat regime restoration on the boundary of a one-dimensional body taking as a basis for computation the principle of *step regularization* when utilizing the additional internal properties of viscosity from numerical schemes. The chapter is outlined as follows.

Section 5.1 gives a description of one of the approaches to a construction of the direct finite-difference method for computation of the heat flux and temperature on the surface of the body using the results of internal temperature measurements. This approach is based on the use of *explicit schemes* of numerical continuation of the temperature field. How to estimate stability of a numerical method of the inverse problem solution is discussed in Sect. 5.2. Various, more concrete aspects of stability concerning IHCP solution using explicit difference schemes as well as the technique of selecting the corresponding steps of approximation will be studied in greater detail in Sect. 5.3.

The difference method of IHCP solution, based on implicit difference schemes of temperature field continuation, is described in section 5.4. It is shown that this method has better stability characteristics compared with explicit schemes.

Section 5.5 completes the chapter. It describes in brief a technique proposed by the author to increase the stability of the IHCP numerical solution by introducing in the heat conduction equation a stabilizing term as the second time derivative of a temperature with low parameter. As a result, the equation artificially becomes hyperbolic, for which a boundary inverse problem is well-posed. A value of the parameter is selected by conforming the temperature residual with the input data error.

5.1 Construction of Difference Algorithms

5.1.1 Problem Statement and Approach to Solution

Let us consider a non-linear boundary inverse problem for heat conduction. Assume that the body has boundaries $[0, b]$ and at one of them $(x = b)$ is known a heat flux $q^*(\tau)$. Given are temperature measurements $f(\tau)$ at some point $0 < d \leqslant b$ and initial distribution $\varphi(x)$. It is required to determine a temperature field in the body and the conditions at boundary $x = 0$ from

$$C(T)\frac{\partial T}{\partial \tau} = \frac{\partial}{\partial x}\left[\lambda(T)\frac{\partial T}{\partial x}\right], \quad 0 < x < b,\ 0 < \tau \leqslant \tau_m ; \tag{5.1}$$

$$T(x, 0) = \varphi(x), \quad 0 \leqslant x \leqslant b ; \tag{5.2}$$

$$T(d, \tau) = f(\tau), \quad 0 \leqslant \tau \leqslant \tau_m ; \tag{5.3}$$

$$-\lambda(T(b, \tau))\frac{\partial T(b, \tau)}{\partial x} = q^*(\tau), \quad 0 \leqslant \tau \leqslant \tau_m , \tag{5.4}$$

where $C(T)$, $\lambda(T)$, $\varphi(x)$, $f(\tau)$, $q^*(\tau)$ are given functions.

In using direct methods of solving the given problem the results are error-sensitive in the assignment of thermophysical properties, the smoothness of results depending on the smoothness of functions $\lambda(T)$ and $a(T) = \lambda(T)/C(T)$. The table form of coefficients presentation in this respect appears not to be suitable, since it leads to supplementary errors as a result of the interpolation operation between nodal points. Also, the values of functions in the nodes of the temperature grid are not always assigned with regard to the necessary smoothness. That is why in solving a non-linear IHCP it is preferable to present thermophysical properties of material in the form of some approximating dependences satisfying the condition of continuous differentiability.

In this case it is convenient to transfer from Eq. (5.1) to another the indentical form

$$\frac{1}{a(T)}\frac{\partial T}{\partial \tau} = \frac{\partial^2 T}{\partial x^2} + \frac{\lambda'(T)}{\lambda(T)}\left(\frac{\partial T}{\partial x}\right)^2, \quad 0 < x < b,\ 0 < \tau \leqslant \tau_m , \tag{5.5}$$

where $\lambda' \equiv \mathrm{d}\lambda/\mathrm{d}T$.

Introduce a uniform rectangular grid $\{x_l = \Delta x l,\ l = 0, 1, \ldots, L;\ \tau_n = \Delta\tau n,\ n = 0, 1, \ldots, m\}$ and write a difference analog of Eq. (5.5), using an implicit six-point scheme (Fig. 5.1) in approximating derivative $\partial^2 T/\partial x^2$ on the upper layer:

$$\left.\frac{\partial^2 T}{\partial x^2}\right|_{ln} = \frac{T_{l+1,n+1} - 2T_{l,n+1} + T_{l-1,n+1}}{\Delta x^2} + O(\Delta x^2);$$

$$\left.\frac{\partial T}{\partial x}\right|_{ln} = \frac{T_{l+1,n} - T_{l-1,n}}{2\Delta x} + O(\Delta x^2);$$

$$\left.\frac{\partial T}{\partial \tau}\right|_{ln} = \frac{T_{l,n+1} - T_{l,n}}{\Delta \tau} + O(\Delta \tau);$$

where

$$T_{ln} = T(\Delta x l, \Delta \tau n).$$

Then, assuming that thermophysical properties are estimated by the temperature of the previous time layer, we obtain

$$-\frac{a_{ln}\Delta\tau}{\Delta x^2} T_{l-1,n+1} + \left[1 + \frac{a_{ln}\Delta\tau}{\Delta x^2}\right] T_{l,n+1} - \frac{a_{ln}\Delta\tau}{\Delta x^2} T_{l+1,n+1}$$

$$= T_{ln} + \frac{a_{ln}\Delta\tau}{\Delta x^2}\frac{\lambda'_{ln}}{4\lambda_{ln}}(T_{l+1,n} - T_{l-1,n})^2, \tag{5.6}$$

where $a_{ln} = a(T_{ln})$, $\lambda_{ln} = \lambda(T_{ln})$, $\lambda'_{ln} = \lambda'(T_{ln})$.

For a discrete presentation of heat fluxes $q(\tau)$ and $q^*(\tau)$ we write difference relationships of the second order of accuracy counting heat accumulation in the selected element of thickness Δx [54, 126, 191]:

$$q_{n+1} = \lambda_{0n}\frac{T_{0,n+1} - T_{1,n+1}}{\Delta x} + C_{0n}\frac{\Delta x}{2}\frac{T_{0,n+1} - T_{0,n}}{\Delta\tau};$$

$$q^*_{n+1} = \lambda_{Ln}\frac{T_{L,n+1} - T_{L-1,n+1}}{\Delta x} + C_{Ln}\frac{\Delta x}{2}\frac{T_{L,n+1} - T_{L,n}}{\Delta\tau}, \tag{5.7}$$

where $C_{ln} = C(T_{ln})$.

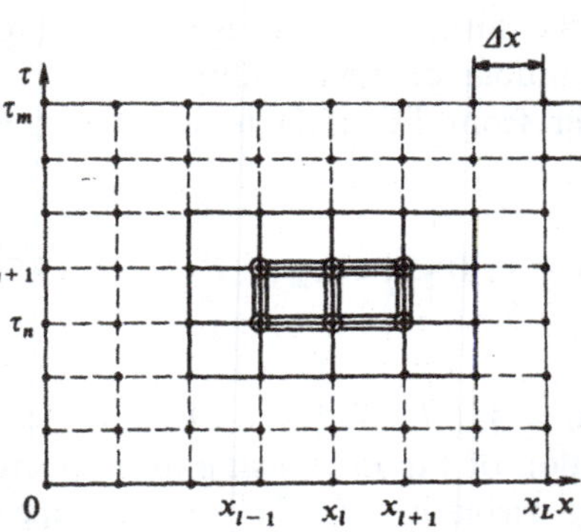

Fig. 5.1. A grid domain and a pattern (position of nodes of the scheme) of numerical solution of a heat conduction equation

The discrete function of the initial distribution of temperature is

$$T_{l0} = \varphi(x_l), \quad l = 0, 1, \ldots, L. \tag{5.8}$$

Relationships (5.6)–(5.8) allow a system of linear algebraic equations to be obtained for defining an unknown heat flux q_{n+1} and unknown temperatures $T_{l,n+1}$ ($l = 0, 1, \ldots, \kappa - 1, \kappa + 1, \ldots, L$; T_κ is the known temperature at point $l = \kappa$ inside the body):

$$\sum_{k=0}^{L} d_{ik} z_k = b_i, \quad i = 0, 1, \ldots, L, \tag{5.9}$$

where

$$z_k = T_{k,n+1}, \quad k = 0, 1, \ldots, \kappa - 1, \kappa + 1, \ldots, L - 1;$$

$$z_L = \frac{2a_{on}\Delta\tau}{\lambda_{0n}\Delta x} q_{n+1} + T_{0n}.$$

Coefficients of this system are null, excluding the following:

$$d_{ii}|_{i=0,1,\ldots,\kappa-1} = d_{i,i-1}|_{i=\kappa+1,\kappa+2,\ldots,L} = 1 + 2\frac{a_{ln}\Delta\tau}{\Delta x^2};$$

$$d_{ii}|_{i=\kappa,\kappa+1,\ldots,L-1} = d_{i,i+1}|_{i=1,2,\ldots,\kappa-2} = d_{i,i-2}|_{i=\kappa+2,\kappa+3,\ldots,L-1}$$

$$= d_{i,i-1}|_{i=1,2,\ldots,\kappa} = -\frac{a_{ln}\Delta\tau}{\Delta x^2};$$

$$d_{01} = -2\frac{a_{0n}\Delta\tau}{\Delta x^2}, \quad d_{L,L-2} = -2\frac{a_{Ln}\Delta\tau}{\Delta x^2}, \quad d_{0L} = -1.$$

The right sides are

$$b_0 = 0, \quad b_i|_{\substack{i=1,2,\ldots,\kappa-2 \\ i=\kappa+2,\kappa+3,\ldots,L-1}} = b_{in}, \quad b_i|_{i=\kappa-1,\kappa+1} = b_{in} + T_{\kappa,n+1}\frac{a_{in}\Delta\tau}{\Delta x^2},$$

$$b_\kappa = b_{\kappa n} - T_{\kappa,n+1}\left[1 + \frac{2a_{\kappa n}\Delta\tau}{\Delta x^2}\right], \quad b_L = T_{Ln} + \frac{2a_{Ln}\Delta\tau}{\lambda_{Ln}\Delta x} q^*_{n+1},$$

where $b_{in} = T_{in} + \dfrac{a_{in}\Delta\tau}{\Delta x^2}\dfrac{\lambda'_{in}}{4\lambda_{in}}(T_{i+1,n} - T_{i-1,n})^2$.

In the result, the problem of determining heat flux $q(\tau)$ and temperature field $T(x, \tau)$ is reduced to solution at every time step of the system of linear algebraic equations. By known temperatures $T_{i,n}$ and $T_{\kappa,n+1}$ we can find temperatures $T_{k,n+1}$ and a heat flux

$$q_{n+1} = \frac{\lambda_{0n}\Delta x}{2a_{0n}\Delta\tau}(z_L - T_{0n}).$$

If necessary, the given technique of numerical solution of IHCP is refined at the expense of iterations by coefficients.

5.1.2 Transition to an Explicit Form of the Difference Algorithm

Analyse the above obtained result. For this let us present system (5.9) in the matrix form

$$Az = b\,, \tag{5.10}$$

where

$$A = \begin{bmatrix}
\beta_{0n} & \gamma_{0n} & 0 & 0 & & & & & & 0 & -1 \\
\alpha_{1n} & \beta_{1n} & \gamma_{1n} & 0 & & & & & & & 0 \\
0 & \alpha_{2n} & \beta_{2n} & \gamma_{2n} & & & & & & & 0 \\
\cdot & & \ddots & & & & & & & & \vdots \\
\cdot & & \alpha_{K-2,n} & \beta_{K-2,n} & \gamma_{K-2,n} & & & & & & \cdot \\
\cdot & & & \alpha_{K-1,n} & \beta_{K-1,n} & & & & & & \cdot \\
\cdot & & & & \alpha_{Kn} & \gamma_{Kn} & & & & & \cdot \\
\cdot & & & & & \beta_{K+1,n} & \gamma_{K+1,n} & & & & \cdot \\
\cdot & & & & & \alpha_{K+2,n} & \beta_{K+2,n} & \gamma_{K+2,n} & & & \cdot \\
\cdot & & & & & & \ddots & & & & \cdot \\
0 & \cdots & \cdots & \cdots & \cdots & \cdots & 0 & \alpha_{Ln} & \beta_{Ln} & & 0
\end{bmatrix}$$

$$z = [T_{0,n+1}; T_{1,n+1}; \ldots; T_{K-2,n+1}; T_{K-1,n+1}; T_{K+1,n+1}; T_{K+2,n+1}; \ldots; T_{L,n+1}; z_L]^T;$$

$$b = [0; b_1; b_2; \ldots; b_L]^T.$$

The coefficients of matrix A are expressed by the relationships

$$\beta_{ln} = 1 + 2\frac{a_{ln}\Delta\tau}{\Delta x^2}, \quad l = 0, 1, \ldots, L;$$

$$\alpha_{ln} = \gamma_{ln} = -\frac{a_{ln}\Delta\tau}{\Delta x^2}, \quad l = 1, 2, \ldots, L-1;$$

$$\alpha_{Ln} = -\frac{2a_{Ln}\Delta\tau}{\Delta x^2}; \quad \gamma_{0n} = -\frac{2a_{0n}\Delta\tau}{\Delta x^2}; \quad \gamma_{Ln} = 0\,.$$

The system shown in Eq. (5.10) can be solved by any known method, in particular, using the Gaussian elimination method. But for computer time expense such an approach is not efficient. Despite the predominance of null elements in the matrix (in the total number of L^2 coefficients, only $3L-2$ are other than zero), the traditional Gaussian elimination method will require $\sim n^3$ operations nevertheless.

That is why we consider another system, identical to the algebraic system shown in Eq. (5.10)

$$By = b\,, \tag{5.11}$$

where $y = [z_L; T_{0,n+1}; T_{1,n+1}; \ldots; T_{\kappa-2,n+1}; T_{\kappa-1,n+1};$
$T_{\kappa+1,n+1}; T_{\kappa+2,n+1}; \ldots; T_{L,n+1}]^T;$

$$
B = \begin{bmatrix}
-1 & \beta_{0n} & \gamma_{0n} & & & & & & & \\
 & \alpha_{1n} & \beta_{1n} & \gamma_{1n} & & & & & 0 & \\
 & & \alpha_{2n} & \beta_{2n} & \gamma_{2n} & & & & & \\
 & & \ddots & \ddots & \ddots & & & & & \\
 & & & \alpha_{\kappa-2,n} & \beta_{\kappa-2,n} & \gamma_{\kappa-2,n} & & & & \\
 & & & & \alpha_{\kappa-1,n} & \beta_{\kappa-1,n} & & & & \\
 & & & & & \alpha_{\kappa n} & \gamma_{\kappa n} & & & \\
 & & & & & & \beta_{\kappa+1,n} & \gamma_{\kappa+1,n} & & \\
 & & & & & & \alpha_{\kappa+2,n} & \beta_{\kappa+2,n} & \gamma_{\kappa+2,n} & \\
 & 0 & & & & & \ddots & \ddots & \ddots & \\
 & & & & & & \alpha_{L-1,n} & \beta_{L-1,n} & \gamma_{L-1,n} & \\
 & & & & & & & \alpha_{Ln} & \beta_{Ln} &
\end{bmatrix}
$$

The identity of matrix Eqs. (5.10) and (5.11) denotes only the identity of unknown components of solutions. Computationally Eq. (5.11) gains the advantage. This is because from the system (5.11) we can single out an independent closed system with a tridiagonal matrix

$$B_2 y_2 = b_2 , \tag{5.12}$$

where

$$y_2 = [T_{\kappa+1,n+1}; T_{\kappa+2,n+1}; \ldots; T_{L,n+1}]^T;$$
$$b_2 = [b_{\kappa+1,n}; b_{\kappa+2,n}; \ldots; b_{Ln}]^T;$$

$$
B_2 = \begin{bmatrix}
\beta_{\kappa+1,n} & \gamma_{\kappa+1,n} & 0 \\
\alpha_{\kappa+2,n} & \beta_{\kappa+2,n} & \gamma_{\kappa+2,n} \\
 & \ddots & \\
0 & \alpha_{Ln} & \beta_{Ln}
\end{bmatrix}
$$

The matrix Eq. (5.12) is in fact an algebraic form of the implicit finite-difference solution of the *direct problem* of heat conduction in domain $[d, b]$. Due to this, the system (5.12) is well-conditioned and effectively (quickly and accurately) solved by a run method.

Knowing vector y_2, we can present a difference solution of the *inverse problem* in domain $[0, d]$ as an algebraic system with a triangular matrix which has only three non-zero diagonals:

$$B_1 y_1 = b_1 \tag{5.13}$$

where

$$y_1 = [z_L; T_{0,n+1}; T_{1,n+1}; \ldots; T_{\kappa-1,n+1}]^T;$$
$$b_1 = [0; b_{1n}; b_{2n}; \ldots; b_{\kappa-1,n}\bar{b}_{\kappa n}]^T$$
$$\bar{b}_{\kappa n} = b_{\kappa n} - \beta_{\kappa n} T_{\kappa+1,n+1},$$

$$B_1 = \begin{bmatrix} -1 & \beta_{0n} & \gamma_{0n} & & & & 0 \\ & \alpha_{1n} & \beta_{1n} & \gamma_{1n} & & & \\ & & \alpha_{2n} & \beta_{2n} & \gamma_{2n} & & \\ & & & \ddots & \ddots & \ddots & \\ & & & & \alpha_{\kappa-2,n} & \beta_{\kappa-2,n} & \gamma_{\kappa-2,n} \\ & 0 & & & & \alpha_{\kappa-1,n} & \beta_{\kappa-1,n} \\ & & & & & & \alpha_{\kappa n} \end{bmatrix}$$

The system (5.13) *is solved explicitly*, i.e. the components of vector y_1 are calculated using the recurrent technique, starting from $T_{\kappa-1,n+1}$. Here, a large number of zero coefficients leads us to a significant reduction of the volume of calculations.

It should be noted that in the case of constant thermophysical properties ($a =$ const, $\lambda' \equiv 0$), a six-point difference scheme (Fig. 5.2a) is degenerated into a T-like four-point scheme (Fig. 5.2b). We could have made use of a T-like scheme in the case of varying properties of the body, approximating the right-hand side of Eq. (5.1) as follows:

$$\frac{\partial}{\partial x}\left(\lambda \frac{\partial T}{\partial x}\right)\bigg|_{l,n+1}$$
$$\simeq \frac{\lambda^n_{l+1/2}\left(\dfrac{T_{l+1,n+1} - T_{l,n+1}}{\Delta x}\right) - \lambda^n_{l-1/2}\left(\dfrac{T_{l,n+1} - T_{l-1,n+1}}{\Delta x}\right)}{\Delta x}$$
$$= \frac{\lambda^n_{l+1/2}}{\Delta x^2} T_{l+1,n+1} - \left(\frac{\lambda^n_{l+1/2} + \lambda^n_{l-1/2}}{\Delta x^2}\right) T_{l,n+1} + \frac{\lambda^n_{l-1/2}}{\Delta x^2} T_{l-1,n+1},$$

where $\lambda^n_{l\pm 1/2} = \frac{1}{2}(\lambda^n_l + \lambda^n_{l\pm 1})$, $\lambda^n_l = \lambda(T_{ln})$.

The difference methods can be used for solving the boundary IHCP in domains with moving boundaries. One such algorithm, in which an explicit

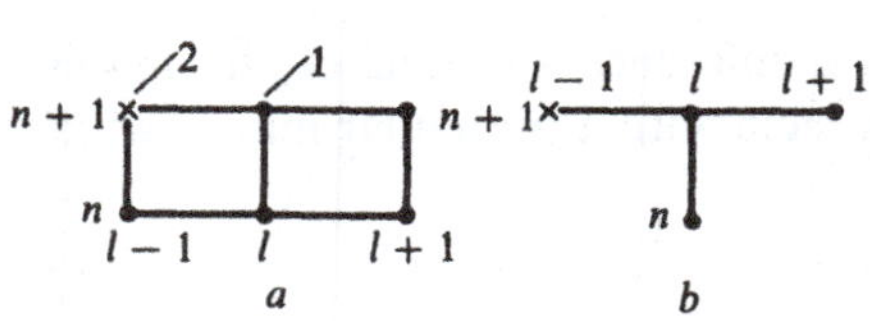

Fig. 5.2. Difference schemes: *1* – the grid nodes with known temperature values; *2* – a node with unknown value of temperature

T-type scheme of the temperature field numerical continuation is used, will be considered in Sect. 7.7 for a generalized non-linear heat conduction equation.

5.1.3 Some Notes About Explicit Schemes of the Temperature Field Numerical Continuation

The algorithmic construction, considered in Sect. 5.1.1, can be interpreted as a numerical solution of IHCP in the boundary-value statement. In fact, an implicit difference scheme has been taken as the initial material, designed for solving a direct problem, and in every time step a corresponding computational algorithm was transformed for solving an inverse problem – a known temperature T_κ was "introduced" into the right part of the system, and the unknown value of boundary condition was included into the number of unknowns. However, on the other hand, the solution of IHCP, written in the form of Eq. (5.11), was actually reduced to the Cauchy statement, i.e. to a *numerical continuation of the temperature field*. Such an approach seems to be less labour-consuming as compared with Eq. (5.10).

The general method for constructing a numerical procedure for solving an inverse problem with the Cauchy statement includes dividing the initial problem in two. The first is to define the grid function of heat flux $q_{\kappa n}$ at the point of temperature measurement (or the temperature itself in the adjacent layer $T_{\kappa+1,n}$). Hence, it is a direct problem, which is solved in domain $D_1 = \{d \leqslant x \leqslant b,\ 0 \leqslant \tau \leqslant \tau_m\}$. The second, being an inverse problem, is considered in domain $D_2 = \{0 \leqslant x \leqslant d,\ 0 \leqslant \tau \leqslant \tau_m\}$.

A six-point difference scheme for the combined Cauchy statement (in which simultaneously with the conditions on boundary $x = d$, the initial distribution of temperature are given) allows one to determine the temperature values in all nodes of the rectangular domain D_2.

In practice with heat experiments we can observe a situation when the initial temperature conditions of the body are not known.[1] In this case a T-type scheme of the heat conduction equation approximation can be applied, trading for a part of domain D_2 (Fig. 5.3a). Domain ABCD will be a domain of definiteness of the difference equation. Its dimensions, depending on angle $<$ BAD, vary by changing the ratio of steps Δx and $\Delta\tau$: $tg <$ BAD $= \Delta x/\Delta\tau$. Thus, the decrease of Δx and increase of $\Delta\tau$ result in the reduction of a definiteness domain.

Using the scheme, a pattern of which is shown in Fig. 5.3b (an explicit scheme for a direct heat conduction problem), we can solve a boundary inverse problem with known initial condition in domain AOBD. To calculate a temperature field in domain BCD it is necessary to gain information about the final temperature distribution in the body $T(x, \tau_m)$. Such data are usually not known

---

[1] For IHCP in Cauchy statement the assignment of the initial condition is, generally speaking, not an obligatory requirement see, e.g., solution (2.11), but additional restrictions should be observed (see Sect. 2.3.2).

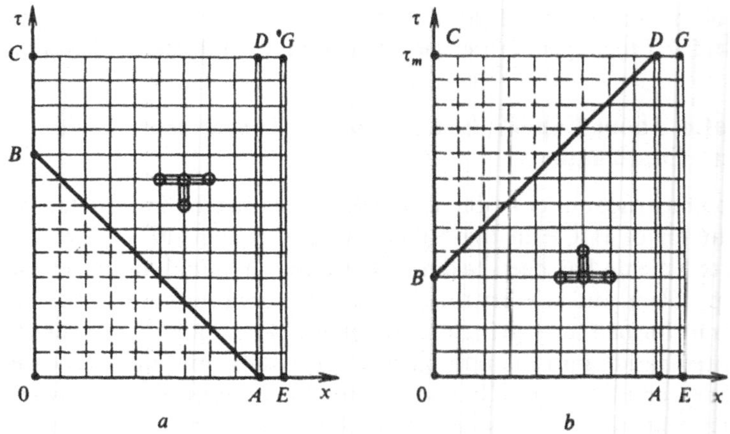

Fig. 5.3. Grid domains and patterns of numerical solution of IHCP (the initial temperature values are given on lines EG and AD)

and in order to expand the domain of temperature field recoverability till the moment $\tau = \tau_m$ we must have values of the heat flux and temperature at point $x = d$ not only on segment $[0, \tau_m]$, but at $\tau > \tau_m$ as well (Fig. 5.4). An additional section DG must include the number of time steps equalling the number of points of discretization of the interval $[0, d]$, $\tau_M - \tau_m = \kappa \Delta\tau$. A similar technique can also be used for IHCP without initial condition (see Fig. 5.2b), provided the Cauchy conditions are obtained at $\tau < 0$.

Also, using the technique for numerical extension of the heat conduction equation, there is another possibility (may be formal) for expanding the domain of temperature field recoverability, if neither the initial nor final distribution of temperatures in the body are known. It includes a combination of schemes, shown on Fig. 5.3.

Finally, we can postulate a condition on one of the boundaries $\tau = 0$ and $\tau = \tau_m$, taking into account that a solution of the inverse problem will be distorted by this condition in a time domain $0 < \tau < \tau_1^*$ or $\tau_2^* < \tau < \tau_m$, respectively.

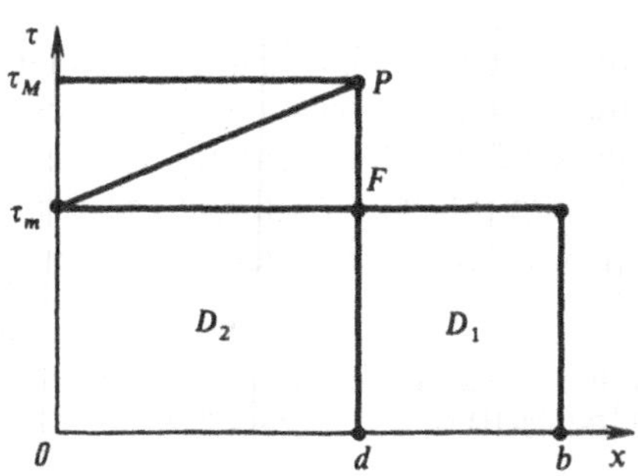

Fig. 5.4. Domains D_1, D_2 of solutions of direct and inverse heat conduction problems, respectively

Note in conclusion that we can construct other explicit schemes for the numerical extension of solution of a heat conduction equation, such as, for example, a "cross" scheme.

The central criterion in selecting a suitable difference approximation of a heat conduction equation when solving inverse problems will be that of *stability*.

5.2. Stability Criterion of the Difference Method for Solving a Boundary Inverse Problem

When using this method it is necessary to keep in mind the initial instability of IHCP, which under definite conditions must inevitably develop in the form of oscillating solutions.

Since the matrix of the system (5.13) is triangular, its determinant is equal to a product of diagonal coefficients

$$\det \boldsymbol{B}_1 = -\prod_{i=1}^{\kappa} \alpha_{in}, \quad \text{where } \alpha_{in} = -a_{in}\Delta\tau/\Delta x^2 .$$

Therefore, an increase of the number of steps κ under given values Δx and $\Delta\tau$ results in the continuous decrease of the determinant value, and if $\kappa \to \infty$ $\det \boldsymbol{B}_1 \to 0$. At the same time one of the elements, $a_{11} = -1$, does not change its value by the change in κ. One can conclude (naturally considering the instability of an inverse problem) that the increased distance d from the boundary f the body with an unknown condition, keeping $\Delta\tau$ and Δx constant, is accompanied by worsening problem conditionality, i.e. reduced accuracy in the temperature field recovery.

For the direct algebraic method for solving integral forms of IHCP, it was shown in Chap. 4 that the influence of perturbations in the temperature data on the results sought could be significantly smoothed due to definite selection of the discretization step of a function to be found. Here, on the one hand, the finite in value errors of input data may cause finite disturbances in unknown boundary conditions but, on the other hand, these disturbances may be attenuated in the course of time, i.e. a condition of asymptotic stability in the solution is satisfied. We can expect that a similar behaviour in the results may also be observed when using finite-difference methods of solution.

After approximation of a differential problem we now consider the *stability of the system of difference equations*. This condition is completely defined by the properties of a difference problem obtained. It should be remembered that by stability of a finite-difference scheme $A_h u_h = f_h$ is meant a property defined by the inequality

$$\|u_h\| \leqslant C \|f_h\| , \tag{5.14}$$

where A_h is a linear difference operator, dependent upon approximation parameters $\Delta\tau$ and Δx; u_h, f_h are grid functions u_{kn} and f_{kn}; $\|u_h\|$ and $\|f_h\|$ are the accepted norms for difference solution and the right-hand side and C is a positive constant.

It is important to note that the constant C in inequality (5.14) must not depend on steps Δx and $\Delta\tau$. Also, it is desirable to have C not too large.

The norm of the grid function can be given, for example, as the maximum of the modulus of function values $\|u_h\| = \max\limits_{k,n} |u_{kn}|$.

For a numerical solution of direct heat conduction problems, when the initial differential statement is well-posed, its change using the difference analogue must ensure computational stability, i.e. exclude any distortion of solution as a result of gaining the input data error and round-off errors.

It is obvious that the type of approximation scheme and its parameters must exert an influence on the character of IHCP numerical solution as well, leading either to a crucial distortion of results due to the rapid accumulation of errors or, on the contrary, damping the oscillating character of solution.

A possibility of artificial smoothing is studied below for a difference solution of IHCP by selecting the approximation schemes and steps of the grid. Since the initial problem is ill-posed, one should discard the definition given above for computational stability. In fact, if by reducing the grid steps, a difference scheme approximates the heat conduction equation more accurately, the solution of IHCP would tend to a "precise" (non-stable) solution, corresponding to the disturbed input data. In this case at $\Delta\tau$, $\Delta x \to 0$ we shall have $C \to \infty$.

Introduce a concept of *weak stability* of the difference scheme [126], defined by inequality

$$\|u_h\| \leqslant C_h \|f_h\| , \tag{5.15}$$

where C_h is a positive constant, depending on parameters of approximation.

The inequality (5.15) shows that there is a region of step variation of the grid in which a stability of difference equations is observed.

Therefore, our task is to obtain the approximate conditions of weak stability of difference schemes when solving poorly based inverse problems. It should be noted that the above concerns the *stability on the finite interval* of spatial variable change.

5.3 Investigation into the Stability of Numerical Solution for Inverse Problems

Let us discuss the main distinction between a difference procedure for boundary condition reconstruction and a semi-analytical one, considered in Chap. 4. An algebraic solution of integral equations of the first kind, corresponding to a boundary-value IHCP statement in the region with stationary boundaries, was in line with a minimum number of discrete steps, when only boundary conditions have been approximated. The heat conduction equation itself was not subjected to a finite-difference approximation. At the same time numerical algorithms of IHCP solution assume a discretization of conditions on the boundary of a space-time domain considered as well as the difference representation of a heat conduction equation both by coordinate and time.

A question arises: could difference schemes of IHCP solution possess better characteristics of stability compared with a semi-analytical methodology? Let us try to give an answer to this question after analysing a numerical solution of IHCP in linear statement.

5.3.1 A Point Error Technique

Consider a T-type scheme for the case with constant thermophysical properties of the body:

$$T_{l+1,n+1} = (2+p)T_{l,n+1} - pT_{ln} - T_{l-1,n+1}, \; l = 1, 2, \ldots, L, \quad n = 0, 1, \ldots, m, \tag{5.16}$$

where $p = \Delta x^2/\Delta\tau$.

The input data are known on two spatial layers

$$l = 1, 2\colon \; T_{1n} = f_n, \quad T_{2n} = \varphi_n .$$

It is supposed that the direction of numerical integration of the heat conduction equation by spatial variable x coincides with the direction of this coordinate.

Let the input information be given with errors $\tilde{f}_n = f_n + \delta_{fn}$, $\tilde{\varphi}_n = \varphi_n + \delta_{\varphi n}$. Then the disturbed grid functions $\tilde{T}_{ln}$ will take on the form $\tilde{T}_{l+1,n+1} = (2+p)\tilde{T}_{l,n+1} - p\tilde{T}_{ln} - \tilde{T}_{l-1,n+1}$, $l = 1, 2, \ldots, L$, $n = 0, 1, \ldots, m$, $\tilde{T}_{1n} = \tilde{f}_n$, $\tilde{T}_{2n} = \tilde{\varphi}_n$.

The errors in solution $\mathscr{E}_{ln} = T_{ln} - \tilde{T}_{ln}$ are determined by formula

$$\mathscr{E}_{l+1,n+1} = (2+p)\mathscr{E}_{l,n+1} - p\mathscr{E}_{ln} - \mathscr{E}_{l-1,n+1}, \quad l = 1, 2, \ldots, L, \quad n = 0, 1, \ldots, m,$$

where $\mathscr{E}_{1n} = \delta_{fn}$, $\mathscr{E}_{2n} = \delta_{\varphi n}$.

Assuming for simplicity that a disturbance is of the point form $\delta_{f_n} = \delta_{\varphi_n} = 1$, $n = k$; $\delta_{f_n} = \delta_{\varphi_n} 0$, $n \neq k$, let us follow the behaviour of errors $\mathscr{E}_{ln}$ while starting from the points of disturbance in the time and coordinate axes. The results of calculations for different values of parameter p are shown on the so-called $\mathscr{E}$-schemes (Fig. 5.5) – a corresponding error value is set in every node of the grid. Values $\Delta Fo = a\Delta\tau/d^2$ are given in the same graphs, where distance d is counted from layer $l = 2$ deep into the body. Attention should be paid to some specific features of errors distribution over the grid domain.

1. The region of influence for initial errors $\delta_{\varphi k} = \delta_{fk} = 1$ on the temperature field for a T-type scheme is an equilateral right-angled triangle, taking up half of the grid domain $\{l = 2, 3, \ldots, L; n = k, k+1, \ldots, k+L-2\}$.

2. The errors grow heavily as parameter p increases; this is equivalent to a consideration of the range of smaller values of number $\Delta Fo = 1/p(l-2)^2$, $l = 3, 4, \ldots, L$.

3. For one time layer the error is sign constant and is a continuously increasing function of parameter l ($l \geqslant 2$).

4. In every time layer, with n ($n = k, k + 1, \ldots$) increasing, the error absolute value can grow at the start and then diminishes to zero.

Now, if we compare the above specific features of errors in numerical solution with similar errors in the semi-analytical method, we can observe characteristic distinctions. The algebraic solution of integral equation remains disturbed at any moment of time, beginning with emergence of a pulsed error in the input temperature. Here, providing $\Delta Fo > \Delta Fo_{cr}$ in the domain of stability, the error of solution subsides quickly with time, and an intense rise of errors at $\Delta Fo \leqslant \Delta Fo_{cr}$ occurs. At the same time a numerical continuation of the temperature field is characterized by a finite region of influence for the initial error on the boundary condition under reconstruction. That is why we can expect that, due to discretization of the heat conduction equation by a time variable, the practical stability of the difference solution of IHCP will be observed with smaller values of ΔFo than in the case with an integral form of the problem. This conclusion is proved by the results from solving methodical examples.

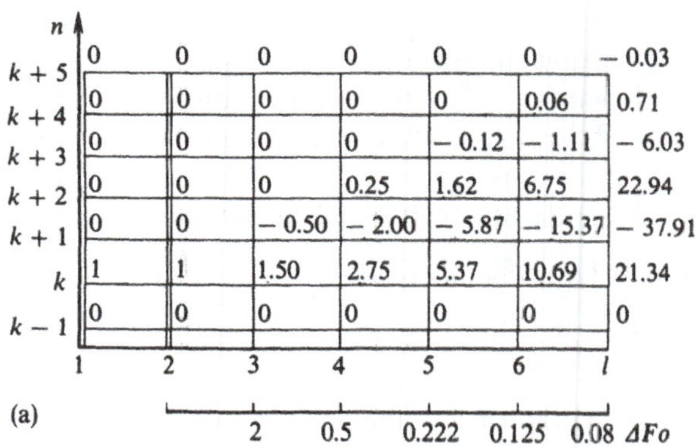

(a)

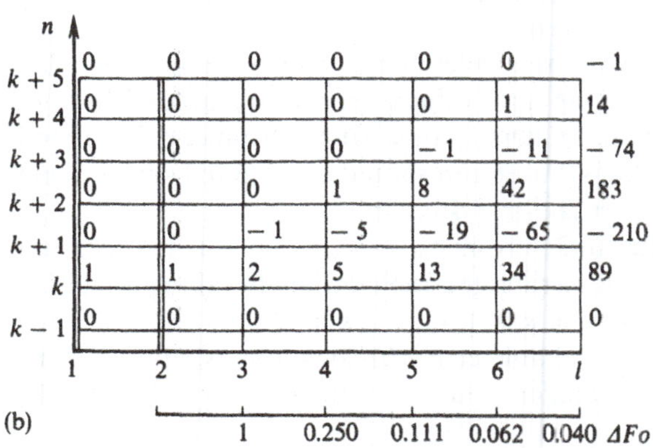

(b)

Fig. 5.5 a–b continued

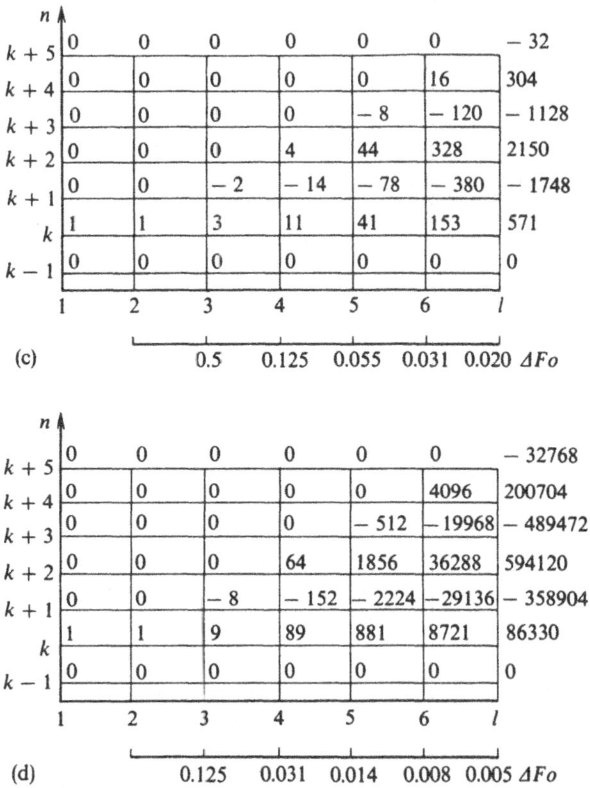

Fig. 5.5a–d. Distribution of errors in the nodes of the grid domain. **a** – $p = 0.5$; **b** – $p = 1$; **c** – $p = 2$; **d** – $p = 8$.

5.3.2 Scheme Viscosity as a Regularizing Factor

To understand the cause of *supplementary natural regularization* of IHCP solving, concerned with a discrete representation of the heat conduction equation, let us consider the first differential approximation to a difference scheme to be studied. With this in mind, expand the grid functions inherent to Eq. (5.16) by their Taylor series expansion in terms of the parameters Δx and $\Delta\tau$:

$$T_{l,n+1} = T_{ln} + \frac{\Delta\tau}{1!}\frac{\partial T}{\partial\tau}\bigg|_{ln} + \frac{\Delta\tau^2}{2!}\frac{\partial^2 T}{\partial\tau^2}\bigg|_{ln} + \dots$$

$$T_{l\pm1,n+1} = T_{ln} \pm \frac{\Delta x}{1!}\frac{\partial T}{\partial x}\bigg|_{ln} + \frac{\Delta\tau}{1!}\frac{\partial T}{\partial\tau}\bigg|_{ln} + \frac{\Delta x^2}{2!}\frac{\partial^2 T}{\partial x^2}\bigg|_{ln}$$

$$+ \frac{\Delta\tau^2}{2!}\frac{\partial^2 T}{\partial\tau^2}\bigg|_{ln} \pm \Delta x\Delta\tau\frac{\partial^2 T}{\partial x\,\partial\tau}\bigg|_{ln} + \dots$$

and reject in the final expression the components with the second-order and above-order of smallness. In the result we obtain

$$\left.\frac{\partial T}{\partial \tau}\right|_{ln} + \frac{\Delta\tau}{2}\left.\frac{\partial^2 T}{\partial \tau^2}\right|_{ln} = a\left.\frac{\partial^2 T}{\partial x^2}\right|_{ln}. \tag{5.17}$$

Thus, in spite of the fact that a difference equation (5.16) is an approximation of the parabolic heat conduction equation, *the first differential approximation* corresponds to a *hyperbolic heat conduction equation*, where coefficient $\Delta\tau/2$ is a fictitious relaxation time. Since the Cauchy problem for a hyperbolic equation is well-posed we can suppose that the influence of the term $(\Delta\tau/2)(\partial^2 T/\partial\tau^2)$ will improve stability in the inverse problem numerical solution, the smoothness of results increasing as $\Delta\tau$ rises.

Such a hypothesis is based on the following physical considerations. The Fourier heat conduction equation is deduced on the assumption that there is an infinitely high speed of heat propagation $w \to \infty$. Setting w as a finite value, the process of heat conduction can be described by a hyperbolic equation [93]

$$\frac{w^2}{a}\frac{\partial T}{\partial \tau} + \frac{\partial^2 T}{\partial \tau^2} = w^2\frac{\partial^2 T}{\partial x^2}. \tag{5.18}$$

As w increases, the time lag (time between heat action start and temperature reaction at some point inside the body) decreases. Component $w^2/a\,\partial T/\partial\tau$, playing the role of "temperature friction", enhances the smoothing effect of the boundary conditions. With $w \to \infty$ there is no lag and the smoothing effect is the greatest. This brings us to a situation when large disturbances in the boundary functions cause small deviations of temperature values in the inner points of the body, and vice versa, small deviations of temperature values at some inner point correspond to large deviations of surface temperature or boundary heat flux.

The speed of heat pulse propagation being decreased, a shift between the time of heat disturbance emergence and temperature change in the inner point of the body increases and, especially important, the amplitude of temperature response rises, i.e. the smoothing effect of specific features of the boundary condition becomes less.

5.3.3 An Approximate Estimate of Admissible Steps of Time Approximation

We continue with the stability analysis, making use of the obtained differential approximation [see Eq. (5.17)], which we shall study not at the nodes of the grid but at all points of the space-time domain:

$$\frac{\partial T}{\partial \tau} + \frac{\Delta\tau}{2}\frac{\partial^2 T}{\partial \tau^2} = a\frac{\partial^2 T}{\partial x^2}, \quad 0 < x < b,\ 0 < \tau \leqslant \tau_m .$$

Introduce new variables $Fo = a\tau/d^2$, $\bar{x} = x/d$ (d is a fixed point of the body). Then

$$\frac{\partial T}{\partial Fo} + \frac{\Delta Fo}{2}\frac{\partial^2 T}{\partial Fo^2} = \frac{\partial^2 T}{\partial \bar{x}^2}, \quad 0 < \bar{x} < \bar{b}, \quad 0 < F_0 \leqslant Fo_m , \tag{5.19}$$

where

$$\Delta Fo = a\Delta\tau/d^2, \quad \bar{b} = b/d, \; Fo_m = \frac{a\tau_m}{d^2}.$$

Consider for simplicity of analysis a semi-infinite body ($\bar{b} \to \infty$). In this case the solution of the first boundary-value problem for Eq. (5.19) at point $\bar{x} = 1$ assumes the form (see [94]):

$$T(1, Fo) = T_w\left(Fo - \sqrt{\frac{\Delta Fo}{2}}\right)\exp\left[-\frac{1}{\sqrt{2\Delta Fo}}\right]$$

$$+ \frac{1}{\sqrt{2\Delta Fo}} \int\limits_{\sqrt{\Delta Fo/2}}^{Fo} T_w(\xi)\exp\left[-\frac{Fo-\xi}{\Delta Fo}\right]\frac{I_1\left(\frac{\sqrt{(Fo-\xi)^2 - \Delta Fo/2}}{\Delta Fo}\right)}{\sqrt{(Fo-\xi)^2 - \Delta Fo/2}}\,d\xi\,,$$

where $T_w(Fo)$ is a temperature on the boundary of the body $[T(0, Fo) = T_w(Fo)]$; $I(z)$ is a modified Bessel's function of the first kind.

Now, state the problem: by the known inside temperature $T(1, Fo) = f(Fo)$ find temperature $T_w(Fo)$ on the boundary of the body. Such a statement of the inverse problem corresponds to a solution of the following integral equation with a variable upper limit:

$$\alpha(\Delta Fo)T_w\left(Fo - \sqrt{\frac{\Delta Fo}{2}}\right) + \frac{1}{\sqrt{2\Delta Fo}}$$

$$\times \int\limits_{\sqrt{\Delta Fo/2}}^{Fo} T_w(\xi)K(Fo-\xi, \Delta Fo)d\xi = f(Fo)\,, \tag{5.20}$$

where

$$\alpha(\Delta Fo) = \exp[-1/\sqrt{2\Delta Fo}]\,;$$

$$K(Fo-\xi, \Delta Fo) = \exp\left[-\frac{Fo-\xi}{\Delta Fo}\right]\frac{I_1(\sqrt{(Fo-\xi)^2 - \Delta Fo/2}/\Delta Fo)}{\sqrt{(Fo-\xi)^2 - \Delta Fo/2}}.$$

There is a term in this equation which contains an unknown function outside the integral sign, this formally makes the task of its definition as being well-posed. However, a contribution of this term towards the solution depends upon the value of parameter ΔFo. By small ΔFo, the solution will not then be of a regular nature, it will become even more smooth, as ΔFo increases.

Note that smoothing of the solution of Eq. (5.20) is determined not only by the influence of the coefficient $\alpha(\Delta Fo)$ but by the form of kernel $K(Fo, \Delta Fo)$ of the integral equation as well. By a sufficiently small ΔFo function $K(Fo)$, beginning from value $K(\sqrt{\Delta Fo/2})$ increases then passes through the maximum and decreases. Using de L'Hospital's rule for removing uncertainty, we obtain

$$\lim_{Fo \to \sqrt{\Delta Fo/2}} K(Fo) = 1/2\Delta Fo\,.$$

Taking the kernel as a temperature response at point $x = d$ to a pulse changing of the boundary heat regime and following the *principle of utmost effect* described in Sect. 4.3.3, we can find such a minimal value of parameter ΔFo by which $K(Fo)$ for all $Fo > \sqrt{\Delta Fo/2}$ becomes a non-increasing function. It can be shown that value $\Delta Fo \simeq 0.04$ satisfies this condition.

Since the differential Eq. (5.19) coincides with a T-type difference scheme (see Fig. 5.3a) with the accuracy of up to the terms of the second order of smallness we can suppose that the properties for Eq. (5.20) solution will in the main coincide with those for numerical solution [see Eq. (5.16)]. Thus, it is possible to forecast the stability of a given process of numerical extension of temperature field up to the values $\Delta Fo_{min} \simeq 0.04$, i.e. by steps, approximately four times less than in the case of an algebraic solution of IHCP in its integral form.

5.3.4 The Analysis of Results of Numerical Simulation

The errors in temperature are depicted in Fig. 5.6, which have been reconstructed through a numerical procedure and by a semi-analytical method for two values of ΔFo. The calculations correspond to the action at the k-th moment of a unit-equal disturbance in the temperature data. It is seen that errors of numerical solution are considerably less for $\Delta Fo = 0.04$.

We should also take notice of the fact that for such ΔFo values as used in these calculations, for which we expect a good enough behaviour of numerical solution with smooth input data, the initial error, nevertheless, rises noticeably in spite of the general tendency towards a subsequent decrease. That is why a difference scheme under study cannot give satisfactory results using small ΔFo when the Cauchy data are burdened by noticeable fluctuating errors.

Note, in addition, that a transition from temperature grid values to values of heat flux is related with additional loss of accuracy due to numerical differentiation. If we concentrate on the analysis of solutions of IHCP integral forms counting upon the smoothing action of step ΔFo, we can expect a sufficiently

n	a	b	c	d
$k+4$	~ 0	~ 0	10	− 559700
$k+3$	− 0.04	0.04	− 70	20980
$k+2$	0.56	− 0.05	180	− 780
$k+1$	− 2.27	0.62	− 210	30
k	2.86	3.79	90	2460

Fig. 5.6. Variation of errors in time: **a** and **c** are for the numerical scheme at $p = 0.1$; $l = 7$, $\Delta Fo = 0.4$ and $p = 1$, $l = 7$, $\Delta Fo = 0.04$ respectively, **b** and **d** are for the solutions of the integral equation at $\Delta Fo = 0.4$ and 0.04, respectively

smooth reconstruction of heat flux $q(\tau)$ with values of ΔFo approximately 2–3 times larger than the above found value, 0.04.

Numerical simulation on the whole has proved this to be so. Here are the results to the solution of one of the examples. It was required to reconstruct the heat flux and temperature on the surface of the slab. Thermophysical properties of the material were taken as constant and equalling $\lambda = 1.3 \cdot 10^{-4}$ Kw/m·K; $a = 1.2 \cdot 10^{-7}$ m^2/s. Thickness of the slab was $b = 0.27 \cdot 10^{-2}$ m. Temperature was set at point $d = 0.8 \cdot 10^{-3}$ m; the heat flux on the boundary $x = b$ was considered to be equal to zero.

Nominal values of input temperature were determined for the dependence of the heat flux $\bar{q} \equiv qd/\lambda = 48 + 88\, Fo_d + 27 Fo_d^2$ from the solution of a direct problem with accuracy $\delta \lesssim 10^{-3}$ K.

The results of IHCP solution are given in Fig. 5.7. Within the considered range $\Delta Fo = 0.06 \div 0.32$ the curves obtained are stable and close enough to exact. However, random disturbances of initial temperature values cause an instability of the calculation process by small values of ΔFo. The *smoothing* in advance of input information $T_\delta(\tau)$ by a *regularization method* enables sufficiently smooth and accurate results to be obtained when solving IHCP (Fig. 5.8). At the same time, smoothing by one of the variants of the least-squares technique – using fourth differences [187] (by "five points") – can lead to a considerable distortion of the view of the curve sought.

For numerical solution of an inverse problem in the Cauchy statement we can use other difference schemes. In particular, the results of calculations by the scheme "cross" are given on Fig. 5.9 [8]. For undisturbed input data they turned out to be close to the results given earlier, obtained by a T-type scheme. Note that in both cases the stability of problem solving did not almost depend on the step in spatial coordinate on the considered bounds of the ratio of steps satisfying the condition $a\Delta\tau/\Delta x^2 = 1.6 \cdot 10^0 \div 1.3 \cdot 10^3$.

The principle of step regularization stated above for numerical IHCP solution presupposes "roughening" of results at the expense of the smoothing effect of viscous terms in the corresponding differential approximation equation. As the domain of temperature field continuation becomes larger by coordinate

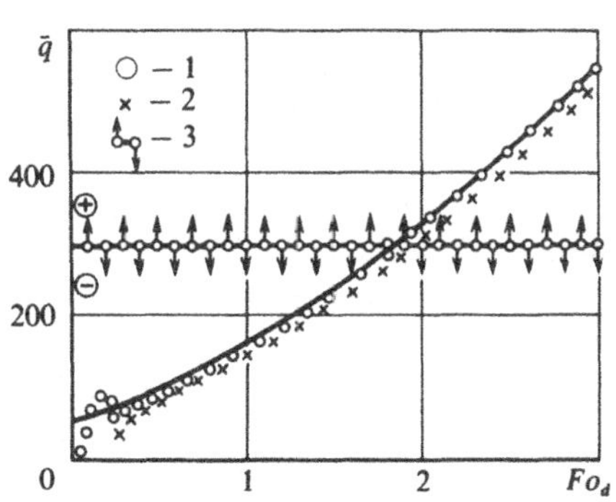

Fig. 5.7. Results of the inverse problem solution for exact and disturbed values of input temperature (full value – true solution); 1 – $\delta T \sim 0$, $\Delta Fo_d = 0.06$, $\kappa = 10$, $L = 40$; 2 – $\delta T \sim 0$, $\Delta Fo_d = 0.32$, $\kappa = 5$, $L = 20$; 3 is with a normal law of errors distribution (Gaussian random variables) in temperature with a spread $3\sigma(\tau) = 0{,}05\, T_{\kappa\,\max}$, $\Delta Fo_d = 0.06$.

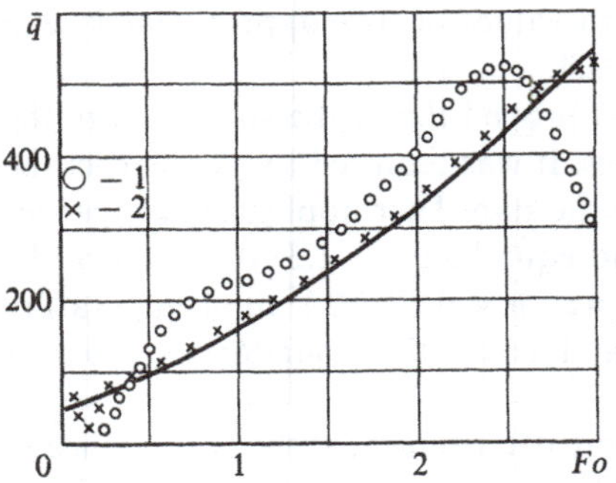

Fig. 5.8. Results of the inverse problem solution for smoothed temperature values (the initial data contained errors, distributed by normal law with a spread $3\sigma(\tau) = 0.1\ T_{\kappa\,max}$) $\Delta Fo_d = 0.06$, $\kappa = 5$ (full curve – true solution): 1 is for smoothing $T_\kappa(\tau)$ by five points, the number of such smoothings $N = 1000$; 2 is for smoothing $T_\kappa(\tau)$ by a regularization method of the second order.

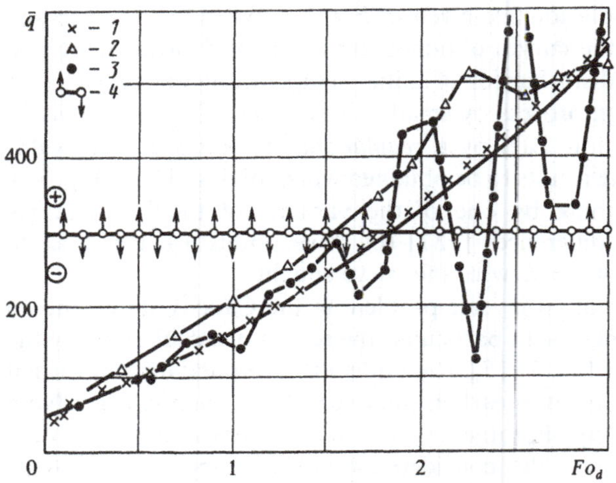

Fig. 5.9. Results of the inverse problem solution for exact and disturbed temperature values (full curve – true solution): 1 – $\delta T \sim 0$, $\Delta Fo_d = 0.06$, $\kappa = 8$; 2 is with a normal law of errors distribution in temperature with a spread $3\sigma(\tau) = 0.1\,T_\kappa(\tau)$, $\Delta Fo_d = 0.32$, $\kappa = 6$; 3 – $3\sigma(\tau) = 0.1\,T_\kappa(\tau)$, $\Delta Fo_d = 0.15$, $\kappa = 6$; 4 – $3\sigma(\tau) = 0.05\ T_{\kappa\,max}$, $\Delta Fo_d = 0.06$.

x the condition of weak stability [see inequality (5.15)] requires still greater increase of the time discretization step. The accuracy of causal characteristic recovery inevitably suffers here because of approximation worsening of the heat conduction equation, and the numerical solution obtained might not correspond to the actual conditions of a heat exchange process.

The analysis performed showed that under definite conditions there is a possibility of quite effective use of direct numerical methods for IHCP solving. This possibility is based on the natural regularizing properties of difference algorithms which are displayed more clearly than in the case of approximate analytical methods for solving integral forms of IHCP. It should be added that numerical continuation algorithms of the temperature field, using explicit difference schemes, are simple and convenient in programming and do not require much computer time spent on their realization. However, the explicit schemes considered have rigid limitations on the minimum admissible steps of time coordinate discretization.

5.4 An Implicit Scheme for Inverse Problem Numerical Solution

The necessity to increase the viscous properties of the computational algorithm and thus broaden the area of applicability of direct numerical methods brings us to an *implicit scheme* for the approximation of the heat conduction equation in the problem of temperature field continuation using Cauchy data. In this case the values of temperature T_{ln}, $n = 1, 2, \ldots, m$ on some spatial layer l are found simultaneously from a solution of the algebraic system with the input data in the form of two time series of temperatures $T_{l-1,n}$ and $T_{l-2,n}$, $n = 1, 2, \ldots, m$.

5.4.1 The Computational Algorithm Construction

Consider an implicit grid approximation of the Cauchy problem or Eq. (5.5) with conditions (5.2)–(5.4), taking $d = b$. Having selected a rectangular grid, set, as earlier, $x_l = l\Delta x$, $l = 0, 1, \ldots, L$; $\tau_n = n\Delta\tau$, $n = 0, 1, \ldots, m$ where Δx and $\Delta\tau$ are steps in coordinate and time, respectively, replace the time and spatial differential operators by the following difference relationships (Fig. 5.10):

$$\left.\frac{\partial T}{\partial \tau}\right|_{ln} \simeq \frac{T_{l,n+1} - T_{l,n-1}}{2\Delta\tau}\,; \quad \left.\frac{\partial T}{\partial x}\right|_{ln} \simeq \frac{T_{l+1,n} - T_{ln}}{\Delta x}\,;$$

$$\left.\frac{\partial^2 T}{\partial x^2}\right|_{ln} \simeq \frac{T_{l+2,n} - 2T_{l+1,n} + T_{ln}}{\Delta x^2}\,.$$

In order to calculate a grid temperature function in all nodes of the rectangular domain, it is necessary to set a supplementary condition on its upper boundary with $\tau = \tau_m$. As computational experiments showed, we can take one of the following a priori conditions:

$$\frac{\partial T(x, \tau_m)}{\partial \tau} \simeq \frac{T_{lm} - T_{l,m-1}}{\Delta\tau} = c_1\,; \tag{5.21}$$

$$\frac{\partial^2 T(x, \tau_m)}{\partial \tau^2} \simeq \frac{T_{l,m+1} - 2T_{lm} + T_{l,m-1}}{\Delta\tau^2} = c_2, \quad l = L-1, L-2, \ldots, 0\,, \tag{5.22}$$

where $c_1 = \dfrac{\mathrm{d}f(\tau_m)}{\mathrm{d}\tau} = \text{const}$, $c_2 = \dfrac{\mathrm{d}^2 f(\tau_m)}{\mathrm{d}\tau^2} = \text{const}$.

A deviation of solution from the unknown, conditioned by introducing constraints on the derivatives [see (5.21) and (5.22)], is observed in a comparatively small domain $\bar{\tau} < \tau < \tau_m$, the condition (5.22)) giving a lesser distortion of results. We can avoid the above worsening of solution in the vicinity of $\tau = \tau_m$, if

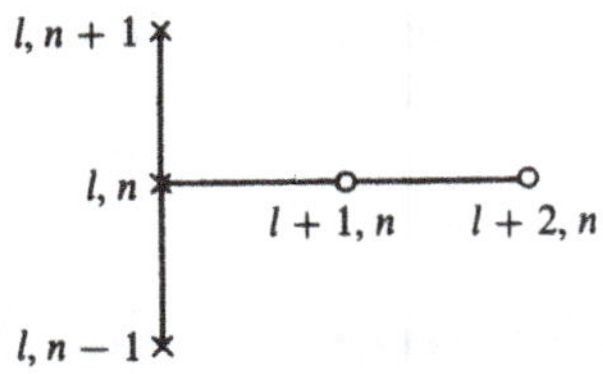

Fig. 5.10. A difference scheme

the values of temperature $f(\tau)$ are known not only at time interval $[0, \tau_m]$ but at $\tau > \tau_m$ as well. In this case, as has been already stated in Sect. 5.1.3, there is no need to set a supplementary condition and the area of numerical IHCP solution becomes trapezoidal.

Another possibility consists in the approximation of a heat conduction equation in the boundary grid nodes ($n = m$) using the explicit scheme. For this, let us make use of the following relationship:

$$\frac{\partial T(x, \tau_m)}{\partial \tau} \simeq \frac{T_{lm} - T_{l,m-1}}{\Delta\tau}. \tag{5.23}$$

A finite-difference analogue of the problem given by Eq. (5.5) and by the conditions (5.2)–(5.4)) for the l-th spatial layer corresponds to a system of nonlinear algebraic equations with a tridiagonal matrix ($l = L - 2, L - 3, \ldots, 0$)

$$T_{lo} = \varphi_l; \tag{5.24}$$

$$A_{ln} T_{l,n+1} + B_{ln} T_{ln} + D_{ln} T_{l,n-1} = F_{ln}, \quad n = 1, 2, \ldots, m - 1, \tag{5.25}$$

$$\alpha_{lm} T_{lm} + \gamma_{lm} T_{l,m-1} = \beta_{lm}, \tag{5.26}$$

where

$$A_{ln} = \frac{1}{2\Delta\tau}; \; B_{ln} = -\frac{a_{ln}}{\Delta x^2}; \; D_{ln} = -\frac{1}{2\Delta\tau};$$

$$F_{ln} = a_{ln} \frac{T_{l+2,n} - 2T_{l+1,n}}{\Delta x^2} - \frac{\lambda'_{ln} a_{ln}}{\lambda_{ln}} \left(\frac{T_{l+1,n} - T_{ln}}{\Delta x}\right)^2.$$

Values α_{lm}, γ_{lm} and β_{lm} are determined by the form of the a priori condition at the end of time interval. Setting the first derivative in the form of condition (5.21), we have $\alpha_{lm} = -\gamma_{lm} = 1$, $\beta_{lm} = c_1 \Delta\tau$. For the other two conditions the corresponding expressions are obtained subject to the main difference Eq. (5.25).

Condition (5.22):

$$\alpha_{lm} = 1 - \frac{a_{lm}\Delta\tau}{\Delta x^2}; \; \gamma_{lm} = -1; \; \beta_{lm} = F_{lm}\Delta\tau - \frac{\Delta\tau^2}{2} c_2.$$

Condition (5.23):

$$\alpha_{lm} = 1 - \frac{a_{lm}\Delta\tau}{\Delta x^2}; \; \gamma_{lm} = -1; \; \beta_{lm} = F_{lm}\Delta\tau.$$

The coefficients and right-hand sides of the system (5.24)–(5.26) are functions of unknown temperatures. Hence, for every step by coordinate x it is necessary to repeat the process of iterative approximation till the moment when, in the two successive iterative cycles, the noticeable difference between the grid functions T_l would disappear:

$$\max_n |T_{ln}^{(p)} - T_{ln}^{(p-1)}| \leqslant \varepsilon$$

where p is the number of iteration and $\varepsilon > 0$ is a small value.

Having satisfied the conditions of convergence we can perform a transition to the next spatial layer. If function $\lambda(T)$ is given in a table form, it is preferable to use a somewhat varified algorithm, in which the derivative $\lambda'_{ln} = \left.\frac{\mathrm{d}\lambda}{\mathrm{d}T}\right|_{ln}$ is approximated by a difference relationship

$$\lambda'_{ln} \simeq \frac{\lambda_{l+1,n} - \lambda_{ln}}{T_{l+1,n} - T_{ln}} .$$

Then in the system shown in Eqs (5.24)–(5.26) the coefficients B_{ln} and right-hand sides F_{ln} will be written as follows:

$$B_{ln} = \frac{a_{ln}}{\Delta x^2}\left(\frac{\lambda_{l+1,n}}{\lambda_{ln}} - 2\right); \; F_{ln} = \frac{a_{ln}}{\Delta x^2}\left[T_{l+2,n} + T_{l+1,n}\left(\frac{\lambda_{l+1,n}}{\lambda_{ln}} - 3\right)\right].$$

To begin computations we have to know temperatures T_{Ln} and $T_{L-1,n}$. The first time series $\{T_{Ln}\}$ is gained by a corresponding discretization of the input function $f(\tau)$. Temperatures $\{T_{L-1,n}\}$ are calculated by an expression for difference approximation of the boundary condition of, for example, the type shown in Eq. (5.7).

The approximation scheme considered [Eqs. (5.24)–(5.26)] has the second order in time and the first one in space. We can obtain difference equations which will have the second order in spatial variable as well.

5.4.2 The Analysis of the Procedure Stability

In spite of the implicit nature of approximation, the difference schemes described above are not stable in classical sense. This can be explained by the specific nature of inverse heat conduction problems. To prove this conclusion let us analyse the stability of scheme shown in Eq. (5.25) with constant coefficients. In this case the main computational relationship (for simplicity assumed $a = 1$) is

$$\frac{1}{2\Delta\tau}(T_{l,n+1} - T_{l,n-1}) - \frac{1}{\Delta x^2}(T_{l+2,n} - 2T_{l+1,n} + T_{ln}) = 0 . \tag{5.27}$$

An IHCP solution with a Cauchy statement is reduced to a search for successive solutions at the spatial points x_l, the solutions on three adjacent spatial layers being related via a definite operator. If this operator does not increase errors in the Cauchy data in going to the next spatial step, the scheme will be stable. And, vice versa, for an unstable scheme the errors will increase without limit.

For stability analysis we can use Neumann's method, assuming that the solution of problem (5.27) is like

$$T_{ln} = p^l e^{i\beta n}\,, \tag{5.28}$$

where i is an imaginary unit, β is a real-valued number and $p = p(\beta)$.

Substituting the solution (5.28) into Eq. (5.27) and determining the roots $p_{1,2}$ of the equation obtained it is easy to see that a necessary Neumann condition of stability $|p_{1,2}| \leqslant 1$ for the scheme shown in Eq. (5.27) is not satisfied under any of the ratio of steps Δx and $\Delta\tau$.

Thus, just as in the case with an explicit scheme of integrating the heat conduction equation with Cauchy data, we can only discuss the *weak stability* of the implicit difference scheme on a finite interval using the coordinate x. The following inequality will be taken as a condition of weak stability

$$\Delta Fo \geqslant \Delta Fo_{adm}\,. \tag{5.29}$$

In the case of non-linear heat conduction problems the condition (5.29) is a local stability condition and must be satisfied at all nodes on the spatial-time grid.

Practical analysis of the stability condition was performed for the algorithms described above. As expected, implicit difference schemes, compared with explicit ones, possess higher viscous properties exerting a stabilizing action on the development of instability in the results. Violation of the smooth nature of the solution reconstructed occurs through smaller time steps. For sufficiently smooth input functions $f(\tau)$ a stable computation of the heat flux $q(\tau)$ using the implicit scheme is possible at $\Delta Fo \geqslant 0.01 \div 0.02$. The results of one of the methodical examples are given in Fig. 5.11.

Thus the implicit numerical scheme of IHCP solution for stability takes precedence over the explicit one. But the implicit scheme is not recurrent thus slightly complicating the algorithms and increasing the computing time.

The algorithms are considered in [24] for solving non-linear IHCP using such direct methods for more complex models, including the motion of body boundaries, convective heat transfer and distributed sources.

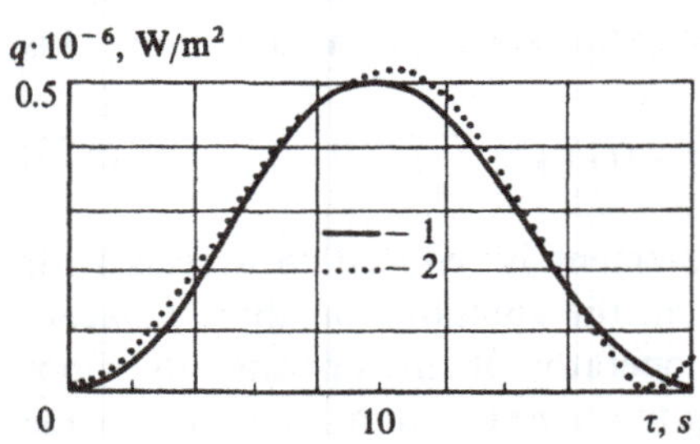

Fig. 5.11. Results of the heat flux recovery in the case of non-disturbed input data $T^*(\tau) = f^*(\tau)$, $q^*(\tau) = 0$, $b = 0.002$ m, $a = (0.4 \cdot 10^{-6} - 0.143 \cdot 10^{-9}\,T + 0.408 \cdot 10^{-12}\,T^2)\,\frac{\text{m}^2}{\text{s}}$, $\lambda = (0.721 + 0.288 \cdot 10^{-3}\,T)$ KW/(m · K); 1 is a true solution $q_T = 0.25 \cdot 10^6 \{\sin[0.1\pi(\tau - 5)] + 1\}\,\frac{\text{W}}{\text{m}^2}$; 2 is a numerical solution, $\Delta\tau = 0.4$ s

5.5 Artificial Hyperbolization of the Heat Conduction Equation in Solving a Boundary Inverse Problem

A question was discussed in Sect. 5.3 about a certain improvement of numerical stability in solving IHCP due to natural hyperbolization of differential approximation of the difference scheme. Developing this idea, we can try to reformulate an inverse problem for parabolic equation into an inverse problem for hyperbolic equation of heat conduction.[2]

Consider a hyperbolic equation

$$\frac{\partial T}{\partial \tau} + \alpha \frac{\partial^2 T}{\partial \tau^2} = a \frac{\partial^2 T}{\partial x^2}, \ 0 < x < b, \ 0 < \tau \leqslant \tau_m, \tag{5.30}$$

where parameter α corresponds to a relaxation time in Eq. (5.18) $t_r = a/w^2$.[3] In selecting the best approximation to a boundary condition sought for parameters α acts as a fictitious relaxation time. For exact input data its value will be close to a true relaxation time. When errors of temperature measurements as well as errors of approximation and rounding-off increase, the effective value of α must be increased in order to smooth the results of IHCP solving. It is necessary, therefore, in some definite way to match α with the accuracy of the initial data.

Consider the following inverse problem, corresponding to a combined statement of the boundary problem for a hyperbolic equation. It is required to find a twice continuously differentiable function $T(x, \tau)$ which in domain $D = \{0 < x < b, 0 < \tau < \tau_m\}$ satisfies equation (5.30), and on boundaries $\tau = 0$, $\tau = \tau_m$, $x = b$ and at point $x = d$, $0 < d < b$ – the conditions $T(x, 0) = 0$; $0 \leqslant x \leqslant b$; $\partial T(x, \tau_m)/\partial \tau = 0$, $0 \leqslant x \leqslant d$;[4]

$$T(d, \tau) = T^*(\tau), \quad -\lambda \frac{\partial T(b, \tau)}{\partial x} = q^*(\tau), \ 0 \leqslant \tau \leqslant \tau_m, \tag{5.31}$$

where $T^*(\tau)$ and $q^*(\tau)$ are known functions. By finding field $T(x, \tau)$ we can reconstruct the heat flux $q^\alpha(\tau) = -\lambda \left.\frac{\partial T}{\partial x}\right|_{x=0}$ at boundary $x = 0$ under the given value of parameter α.

The difference analogue of Eq. (5.30) can be constructed, for instance, by the "cross" explicit scheme on the grid $\{x_l = l\Delta x,\ l = 0, 1, \ldots, \kappa;\ \tau_n = n\Delta\tau,\ n = 0, 1, \ldots, m\}$.

---

[2] This approach was proposed by the author and, defended in his thesis in 1971, and later, described in *Identification of Heat Transfer Processes of Flying Vehicles*, Moscow, Mashinostroenie Publishing Agency, 1979 (in Russian).

[3] This value for solids is very small (e.g. for aluminium $t_r \sim 10^{-11}$ c) and the influence of term $t_r \partial^2 T/\partial \tau^2$ on solution of a direct heat conduction problem is negligible.

[4] Conditions of other types can be assigned as well.

In the result we get

$$T_{l+1,n} = 2T_{ln} - T_{l-1,n} + \frac{\Delta x^2}{2a\Delta\tau}(T_{l,n+1} - T_{l,n-1})$$
$$+ \frac{\alpha}{a}\left(\frac{\Delta x}{\Delta\tau}\right)^2 (T_{l,n+1} - 2T_{ln} + T_{l,n+1}). \tag{5.32}$$

A discrete approximation to Cauchy conditions at point $x = d$ by numerical solution of the problem can be presented as $T_{\kappa n} = T_n^*$, $T_{\kappa+1,n} = f_n$ where the grid function f_n is obtained from a solution of the direct heat conduction problem in domain $d \leqslant x \leqslant b$, $0 \leqslant \tau \leqslant \tau_m$ by known boundary conditions (5.31).

Let us refer to the resuts of solving a methodical example similar to that considered in Sect. 5.3.4. It was established in the process of computations that when α increases, the oscillation amplitudes of function $q^\alpha(\tau)$ computed using the disturbed temperature $T^*(\tau)$ diminish; for the best approximations (within the limits of the "cross" scheme) the amplitudes diminish 3–5 times from their previous value compared with $\alpha = 0$. The oscillations occur near some "weighted average" curve $\tilde{q}(\tau)$, comparatively close to a true one (Fig. 5.12).

If we solve this problem using an implicit approximation scheme with a pattern of the same type as shown on Fig. 5.10, the results do not change qualitatively, i.e. the smoothing effect rises as α increases. However, on the other hand, in the result of ever growing violation of the heat balance equation a noticeable distortion of the boundary condition could take place. Usually, the curve, not being properly smoothed, begins in the main "to drift" from the unknown solution.

The main conclusion of the analysis is that a technique described to stabilize the solution of IHCP enables the stability of results to be increased, the

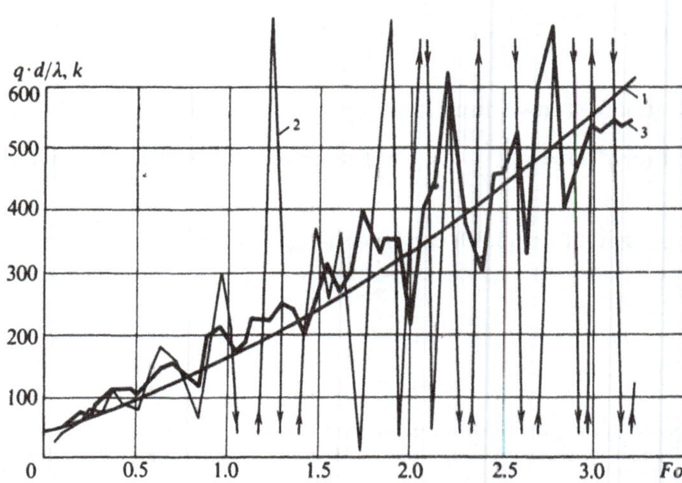

Fig. 5.12. Results of the heat flux density recovery for normal law of error distribution in the initial temperature values $\Delta Fo_d = 0.065$, $3\sigma = 0.05T^*_{max}$: 1 is an exact solution; 2 is for $\alpha = 0$; 3 is for $\alpha = 0.7$.

parameter α value being such, by which a deviation of the solution obtained from a true one becomes the smallest. This value can be approximately found if we match the computed function of temperature $T^{\alpha}_{\kappa n}$ with the assigned function in accordance with condition $\left| \sum_{n=1}^{m} (T^{\alpha}_{\kappa n} - T^{*}_{n})^2 - \delta^2 \right| = \min$, where $T^{\alpha}_{\kappa n}$ are temperature values, gained from a solution of the Fourier heat conduction equation in domain D using a known boundary condition $q^*(\tau)$ and condition $q^{\alpha}(\tau)$, computed in the result of IHCP solution for the hyperbolic equation where δ is the known error of initial temperature T^*. It is assumed here that value $q^*(\tau)$ is assigned precisely, and error δ is much larger than the error of approximation when defining $T^{\alpha}_{\kappa n}$.

Note that a close approach to solution of IHCP based on the approximate substitution of a heat conduction equation into an equation of hyperbolic type has been used later in [205].

5.6 Summary

1. Difference algorithms are suggested and investigated for solving non-linear boundary IHCP, based upon explicit and implicit schemes of numerical continuation of the temperature field when selecting approximation steps from the condition of stability of results.
2. The property of the scheme's viscosity is set as a supplementary regularizing factor in solving inverse problems. The estimates are gained for the minimum admissible steps of time approximation of a heat conduction equation.
3. It is shown that an artificial hyperbolization of the heat conduction equation allows stability of numerical solution of boundary IHCP to be increased.

Earlier, in Chaps. 4 and 5, for solution of boundary IHCP a step regularization principle, was used which, as observed, allows the construction of simple and practical algorithms but it still has a limited area of application. This can be considerably extended by such methods of IHCP solution which are based on the iterative and variational regularization principles, a study of which can be found in Chaps. 6 and 7.

Chapter 6

The Extremal Formulations and Methods of Solving Inverse Heat Conduction Problems

Introduction

One of the more promising directions in solving inverse heat conduction problems is to reduce them to *extremal formulations* and apply numerical methods of the optimization theory. Two cases are thereby possible:

1. a solution is sought in the space of parameters;
2. the task is solved in the functional space.

In the first case, the function of the finite number of parameters is extremized, i.e. it is a *parametric optimization*, and in the second case, functions are defined by extremizing the functionals of these functions, i.e. it is a *functional optimization*.

These two approaches will be applied here to IHCP solution with the help of one class of iterative methods known as *gradient methods*. The solution of ill-posed IHCP using gradient algorithms heightens specificity, which will be considered in this chapter and in Chap. 8. The iterative method of inverse problem solution will be developed as treated by the present author in [11, 15]. The structure of this chapter is outlined as follows. In Sect. 6.1, leaving for the time being the instability of boundary IHCP, there are descriptions of the possible extreme problem statements in the search for heat flux at the boundary of a one-dimensional body using temperature measurements inside the body. Possible criterion for obtaining stability of IHCP solutions while using an iterative procedure of optimization is discussed in Sect. 6.2.

Section 6.3 contains a general description of a parametric form of identification of the unknown heat flux with appropriate geometric interpretation of the gradient method. Concrete definition of iterative algorithms of gradient type as a steepest descent method and that of conjugate gradient method is given in Sect. 6.4. Functional optimization for a solution of the inverse problem using methods of steepest descent and conjugate gradients is touched upon in Sect. 6.5. An approach is also considered as to the determination of a gradient for a residual functional, via a solution of *the adjoint boundary-value problem*. A crucial question is discussed in Sect. 6.6 on the conditions necessary for completing an iterative process in solving ill-posed IHCP. It is in this section that a short comparative analysis of iterative methods is made for a solution of unstable inverse problems. The principles of constructing algorithms while using parametric and functional forms of optimization are considered in Sect. 6.7.

Sections 6.8 and 6.9 are devoted to a numerical investigation of algorithms constructed in solving model problems as well as drawing up appropriate practical recommendations. Some features of boundary IHCP solution for another type of resulting functional are pointed out in Sect. 6.10. Finally, in Sect. 6.11, it is shown how to approach a solution of two-dimensional boundary IHCP by means of an iterative method.

6.1 A Boundary Inverse Problem in the Extremal Statement

Consider an inverse problem for a one-dimensional heat conduction equation

$$C\frac{\partial T}{\partial \tau} = \frac{\partial}{\partial x}\left(\lambda \frac{\partial T}{\partial x}\right).$$

The initial temperature distribution and thermal-physical coefficients are known quantities. It is required to determine a temperature field $T(x, \tau)$ in the domain $D = \{0 \leqslant x \leqslant b,\ 0 < \tau \leqslant \tau_m\}$ and a heat flux $q(\tau) = -\lambda \left.\frac{\partial T}{\partial x}\right|_b$ at the boundary of the body.

With the input information we can set:

1. conditions of the fourth kind, i.e. the heat flux $q^*(\tau)$ and temperature $T^*(\tau)$ at a certain fixed point of the body $x = d$;
2. temperature $T^*(\tau)$ at the point $x = d$ and any boundary condition (of the 1st, 2nd, and 3rd kind), at some other point $x = l$. It is assumed here, that $0 < d \leqslant b$, $0 < l \leqslant d$, $d < l$.

These additional conditions give rise to the occurrence of two different IHCP statements which we shall call problems I and II, respectively. In spite of the fact that the second statement is more common and problem I is realized in its particular case, such a division is justified for solution methods. This will be made clear from further considerations.

Problem 1. Introduce the given problem as a mathematical representation

$$C\frac{\partial T}{\partial \tau} = \frac{\partial}{\partial x}\left(\lambda \frac{\partial T}{\partial x}\right), \quad 0 < x < d,\ 0 < \tau \leqslant \tau_m;$$

$$T(x, 0) = \varphi(x),\ 0 \leqslant x \leqslant d; \tag{6.1}$$

$$\frac{\partial T(d, \tau)}{\partial x} = -q^*(\tau)/\lambda; \quad T(d, \tau) = T^*(\tau), \qquad 0 \leqslant \tau \leqslant \tau_m.$$

Let the heat flux $q(\tau) = -\lambda\,\partial T(0, \tau)/\partial x$ be the desired function. Knowing $q(\tau)$, one can find a temperature field in the domain studied from a solution of the direct heat conduction problem.

Note that a similar problem can be stated as well in the domain $d \leqslant x \leqslant b$, $0 < \tau \leqslant \tau_m$. By a set of these two problems, the initial IHCP statement will be fully reproduced.

To treat the problem as shown in Eqs. (6.1) in the form of optimal control, it is necessary to formulate the *optimization criterion.* Let us select the control $q(\tau)$ from the condition of definite conformity of the known temperature $T^*(\tau)$ with temperature $\tilde{T}(q, d, \tau)$ corresponding to the given control.

In the exact extremal statement, the definition of function $q(\tau)$ corresponds to the minimization of residual $\rho_F^2(\tilde{T}(q, d, \tau), T^*)$ characterizing the deviation of temperature $\tilde{T}(q, d, \tau)$ calculated for a certain density $q(\tau)$ from known temperature $T^*(\tau)$ in the metric of space of the input data F. Select the mean-square residual as a deviation measure

$$J_1(q) = \int_0^{\tau_m} [\tilde{T}(q, d, \tau) - T^*(\tau)]^2 \, d\tau. \tag{6.2}$$

Value $J_1(q)$ is a functional in the space of functions $q(\tau)$ and its numerical value specifies the distance in the functional space L_2 between the given $T^*(\tau)$ and calculated $\tilde{T}(q, d, \tau)$ temperatures. The degree of proximity of functions $T^*(\tau)$ and $\tilde{T}(q, d, \tau)$ can also be judged by some other criteria which correspond to distances in other functional spaces. However, a quadratic measure of residual in the form of the functional (6.2), both from mathematical and practical points of view carries a number of useful properties enabling effective computational algorithms of minimization to be constructed for solving different inverse problems.

Problem 2. Suppose that a condition of the second kind is known at the boundary $x = l$. Then, the temperature field $T(x, \tau)$ must satisfy the equations

$$\begin{aligned} &C\frac{\partial T}{\partial \tau} = \frac{\partial}{\partial x}\left(\lambda \frac{\partial T}{\partial x}\right), \quad 0 < x < l, \ 0 < \tau \leqslant \tau_m; \\ &T(x, 0) = \varphi(x), \quad 0 \leqslant x \leqslant l; \\ &\frac{\partial T(l, \tau)}{\partial x} = -g^*(\tau)/\lambda; \quad T(d, \tau) = T^*(\tau), \qquad 0 \leqslant \tau \leqslant \tau_m, \end{aligned} \tag{6.3}$$

where $0 < d < l$.

Formally, this statement can be reduced to problem I. For this it is necessary to determine the temperature field in the region $d \leqslant x \leqslant l, 0 < \tau \leqslant \tau_m$ from a solution of the problem

$$\begin{aligned} &C\frac{\partial T}{\partial \tau} = \frac{\partial}{\partial x}\left(\lambda \frac{\partial T}{\partial x}\right), \quad d < x < l, \quad 0 < \tau \leqslant \tau_m; \\ &T(x, 0) = \varphi(x), \quad d \leqslant x \leqslant l; \\ &T(d, \tau) = T^*(\tau); \quad \frac{\partial T(l, \tau)}{\partial x} = -g^*(\tau)/\lambda, \qquad 0 \leqslant \tau \leqslant \tau_m. \end{aligned} \tag{6.4}$$

By the known function $T(x, \tau)$ we calculate a heat flux at point $x = d$ $q^*(\tau) = -\lambda \dfrac{\partial T(d, \tau)}{\partial x}$ thus formulating conditions of the fourth kind.

Parallel with this statement, let us consider another two extremal forms of the inverse problem (6.3) which differ in selecting a measure of proximity of already measured and computed temperatures.

In the first case, after the analogy of problem I, introduce the following criterion:

$$J_2(q) = \int_0^{\tau_m} [\tilde{T}(q, d, \tau) - T^*(\tau)]^2 \, d\tau. \tag{6.5}$$

Here, temperature $\tilde{T}(q, d, \tau)$ corresponds to heat fluxes $q(\tau)$ and $g^*(\tau)$, acting at boundaries $x = 0$ and $x = l$.

Now let us consider the other statement. Suppose that a certain approximation to the desired function $q(\tau) \equiv -\lambda \dfrac{\partial T(0, \tau)}{\partial x} = \tilde{q}(\tau)$ is given. Solving the direct problem in the domain $D_1 = \{0 \leqslant x \leqslant l, 0 < \tau \leqslant \tau_m\}$, we obtain a temperature field $\tilde{T}(x, \tau)$ corresponding to this approximation. Function $\tilde{T}(x, \tau)$ in domain $D_2 = \{d \leqslant x \leqslant l, 0 < \tau \leqslant \tau_m\}$ will somehow deviate from the field $T(x, \tau) = T^*(x, \tau)$ found from the solution of problem Eq. (6.4) by the known conditions $T^*(\tau)$ and $q^*(\tau)$. For numerical expression of this deviation, use functional

$$J_3(q) = \int_0^{\tau_m} \int_d^{l} [\tilde{T}(q, x, \tau) - T^*(x, \tau)]^2 \, dx \, d\tau \,. \tag{6.6}$$

Varying a control function $q(\tau)$, we shall obtain different values $J_3(q)$.

Note. Problem II with the criterion (6.6) admits generalization for several points of temperature in domain D_2. Let the temperatures $T_1^*(\tau), T_2^*(\tau), \ldots, T_r^*(\tau)$ be known corresponding to points $x = d_1, d_2, \ldots, d_r$ where $d_1 < d_2 < \cdots < d_r = l$. By these data we reconstruct a temperature field in the whole domain D_2. Depending on the number of points r and quality of measurements, we can use interpolating and smoothing of the initial information for this purpose, as well as the solution of assemblage of direct problems with the boundary conditions of the first kind, i.e. functions $T_i^*(\tau)$, $i = 1, 2, \ldots, r$. In the result, function $T^*(x, \tau)$ is formed which is considered as a given temperature field in the criterion (6.6).

6.2 The Iterative Regularization Principle

A mathematical description of heat conduction processes with corresponding boundary conditions was given above as well as the control actions and optimization criteria being selected. To complete any one of these statements of an extremal problem, it is necessary to determine an admissible set Q of functions of control and consider all control actions as belonging to this very set ($q \in Q$).

Supposing that the initial problem for which the extremal statement is formulated is well-posed, the domain of admissible solutions can be selected mainly from pure physical considerations which will vary for each particular problem. For example, in some heat tests, it is known in advance that during definite time intervals the desired function or its derivative are sign-definite. To a certain degree of accuracy, one can also predict the highest and the least values $q(\tau)$, the approximate times of extremum realization, etc.

However, inverse heat conduction problems by their very matter are unstable and such restrictions could be insufficient to formulate a mathematically well-posed problem of optimal control. Moreover, a priori assignment of pure physical restrictions onto the class of desired functions is at times impossible to carry out for the accuracy required. As a result one has to accept this uncertainty and try to solve the problem without any restrictions. It is quite obvious that the optimal solution $q(\tau)$ obtained could have nothing in common with a true one.

First, it occurs owing to the fact that initial functions in actual conditions are always known to have some error which is often sufficiently large and of a pulsating nature. Second, to solve the given problem using a computer it is necessary to transfer to a finite-difference analogue. The discretization process inevitably leads to errors conditioned by the approximation of continuous functions through piecewise-polynomial dependences, as well as by the approximation of differential and integral operators through difference and summation operators. Moreover, errors of rounding-off occur when performing arithmetic operations. All these errors give instability of results in approaching the point of minimum of the residual functional.

Thus, if specific restrictions to a set of admissible functions $q(\tau)$ are not introduced, a stable solution of IHCP can be based only on the use of natural regularizing properties of the heat conduction process and computational algorithms. Earlier, in chapters 4 and 5, one parameter was selected for this purpose, a step in time. One can also use such a technique for a extremal problem statement. However, as has been already shown, such a technique is not always usable and so another possibility of IHCP regularization in the extreme statement will be considered below.

This possibility is implied by some iterative optimization algorithms in which a sequential improvement of the solution takes place in accordance with the formula:

$$q^{k+1}(\tau) = q^k(\tau) + \Delta q^k(\tau), \quad k = 0, 1, \ldots, \tag{6.7}$$

where $q^0(\tau)$ is the initial approximation. The correction $\Delta q^k(\tau)$ at every iteration is computed from the condition of a decrease of the objective functional $J(q^{k+1}) < J(q^k)$.

It is possible to suppose that methods enabling the effective initiation of the iterative process from a distant approximation $q^0(\tau)$ and a sharp slowing down in approaching the functional minimum would appear useful when solving inverse heat conduction problems. Such a method of damping the instability when specifying an approximate solution for a ill-posed problem is based on "viscous" properties of numerical algorithms of optimization.

It is necessary to keep in mind that as the number of iterations increases an IHCP solution can worsen, gradually losing its smooth character. Any waviness appearing in $q^k(\tau)$ will gain in strength as fast as the increasing fluctuating errors burden the initial temperature data and the greater would be the distance between the boundary with $q(\tau)$ condition sought and the point of temperature measurement.

For this, a suggestion arises to halt the iterative process at a certain iteration $k = \bar{k}$, without admitting a "shake up" of the solution. We shall discuss the *admissible minimization level* of the integral temperature residual further.

The main question in such an approach is to select a *halt criterion*. With this in mind let us introduce a complementary condition specifying the admissible degree of proximity of the approximation sought to the "exact" function $q_{\delta T}(\tau)$ corresponding to the disturbed input data. A "discipline" condition as such can be a restriction to the residual level given as an aggregate error δ including an error δ_T in temperature data and an approximation error δ_a in the boundary-value heat conduction problem.

Thus, let us bind the iterative sequence (6.7) according to the condition

$$J(q) \simeq \delta^2, \tag{6.8}$$

where

$$\delta^2 = \delta_T^2 + \delta_a^2. \tag{6.9}$$

Having outlined the iterative solution of an ill-posed inverse problem, we can proceed to concrete implementation of the given method.

6.3 Parametric Optimization in Solving Inverse Problems

Instead of the continuous function $q(\tau)$ let us consider its finite-dimensional analogue as vector $\boldsymbol{q} = [q_1, q_2, \ldots, q_m]^T$, the components of which are obtained by sampling in the chosen time grid $q_n = q(\tau_n)$, $n = 1, 2, \ldots, m$. Assume the time series $\{q_n\}_1^m$ be used for a piecewise-polynomial approximation of the desired function $q(\tau)$: $\hat{q}(\boldsymbol{q}, \tau) \simeq q(\tau)$. Then, the initial extremal statement of IHCP can be re-formulated as a problem of parametric optimization.

To be precise, let us consider the second extremal problem with the functional (6.5). In this case, problem I will follow from it as a particular statement.

Keeping aside for the time being the instability of IHCPs and hence assuming the minimization of the functional (6.5) with no special constraints, we shall obtain a *parametric equivalent* of the initial problem: define such a set of parameters $\{q_n\}_1^m$ from a certain domain G which could minimize the criterion $J(\boldsymbol{q}) \equiv J_2(\boldsymbol{q})$:

$$\boldsymbol{q} = \arg\inf_{\boldsymbol{q} \in G} J(\boldsymbol{q}), \tag{6.10}$$

$$J(\boldsymbol{q}) = \int_0^{\tau_m} [\tilde{T}(\hat{q}(\boldsymbol{q}, \tau), d, \tau) - T^*(\tau)]^2 \, d\tau, \tag{6.11}$$

where $\tilde{T}(\hat{q}(\boldsymbol{q}, \tau), d, \tau)$ is the temperature corresponding to a solution of the direct heat conduction problem with a given approximation $\hat{q}(\boldsymbol{q}, \tau)$ to a function sought.

Note that the statement (6.10) generally is related to *non-linear programming*, that is, the quality criterion $J(\boldsymbol{q})$ represents a non-linear function of the parameters and the G set can be extracted by non-linear constraints.

The graph of dependence of the objective function $J(\boldsymbol{q})$ on variables of control $\boldsymbol{q}$ is called a *response surface* and the lines of equal values of the function, *level lines*.

There are many situations in practice when a priori information concerning the G set is not available or it is assumed in advance that the extremum point satisfies all possible constraints. Then the problem is solved in the open domain (with no constraints).

This statement, of course, is not the only possible one. Other values can also act as the parameters sought such as, for example, the coefficients a_k in truncated expansion into a Fourier series by some orthonormal system of functions $\{\psi_k\}: q(\tau) \approx \sum_{k=1}^{r} a_k \psi_k(\tau)$ or the coefficients C_k in approximation by B-splines $q(\tau) \simeq \sum_{k=0}^{l} C_k B_k(\tau)$, where $B_k(\tau)$ is the B-spline, corresponding to the kth time interval [130].

In practical realization, using a computer, the integral quality criterion is substituted by a finite sum, e.g. the function (6.11) can be roughly presented as

$$J(\boldsymbol{q}) \simeq \Delta\tau \sum_{n=1}^{m} [\tilde{T}(\boldsymbol{q}, d, \tau_n) - T^*(\tau_n)]^2 \equiv \Delta\tau \, \| A[\boldsymbol{q}] - \boldsymbol{f} \|^2,$$

where τ_n are the points of uniform sampling of the continuous functions $\tilde{T}(\tau)$ and $T^*(\tau)$, $\Delta\tau$ is a step in time, $A[\boldsymbol{q}] \equiv \tilde{\boldsymbol{T}} = [\tilde{T}_1, \tilde{T}_2, \ldots, \tilde{T}_m]^T$ is the temperature vector at point d computed by known $\boldsymbol{q}$, A is generally a non-linear operator, $\boldsymbol{f} = [T_1^*, T_2^*, \ldots, T_m^*]^T$ is a vector of temperature measurements and $\|\boldsymbol{y}\|$ is the Euclidean norm of some $\boldsymbol{y}$ vector.

Note that the sampling grids of unknown quantity $q(\tau)$ and temperature $\tilde{T}(\tau)$ may not coincide.

Thus, the inverse problem in the extreme statement can be reduced to a minimization of the temperature residual in the Euclidean metric

$$\min_{\boldsymbol{q}} \| A[\boldsymbol{q}] - \boldsymbol{f} \|^2. \tag{6.12}$$

To solve problem (6.12) let us make use of the discrete iterative process analogous to (6.7):

$$\boldsymbol{q}^{k+1} = \boldsymbol{q}^k + \Delta\boldsymbol{q}^k, \tag{6.13}$$

where $\boldsymbol{q}_0$ is the initial estimation of the unknown vector.

As an increment $\Delta\boldsymbol{q}^k$ in proceeding to the next approximation we use vector

$$\Delta\boldsymbol{q}^k = -\beta_k \boldsymbol{\xi}^k, \tag{6.14}$$

where $\boldsymbol{\xi}^k$ is the direction from point $\boldsymbol{q}^k$ and β^k is a scalar quantity determining the step length along this direction.

Let us introduce into our consideration a *gradient* of criterion function $J'(q)$, which by definition is a vector orthogonal to the level line pointing the direction of the most rapid increase of the function. Its components are partial derivatives of $J(q)$ for the unknown parameters

$$J'(q) = [\partial J/\partial q_1, \partial J/\partial q_2, \ldots, \partial J/\partial q_m]^T.$$

The admissible direction of descent $-\xi^k$ must be assigned so that the angle between it and direction opposite to gradient $(-J'(q^k))$ in point q^k satisfies the condition $\cos \gamma_k = \dfrac{(J'^k, \xi^k)}{\|J'^k\| \, \|\xi^k\|} > 0$, where (J'^k, ξ^k) is an inner product in Euclidean space. In optimization methods, the depth of descent β_k characterizing the value of step in going to the next approximation is selected as a non-negative value β locally minimizing the criterion function in direction ξ^k.

The descent method considered allows us to correct step by step the control $q(\tau)$, sequentially decreasing the function value, i.e. $J(q^{k+1}) < J(q^k)$. Assuming a uniqueness of IHCP solution function $J(q)$ would have one minimum and, if the initial functional is convex (this condition is satisfied in advance for linear problems), the sequence thus constructed [see formulae (6.13)–(6.14)] would converge towards this extremal point.

An evident geometric interpretation of the discrete iterative process for inverse problem solving can be presented for the case of a two-dimensional

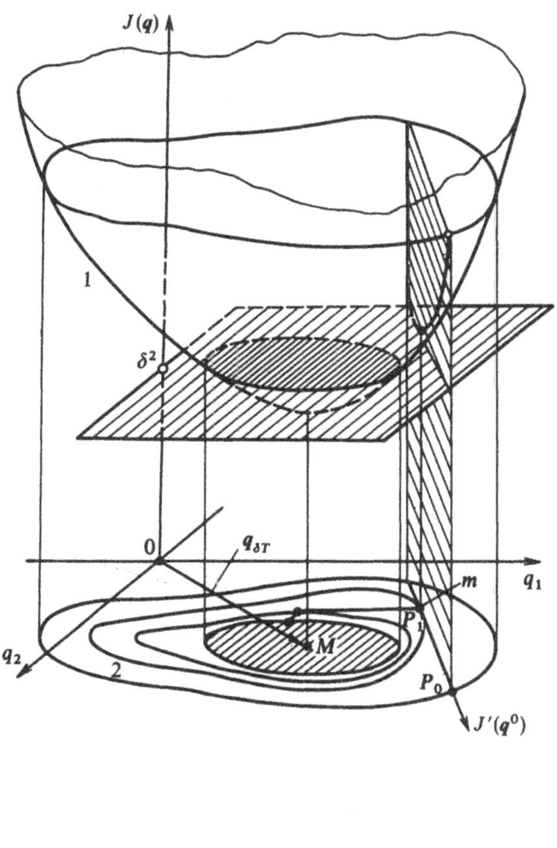

Fig. 6.1. Schematic illustration of the gradient method of IHCP solving with an optimal selection of descent depth (M = the point of minimum of criterion function, corresponding to an exact IHCP solution with disturbed data vector $q_{\delta T}$ is the point of minimum at a given iterative step) 1, response surface; 2, level line

vector $\boldsymbol{q} = [q_1, q_2]^T$. The response surface and level-lines are shown in Fig. 6.1. The initial approximation is designated by point P_0; to it corresponds some value of the $J(\boldsymbol{q}^0)$ function.

The path of movement from P_0 into optimal point M is a broken line, each section of which corresponding to one iterative step.

6.4 Gradient Methods of Parametric Optimization

The iterative sequence for minimizing the criterion function will be constructed by two methods, namely the *steepest descent* method and the *conjugate gradient* method.

The steepest descent method. In this case the strategy of iterative IHCP solution is depicted in Fig. 6.2. The process begins from some initial approximation (point P_0). To go onto the next point P_1 the direction of antigradient is used. The movement in this direction corresponds to the fastest decrease of the criterion function. Step β_K determines such point at which function $J(\boldsymbol{q}^{k+1})$ acquires the minimum value. Thus, we have an algorithm

$$\boldsymbol{q}^{k+1} = \boldsymbol{q}^k - \beta_k \boldsymbol{J}'^k, \quad k = 0, 1, \ldots; \tag{6.15}$$

$$\beta_k : J(\boldsymbol{q}^{k+1}) = \min_{\beta > 0} J(\boldsymbol{q}^k - \beta \boldsymbol{J}'^k),$$

where $\boldsymbol{q}^0$ is a pre-set approximation.

The fact that a one-dimensional minimization of function on beam $\boldsymbol{q}^k - \beta \boldsymbol{J}'^k$ is executed on every iteration denotes orthogonality of a new direction with respect to the previous direction. Note that the steepest descent method must not lead us astray as to the optimality of the whole process of moving to the extremal point. It is only concerned with the size of the step in the direction of descent being chosen on every iteration from the condition of function minimum (see Fig. 6.1). On the whole, however, using this method it is impossible to make

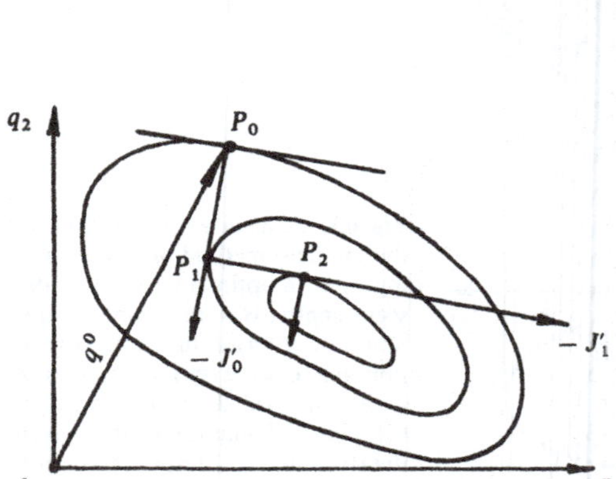

Fig. 6.2. An optimization scheme by the steepest descent method

a search for an optimal point not only with minimum but even for a finite number of steps. For this a method using conjugate gradients has better characteristics.

The conjugate gradient method. We shall explain this method using the example of linear IHCP with constant coefficients when using the criterion function (6.12). In this case vector $\boldsymbol{q}$ is a solution of the system of linear algebraic equations (see Chap. 4)

$$A_{\Delta}\boldsymbol{q} = \boldsymbol{f}, \tag{6.16}$$

where A_{Δ} is a square matrix, approximating the integral operator of heat conduction [see Eq. (3.1)].

In the extremal statement the problem (6.16) corresponds to a quadratic function

$$J(\boldsymbol{q}) = \|A_{\Delta}\boldsymbol{q} - \boldsymbol{f}\|^2. \tag{6.17}$$

We can write

$$\|A_{\Delta}\boldsymbol{q} - \boldsymbol{f}\|^2 = (A_{\Delta}\boldsymbol{q} - \boldsymbol{f})^T(A_{\Delta}\boldsymbol{q} - \boldsymbol{f}) = \boldsymbol{q}^T\boldsymbol{B}\boldsymbol{q} - 2\boldsymbol{q}^T\boldsymbol{b} + \boldsymbol{f}^T\boldsymbol{f},$$

where $\boldsymbol{B} = A_{\Delta}^T A_{\Delta}$ is a symmetric and positively definite matrix and $\boldsymbol{b} = A_{\Delta}^T\boldsymbol{f}$, $\boldsymbol{f}^T\boldsymbol{f}$ are fixed elements.

Based on the initial approximation of $\boldsymbol{q}^0$, let us construct a sequence of vectors $\boldsymbol{q}^1, \boldsymbol{q}^2, \ldots$ using formulae (6.13) and (6.14).

$$\boldsymbol{q}^1 = \boldsymbol{q}^0 - \beta_0\boldsymbol{S}^0;\ \boldsymbol{q}^2 = \boldsymbol{q}^1 - \beta_1\boldsymbol{S}^1 = (\boldsymbol{q}^0 - \beta_0\boldsymbol{S}^0) - \beta_1\boldsymbol{S}^1; \ldots, \boldsymbol{q}^n$$

$$= \boldsymbol{q}^0 - \sum_{k=0}^{n-1} \beta_k\boldsymbol{S}^k, \tag{6.18}$$

where $\boldsymbol{S}^k$ are some vectors.

Numerical parameters β_k for every iterative step are specified from the condition of function minimum at the k-th iteration:

$$\min_{\beta \geqslant 0} J(\boldsymbol{q}^k - \beta\boldsymbol{S}^k).$$

Choose directions $\boldsymbol{S}^k$ so that

$$\boldsymbol{S}^{kT}\boldsymbol{B}\boldsymbol{S}^j = 0, \quad k \neq j \tag{6.19}$$

would be satisfied. Vectors $\boldsymbol{S}^k$, satisfying condition (6.19), are called conjugate vectors (generalized-orthogonal) with respect to the $\boldsymbol{B}$ matrix.

The conjugate directions can be chosen in a different way [55, 74, 197]. In particular, it is proven that if vector $\boldsymbol{S}^k$ is the linear combination of the residual gradient at the given point and that of the previous vector:

$$\boldsymbol{S}^k = \boldsymbol{J}'(\boldsymbol{q}^k) + \gamma_k\boldsymbol{S}^{k-1}, \tag{6.20}$$

where $\gamma_k = \|\boldsymbol{J}'(\boldsymbol{q}^k)\|^2/\|\boldsymbol{J}'(\boldsymbol{q}^{k-1})\|^2, \gamma_0 = 0$, then directions $\boldsymbol{S}^k$ and $\boldsymbol{S}^{k-1}$ are conjugate.

Another variant of the method is

$$\gamma_k = -\frac{(J'(q^k), J'(q^{k-1}) - J'(q^k))}{\|J'(q^{k-1})\|^2}.$$

The computation procedure using the conjugate gradient method is as follows: beginning from some point q^0 the motion is in the direction of anti-gradient to the minimum which is reached at point q^1. Then the direction is determined, conjugate to the previous one, along which a transition to the next minimal point takes place. In the result we can specify a new approximation q^2, and then in the same way $-q^3$, etc. (Fig. 6.3).

An important feature of conjugate gradients is that the number of iterations for gaining an exact minimum of the quadratic function (6.17) is theoretically equal to the m, i.e., dimensionality of the unknown vector. In practice, however, because of errors which occur in the course of computations this condition is not satisfied and the convergence of the method becomes worse. To "revitalize" the iterative process it is recommended, after the expiration of m iterations, to perform one step using the steepest descent method ($\gamma_{m+1} = 0$) and then continue computations with the help of conjugate gradients. This approach, known as a renewal of iterations, decreases errors.

For a non-linear IHCP, when thermophysical properties of the body depend on temperature, the criterion function ceases to be quadratic. In this case the application of the conjugate gradient method is based on the assumption that $J(q)$ has properties nearing those of the quadratic function.

In the methods considered above it has been necessary to choose descent depth β_k at every iteration, that is scalar quantities in formulae (6.16) and (6.18). For this, one has to perform function minimization of one variable. In the case with a linear heat conduction problem the expression β_k is gained in the final form (see Sect. 6.7.1). In solving non-linear IHCP a linear estimation β_k can give no converging sequence of approximations and so it is necessary to have

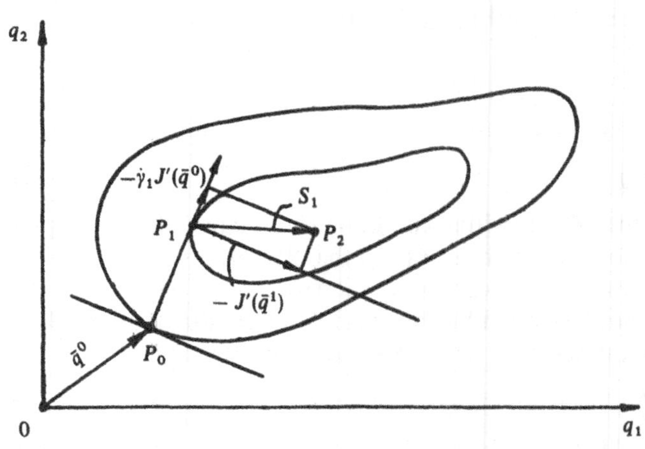

Fig. 6.3. An optimization diagram using the conjugate gradient method

a supplementary iterative computation procedure for this parameter. A method of golden section (see [55]) and an algorithm based on quadratic approximation of the criterion function are effectively used for this purpose [201].

One of the major problems in practice for methods of gradient type when solving inverse problems is the construction of an effective procedure for computing the gradient of a criterion function. Not only for the effectiveness of the whole design procedure but in a whole number of cases does the principal possibility of using iterative algorithms for IHCP solving depend on the successful solution of this problem.

The numerical technique is in difference approximation of the partial derivative $\partial J/\partial q_i$ in the vicinity of $\boldsymbol{q}$ approximation should be recognized as the least satisfactory technique of gradient computation. In the simplest case it has the form

$$\frac{\partial J}{\partial q_n} = \frac{J(q_1, q_2, \ldots, q_{n-1}, q_n + \Delta q_n, q_{n+1}, \ldots, q_m)}{\Delta q_n} \tag{6.21}$$

where Δq_n is a small increment in the n-th component of vector $\boldsymbol{q}$.

Thus, to compute all components of the gradient on the given iterative step it is necessary to solve m times the direct problem of heat conduction. Such an approach uses a large amount of computing time.

Another difficulty of this technique concerns the selection of the value of increments Δq_n. With very small Δq_n, high inaccuracies in determining the increment of the functions occur, conditioned by approximation and rounding-off errors. On the other hand, the increase of Δq_n can result in a sufficiently large violation of the assumption concerning linearity of the response surface at the point studied.

That is why we continue with gradient computations using other and more reasonable methods.

6.5 Functional Optimization in Inverse Problems

6.5.1 The Construction of the Iterative Sequence

In Sects. 6.3 and 6.4 the unknown dependence of heat flux on time $q(\tau)$ has been substituted by a finite-dimensional vector, minimizing the criterion function. Now consider a more complicated problem of finding function $q(\tau)$ on which the selected functional takes the minimal value. As before, we shall not discuss the question of stability of IHCP solving in the form of optimal control, leaving the analysis for Sect. 6.6. Taking this into account, the problem statement will look like the following: among functions of one variable $q = q(\tau)$ in the class of piecewise-continuous functions set in segment $[0, \tau_m]$ one is to find a function on which the functional

$$J(q) = \int_0^{\tau_m} [T^*(q, d, \tau) - T^*(\tau)]^2 \, d\tau \tag{6.22}$$

reaches the minimum value under conditions specified by a corresponding boundary heat condition problem statement.

Assuming that the unknown function can be represented by an infinite series using a certain system of functions $\{f_k(\tau)\}$: $q(\tau) = \sum_{k=1}^{\infty} a_k f_k(\tau)$, where a_k are unknown coefficients, it is possible formally to reduce the given problem of functional optimization to a parametric one. In this case the number of variables will be infinite. On the other hand, it is always possible to confine them to within a large enough but finite number of parameters. Moreover, computations in the computer call for a discrete representation of the problem. A question arises whether the transition to a more complex problem of optimal control is of reason.

The answer to this question in particular is that with a finite-dimensional approximation the initial problem of optimization is not solved exactly and, even providing the convergence of finite-dimensional approximation to optimal control in the initial infinite-dimensional space, this convergence may be quite irregular. Also, there is a danger that the process of approximation will be unstable regarding computation errors.

Note additionally that investigation into continuous problems of control as against discrete problems is of more interest to experts dealing with the solution of equations in partial derivatives.

Therefore, the use of a more general form – functional optimization – is quite substantiated.

Introduce a notion of the gradient of functional (the first Frechnet derivative [83]) at some point q. $q = q(\tau)$, $\tau \in [0, \tau_m]$ is the function considered as an element of the functional space $q \in Q = L_2$. If the functional increment at the point can be presented as

$$J(q + \Delta q) - J(q) = \int_0^{\tau_m} J'(\tau)\Delta q(\tau)\,d\tau + o(\|\Delta q\|),$$

where $q + \Delta q \in Q$, then function $J'(\tau)$ is called *a gradient* of functional at point $q \in Q$. The gradient $J'(\tau)$ as distinct from the one considered earlier determines the direction not in the space of unknown parameters but in the functional space.

As a numerical method of functional optimization take the gradient descent method in which the iterative sequence is constructed by the scheme similar to a finite-dimensional case

$$q^{k+1}(\tau) = q^k(\tau) - \beta_k \xi^k(\tau), \quad k = 0, 1, \ldots, \tag{6.23}$$

where $q^0(\tau)$ is the initial approximation and $\xi^k(\tau)$ is the admissible direction of descent.

In the *steepest descent method* in order to obtain $q^{k+1}(\tau)$ one should determine the gradient of residual at the $q^k(\tau)$ point and minimize $J(q)$ in the direction of the most rapid decrease of the functional (antigradient) coming from this point:

$$q^{k+1}(\tau) = q^k(\tau) - \beta_k J'^k(\tau), \quad k = 0, 1, \ldots,$$
$$\beta_k : J(q^{k+1}) = \min_{\beta \geq 0} J(q^k - \beta J'^k). \tag{6.24}$$

The *conjugate gradient method* corresponds to the following iterative process

$$q^{k+1}(\tau) = q^k(\tau) - \beta_k S^k(\tau), \quad k = 0, 1, \ldots; \tag{6.25}$$

$$S^k(\tau) = J'^k(\tau) + \gamma_k S^{k-1}(\tau), \quad \gamma_0 = 0; \tag{6.26}$$

$$\beta_k : J(q^{k+1}) = \min_{\beta \geq 0} J(q^k - \beta S^k).$$

As seen earlier in parametric optimization, different variants of the conjugate gradient method are specified by selection of the γ_k parameter [55]. Assuming that within one iteration the Frechnet second derivative varies little, we determine γ_k using the expression

$$\gamma_k = - \frac{\int_0^{\tau_m} J'^k(\tau)[J'^{k-1}(\tau) - J'^k(\tau)\, d\tau}{\int_0^{\tau_m} [J'^{k-1}(\tau)]^2\, d\tau}.$$

If we assume the functional minimized is close to *a quadrctic* one, then it is possible to use the formula:

$$\gamma_k = \int_0^{\tau_m} (J'^k)^2\, d\tau \Big/ \int_0^{\tau_m} (J'^{k-1})^2\, d\tau.$$

6.5.2 Derivation of the Formula for the Gradient

The basic question of gradient minimization is in determining derivative $J'(q, \tau)$. An effective method of solving this problem is stated below.

Let thermophysical properties of the body vary with temperature, i.e. the inverse problem, corresponding to Eq. (6.1), is a non-linear one. Write the direct problem for model temperature assigned by the Kirchhoff transformation $\left(\theta = \frac{1}{\lambda_0}\int_0^T \lambda(T)\, dT, \text{ where } \lambda_0 \text{ is some constant}\right)$. Supposing for simplicity $d = b$, we obtain

$$\frac{\partial \theta}{\partial \tau} = a(\theta)\frac{\partial^2 \theta}{\partial x^2}, \quad 0 < x < b, 0 < \tau \leqslant \tau_m; \tag{6.27}$$

$$\theta(x, 0) = \xi(x); \tag{6.28}$$

$$\frac{\partial \theta(b, \tau)}{\partial x} = -\frac{1}{\lambda_0} q^*(\tau) \equiv z(\tau); \tag{6.29}$$

$$\frac{\partial \theta(0, \tau)}{\partial x} = -\frac{1}{\lambda_0} q(\tau) \equiv p(\tau), \tag{6.30}$$

where

$$\xi(x) = \frac{1}{\lambda_0} \int_0^{\varphi(x)} \lambda(T)\,\mathrm{d}T .$$

The measure of the residual for selection of a controlling function will be the form

$$J(p) = \int_0^{\tau_m} [\theta(b,\tau) - f(\tau)]^2\,\mathrm{d}\tau , \tag{6.31}$$

where $f(\tau)$ is the known function $\left(f(\tau) = \frac{1}{\lambda_0} \int_0^{T^*(\tau)} \lambda(T)\,\mathrm{d}T\right)$.

Give $p(\tau)$ an increment $\Delta p(\tau)$. The corresponding changes will be acquired from the temperature at every point of the body:

$$\frac{\partial\theta(p)}{\partial\tau} - \frac{\partial\Delta\theta}{\partial\tau} = (a(\theta) + \Delta a)\left[\frac{\partial^2\theta(p)}{\partial x^2} + \frac{\partial^2\Delta\theta}{\partial x^2}\right].$$

Taking into account that $\Delta a \simeq \frac{\mathrm{d}a}{\mathrm{d}\theta}\Delta\theta$, and neglecting the value of the second order of smallness $\Delta a \frac{\partial^2\Delta\theta}{\partial x^2}$, we have an equation for the increment

$$\frac{\partial\Delta\theta}{\partial\tau} = \frac{\mathrm{d}a}{\mathrm{d}\theta}\Delta\theta\frac{\partial^2\theta(p)}{\partial x^2} + a(\theta(p))\frac{\partial^2\Delta\theta}{\partial x^2} . \tag{6.32}$$

Similarly, consider the boundary conditions which increment $\Delta\theta$ must satisfy:

$$\Delta\theta(x,0) = 0 ; \tag{6.33}$$

$$\frac{\partial\Delta\theta(b,\tau)}{\partial x} = 0 ; \tag{6.34}$$

$$\frac{\partial\Delta\theta(0,\tau)}{\partial x} = \Delta p(\tau) . \tag{6.35}$$

The functional (6.31) will have the increment $\Delta J \equiv J(p + \Delta p) - J(p) = J(\theta + \Delta\theta) - J(\theta)$. From here, neglecting the member of the second order of smallness,

$$\Delta J \approx 2\int_0^{\tau_m} \Delta\theta(b,\tau)[\theta(b,\tau) - f(\tau)]\,\mathrm{d}\tau .$$

To determine the functional gradient, let us introduce into our consideration a boundary-value problem which is called an adjoint problem to that of (6.27)–(6.30)

$$-\frac{\partial\psi}{\partial\tau} = \frac{\partial^2}{\partial x^2}[a(\theta)\psi] + \psi\frac{\mathrm{d}a}{\mathrm{d}\theta}\frac{\partial^2\theta}{\partial x^2}, \quad 0 < x < b,\ 0 < \tau < \tau_m ; \tag{6.36}$$

$$\psi(x, \tau_m) = 0; \tag{6.37}$$

$$\frac{\partial}{\partial x}[a(\theta(b, \tau))\psi(b, \tau)] = 2[\theta(b, \tau) - f(\tau)]; \tag{6.38}$$

$$\frac{\partial}{\partial x}[a(\theta(0, \tau))\psi(0, \tau)] = 0. \tag{6.39}$$

In this case the increment of the functional can be expressed in the form

$$\Delta J = \int_0^{\tau_m} \Delta\theta(x, \tau)\frac{\partial}{\partial x}[a(\theta(x, \tau))\psi(x, \tau)]|_0^b \,\mathrm{d}\tau$$
$$= \int_0^b \left\{\frac{\mathrm{d}}{\mathrm{d}x}\int_0^{\tau_m}\frac{\partial}{\partial x}[a(\theta(x, \tau))\psi(x, \tau)]\Delta\theta(x, \tau)\,\mathrm{d}\tau\right\}\mathrm{d}x$$

or

$$\Delta J = \int_0^b I_1(x)\,\mathrm{d}x + \int_0^b I_2(x)\,\mathrm{d}x,$$

where

$$I_1(x) = \int_0^{\tau_m} \Delta\theta(x, \tau)\frac{\partial^2}{\partial x^2}[a(\theta(x, \tau))\psi(x, \tau)]\,\mathrm{d}\tau;$$

$$I_2(x) = \int_0^{\tau_m} \frac{\partial\Delta\theta(x, \tau)}{\partial x}\frac{\partial}{\partial x}[a(\theta(x, \tau))\psi(x, \tau)]\,\mathrm{d}\tau.$$

For I_1, taking into account Eqs. (6.33), (6.36) and (6.37) and after integration by parts we shall obtain

$$I_1(x) = -\int_0^{\tau_m} \Delta\theta(x, \tau)\psi(x, \tau)\frac{\mathrm{d}a(\theta(x, \tau))}{\mathrm{d}\theta}\frac{\partial^2\theta(x, \tau)}{\partial x^2}\,\mathrm{d}\tau$$
$$+ \int_0^{\tau_m} \psi(x, \tau)\frac{\partial\Delta\theta(x, \tau)}{\partial\tau}\,\mathrm{d}\tau.$$

Using the equation for increments [Eq. (6.32)] we can write

$$\int_0^b I_1(x)\,\mathrm{d}x = \int_0^b\left\{\int_0^{\tau_m}\psi(x, \tau)a(\theta(x, \tau))\frac{\partial^2\Delta\theta(x, \tau)}{\partial x^2}\,\mathrm{d}\tau\right\}\mathrm{d}x.$$

Having changed the order of integration and integrating by parts the inner integral with the use of conditions (6.34) and (6.35) we will have

$$\Delta J = -\int_0^{\tau_m}\psi(0, \tau)a(\theta(0, \tau))\Delta p(\tau)\,\mathrm{d}\tau.$$

Since $\Delta J \simeq \int_0^{\tau_m} J'\Delta p\,\mathrm{d}\tau$, we finally obtain

$$J'(\tau) = -\psi(0, \tau)a(\theta(0, \tau)). \tag{6.40}$$

Thus, to determine the gradient of the functional it is necessary in accordance with the formula obtained to solve an adjoint problem and therefore find function $\psi(0, \tau)$.

The method described for obtaining the formula for the gradient can be also expanded onto a whole number of other statements of inverse problems, in particular, for a generalized heat conduction equation [13, 33, 34].

6.6 The Selection of Approximate Solution and the General Appraisement of Gradient Methods

6.6.1 About the Rule of Stoppage of Iterations

So far when considering parametric and functional optimizations it was assumed that the IHCP solution sought will minimize the performance criterion. However, if the input data are known to have some errors and the intervals of sampling of $q(\tau)$ are so small that a step self-regularization is actually absent, the approximate solution thus gained will acquire rather an oscillating nature. That is why, as mentioned in Sect. 6.2, there is no need in trying to reach an optimal point with an accuracy higher than the accuracy of the input data assignment. It is reasonable to interrupt the iterative process using condition (6.8), i.e. as soon as the function (6.11) (or the functional (6.22)) reaches the value of the summary error δ^2 in the temperature measurements and computations.

In particular, for a parametric IHCP statement with disturbed initial data for the admissible level of minimization we will take

$$J(\boldsymbol{q}) \simeq \delta^2 , \tag{6.41}$$

where $\delta^2 = \delta_T^2 + \delta_a^2$.

Value δ_T^2 is estimated by the formula $\delta_T^2 = \int\limits_0^{\tau_m} \sigma^2(\tau)\,\mathrm{d}\tau$, where $\sigma^2(\tau)$ is a function $T^*(\tau)$ variance. It is rather a complicated matter to determine error δ_a and that is why the method indicated is useful when the condition $\delta_a \ll \delta_T$ is satisfied. In this case we can assume $\delta \simeq \delta_T$.

For parametric optimization in the case of two variables the level-line corresponding to condition (6.41), below which it is not supposed to descend, is shown in Fig. 6.1. In the given problem it confines the shaded region of non-feasible solutions. The constraint formulated is not tough, but it should be kept in mind that for large fluctuating errors at T^*, the quality of the reconstructed function can rapidly worsen while moving towards the optimal point.

To choose vector $\bar{q}_\delta$ or function $q_\delta(\tau)$ satisfying the matching condition with an error in the input data one can use some limiting number of iterations k, i.e. the number of iterative cycles is considered as a *regularization parameter* of gradient methods in IHCP solving.

The computational process for the given initial approximation is completed when performing the inequality $J^k \leqslant \delta^2$.

A supplementary regularizing parameter in gradient methods is the quantity β_k. One can demand quite an exact approach of the residual level assigned at the expense of variation of the descent depth in the last iteration, i.e. one has to choose a variable $\beta_{\bar{k}}$ by condition $\min_{\beta}|J^{\bar{k}} - \delta^2|$. Here $\bar{k}$ is the number of iteration at which the following inequality is performed (Fig. 6.4)

$$J^{\bar{k}} > \delta > J^{\bar{k}+1}.$$

In the statement suggested for IHCP solution, which includes a regularizing condition of matching parameters for the iterative process with the value of summary error, there is a problem of the many approximate solutions available which correspond to different initial elements, i.e. the problem, strictly speaking, turns out to be indefinite. If a set of functions $q(\tau) \in Q_\delta$ found by iteration starting from different sufficiently smooth trial functions $q^0(\tau)$ has a large diameter $d(Q_\delta) = \sup_{q', q'' \in Q_\delta} \rho(q', q'')$,

where

$$\rho(q', q'') = \int_0^{\tau_m} [q'(\tau) - q''(\tau)]^2 \, d\tau,$$

then a need for refinement of the IHCP initial statement would appear in order to reveal the required representative $q(\tau)$. So the question arises, how great could a solution be distorted when found by the condition (6.41) with zero initial approximation $q^0(\tau) = 0$ involving no supplementary constraints?

As the results of solving methodical examples have shown, the functions reconstructed turned out to be close enough to standard functions when these have had zero values of function and its derivative ($q(\tau_m) = q'(\tau_m) = 0$) at the end of the time interval. That is why we may suppose that if there is a possibility to a priori point out the approximate values of heat flux reconstructed as well as its derivative at the end of the time interval and to choose the initial approximation, these conditions taken into account, then a solution thus gained will be close enough to the solution sought.

The selection of the objective functional $J(q)$ in the form of a mean-square residual assumes solutions to be close to actual ones only in the mean. In particular, gradient $J'(\tau)$ equals zero in advance at the boundary of the time interval with $\tau = \tau_m$. In the result value $q(\tau_m)$ in the process of iterations would not change. Therefore, if one does not manage to predict exactly the behaviour

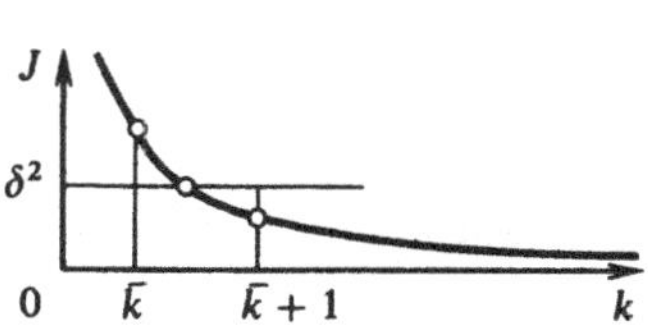

Fig. 6.4. A change of residual in the process of iterative minimization

of the unknown function at the end of the time interval and count it while assigning the initial approximation, then at some vicinity of $\tau = \tau_m$ the curve $q(\tau)$ declines from the actual solution.

6.6.2 Methods of Gradient Minimization

Consider the main reasons for using the steepest descent and the conjugate gradient method for IHCP iterative solution.

1. These methods largely satisfy the requirements considered in Sect. 6.2. The process of approximate solution refinement proceeds rapidly at first. It depends weakly on errors of input data and computational errors; it slows down sharply as the optimal point is approached. As a result, on the one hand, one of the principal difficulties of optimization is overcome that which has been connected with the selection of a sufficiently close initial approximation, and, on the other hand, there appears a real possibility of using viscous properties of algorithms to obtain smooth IHCP solutions. *Thus, the condition of slow convergence of methods near the minimum point being unwanted while solving well-posed problems of control turns out to be useful in solving ill-posed problems.*
2. Methods are applicable for solution of both linear and non-linear problems. They are universal and easily generalized onto other types of inverse heat transfer problems.
3. Realization of the corresponding gradient algorithms, as shown below, leads to effective computational processes and due to their simplicity, and by their ability to determine sufficiently close approximation to unknown functions little computer time is spent. It is also important to note that these methods are realized effectively not only in the digital computer but in the hybrid computational system as well thus enabling still greater economies of computing time.
4. As will be shown in Chap. 7, methods of inverse problem solution applying the procedures of steepest descent and conjugate gradients enable us to formulate an effective regularized form of the algorithm with uniform approximation to an unknown solution. Here, it becomes very important that any increase of calculations in comparison with the initial algorithm, providing the convergence in the mean, is kept small.

Comparing the possibilities of gradient methods we can obtain better characteristics of the computational process using the *conjugate gradient method.* Primarily, this is referred to problems with smooth input data. Let us explain this. The steepest descent method is the first order method based on linear approximation of the criterion assumed. The quality of such an approximation worsens sharply while nearing a stationary point. The method has no "memory", since a complete renewal of information occurs at every iteration. In the result a bad convergence of the method begins to manifest itself already at a sufficiently large distance from the optimal solution.

We may assume, however, that in the case of large fluctuating errors in measurements this drawback would not be so noticeable, for it is planned to interrupt the relaxation process by the condition (6.8) with a greater admissible level of the residual δ^2. As a result, the required number of iterations would not appear to be large and there would be no noticable moderation of convergence so far.

Methods of minimization of the *second order* (of Newton type) are *unsuitable* for the above considered iterational regularization. This can be explained by the fact that in the linear case they are reduced to a direct transformation of operator A of the inverse problem $Au=f$, and in the non-linear case an inversion of the derivative of this operator is required. At the same time, operator A and its derivative do not have continuous inverses, thus excluding them from such an operation. That is why *Newton type* methods can be used only for a step regularization of the solution.

The conjugate gradient method combines the advantages of first and second order methods. At every current iteration the information about the previous step is also used, since the gradient is required not only at the given point but also the one proceeding in order to compute a new conjugate direction. This method enables closer and with a lesser number of iterations to reach a required vicinity of optimal solution starting from a distant initial approximation. At the same time difficulties inherent to Newton's optimization methods are in fact absent in the conjugate gradient method, because the matrix of second derivatives does not need to be computed or inverted. It will be shown below (see 6.8, 6.9 and Chap. 8) that the conjugate gradient method is an effective computational procedure for solving boundary IHCPs. This method can be successfully used for solving other inverse problems, such as coefficient problems (see Chap. 8).

6.7 Iterative Algorithms for Solving a Linear Inverse Problem

Consider problem II, presented by statement (6.3), with functional (6.5) in the case of constant thermophysical properties and uniform initial distribution of temperatures:

$$\frac{\partial T}{\partial \tau} = a\frac{\partial^2 T}{\partial x^2}, \quad 0 < x < l, \quad 0 < \tau \leqslant \tau_m;$$

$$T(x,0) = 0, \quad 0 \leqslant x \leqslant l; \tag{6.42}$$

$$-\lambda\frac{\partial T(l,\tau)}{\partial x} = g^*(\tau); \; T(d,\tau) = T^*(\tau), \quad 0 \leqslant \tau \leqslant \tau_m, \quad 0 < d \leqslant l;$$

$$-\lambda\frac{\partial T(0,\tau)}{\partial x} = q(\tau) - ? \tag{6.43}$$

The temperature field for heat fluxes of assigned density $q(\tau)$ and $g^*(\tau)$ is determined by the integral formula

$$T(x, \tau) = \int_0^\tau q(\xi) \frac{\partial \theta(x, \tau - \xi)}{\partial \tau} \mathrm{d}\xi + \int_0^\tau g^*(\xi) \frac{\partial \theta(l - x, \tau - \xi)}{\partial \tau} \mathrm{d}\xi, \tag{6.44}$$

where $\theta(x, \tau)$ is a solution of the boundary-value heat conduction problem with the unit density of heat flux at one boundary ($x = 0$) and zero density at the other boundary ($x = 1$) (see Sect. 2.1.2).

In practice the parametric and functional forms of solving the inverse problem shown in Eqs. (6.42) and (6.43) would differ in approach to determine the direction of steepest descent. In the first case the simplest technique of determining the gradient components is used by calculating the partial derivatives of the surface response via the variables of control but without the difference approximation of the type given by formula (6.21). The corresponding quadrature formula for the expression (6.44) provides such a possibility. In the case of functional optimization the gradient cannot be determined in such a direct way, and an adjoint problem would then be used.

6.7.1 A Parametric Method

Instead of functions $q(\tau)$ and $g^*(\tau)$ let their piecewise-polynomial analogues be introduced, a stepwise approximation being used here $\hat{q}_i = \frac{1}{2}(q_{i-1} + q_i)$, $\hat{g}_i^* = \frac{1}{2}(g_{i-1}^* + g_i^*)$. Then, for a uniform grid with step $\Delta\tau$ we receive an approximate formula, specifying the temperature at n moment of time (see Sect. 4.1):

$$T_n(x) = \sum_{i=1}^{n} \hat{q}_i \Delta \vartheta_{n-i}(x) + \sum_{i=1}^{n} \hat{g}^* \Delta \vartheta_{n-i}(l - x), \quad n = 1, 2, \ldots, m, \tag{6.45}$$

where $\Delta \vartheta_{n-i}(x) = \vartheta(x, \Delta\tau(n - i + 1)) - \vartheta(x, \Delta\tau(n - i))$.

Supposing the criterion function $J = \sum_{n=1}^{m} [T_n(d) - T_n^*]^2$, the variables of control being values $\hat{q}_k$, $k = 1, 2, \ldots, m$, find the components of the gradient

$$\frac{\partial J(\boldsymbol{q})}{\partial \hat{q}_k} = 2 \sum_{n=k}^{m} [T_n(d) - T_n^*] \psi_{kn}, \quad k = 1, 2, \ldots, m, \tag{6.46}$$

where $\psi_{kn} = \dfrac{\partial T_n(d)}{\partial \hat{q}_k} = \Delta \vartheta_{n-i}(d)$; $\boldsymbol{q} = [\hat{q}_1, \hat{q}_2, \ldots, \hat{q}_m]^T$.

Using the expression (6.46) at every iteration we can find a desired direction ξ^k in the methods of both steepest descent and conjugate gradients and define the next approximation $\boldsymbol{q}^{k+1}$ by formula $\boldsymbol{q}^{k+1} = \boldsymbol{q}^k - \beta_k \xi^k$. The value of step β_k will be determined from the condition of minimum of function $J(\boldsymbol{q}^{k+1})$:

$$\frac{\partial J(\boldsymbol{q}^{k+1}, d)}{\partial \beta} = 2 \sum_{n+1}^{m} [T_n(\boldsymbol{q}^{k+1}, d) - T_n^*] \frac{\partial T(\boldsymbol{q}^{k+1}, d)}{\partial \beta} = 0. \tag{6.47}$$

The derivative is

$$\frac{\partial T_n(\boldsymbol{q}^{k+1}, d)}{\partial \beta} = \sum_{i=1}^{n} \frac{\partial \hat{q}_i^{k+1}}{\partial \beta} \Delta\vartheta_{n-i}(d) = -\sum_{i=1}^{n} \xi_i^k \Delta\vartheta_{n-i}(d). \tag{6.48}$$

A formula follows from expressions (6.45), (6.47) and (6.48) to determine an optimal value of

$$\beta_k = \frac{\sum_{n=1}^{m} [T_n^k(d) - T_n^*] \sum_{i=1}^{m} \xi_i^k \Delta\vartheta_{n-i}(d)}{\sum_{n=1}^{m} \left[\sum_{i=1}^{m} \xi_i^k \Delta\vartheta_{n-i}(d)\right]^2}. \tag{6.49}$$

6.7.2 The solution in the infinite-dimensional space

In this case the functional (6.22) has to be minimized. Suppose that $d = l = b$, then the *functional gradient* is given by expression $J'(\tau) = -\psi(0,\tau)a(\theta(0,\tau))$.

Introduce the inverse time $t = \tau_m - \tau$ and rewrite the adjoint problem as in Eqs. (6.36)–(6.39) with the account of constancy of thermophysical properties ($a = \text{const}, \lambda = \lambda_0 = \text{const}$):

$$\frac{\partial \bar{\psi}}{\partial t} = a\frac{\partial^2 \bar{\psi}}{\partial x^2}, \quad 0 < x < l, 0 < t < \tau_m; \tag{6.50}$$

$$\bar{\psi}(x, 0) = 0;$$

$$\frac{\partial \bar{\psi}(l, t)}{\partial x} = 2[T(l, t) - T^*(t)] \equiv e(t); \quad \frac{\partial \bar{\psi}(0, t)}{\partial x} = 0,$$

where $\bar{\psi} = a\psi$.

Solution of the problem (6.50) is of the form

$$\bar{\psi}(x, t) = -\lambda \int_0^t e(t)\frac{\partial \vartheta(l - x, t - \xi)}{\partial t} d\xi .$$

Thus

$$J'(t) = \bar{\psi}(0, t) = -\lambda \int_0^t e(t)\frac{\partial \vartheta(l, t - \xi)}{\partial t} d\xi .$$

The coefficient β_k, determining a value of the step proceeding from $p^k(\tau)$ to $p^{k+1}(\tau)$ in the direction $\xi(\tau)$, is found from the condition

$$\min_{\beta \geq 0} \int_0^{\tau_m} [T(p^k - \beta\xi^k) - T^*]^2 d\tau$$

$$= \min_{\beta \geq 0} \int_0^{\tau_m} [-\beta \Delta T(\xi^k) + T(p^k) - T^*]^2 d\tau ,$$

where

$$\Delta T(\xi^k) = \Delta T(\xi^k, l, \tau) .$$

Setting the derivative (with respect to β) to zero, we receive

$$\beta_k = \frac{\int_0^{\tau_m} \Delta T(\xi^k)[T(p^k) - T^*]\,\mathrm{d}\tau}{\int_0^{\tau_m} \Delta T^2(\xi^k)\,\mathrm{d}\tau} . \tag{6.51}$$

An increment in temperature $\Delta T(\xi)$ is found from a solution of the boundary-value problem

$$\frac{\partial \Delta T}{\partial \tau} = a \frac{\partial^2 \Delta T}{\partial x^2}, \quad 0 < x < l, \quad 0 < \tau \leqslant \tau_m;$$

$$\Delta T(x, 0) = 0 ;$$

$$\partial \Delta T(l, \tau)/\partial x = 0; \quad \partial \Delta T(0, \tau)/\partial x = \xi(\tau) ,$$

from which

$$\Delta T(\xi) = \Delta T(\xi, l, \tau) = -\lambda \int_0^{\tau} \xi(\tau) \frac{\partial \vartheta(l, \tau - \xi)}{\partial \tau}\,\mathrm{d}\xi .$$

In constructing the sequence of elements $q^k(\tau)$ by formulae (6.24) or (6.25) it is necessary at every iteration to solve *three* problems on determining temperature $T(l, \tau)$, functions $\bar{\psi}(0, \tau)$ and $\Delta T(l, \tau)$. For this we can use a discrete representation (6.45), the values of coefficients $\Delta\theta_{n-i}$ for all three problems all being the same provided that these problems are approximated with the same time step.

6.8 Experimental Investigation of Algorithms

After a short description of algorithms let us answer the following questions:

1. How do computational and temperature measurements errors affect the convergence of iterative processes and the quality of approximations being obtained?
2. By what condition should one end iterations?
3. How great could a distortion of solution be at the expense of inaccurate assignment of the initial approximation in the vicinity $\tau = \tau_m$?

To answer these questions we have made a practical analysis of algorithms efficiency with different conditions of application. All methodical calculations have been made with steps using Fourier number $\Delta Fo = a\Delta\tau/l^2$, which were sufficiently less than those marginally admissible values which corresponded to the stability limit of direct algebraic methods for solving integral forms of IHCP (see Chap 4). Thus, a step regularization did not, in fact, show itself. So, it was possible to investigate the regularizing properties of the suggested iterative methods. Let us consider the results obtained.

6.8.1 Stability with respect to Errors of Approximation and Rounding-off. Accuracy of Reconstruction of the Boundary Condition Using Smooth Data

It has been established that the process of iterative refinement of heat flux $q(\tau)$ is stable with respect to errors connected with realization of computational algorithms in the computer. This conclusion can refer to both algorithms of parametric and functional optimization constructed using steepest descent and conjugate gradient methods. A methodical account was carried out in which distant values $\boldsymbol{q}^0 = 0$ or $q^0(\tau) = 0$ were selected as initial approximations. The iterative processes converged rapidly at first and slowed down abruptly after *three to six iterations. In most cases* the form of heat flux $q(\tau)$ was revealed.

Viscous properties of gradient methods proved to be sufficient so that for any reasonable number of iterations *no significant oscillations* in the solution emerged provided that the non-disturbed data had been used as input information (Figs 6.5 and 6.6, see also Figs. 6.9 and 6.10). At the same time, the accuracy of reconstruction of a boundary condition of the complex structure turned out to be quite suitable for the majority of practical applications.

Such behaviour of the iterative process avoids introducing an admissible level of minimization and terminates the search using common methods, e.g. by "adhesion" of two successive approximations to the heat flux:

$$\max_{\tau_n} |q^{k+1}(\tau_n) - q^k(\tau_n)| \leqslant \varepsilon, \; n = 1, 2, \ldots, m.$$

The absence of instability of the iterative solution of IHCP with exact data suggests that a similar behaviour of the iterative process is also possible when using smoothed temperature information. The calculations carried out proved

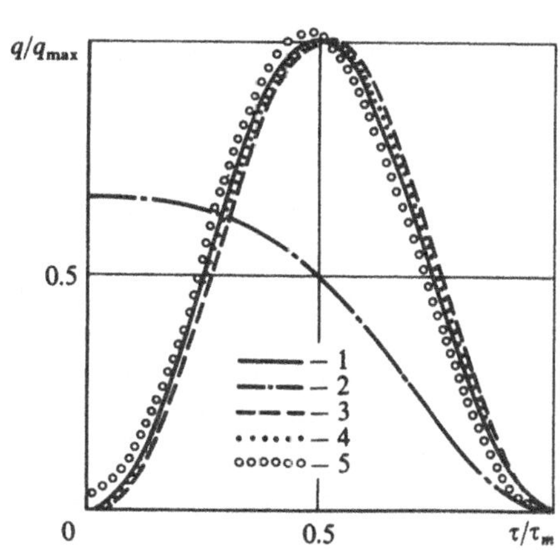

Fig. 6.5. Results of the heat flux recovery using the steepest descent and conjugate gradient methods for non-disturbed data, $b = d$, $\Delta Fo_b = 0.01$: 1 is a true solution $q_T = \frac{q_{T\max}}{2}\left\{\sin\left[\frac{2\pi}{\,}\left(\tau - \frac{\tau_m}{4}\right)\right] + 1\right\}$; 2, 3, 4 are the 1st, 9th and 50th approximations, respectively, using conjugate gradients; 5 is the 50th approximation using steepest descent.

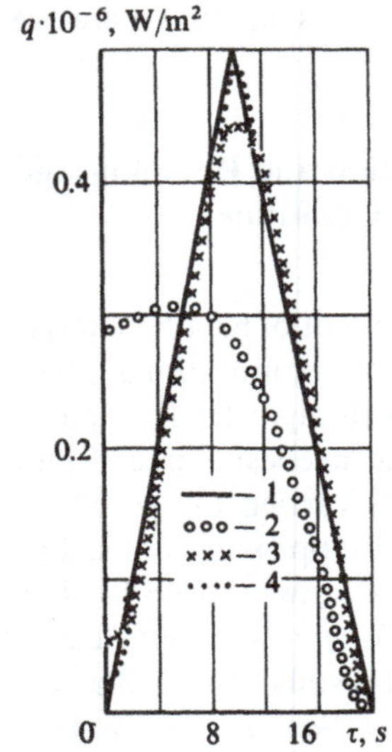

Fig. 6.6. Results of the heat flux recovery using conjugate gradients for non-disturbed data, $b = 0.003$m, $d = 0.002$m, $a = 0.35 \cdot 10^{-6} \frac{m^2}{s}$, $\lambda = 0.721$ KW/(m·K), $\Delta\tau = 0.4$ s:
1 is a true solution $q_T, \frac{W}{m^2} = \begin{cases} 0.5 \cdot 10^5 \tau, & \tau, s = [0, 10] \\ (1 - 0.05) \cdot 10^6 \tau, & \tau, s = [10, 20] \end{cases}$;
2, 3 are the 1st, 3rd and 100th approximations.

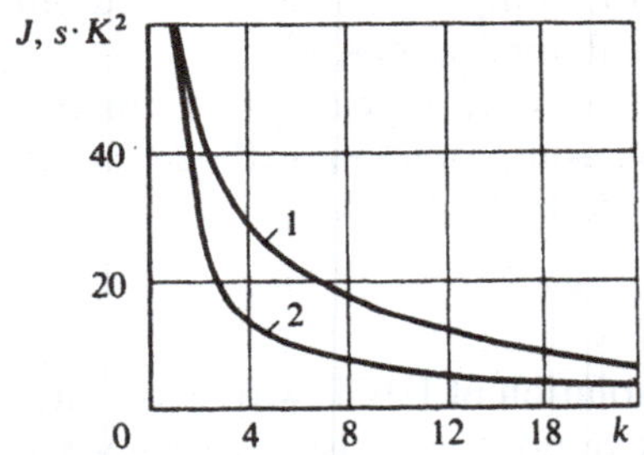

Fig. 6.7. An example of convergence of iterative successions on non-disturbed data: 1, the steepest descent method; 2, the conjugate gradient method

this suggestion. An algorithm uniformly approximating both the function and its derivative has been used in them as a smoothing algorithm [24].

An **important conclusion** follows: to solve ill-posed inverse problems one can use a traditional formula of gradient minimization, provided that the results of temperature measurements are *smoothed* in advance. In a number of cases such an approach to the treatment of heat test data appears to be not only quite suitable but the most reasonable method.

Finally, as expected, the sequence of approximate solutions, constructed using the conjugate gradients method, converged more quickly with the possibility of approaching the minimum of accepted criterion more closely than in the case of steepest descent. For one of the examples such a situation is shown on Fig. 6.7. The algorithm of IHCP solution based on conjugate gradients is slightly more complicated than the algorithm of steepest descent and has an advantage in the overall expenditures of computing time in obtaining the unknown solution with sufficiently high accuracy. Therefore this algorithm is recommended for practical application in the treatment of experimental data.

6.8.2 Stability with respect to Errors in Input Data and Selection of Approximate Solutions

The emergence of fluctuating errors as a result of temperature measurements violates the above described behaviour of the iterative process. Difference is also revealed in the properties of algorithms of parametric optimization from those with optimal control.

The rate of convergence of methods, in which the gradient is calculated by means of an adjoint problem, in the first iterations is comparable with the rate used in the case of exact assignment of input information. The course of the iterative process with errors in the input data could be tentatively divided into **two stages**: a stage needed to reveal and refine the basic structural features of the unknown function and a stage for the gradual development of oscillations in the solution. It is important to note that the first stage in the algorithms of optimal control for curves $q(\tau)$ of reasonable complexity has managed on the whole to end before the stage of problem instability was revealed. The course of solution refinement at the first stage depended only slightly upon the level and distribution of high-frequency component errors in the temperature data.

On Fig. 6.8(a), for sufficiently large fluctuating disturbances of input data ($T^*_{\delta n} = T^*_n + \delta_0 \omega_n$ where $\delta_0 = 10\% \ T^*_{\max}$; ω_n is a random value, distributed by uniform law at segment $[-1, 1]$), are shown the results for three first iterations obtained using the conjugate gradient method. The last approximation is already quite close to the unknown curve $q(\tau)$. If we continue the computational process still further, then during the following three iterations the results would vary relatively little. Starting from the 7th–8th iteration we can observe a gradual "build-up" of the solution (Fig. 6.8b depicts the 50th approximation). Approximately the same behaviour is observed in other cases as well.

The convergence of the parametric optimization algorithm turned out to be sufficiently less regular and depended upon the nature of the value of errors in the data of measurements. In both cases we introduced an admissible level of minimization, i.e. the number of iterations was conformed reasonably with the accuracy in the input data assignment so that the iterative process could be stopped somewhere on the joint of the first and second stages. As computations show, an oscillating nature of solutions develops gradually and a build-up of the amplitude of oscillations unfolds slowly. This contributes to a sufficiently accurate determination of the approximation to the unknown function.

An approximate analytical form (6.45) representing the solution of the direct heat conduction problem with a corresponding step selection of the discretization process $\Delta\tau$ enables a good quality of approximation of the initial continuous problem to be achieved. In total, with appreciable errors in temperature measurements, inequality $\delta_T \gg \delta_a$ is satisfied and one can demand to

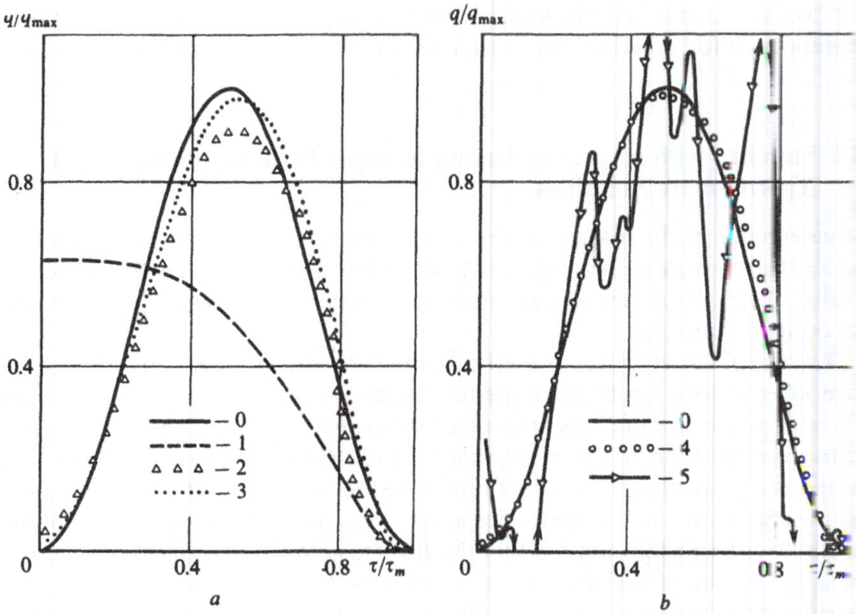

Fig. 6.8. The iterative solution of IHCP using the conjugate gradient method with disturbed data, $d = b$, $\Delta Fo_b = 0.1$: 0 is a true solution (see Fig. 6.5); 1, 2, 3 are the 1st, 2nd, and 3rd approximations, respectively; 4 is an approximation chosen by the residual criterion (the 4th iteration); 5 is the 50th approximation.

terminate the search by condition

$$J \simeq \delta^2 , \tag{6.52}$$

where $\delta = \delta_T$.

Suppose that as a result of the iterative process the condition $J(q^{\bar{k}-1}) = \delta^2$ has been satisfied at some step $k = \bar{k}$ with given accuracy. Then, function $q^{\bar{k}-1}$ can be taken for the approximation sought to a solution. If, however, on the two successive iterations $J(q^{\bar{k}-2}) > \delta^2$ and $J(q^{\bar{k}-1}) < \delta^2$ then $q^{\bar{k}-2}$ could appear as "oversmoothed" and $q^{\bar{k}-1}$ as insufficiently smooth approximations. In this case it is reasonable [11, 15] to correct the $q^{\bar{k}-2}$ by sampling in some way a new value of parameter β, for example, by substituting the optimal value $\beta_{\bar{k}-2}$ by value $\tilde{\beta}_{\bar{k}-2}$, defined from equality

$$\tilde{\beta}_{\bar{k}-2}\colon J(q^{\bar{k}-1}(\beta)) \simeq \delta^2 , \tag{6.53}$$

where

$$0 < \beta < \beta_{\bar{k}-2} .$$

Thus, at the last iteration we will receive the required approximation

$$q^{\bar{k}-1} = q^{\bar{k}-2} - \tilde{\beta}_{\bar{k}-2}\xi^{\bar{k}-2} . \tag{6.54}$$

We can show that condition (6.53) is satisfied by value

$$\tilde{\beta}_{\bar{k}-2} = \beta_{\bar{k}-2} - \sqrt{\beta^2_{\bar{k}-2} - \frac{J'(q^{\bar{k}-2}) - \delta^2}{\int_0^{\tau_m} \Delta T^2(\xi^{\bar{k}-2}, l, \tau)\,\mathrm{d}\tau}}. \tag{6.55}$$

The algorithm, realizing the solution of IHCP using the gradient method with a halt in the iterative process by the admissible level of the residual, is as follows ($k = 1, 2, \ldots, \bar{k}$, the initial approximation $q^0(\tau)$ is given):

1. We calculate the value of the criterion J at point q^k

   $$J^k = J(q^k).$$

2. We check the condition of termination of optimal search by the admissible level of minimization

   $$J^k \leqslant \delta^2. \tag{6.56}$$

   If condition (6.56) is satisfied, we perform a transition to item 7.
3. We solve an adjoint problem and find the gradient J'^k – eq. No. ?
4. In the conjugate gradient method we determine a conjugate direction $\xi^k = S^k$ by formula (6.26), while in the steepest descent method $\xi^l = J'^k$.
5. We calculate the value of step $\beta_k \geqslant 0$ by formula (6.51).
6. We determine the next approximation to the heat flux by formula (6.23) and perform a transition to item 1.
7. By formula (6.55) we calculate value $\tilde{\beta}_{\bar{k}-2}$.
8. We determine the final approximation to the heat flux as in (6.54).

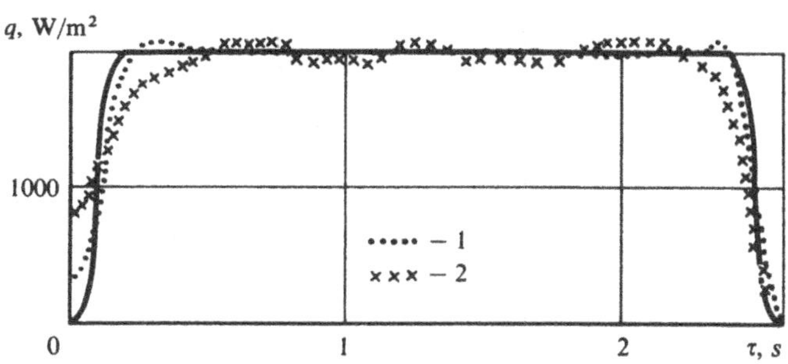

Fig. 6.9. Results of iterative solution of IHCP using the conjugate gradient method, $b = d = 0.015$ m, $a = 10^{-4}$ m²/s, $\lambda = 360$ W/(m·K), $\Delta\tau = 0.025$ s:

full curve is a true solution: $q_T = \begin{cases} 1000\{\sin[5\pi(\tau - 0.1)] + 1\}, \tau = [0, 0.2] \\ 2000, \tau = [0.2, 2.3] \\ 1000\{\sin[5\pi(\tau - 2.2)] + 1\}, \tau = [2.3, 2.5]; \end{cases}$

1 is the 100th approximation for exact data; 2 is the normal law of error distribution with a spread $3\sigma(\tau) = 0.02\ T^*$ max (the halt chosen by the residual criterion).

In Figs 6.8 and 6.9 are shown approximate solutions $q(\tau)$, obtained by means of the above described algorithm with a halt by condition (6.52). These results can be estimated as satisfactory.

6.8.3 The Absence of Uniform Convergence

In Sect. 6.6.1 mention was made of the absence of uniform convergence of solution using the method stated. Thus it is of interest to clarify how the "nulling" of gradient at $\tau = \tau_m$ would affect the solution of inverse problem. Note that in the parametric optimization at the expense of difference approximation of the criterion function the "last" component of gradient $\partial J / \partial q_m$ is not equal to zero but its value is continuously diminishing by a more exact approximation of the criterion function. In any case a reconstructed solution will deviate from a true one in some vicinity of the end point $\tau = \tau_m$, this resulting in a distortion of the solution.

Figure 6.10(a) depicts the results of calculations of the heat flux using the conjugate gradient method provided that a reconstructed function is not equal to zero at $\tau = \tau_m$ with initial condition $q^0(\tau) = 0$, $\tau \in [0, \tau_m]$. For another initial estimation, the results are shown on Fig. 6.10(b). It can be seen that a more accurate account of the behaviour of the unknown function under the assignment of initial approximation gives better results.

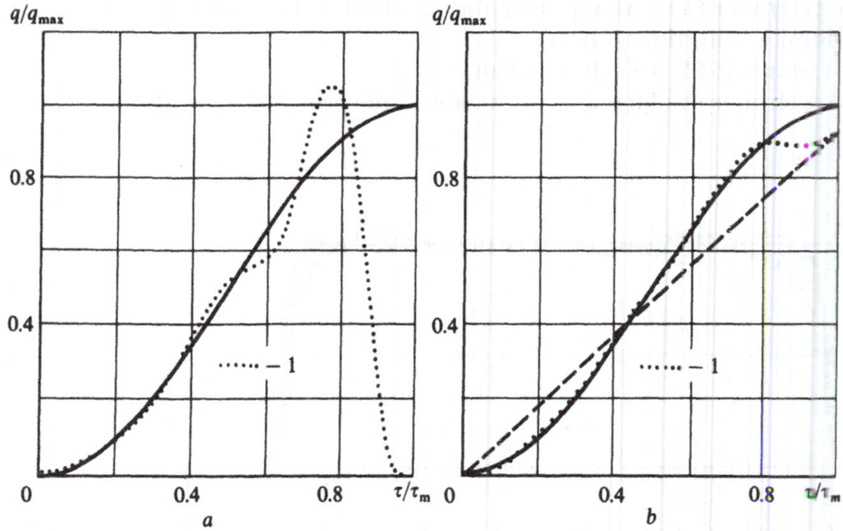

Fig. 6.10. A methodical example, $d = b$, $\Delta Fo_b = 0.1$ (the input data are non-disturbed; full curve is a true solution

$$q_T = \frac{q_{T\max}}{2}\left\{\sin\left[\frac{\pi}{\tau_m}\left(\tau - \frac{\tau_m}{2}\right)\right] + 1\right\}:$$

a – the initial approximation $q^0(\tau) = 0$; 1 is the 100th approximation; *b* – the initial approximation is dashed line, 1 is the 25th approximation.

In varying the initial approximation $q^0(\tau)$ in the class of constant values $q^0(\tau) = C_1 = \text{const}$, $0 \leqslant \tau \leqslant \tau_m$ or constant values of the derivative $dq^0(\tau)/d\tau = C_2 = \text{const}$ and $q^0(0) = 0$, $0 \leqslant \tau \leqslant \tau_m$ in the obtained set of solutions $q(\tau)$ there is an approximation closest to a true solution.

In selecting this approximation for the case of smooth input data it is possible to look at the value of minimized residual of the temperatures after 10–20 iterations: through a closer assignment of the initial approximation in the vicinity of the end point the value of residual is less. In particular, for a problem, the results of which solution are given on Fig. 6.10(a), where $q^0 = 0$, the residual $J^{10} \simeq 6$, while a closer initial estimate gives a lesser residual $J^{10} \simeq 2$ (see Fig. 6.10b).

6.9 Numerical Determination of Heat Loads Under Varying Thermophysical Properties of the Body

Methods available at present do not generally allow construction of an analytical solution of the boundary-value heat conduction problem counting the problem as being non-linear by thermophysical properties That is why, for practical realization of the gradient search of heat loads given that the properties of the body depend on temperature, it is necessary to use numerical methods in computing a temperature field. To solve an adjoint boundary-value problem and the problem for increments of temperature we shall apply a numerical approach as well.

Let us present the problems (6.27)–(6.30), (6.32)–(6.35) and (6.36)–(6.39) in a finite-difference form on the uniform space-time grid

$$\{x_l = l\Delta x,\ l = 0, 1, \ldots, L;\ x_0 = 0,\ x_L = b\},$$

$$\{\tau_n = n\Delta\tau,\ n = 0, 1, \ldots, m;\ \tau_0 = 0\}\,.$$

The corresponding algorithms being available, we can construct a procedure of iterative refinement for the initial estimate $p^0(\tau) = q^0(\tau)/\lambda_0$ by means of a gradient method. The approximation of problems will be performed on the implicit six-point pattern described in Sect. 5.1.

As a result, for the numerical determination of the model temperature $\theta(x, \tau)$, adjoint variable $\psi(x,\tau)$ and increment $\Delta\theta(x, \tau)$ in every time layer, we shall have corresponding three-diagonal systems of algebraic equations: non-linear for θ and linear for ψ and $\Delta\theta$. The solution of these systems is performed by a "run" method (in combination with iterations for coefficients in computing θ). Note that a numerical integration of adjoint equation (6.36) is performed in the direction from $\tau = \tau_m$ to $\tau = 0$. That is why an implicit scheme of approximation will be written in an "upside" form as well (as compared to approximation of a heat conduction equation).

In performing the computations we can preset thermal diffusivity $a(T)$ as a formula or a table. The use of model temperature $\theta(x, t)$ in the formulation of the problem as in Eqs. (6.27)–(6.30) suggests a transition from dependence $a(T)$ to dependence $a(\theta)$ which can be performed numerically or analytically. It is especially convenient when properties $\lambda(T)$ and $a(T)$ are given by simple formula enabling us to find an analytical expression for $a(\theta)$. For example, in many practical applications it is quite possible to approximate thermal conductivity $\lambda(T)$ by a linear dependence and thermal diffusivity $a(T)$ by a quadratic dependence:

$$\lambda = \lambda_0 + \lambda_1 T; \tag{6.57}$$

$$a = a_0 + a_1 T + a_2 T^2 . \tag{6.58}$$

In this case, assuming $T \geqslant 0$, it is not difficult to get the following expression for $a(\theta)$:

$$a(\theta) = C_1 + C_2\sqrt{1 + C_3\theta} + C_4\theta, \quad \lambda_1 \neq 0 , \tag{6.59}$$

where

$$C_1 = a_0 - \frac{\lambda_0}{\lambda_1}\left(a_1 - \frac{2a_2\lambda_0}{\lambda_1}\right); C_2 = \frac{\lambda_0}{\lambda_1}\left(a_1 - \frac{2a_2\lambda_0}{\lambda_1}\right); C_3 = \frac{2\lambda_1}{\lambda_0}; C_4 = \frac{2a_2\lambda_0}{\lambda_1} .$$

Besides, analytical notation $\lambda(T)$ usually allows us to avoid approximate integration when computing time series, corresponding to the known functions of experimental temperature $f(\tau)$ and initial condition $\xi(x)$:

$$f_n = \frac{1}{\lambda_0}\int_0^{T_n^*} \lambda(T)\,\mathrm{d}T, \quad n = 0, 1, \ldots, m;$$

$$\xi_l = \frac{1}{\lambda_0}\int_0^{T_{ol}} \lambda(T)\,\mathrm{d}T, \quad l = 0, 1, \ldots, L .$$

So, for dependence (6.57) we have

$$f_n = T_n^* + \frac{\lambda_1}{2\lambda_0}(T_n^*)^2, \quad n = 0, 1, \ldots, m;$$

$$\xi_l = T_{ol} + \frac{\lambda_1}{2\lambda_0} T_{ol}^2, \quad l = 0, 1, \ldots, L .$$

Coefficients in difference relationships, approximating an adjoint problem and also one for increments, are calculated using field $\theta^k(x_l, \tau_r)$ at every k-th iteration. Values $a_{ln}'^k \equiv \left.\frac{\mathrm{d}a}{\mathrm{d}\theta}\right|_{ln}^k$ are determined either from formula $a'(\theta)$ or by numerical differentiation of table function $a_j(\theta_j)$ with a pre-interpolation of initial values $a_j(\theta_j)$. Naturally, it is better, if possible, to use the first technique. In particular, for expression (6.59) we have $a_{ln}' = C_5/\sqrt{1 + C_3\theta_{ln}} + C_4, \quad \lambda \neq 0,$ $C_5 = a_1 - \dfrac{2a_2\lambda_0}{\lambda_1}$.

Having determined the field $\theta(x, \tau)$ at the last iteration, we can transfer to temperature T at any point of the body, including the temperature on the boundary. For this we use the formula of inverse transformation $T(x, \tau) = \lambda_0 \int_0^{\theta(x,\tau)} d\theta/\lambda(\theta)$.

For a heat conduction coefficient of the type seen in expression (6.57) the required relation is easily obtained from formula

$$T(x, \tau) = \frac{\lambda_0}{\lambda_1}\left(\sqrt{1 + \frac{2\lambda_1}{\lambda_0}\theta(x, \tau)} - 1\right).$$

Turn now to the analysis of the results of numerical simulation. Consider the following two examples, corresponding to different thermophysical properties of the body ($\lambda[\mathrm{W/(m \cdot K)}]$, $a[\mathrm{m^2/s}]$):

1. $\lambda = 0.721$; $a = 0.4 \cdot 10^{-6} - 0.143 \cdot 10^{-9} T + 0.408 \cdot 10^{-12} T^2$;
2. $\lambda = 0.721 + 0.288 \cdot 10^{-3} T$; $a = 0.4 \cdot 10^{-6} - 0.143 \cdot 10^{-9} T + 0.408 \cdot 10^{-12} T^2$.

As parameter β_k, characterizing the depth of descent in these examples a linear estimate has been taken [see formula (6.51)]. The numerical solution of the problems below given has been performed on the difference grids with 50 nodes in space and 50 in time. Value $q^0 = 0$ has been taken for the initial estimate. To save computing time, the computations both for direct and for inverse problems were performed without iterations for thermophysical coefficients.

Let us begin with the case of undisturbed input data. These results are given in Fig. 6.11 (example 1) and Figs 6.12 and 6.13 (example 2). The curve 3 (Fig. 6.11a) was obtained from a solution of IHCP when changing the actual thermophysical properties from example 2 by constant values $\lambda = 0.721$ W(m·K), $a = 0.4 \cdot 10^{-6}$ M²/S. It is seen that a neglect of the change of properties could lead to considerable distortions of the unknown boundary condition.

From the results of the calculations we can come to conclusions coinciding on the whole with the results obtained for a semianalytical methodology of iterative determination of heat loads (see Sect. 6.8).

First, both gradient methods – steepest descent and conjugate gradients – can be used for data processing of experimental investigations. However, the conjugate gradient method, being more complicated in comparison with the steepest descent method, has given better results both by the rate of convergence and by the accuracy of boundary condition reconstruction.

Second, for non-disturbed data there have been no signs of instability of IHCP iterative solution, that allowing to use standard techniques of defining a required approximation to the unknown function.

On Fig. 6.14 (example 1) are given the results of reconstruction of the heat flux obtained with the use of disturbed input data:

$$T_\delta^*(\tau_n) = T^*(\tau_n) + \delta_0 \omega_n / 3\,,$$

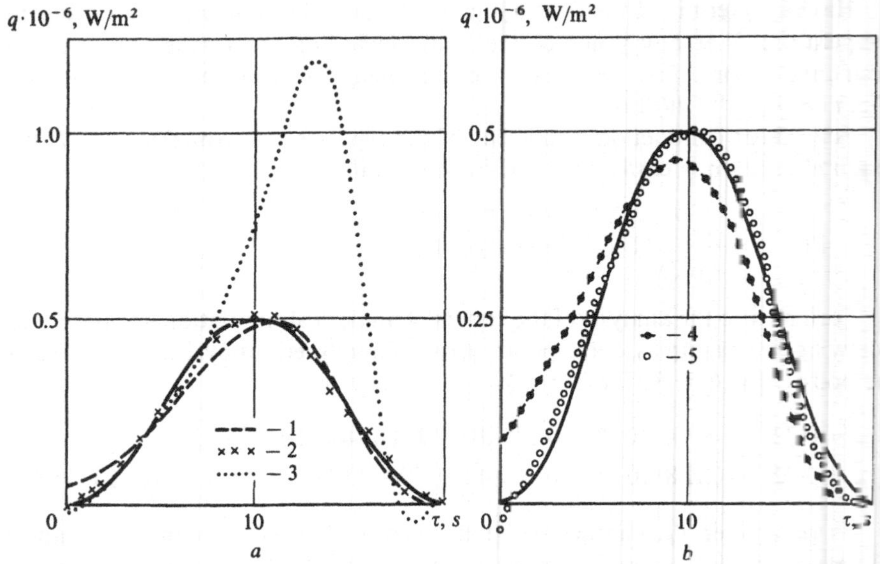

Fig. 6.11. a, b. Examples on reconstruction of the heat flux with undisturbed input data $T^*(\tau) = f(\tau)$, $q^*(\tau) = 0$ and initial approximation $q^0(\tau) = 0$ (for $b = 0.002$ m (**a**) and $b = 0.004$ m (**b**)); full curve is a true solution – see Fig. 6.5: 1 is a calculation using the steepest descent method with varying thermo-physical properties (50th iteration); 2 is a calculation using the conjugate gradient method (50th iteration); 3 is a calculation using the conjugate gradient method on the assumption that $a(T) = \text{const}$ (50th iteration); 4, 5 are the calculations using the conjugate gradient method, 3rd and 20th iterations, respectively.

where $\delta_0 = 10\%\ T^*_{\max}$ is the maximum error possible and ω_z is a random variable, distributed by normal law (Gaussian random variable) with a unit-equal variance and a zero mean.

It is seen that as the number of iterations increases, the computed values of $q(\tau)$ could greatly differ from the unknown values. At the same time the choice of approximation (number of iteration $k = \bar{k}$) by the admissible level of residual [see equality (6.8)] allows a suitable, sufficiently accurate, solution to be obtained. It should be added that allowing for $\delta_a \neq 0$ in condition (6.9) (this error was determined in a practical way) we could gain better results as compared with the use of value δ_T as a total error (see curves 2 and 3 on Fig. 6.14).

Another possibility of reconstructing function $q(\tau)$ is not directly connected with the principle of admissible level of residual and consists in a *pre-smoothing* of temperature data. The corresponding results obtained with the help of regularized procedure of smoothing is given in Fig. 6.13. Note that such an approach to determination of heat loads in a whole number of cases appears to be more reasonable. As calculations showed a pre-smoothing of input information excludes emergence of indications to instability in the solution of IHCP when increasing the number of iterations.

Thus we can terminate the iterative process by a common condition of proximity of the previous and present approximations. In the result we exclude

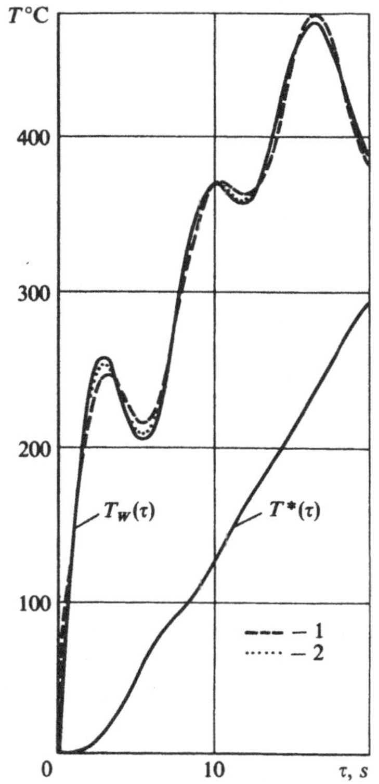

Fig. 6.12. Results on temperature recovery of the body surface, $b = 0.003$ m, full curves are exact functions of the input and reconstructed temperature; they are obtained through a numerical solution of the direct problem for the function given $q_T(\tau)$, as shown in Fig. 6.13: *1* and *2* are the 30th and 60th approximations, respectively, for undisturbed data using the conjugate gradient method.

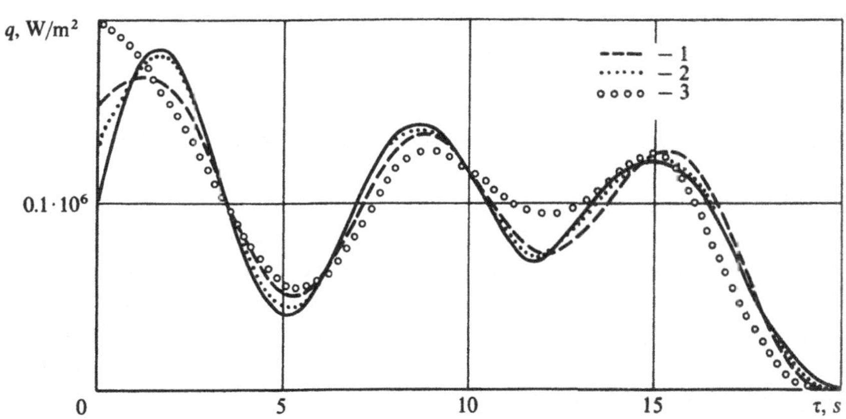

Fig. 6.13. Results on reconstruction of heat flux, $b = 0.003$ m; full curve is a true solution (it was assumed tabulated): *1* and *2* are the 30th and 60th approximations, respectively, for undisturbed data using the conjugate gradient method; *3* is the 60th approximation on smoothed data (initial values of temperature included errors distributed by the normal law with an amplitude $3\sigma = 0.05\ T^*$ max).

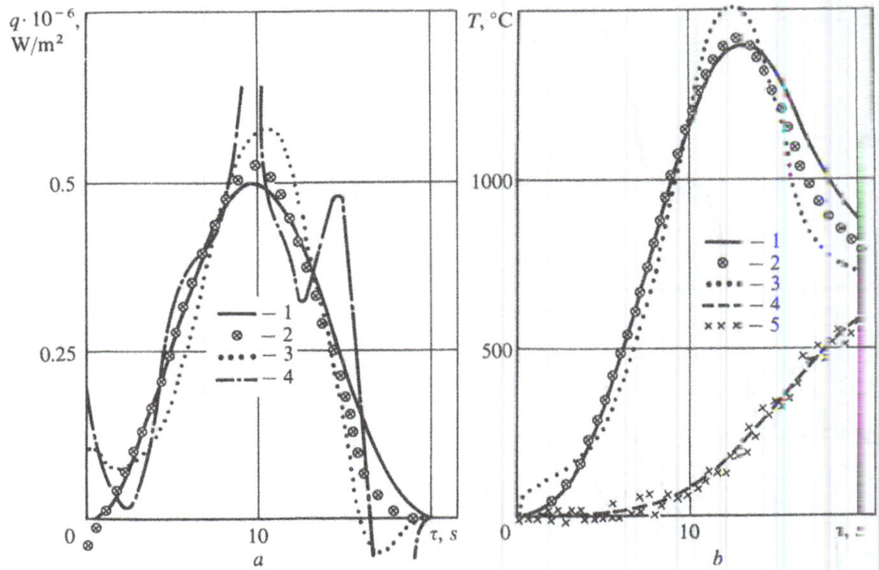

Fig. 6.14. Results on reconstruction of boundary conditions using the conjugate gradient method for disturbed input temperature $T^*(\tau) = f_\delta(\tau)$ and $q^*(\tau) = 0$, $q^0 = 0$, $b = 0.004$ m; *1* are the unknown functions $\left(q_T = \frac{q_{T\max}}{2}\left\{\sin\left[\frac{2\pi}{\tau_m}\left(\tau - \frac{\tau_m}{4}\right)\right] + 1\right\}\right)$, $T_W(\tau)$ was found through a numerical solution of the direct problem for this function $q_T(\tau)$): *2* is an approximation, chosen by the condition (6.8), the 9th iteration; *3* is an approximation, corresponding to a condition $\delta = \delta_T$, the 20th iteration; *4a* is the 30th iteration; *4b* are the exact input temperatures; *5b* are the disturbed input temperatures.

two somewhat indefinite moments. The first is connected with the estimate of error δ_a, the second with the ambiguity of approximation to the unknown function, corresponding to the given level of residual.

6.10 Solution of a Non-Linear Inverse Problem in Statement II

So far the methodical developments and the results of numerical simulation concerned the solution of IHCP in statements I and II when functional (6.2) has been taken as the criterion of optimality. An inverse problem in statement II has been also formulated in Sect. 6.1, but it presupposed the use of criterion (6.6). Transfer now to a consideration of this problem on the basis of principles of functional optimization. Let the temperature field in a form transformed by Kirchoff answer the following conditions:

$$\frac{\partial \theta}{\partial \tau} = a(\theta)\frac{\partial^2 \theta}{\partial x^2}, \quad 0 < x < b,\ 0 < \tau \leqslant \tau_m; \qquad \theta(x, 0) = \xi(x), \quad 0 \leqslant x \leqslant b; \tag{6.60}$$

$$\frac{\partial \theta(0, \tau)}{\partial x} = p(\tau), \quad 0 \leqslant \tau < \tau_m; \qquad \frac{\partial \theta(b, \tau)}{\partial x} = z(\tau), \quad 0 \leqslant \tau \leqslant \tau_m.$$

Function $p(\tau)$ is a variable characteristic selected by the value of the integral error

$$J(p) = \int_0^{\tau_m} \int_d^b [\theta(p(\tau), x, \tau) - f(x, \tau)]^2 \, dx \, d\tau \,, \tag{6.61}$$

where $f(x, \tau) = \frac{1}{\lambda_0} \int_0^{T^*(x,\tau)} \lambda(T) \, dT$; $T^*(x, \tau)$ is the known temperature, preset in domain $0 < d \leqslant x \leqslant b, 0 \leqslant \tau \leqslant \tau_m$.

Suppose that function $p(\tau)$ has received an increment $\Delta p(\tau)$. The response reaction of the body, herein, as its temperature history, will be defined from the solution of the problem

$$\begin{aligned} &\frac{\partial \Delta\theta}{\partial \tau} = a(\theta(p)) \frac{\partial^2 \Delta\theta}{\partial x^2} + D(\theta(p))\Delta\theta, \quad 0 < x < b, \; 0 < \tau \leqslant \tau_m ; \\ &\Delta\theta(x, 0) = 0 ; \\ &\frac{\partial \Delta\theta(b, \tau)}{\partial x} = 0; \quad \frac{\partial \Delta\theta(0, \tau)}{\partial x} = \Delta p(\tau) , \end{aligned} \tag{6.62}$$

where $D(\theta(p)) = \frac{da}{d\theta} \frac{\partial^2 \theta(p)}{\partial x^2}$.

The functional (6.61) will get a corresponding increment. The linear part of this increment is

$$\Delta J \simeq 2 \int_0^{\tau_m} \int_d^b [\theta(p(\tau), x, \tau) - f(x, \tau)] \Delta\theta(x, \tau) \, dx \, d\tau \,. \tag{6.63}$$

Express the gradient of the functional via the solution of an adjoint problem:

$$\begin{aligned} &-\frac{\partial \psi}{\partial \tau} = \frac{\partial^2}{\partial x^2}(a(\theta(p))\psi) + D(\theta(p))\psi + F, \quad 0 < x < b, 0 < \tau < \tau_m ; \\ &\psi(x, \tau_m) = 0 ; \\ &\frac{\partial}{\partial x}(a(\theta(p))\psi(0, \tau)) = 0; \quad \frac{\partial}{\partial x}(a(\theta(p))\psi(b, \tau)) = 0 , \end{aligned} \tag{6.64}$$

$$\text{where} \quad \left. \begin{aligned} &F = 2[\theta(p(\tau), x, \tau) - f(x, \tau)], d < x < b, 0 < \tau \leqslant \tau_m , \\ &F = 0, \; 0 < x < d, 0 < \tau \leqslant \tau_m . \end{aligned} \right\} \tag{6.65}$$

With this purpose consider the value

$$\begin{aligned} \rho = \int_0^{\tau_m} d\tau \int_0^b \Bigg\{ &\psi \left(a \frac{\partial^2 \Delta\theta}{\partial x^2} + D\Delta\theta - \frac{\partial \Delta\theta}{\partial \tau} \right) \\ &- \Delta\theta \left[\frac{\partial^2 (a\psi)}{\partial x^2} + D\psi + F + \frac{\partial \psi}{\partial \tau} \right] \Bigg\} dx \,. \end{aligned}$$

Evidently, $\rho = 0$. Note also that

$$\psi\left(a\frac{\partial^2 \Delta\theta}{\partial x^2} + D\Delta\theta\right) - \Delta\theta\left[\frac{\partial^2}{\partial x^2}(a\psi) + DF + F\right]$$

$$= \frac{\partial}{\partial x}\left[a\psi\frac{\partial \Delta\theta}{\partial x} - \Delta\theta\frac{\partial(a\psi)}{\partial x}\right] - \Delta\theta F\,,$$

$$-\psi\frac{\partial \Delta\theta}{\partial \tau} - \Delta\theta\frac{\partial \psi}{\partial \tau} \equiv \frac{\partial}{\partial x}(-\psi\Delta\theta)\,.$$

Consequently, $$\rho = \int\limits_0^{\tau_m} \mathrm{d}\tau \int\limits_0^b \left\{\frac{\partial}{\partial x}\left[a\psi\frac{\partial \Delta\theta}{\partial x} - \Delta\theta\frac{\partial(a\psi)}{\partial x}\right] - \frac{\partial(\psi\Delta\theta)}{\partial \tau} - F\Delta\theta\right\}\mathrm{d}x$$

$$= \int\limits_0^{\tau_m}\left[a\psi\frac{\partial \Delta\theta}{\partial x} - \Delta\theta\frac{\partial(a\psi)}{\partial x}\right]_0^b \mathrm{d}\tau - \int\limits_0^b (\psi\Delta\theta)\Big|_0^{\tau_m}\mathrm{d}x$$

$$- \int\limits_0^{\tau_m}\int\limits_0^b F\Delta\theta\,\mathrm{d}x\,\mathrm{d}\tau\,.$$

Substituting the limits with initial and boundary conditions of problems (6.62) and (6.64), we obtain

$$\rho = -\int\limits_0^{\tau_m} \Delta p(a\psi(0,\tau))\,\mathrm{d}\tau - \int\limits_0^{\tau_m}\int\limits_0^b \Delta\theta F\,\mathrm{d}x\,\mathrm{d}\tau\,.$$

Since $\rho = 0$ and F is defined by expression (6.65), we have

$$-2\int\limits_0^{\tau_m}\int\limits_0^b (\theta - f)\Delta\theta\,\mathrm{d}x\,\mathrm{d}\tau = \int\limits_0^{\tau_m} \Delta p a\psi\,\mathrm{d}\tau\,.$$

Or, considering equality (6.63), $\Delta J = \int\limits_0^{\tau_m} \Delta p(-a\psi)\,\mathrm{d}\tau$. Hence, the gradient of functional (6.61) is $J'(p) = -a\psi(0,\tau)$.

Having determined $J'(p)$ and taken into account the fact that the algorithm of searching for $p(\tau)$ by using the criterion functional gradient has been explained earlier, we have completed the solution of a non-linear IHCP in statement II. The approximation of problems (6.60), (6.62) and (6.64) for algorithm numerical realization is performed by means of a difference method. In the case of constant thermophysical properties an iterative solution of the given problem is noticeably simplified at the expense of integral forms of solutions of boundary-value problems.

Note. Generally a sufficiently exact specification of the admissible level of residual minimization [see (6.61)] in the presence of errors in the input data is a complex problem. That is why the use of functional (6.61) in the extremal IHCP statement assumes pre-smoothing of the experimental information.

6.11 The Iteration Technique of Determining Non-Stationary Heat Loads in the Two-Dimensional Case

The gradient methods of solving inverse problems in parametric and functional statements already treated in this chapter assume natural generalization for a multi-dimensional case. Let us show this by the example of functional optimization for the problem with two space coordinates x and y, i.e. when the vector of heat flux at every point of the space is parallel to plane xy and does not vary in z. Suppose that the body has a rectangular section and there is no heat exchange on its bounding planes $x = 0$, $x = b$ and $y = 0$ (Fig. 6.15). It is required to find a heat flux $q(x, \tau)$ at the boundary $y = c$ from the condition of admissible minimum of the residual functional:

$$J(p) = \int_0^{\tau_m} d\tau \int_0^b [T(x, 0, \tau) - f(x, \tau)]^2 \, dx, \tag{6.66}$$

where $f(x, \tau)$ is the assigned function, corresponding to temperature measurements at the boundary $y = 0$. Temperature $T(x, 0, \tau)$ is determined from a solution of the direct heat conduction problem with a preset control $p(x, \tau) = -q(x, \tau)/\lambda$:

$$\frac{\partial T}{\partial \tau} = a\left(\frac{\partial^2 T}{\partial x^2} + \frac{\partial^2 T}{\partial y^2}\right), \quad 0 < x < b,\ 0 < y < c,\ 0 < \tau \leqslant \tau_m;$$

$$T(x, y, 0) = \xi(x, y), \quad 0 \leqslant x \leqslant b, \quad 0 \leqslant y \leqslant c;$$

$$\frac{\partial T(b, y, \tau)}{\partial x} = \frac{\partial T(0, y, \tau)}{\partial x} = \frac{\partial T(x, 0, \tau)}{\partial y} = 0, \quad 0 \leqslant \tau \leqslant \tau_m;$$

$$\frac{\partial T(x, c, \tau)}{\partial y} = p(x, \tau), \quad 0 \leqslant \tau \leqslant \tau_m.$$

Let $p(x, \tau)$ and $p(x, \tau) + \Delta p(x, \tau)$ be two admissible controls; $T(x, y, \tau)$ and $T(x, y, \tau) + \Delta T(x, y, \tau)$ their corresponding temperature fields. Then, the increment of functional (6.66) $\Delta J(p) = J(p + \Delta p) - J(p)$ will have a linear part

$$\Delta J(p) \simeq 2 \int_0^{\tau_m} d\tau \int_0^b \Delta T(x, 0, \tau)[T(x, 0, \tau) - f(x, \tau)] \, dx.$$

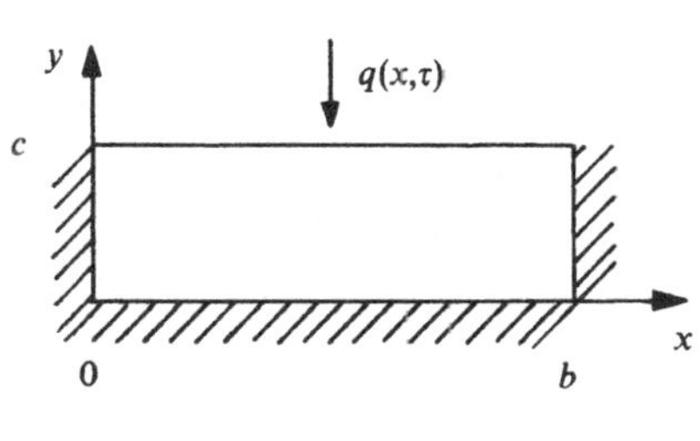

Fig. 6.15. The domain of space coordinates

The function-increment $\Delta T(x, 0, \tau)$ satisfies the conditions of the following boundary-value problem:

$$\frac{\partial \Delta T}{\partial \tau} = a\left(\frac{\partial^2 \Delta T}{\partial x^2} + \frac{\partial^2 \Delta T}{\partial y^2}\right), \quad 0 < x < b,\ 0 < y < c,\ 0 < \tau \leqslant \tau_m; \tag{6.67}$$

$$\Delta T(x, y, 0) = 0; \tag{6.68}$$

$$\frac{\partial \Delta T(b, y, \tau)}{\partial x} = \frac{\partial \Delta T(0, y, \tau)}{\partial x} = \frac{\partial \Delta T(x, 0, \tau)}{\partial y} = 0; \tag{6.69}$$

$$\frac{\partial \Delta T(x, c, \tau)}{\partial y} = \Delta p(x, \tau). \tag{6.70}$$

Then write the *adjoint problem*

$$\frac{\partial \psi}{\partial \tau} = -a\left(\frac{\partial^2 \psi}{\partial x^2} + \frac{\partial^2 \psi}{\partial y^2}\right), \quad 0 < x < b,\ 0 < y < c,\ 0 < \tau < \tau_m; \tag{6.71}$$

$$\psi(x, y, \tau_m) = 0; \tag{6.72}$$

$$\frac{\partial \psi(x, c, \tau)}{\partial y} = \frac{\partial \psi(b, y, \tau)}{\partial x} = \frac{\partial \psi(0, y, \tau)}{\partial x} = 0; \tag{6.73}$$

$$\frac{\partial \psi(x, 0, \tau)}{\partial y} = 2[T(x, 0, \tau) - f(x, \tau)]. \tag{6.74}$$

Consider the value

$$\rho = \int_0^{\tau_m} d\tau \int_0^b dx \int_0^c \{\psi[L(\Delta T) - \partial \Delta T/\partial \tau] - \Delta T[L\psi + \partial \psi/\partial \tau]\}\, dy,$$

where operator $L = a(\partial^2/\partial x^2 + \partial^2/\partial y^2)$. From Eqs. (6.67) and (6.71) we obtain $\rho = 0$.

Evidently we have the following equations:

$$\psi L(\Delta T) - \Delta T L\psi = a\frac{\partial}{\partial x}\left(\psi\frac{\partial \Delta T}{\partial x} - \Delta T\frac{\partial \psi}{\partial x}\right) + a\frac{\partial}{\partial y}\left(\psi\frac{\partial \Delta T}{\partial y} - \Delta T\frac{\partial \psi}{\partial y}\right),$$

$$-\psi\frac{\partial \Delta T}{\partial \tau} - \Delta T\frac{\partial \psi}{\partial \tau} = -\frac{\partial}{\partial \tau}(\psi \Delta T).$$

Introduce designations

$$P(x, y, \tau) = a\psi\frac{\partial \Delta T}{\partial x} - a\Delta T\frac{\partial \psi}{\partial x}; \quad Q(x, y, \tau) = a\psi\frac{\partial \Delta T}{\partial y} - a\Delta T\frac{\partial \psi}{\partial y};$$

$$R(x, y, \tau) = -\psi \Delta T.$$

Functions P, Q and R are continuous in parallelepiped $D = G \times [0 \leqslant \tau \leqslant \tau_m]$ (G is the domain of space variables x, y) and have continuous derivatives of the first order with respect to x, y and τ.

In view of the new designations, value ρ is written as

$$\rho = \int_0^{\tau_m} \mathrm{d}\tau \int_0^b \mathrm{d}x \int_0^c \left(\frac{\partial P}{\partial x} + \frac{\partial Q}{\partial y} + \frac{\partial R}{\partial \tau}\right) \mathrm{d}y. \tag{6.75}$$

Surface ∂D, bounding domain D, is piecewise-smooth. Direction of the external normal n on each face is determined by the cosines $\cos(n, x)$, $\cos(n, y)$ and $\cos(n, \tau)$, out of which one is equal to 1 while the other two to zero, depending on the side being considered (Fig. 6.16).

Applying the Ostrogradsky formula, proceed from the volume integral in expression (6.75) to the surface one

$$\rho = \iint_{\partial D} [P \cos(n, x) + Q \cos(n, y) + R \cos(n, \tau)] \, \mathrm{d}s$$

$$= \int_0^{\tau_m} \mathrm{d}\tau \int_0^c P \, \mathrm{d}y \Big|_{x=0}^{x=b} + \int_0^{\tau_m} \mathrm{d}\tau \int_0^b Q \, \mathrm{d}x \Big|_{y=0}^{y=c} + \int_0^b \mathrm{d}x \int_0^c R \, \mathrm{d}y \Big|_{\tau=0}^{\tau=\tau_m}.$$

Returning to the former designations and taking into account boundary conditions (6.68)–(6.70) and (6.72)–(6.74), we obtain

$$\rho = a \int_0^{\tau_m} \mathrm{d}\tau \int_0^b \{2\Delta T(x, 0, \tau) [T(x, 0, \tau) - f(x, \tau)] + \psi(x, c, \tau) \Delta p(x, \tau)\},$$

or $\rho = a\Delta J + a \int_0^{\tau_m} \mathrm{d}\tau \int_0^b \psi(x, c, \tau) \Delta p(x, \tau) \, \mathrm{d}x$.

Since $\rho = 0$, then

$$\Delta J = -\int_0^{\tau_m} \mathrm{d}\tau \int_0^b \psi(x, c, \tau) \Delta p(x, \tau) \, \mathrm{d}x = \int_0^{\tau_m} \mathrm{d}\tau \int_0^b J' \Delta p \, \mathrm{d}x$$

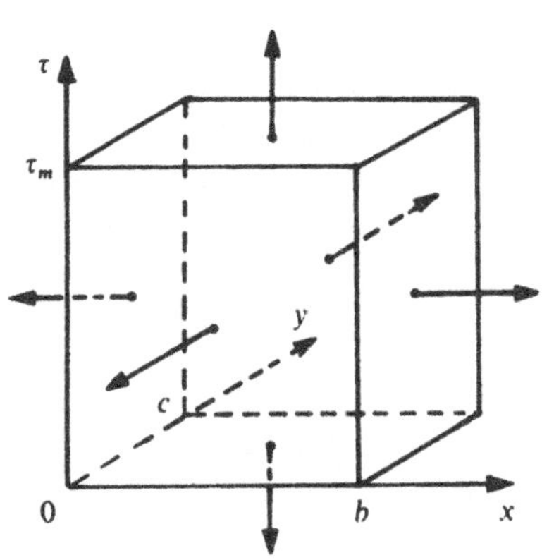

Fig. 6.16. A disposition of external normals

So, the gradient of functional (6.66) will be equal to

$$J' = -\psi(x, c, \tau).$$

Hence, the question of the effective use of gradient methods, that is the determination of the gradient through a solution of the adjoint problem is solved.

A general scheme of IHCP solution within a two-dimensional statement remains the same as in the case of one spatial variable. To calculate functions $T(x, 0, \tau)$, $\psi(x, c, \tau)$ and $\Delta T(x, 0, \tau)$ it is convenient to use an approximate analytical method to solve a two-dimensional heat conduction problem stated in Sect. 4.5. The adjoint problem is reduced thereby to a statement with "inverse" time by changing the variable $t = \tau_m - \tau$. The choice of iterative step (the depth of descent) β in every iteration should be performed by condition $\partial J(p^{k+1}(x, \tau))/\partial\beta = 0$, where $p^{k+1}(x, \tau) = p^k(x, \tau) - \beta\xi^k(x, \tau)$; $\xi^k(x, \tau)$ is the direction of descent.

It is not complicated to show that value β is given by formula

$$\beta_k = \frac{\int\limits_0^{\tau_m} d\tau \int\limits_0^b [T^k(x, 0, \tau) - f(x, \tau)] \Delta T(\xi^k(x, \tau), x, \tau) \, dx}{\int\limits_0^{\tau_m} d\tau \int\limits_0^b \Delta T^2(\xi^k(x, \tau), x, \tau) \, dx}.$$

Without any principal changes a derivation of the formula for the gradient may be extended onto other problem statements, considered for domains with a different geometry, for instance, spherical or cylindrical.

By functional optimization with the use of the adjoint problem for computing the gradient of an integral residual we can also solve other types of inverse problems, namely, coefficient IHCPs and inverse problems for engineering systems.

6.12 Summary

1. An iterative method is suggested for solving inverse problems of heat conduction in which a regularization is performed by matching the number of iterations with the value of error in the initial data. The extremal statements of boundary inverse problems are formulated for one-dimensional and two-dimensional equations of heat conduction and regularized algorithms are developed to solve these problems using the steepest descent and conjugate gradient methods.
2. Regularization and accuracy characteristics (of parametric and functional forms of the given method) have been studied using computational experiment when applied to a solution of linear and non-linear boundary IHCP. It is shown that the method demonstrated allows effective algorithms to solve such problems to be constructed.

The material considered in this chapter can be taken as an introduction to an iterative regularization of inverse problem solution. More detailed and rigorous information together with iterative algorithms for a greater variety of problems can be found in Chap. 8. But before this we shall need to study Chap. 7, where a variational regularization principle will be used for IHCP solution. Further consideration of this approach, including the construction of effective computational procedures in determining heat boundary conditions, will also promote a deeper understanding of another – that of iterative regularization.

Chapter 7

Regularization of Variational Forms of Inverse Heat Conduction Problems

Introduction

As was pointed out in Chap. 2, one of the most general methods used in ill-posed problem solving is the Tikhonov regularization method. At present it has been developed into a broad mathematical direction. This has resulted in a rigorous branched theory which covers a whole number of regular methods within general Tikhonov concept of regularization. The most complete results have been obtained for the major form of this method which is known as *variational principle of regularization*. This principle with its subsequent algorithms has been quite widely applied in practice for solving various inverse problems, including IHCP. The first study in this field was made by Tikhonov and Glasko (see Ref. * on p. 7), where the authors considered a solution of boundary IHCP on determination of the surface temperature of a semi-infinite body. Next followed the studies of Alifanov [4, 7, 17], in which the author elaborated and investigated regularizing algorithms for solving various linear IHCP, mainly for one-dimensional formulations. The expansion of a regularization method onto a solution of non-linear boundary IHCP has been attempted by Alifanov [10, 12] and Alifanov and Artyukhin [21, 22].

Among studies by other scholars there are Beck and Murio [164], in which an algorithm of function specification is regularized. These questions have been also considered in [162].

Methods of solving inverse heat condition problems are studied in this chapter based on the variational regularization principle. The plan of the chapter is as follows.

In Sect. 7.1 one of the most easy-to-grasp interpretations of the variational regularization method obtained by using Lagrange multipliers is considered for the integral form of one-dimensional IHCP. To find extremals of a regularizing functional an appropriate Euler equation is obtained in Sect. 7.2. Section 7.3 is devoted to a construction of simple and economic algorithm of IHCP regularized solution based on a preliminary approximation of the heat conduction integral operator using the scheme stated in Sect. 4.1, with a successive regularization of this algebraic form. Section 7.4 contains a description of the main techniques on selection of *a regularization parameter*; this is a central question as to its correct use for practical purposes. An effective technique is also described here on automatization of parameter selection. The results of computational experiments on the study of accuracy of these algorithms are given in Sect. 7.5.

Sections 7.6 and 7.7 are devoted to an approximate regularized solution of the non-linear boundary IHCP, where a *by-interval regularization scheme and a regularization of the finite-difference continuation of temperature field* are considered. Finally, a regularizing scheme is constructed for solving a two-dimensional IHCP (Sect. 7.8).

7.1 The Regularized Form of Inverse Problems

Let us consider the IHCP with constant coefficients, although the results given below are proved to be extended for a more general case formulated earlier in Sect. 2.6.4. Using the residual principle elaborated for solving operator equations (see [103, 104]) consider a possible interpretation of the regularization method for solution of inverse heat conduction problems (see also [4, 7]).

As was shown in Chap. 3, the integral forms of IHCP lead to an ill-posed operation of solving the integral equation of the first kind in order to determine the unknown boundary condition (e.g. heat flux, temperature of the surface) or some auxiliary function (in particular, thermal potential density) the latter enabling the unknown quantity to be calculated easily. In the case of one-dimensional problems this equation assumes a general form

$$\int_0^\tau u(\xi)K(\tau,\xi)\,\mathrm{d}\xi = f(\tau), \quad 0<\tau\leqslant\tau_m, \tag{7.1}$$

where $u(\tau)$ is the unknown function.

The kernel $K(\tau, \xi)$ of integral equation and free member $f(\tau)$ are defined by a concrete formulation of the problem.

Rewrite equation (7.1) in operator form

$$Au = f; \quad u\in U; \; f\in F, \tag{7.2}$$

where A is a continuous linear operator determined by kernel $K(\tau, \xi)$; U and F are Hilbert spaces.

Assume that the right-hand side of Eq. (7.2) is given with known error δ for the exact data f_T, *namely* $f = f_\delta$, $\| f_T - f_\delta \|_F \leqslant \delta$. That is why instead of Eq. (7.2) consider rather an inequality for the residual

$$\| Au - f_\delta \|_F \leqslant \delta\,. \tag{7.3}$$

The inequality defines all elements u corresponding to deviation of the left-hand side of Eq. (7.2) from its right-hand side within the error limit δ. This condition forms a set of functions D_δ, which also contains, in particular, an unknown solution $u_T(\tau)$, corresponding to the exact right-hand side f_T.

The set D_δ can be called a set of "suspected" solutions of Eq. (7.2). Look for an element $u_\delta \in D_\delta$ which to the minimum deviates from some given element

$u^* \in U$ (u^* may not belong to D_δ):

$$\min_{u \in D_\delta} \| u - u^* \|_U . \tag{7.4}$$

Thus function $u_\delta(\tau)$ can be taken as an approximate solution of the initial equation (7.2).

Define concretely the problem given by conditions (7.3) and (7.4) introducing the Sobolev space $W_2^k[0, \tau_m]$ of generally differentiable functions as a class of admissible solutions. Assume the right-hand side f belongs to $L_2[0, \tau_m]$ space of square integrable functions. The choice $U = W_2^k$ is associated with the fact that an unknown solution has usually a definite smoothness.

For spaces demonstrated the definition of an approximate solution $u_\delta(\tau)$ will be connected with the following *problem on conditional extremum* find function $u_\delta(\tau)$, satisfying the conditions:

$$\| u_\delta - u^* \|_{W_2^k} = \min_{u \in D_\delta} \| u - u^* \|_{W_2^k} ; \tag{7.5}$$

$$D_\delta = \{u : \| Au - f_\delta \|_{L_2} \leqslant \delta\} , \tag{7.6}$$

where an inequality $\| f_T - f_\delta \|_{L_2} \leqslant \delta = \delta_{L_2}$ is valid for the input data. Operator A is continuous and linear and, therefore, the problem formulated is solvable *in a single way*, and a sequence of approximate solutions with decreasing $\delta \to 0$ would converge *uniformly* to an exact solution with all derivatives up to $(k-1)$ order, inclusive (see [103]).

Thus, the initial ill-posed problem, as in Eq. (7.2), is transformed to a correct one. Function $u^*(\tau)$ can be interpreted here as a *trial solution* of the integral equation and may be obtained, for example, from the solution of Eq. (7.2) by any direct method using a natural regularization principle. Null element $u^* = 0$ can be taken as u^* in the cases when trial solution is not available. As computational experiments showed, good accuracy is obtainable for defining the boundary function at null trial function.

A variational form of IHCP has been formulated above with a regularizing constraint on the unknown function. Computationally the solution of problem given by conditions (7.5) and (7.6) has certain difficulties. Hence, the problem on conditional extremum has to be reduced to a *problem on non-conditional extremum* using the Lagrangian multiplier method. For this purpose compose a functional

$$F[u, u^*, \lambda, \gamma] = \| u - u^* \|^2_{W_2^k} + \lambda(\| Au - f \|^2_{L_2} + \gamma^2 - \delta^2_{L_2}) , \tag{7.7}$$

where γ is a supplementary variable; λ is an undetermined multiplier.

The necessary conditions of the extremum for functional (7.7) are written in the form of the Euler equation for $F[u, u^*, \lambda, \gamma]$ and conditions $\partial F[u, u^*, \lambda, \gamma]/\partial \lambda = 0$; $\partial F[u, u^*, \lambda, \gamma]/\partial \gamma = 0$. Minimization by λ and γ gives

$$\| Au - f \|^2_{L_2} = \delta^2_{L_2} - \gamma^2 , \tag{7.8}$$

$$\lambda \gamma = 0 . \tag{7.9}$$

It follows from Eq. (7.9) that either $\lambda = 0$ or $\gamma = 0$. The case $\lambda = 0$ should be excluded from our consideration since the initial problem would then degenerate into an evident solution $u = u^*$, which is not possible due to an arbitrary choice of u^* element.

So, we receive a **final formulation** of the problem: find function $u^\lambda(\tau)$ minimizing functional (7.7) (under the assumption that $\gamma = 0$) when sampling the Lagrangian multiplier λ from the condition

$$\| Au^\lambda - f_\delta \|^2_{L_2} = \delta^2_{L_2} .$$

If we introduce parameter $\alpha = \dfrac{1}{\lambda}$ instead of parameter λ, we shall obtain

$$\begin{gathered}\min_{u \in W_2^k} \{\Phi[u, u^*, \alpha] = \| Au - f_\delta \|^2_{L_2} + \alpha \| u - u^* \|^2_{W_2^k}\} ;\\ \| Au^\alpha - f_\delta \|^2_{L_2} = \delta^2_{L_2} .\end{gathered} \tag{7.10}$$

Evaluating the norms

$$\| z \|^2_{L_2} = \int_0^{\tau_m} z^2(\xi) \, d\xi ;$$

$$\| z \|^2_{W_2^k} = \int_0^{\tau_m} \sum_{j=0}^{k} \left[\frac{d^j z(\xi)}{d\xi^j} \right]^2 r_j(\xi) \, d\xi ,$$

where $r_j(\xi)$ are definite nonnegative and continuous functions ($r_0, r_k > 0$), reduce the above formula to the following variational problem:

$$\begin{aligned}\min_{u \in W_2^k} \Bigg\{ \Phi[u, u^*, \alpha] = \int_0^{\tau_m} d\tau \left[\int_0^{\tau} u(\xi) K(\tau, \xi) \, d\xi - f_\delta(\tau) \right]^2 \\ + \alpha \int_0^{\tau_m} \sum_{j=0}^{k} r_j(\xi) [u(\xi) - u^*(\xi)]^{(j)2} \, d\xi \Bigg\} ,\end{aligned} \tag{7.11}$$

where α is defined from the condition

$$\int_0^{\tau_m} \left[\int_0^{\tau} u^\alpha(\xi) K(\tau, \xi) \, d\xi - f_\delta(\tau) \right]^2 d\tau = \delta^2_{L_2} . \tag{7.12}$$

The k upper limit of summation in expression (7.11) corresponds to the order of regularization. If $k = 0$ ($W_2^0 = L_2$), the mean-square deviation of unknown solution from a prescribed function is minimized. In this case there is no uniform convergence of regularized approximations.

Generally, number k, which shows the degree of smoothness of regularized solution and weights $r_i(\tau)$, which count the influence of every derivative, should be chosen both due to the specific nature of IHCP under study and if using a priori information available on the behaviour of the unknown function. In most cases it is sufficient to take $k = 1$ or 2. We can also assume $r_0 = r_1$

$= r_2 = 1$. Moreover, it is usually sufficient to ask for smoothness of solution by way of minimum of value

$$\int_0^{\tau_m} u'^2(\xi)\,d\xi \quad \text{or} \quad \int_0^{\tau_m} u''^2(\xi)\,d\xi\,. \tag{7.13}$$

The problem statement given by conditions (7.10) corresponds to a *regularization method* [138] when parameter α is chosen according to a residual principle. Functional Φ is called the *Tikhonov functional* (smoothing or regularizing functional), and integral $\Omega = \int_0^{\tau_m} \sum_{j=0}^{k} r_j(\xi)[u(\xi) - u^*(\xi)]^{(j)2}\,d\xi$ is called the *stabilizing functional.*

The *algorithm* (operator) $R[f_\delta, \alpha]$ for obtaining elements u^α from a solution of the problem of Φ minimization in choosing parameter α from Eq. (7.12) is the *regularizing* one for integral Eq. (7.1).

7.2 The Construction of a Regularizing Operator

The most simple regularizing operator for the above problem can be obtained when a solution of the boundary-value problem for the Euler equation is taken as an algorithm of minimization.

Give value $u(\xi)$ an increment as

$$\delta u(\xi) = \gamma\eta(\xi)\,,$$

where $\eta(\xi)$ is an arbitrary function from $W_2^k[0, \tau_m]$ and γ is a numerical parameter. With small values of γ a set of functions $u(\xi) + \delta u(\xi)$ will belong to a certain neighbourhood of function $u(\xi)$. Calculate variation $\delta\Phi$ of the smoothing functional at point $u = u(\xi)$ using a definition

$$\delta\Phi = \frac{\partial}{\partial\gamma}\,\Phi[u(\xi) + \gamma\eta(\xi)]\bigg|_{\gamma=0}\,,$$

and equate $\delta\Phi$ to zero, thus satisfying the necessary condition of extremum.

Then assuming $u^* = 0$ we obtain

$$F + \alpha(u, \eta)_{W_2^k} = 0\,, \tag{7.14}$$

where

$$F = \int_0^{\tau_m} d\tau \left[\int_0^{\tau} K(\tau, \xi)u(\xi)\,d\xi - f_\delta(\tau)\right] \int_0^{\tau} K(\tau, \xi)\eta(\xi)\,d\xi;$$

$$(u, \eta)_{W_2^k} = \int_0^{\tau_m} \left[\sum_{j=0}^{k} r_j(\xi)u^j(\xi)\eta^j(\xi)\right] d\xi$$

is a scalar product in $W_2^k[0, \tau_m]$.

Rewrite F in the form

$$F = \int_0^{\tau_m} d\tau \int_0^{\tau} \left[\int_0^{\tau} K(\tau, \xi)K(\tau, \zeta)u(\zeta)\eta(\xi)\,d\zeta - f_\delta(\tau)K(\tau, \xi)\eta(\xi)\right] d\xi$$

and change the order of integration over the triangular domain in the obtained expression (Fig. 7.1):

$$F = \int_0^{\tau_m} \eta(\xi)\,\mathrm{d}\xi \int_\xi^{\tau_m} \left[\int_0^{\tau} K(\tau,\xi)K(\tau,\zeta)u(\zeta)\,\mathrm{d}\zeta - f_\delta(\tau)K(\tau,\xi) \right] \mathrm{d}\tau\,. \tag{7.15}$$

Before starting transformation of functional $(u,\eta)_{W_2^k}$, note that a variational problem can be formulated by three different statements:

1. with free boundaries, when a search for the extremal is made among all curves $u(\xi)$, the ends of which lie on the prescribed verticals $\xi = 0$, $\xi = \tau_m$;
2. with fixed boundaries, i.e. the extremal is sought among all functions $u(\xi)$ satisfying the conditions $u^{(j)}(0) = a_j$; $u^{(j)}(\tau_m) = b_j$, $j = \overline{0, k-1}$ at points $\xi = 0$, $\xi = \tau_m$;
3. as a combined statement, when one of the boundaries is free and the other is fixed.

In practice the first statement is very often the case since the conditions to be satisfied by the function on the boundaries of segment $[0, \tau_m]$ may be unknown. That is why we consider a variational problem with free boundaries noting that a transition to the other two statements would offer no obstacles.

Consequently using the operation of integration by parts, transform $(u,\eta)_{W_2^k}$ so that under the integration sign the function $\eta(\xi)$ gets rid of differentiation ($u(\xi)$ and $r_j(\xi)$ are assumed to have the wanted number of continuous derivatives)

$$\begin{aligned}(u,\eta)_{W_2^k} &= \int_0^{\tau_m} \eta(\xi) \sum_{j=0}^{k} (-1)^j \frac{\mathrm{d}^j}{\mathrm{d}\xi^j}[r_j(\xi)u^{(j)}(\xi)]\,\mathrm{d}\xi \\ &+ \sum_{j=1}^{k} (-1)^{j+1}\eta^{(j-1)}(\xi) \sum_{i=j}^{k} (-1)^{i+1} \frac{\mathrm{d}^{i-j}}{\mathrm{d}\xi^{i-j}}[r_i(\xi)u^{(i)}(\xi)]\Bigg|_{\xi=0}^{\xi=\tau_m}\,. \end{aligned} \tag{7.16}$$

Substituting expressions (7.15) and (7.16) into Eq. (7.14) we receive

$$\begin{aligned}&\int_0^{\tau_m} \eta(\xi) \left\{ \int_\xi^{\tau_m} \mathrm{d}\tau \int_0^{\tau} K(\tau,\xi)K(\tau,\zeta)u(\zeta)\,\mathrm{d}\zeta - \int_\xi^{\tau_m} f_\delta(\tau)K(\tau,\xi)\,\mathrm{d}\tau \right.\\ &\left. + \alpha \sum_{j=0}^{k} (-1)^j \frac{\mathrm{d}^j}{\mathrm{d}\xi^j}[r_j(\xi)u^{(j)}(\xi)] \right\} \mathrm{d}\xi \\ &+ \alpha \sum_{j=1}^{k} (-1)^{j+1}\eta(\xi) \sum_{i=j}^{k} (-1)^{i+1} \frac{\mathrm{d}^{i-j}}{\mathrm{d}\xi^{i-j}}[r_j(\xi)u^{(i)}(\xi)]\Bigg|_{\xi=0}^{\xi=\tau_m} = 0\,. \end{aligned} \tag{7.17}$$

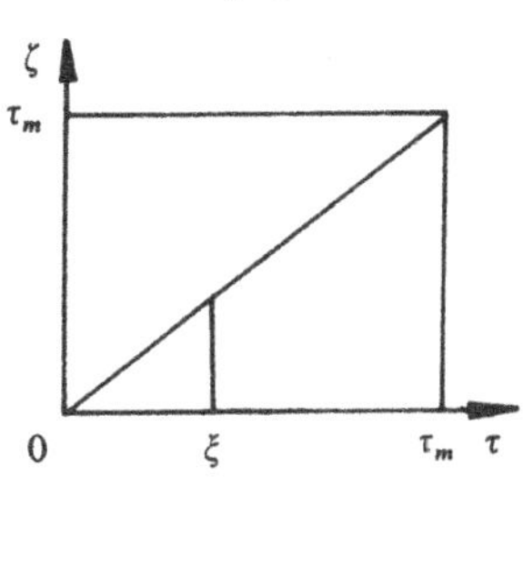

Fig. 7.1. The domain of integration

Owing to arbitrary sampling of function $\eta(\xi)$ including its values on boundaries $\xi = 0$, $\xi = \tau_m$, equality (7.17) will be observed under the following conditions:

$$\int_{\xi}^{\tau_m} d\tau \int_0^{\tau} K(\tau,\xi)K(\tau,\zeta)u(\zeta)\,d\zeta - \int_{\xi}^{\tau_m} f_\delta(\tau)K(\tau,\xi)\,d\tau$$

$$+ \alpha \sum_{j=0}^{k} (-1)^j \frac{d^j}{d\xi^j}[r_j(\xi)u^{(j)}(\xi)] = 0, \quad 0 \leqslant \xi \leqslant \tau_m; \tag{7.18}$$

$$\sum_{i=p}^{k} (-1)^i \frac{d^{i-p}}{d\xi^{i-p}}[r_i(\xi)u^{(i)}(\xi)]\bigg|_{\xi=0;\tau_m} = 0, \quad p = \overline{1,k}. \tag{7.19}$$

The integro-differential Eq. (7.18) is the Euler equation sought and its general solution contains the $2k$ arbitrary constants to be defined by boundary conditions (7.19).

Noting that the domain of integration for the multiple integral is a rectangular trapesoid and changing the order of integration in a proper way we can transform Eq. (7.18) to

$$\int_0^{\xi} u(\zeta)\,d\zeta \int_{\xi}^{\tau_m} K(\tau,\xi)K(\tau,\zeta)\,d\tau + \int_{\xi}^{\tau_m} u(\zeta)\,d\zeta \int_{\zeta}^{\tau_m} K(\tau,\xi)\,K(\tau,\zeta)\,d\tau$$

$$- \int_{\xi}^{\tau_m} f_\delta(\tau)K(\tau,\xi)\,d\tau + \alpha \sum_{j=0}^{k} (-1)^j \frac{d^j}{d\xi^j}[r_j(\xi)u^{(j)}(\xi)] = 0, \quad 0 \leqslant \xi \leqslant \tau_m.$$

Or, finally

$$\int_0^{\tau_m} u(\zeta)\bar{K}(\xi,\zeta)\,d\zeta - \bar{b}(\xi) + \alpha \sum_{j=0}^{k} (-1)^j \frac{d^j}{d\xi^j}[r_j(\xi)u^{(j)}(\xi)] = 0\,, \tag{7.20}$$

where

$$\bar{K}(\xi,\zeta) = \begin{cases} \displaystyle\int_{\xi}^{\tau_m} K(\tau,\xi)K(\tau,\zeta)\,d\tau, & (0 \leqslant \zeta < \xi) \\ \displaystyle\int_{\zeta}^{\tau_m} K(\tau,\xi)K(\tau,\zeta)\,d\tau, & (\xi \leqslant \zeta \leqslant \tau_m) \end{cases}$$

$$\bar{b}(\xi) = \int_{\xi}^{\tau_m} f_\delta(\tau)K(\tau,\xi)\,d\tau.$$

Equation (7.20) can be reduced to a classical form of the Fredholm equation. Rewrite Eq. (7.20) as follows:

$$\alpha \sum_{j=0}^{k} (-1)^j \frac{d^j}{d\xi^j}[r_j(\xi)u^{(j)}(\xi)] = B[\xi, u(\zeta), f_\delta(\tau)]\,, \tag{7.21}$$

where

$$B[\xi, u(\zeta), f_\delta(\tau)] = \bar{b}(\xi) - \int_0^{\tau_m} u(\zeta)\bar{K}(\xi,\zeta)\,d\zeta\,.$$

Regarding $B[\xi, u(\zeta), f_\delta(\tau)]$ as the right-hand side of differential Eq. (7.21) and assuming $G(\xi, S)$ to be the Green's function of the boundary-value problem (7.21)–(7.19), we shall obtain a formal record of solution for this problem via function $G(\xi, S)$:

$$\alpha u(\xi) + \int_0^{\tau_m} u(\zeta)\,\mathrm{d}\zeta \int_0^{\tau_m} G(\xi, S)\bar{K}(S, \zeta)\,\mathrm{d}S = \int_0^{\tau_m} G(\xi, S)\bar{b}(S)\,\mathrm{d}S\,. \tag{7.22}$$

The equation is the Fredholm integral equation of the second kind. Based on the general theory of integral equations, it is possible to illustrate that the solution of Eq. (7.22) and, hence, Eq. (7.20), does exist uniquely and continuously at $[0, \tau_m]$.

Prove now that Eq. (7.20) specifies function $u^\alpha(\xi) \in W_2^k[0, \tau_m]$ which coincides with the problem solving of functional minimization $\Phi_\alpha[u] = \Phi[u, \alpha]$. To avoid bulky expressions, present the Euler equation in the operator form.

As a preliminary, rewrite the stabilizing functional in expression (7.10) taking into account formula (7.16) for an inner product in space W_2^k and also conditions (7.19):

$$\Phi[u, \alpha] = \| Au - f_\delta \|_{L_2}^2 + \alpha(L^k u, u)_{L_2}\,,$$

where

$$L^k = \sum_{j=0}^{k} (-1)^j \frac{\mathrm{d}^j}{\mathrm{d}\xi^j}\left[r_j(\xi)\frac{\mathrm{d}^j}{\mathrm{d}\xi^j}\right]$$

is an operator, which maps $W_2^k[0, \tau_m]$ in $L_2^k[0, \tau_m]$ and defines the smoothness of an approximate solution being obtained.

Introduce an adjoint operator A^* by equality

$$(Au, \vartheta) = (u, A^*\vartheta).$$

In our case it is not difficult to obtain

$$A^*\vartheta = \int_\xi^{\tau_m} K(\tau, \xi)\vartheta(\tau)\,\mathrm{d}\tau\,.$$

This expression having been taken into account, write Eq. (7.20) in the operator form

$$A^*Au + \alpha L^k u = A^* f_\delta\,. \tag{7.23}$$

Consider value $u_1^\alpha = u^\alpha + \vartheta$ (ϑ is any arbitrary element of $W_2^k[0, \tau_m]$) and represent $\Phi_\alpha[u_1^\alpha]$ as follows:

$$\Phi_\alpha[u_1^\alpha] = \Phi_\alpha[u^\alpha] + \| A\vartheta \|_{L_2}^2 + \alpha \| \vartheta \|_{W_2^k}^2 + 2(Au^\alpha - f_\delta, A\vartheta)_{L_2} + 2\alpha(u^\alpha, \vartheta)_{W_2^k}\,.$$

Utilizing the adjoint operator, we obtain

$$\Phi_\alpha[u_1^\alpha] = \Phi_\alpha[u^\alpha] + \| A\vartheta \|_{L_2}^2 + \alpha \| \vartheta \|_{W_2^k}^2 + 2(A^*Au^\alpha - A^*f_\delta, \vartheta)_{L_2} + 2\alpha(u^\alpha, \vartheta)_{W_2^k}.$$

From here, counting Eq. (7.23) and equality $(u, \eta)_{W_2^k} = (\eta, L^k u)_{L_2}$ as followed from expression (7.16) with conditions (7.19) we have

$$\Phi_\alpha[u_1^\alpha] = \Phi_\alpha[u^\alpha] + \| A\vartheta \|_{L_2}^2 + \alpha \| \vartheta \|_{W_2^k}^2 .$$

Noting that $\Phi_\alpha[u_1^\alpha] > \Phi_\alpha[u^\alpha]$, $\alpha > 0$, we come to a **conclusion** that the unknown element u^α found by solving Eq. (7.23) gives minimum to the smoothing functional.

Thus, a regularizing operator for integral equation (7.1) is constructed. Solving problem (7.20)–(7.19) with different values of parameter α we obtain a set of functions $u^\alpha(\xi)$, among which there is a unique function $u_\delta^\alpha(\xi)$ which satisfies the condition of residual in the form of Eq. (7.12). With this we complete the solution of the inverse problem in a regularized statement.

7.3 Regularization of the Inverse Problem Finite-Dimensional Form

The IHCP regularized form in infinite-dimensional spaces of solution and that of initial data has been considered above. Writing a finite difference equivalent for Eq. (7.20) and solving it under known boundary conditions, we can define a grid function $u_{\Delta\tau}^\alpha(\tau)$. A somewhat different approach is possible. Operator A is approximated first by A_Δ and only then the difference analog of integral Eq. (7.1) is regularized. *Computationally the second variant is preferable provided that an approximate calculation of the integral in Eq.* (7.1) *is constructed to include the possibility of analytical integration of kernel* $K(\tau, \xi)$ *in the interval* $[\tau_{i-1}, \tau_i]$. In this case by excluding the difference analogs of integral terms in Eq. (7.20) we have simpler and more effective algorithms. Consider this point in detail, since this is a situation which occurs for the majority of IHCP statements as seen in Chaps. 3 and 4.

Approximate Eq. (7.1) on a uniform grid according to the technique illustrated in Chap. 4. As a result we obtain a system of algebraic equations $\sum_{i=1}^{n} \hat{u}_i \varphi_{ni} = f_{\delta n}$, $n = 1, 2, \ldots, m$ or in matrix representation

$$A_\Delta \boldsymbol{u} = \boldsymbol{f}_\delta \tag{7.24}$$

where A_Δ is the lower triangular matrix.

A regularized form of this problem for the case $k = 2$, $u^* = 0$ and weights $r_0 = 0, r_1(\tau) = r_1 = \text{const} \geqslant 0, r_2(\tau) = r_2 = \text{const} \geqslant 0$ consists in finding of such parameters $\hat{u}_i$, $i = 1, 2, \ldots, m$ which give minimum of function

$$\hat{\Phi}_\alpha[u] = \sum_{n=1}^{m} \left(\sum_{i=1}^{n} \varphi_{ni} \hat{u}_i - f_{\delta n} \right)^2 + \frac{\alpha r_1}{\Delta\tau^2} \sum_n (\hat{u}_n - \hat{u}_{n-1})^2$$

$$+ \frac{\alpha r_2}{\Delta\tau^4} \sum_n (\hat{u}_{n+1} - 2\hat{u}_n + \hat{u}_{n-1})^2 . \tag{7.25}$$

Parameter $\alpha > 0$ is defined from a condition of matching the residual with the error of the right-hand side in Eq. (7.24), given in the metric of Euclidean space:

$$\sum_{n=1}^{m}\left(\sum_{i=1}^{n}\varphi_{ni}\hat{u}_i^{\alpha}-f_{\delta n}\right)^2=\delta_{Em}^2\,. \tag{7.26}$$

Minimizing function (7.25) over all $\hat{u}_n$ under particular conditions we have a system of linear algebraic equations with a symmetrical positive defined matrix

$$\boldsymbol{Bu}=\boldsymbol{D}\,, \tag{7.27}$$

where $\boldsymbol{B}=\boldsymbol{A}_\Delta^T\boldsymbol{A}_\Delta+\alpha\boldsymbol{C}$; $\boldsymbol{A}_\Delta^T$ is a transposed matrix; $\boldsymbol{D}=\boldsymbol{A}_\Delta^T\boldsymbol{f}_\delta+\boldsymbol{g}$.

The elements of normal matrix $\boldsymbol{A}_\Delta^T\boldsymbol{A}_\Delta$ are given by formula

$$b_{kl}=\sum_{n=l}^{m}\varphi_{nk}\,\varphi_{nl},\quad b_{kl}=b_{lk},\quad k,l=1,2,\ldots,m\,.$$

Matrix $\boldsymbol{C}$ is a finite-dimensional analog of the smoothing operator L^2 when $r_0=0$. Vector $\boldsymbol{A}_\Delta^T\boldsymbol{f}_\delta=\left\{\sum_{n=k}^{m}\varphi_{nk}f_{\delta n}\right\}$, $k=1,2,\ldots,m$. Vector $\boldsymbol{g}$ is defined by particular boundary conditions.

A regularization of the first order is very often used. It consists in minimization of the first differences. A regularization of the second order can be also used when only the second differences are minimized.

Consider the **first case**, writing the finite-dimensional analog of regularizing functional in the form

$$\hat{\Phi}_\alpha[u]=\sum_{n=1}^{m}\left(\sum_{i=1}^{m}\varphi_{ni}\hat{u}_i-f_{\delta n}\right)^2+\alpha\sum_{n}(\hat{u}_n-\hat{u}_{n-1})^2$$

and setting the boundary conditions in the following way (e.g. from approximation of the first-order derivatives at the time interval limits): $u_1-u_0=c_{10}$, $u_{m+1}-u_m=c_{1m}$. Then matrix $\boldsymbol{C}$ and vector $\boldsymbol{g}$ are written as

$$\boldsymbol{C}=\begin{bmatrix} 1 & -1 & & & & \\ -1 & 2 & -1 & & & \\ & -1 & 2 & -1 & & \\ & & & \ddots & & \\ & & & -1 & -2 & -1 \\ & & & & -1 & 1 \end{bmatrix} \tag{7.28}$$

$$\boldsymbol{g}=[-\alpha c_{10},0,\ldots,0,\alpha c_{1m}]^T\,. \tag{7.29}$$

In the **second case** we take the criterion function of the following type:

$$\hat{\Phi}_\alpha[u]=\sum_{n=1}^{m}\left(\sum_{i=1}^{n}\varphi_{ni}\hat{u}_i-f_{\delta n}\right)^2+\alpha\sum_{n}(\hat{u}_{n+1}-2\hat{u}_n+\hat{u}_{n-1})^2$$

and boundary conditions in the form of second and third differences corresponding to the approximation of second and third derivatives:

$$u_0 - 2u_1 + u_2 = c_{20}; \quad -u_{-1} + 3u_0 - 3u_1 + u_2 = c_{30};$$

$$u_{m-1} - 2u_m + u_{m+1} = c_{2m}; \quad -u_{m-1} + 3u_m - 3u_{m+1} + u_{m+2} = c_{3m}.$$

In the result we obtain

$$C = \begin{bmatrix} 1 & -2 & 1 & & & & \\ -2 & 5 & -4 & 1 & & & \\ 1 & -4 & 6 & -4 & 1 & & \\ & 1 & -4 & 6 & -4 & 1 & \\ & & & \ddots & & & \\ & & 1 & -4 & 6 & -4 & 1 \\ & & & 1 & -4 & 5 & -2 \\ & & & & 1 & -2 & 1 \end{bmatrix} \tag{7.30}$$

$$g = [\alpha(c_{20} + c_{30}), -\alpha c_{20}, 0, \ldots, 0, \alpha c_{2m}, \alpha(c_{3m} - c_{2m})]^T . \tag{7.31}$$

Solving system (7.27) with a chosen smoothness order of the unknown solution and a given value of parameter α, characterizing the degree of smoothness, we obtain some approximation u^α to a true function $u_T(\tau)$.

It is known that for any self-adjoint operator, such as matrix $\boldsymbol{B}$, there is a system of linearly independent pairwise orthogonal and unit-equal in norm eigenvectors $\{\boldsymbol{e}_n\}$, i.e. a orthonormal basis of Euclidean space of given dimensionality. Following [144] expand elements $\boldsymbol{u}$ and $\boldsymbol{D}$ in this basis: $\boldsymbol{u} = \sum_{n=1}^{m} \xi_n \boldsymbol{e}_n$; $\boldsymbol{D} = \sum_{n=1}^{m} \eta_n \boldsymbol{e}_n$ where ξ_n and η_n are the coordinates of vectors $\boldsymbol{u}$ and $\boldsymbol{D}$ with respect to the basis $\{\boldsymbol{e}_n\}_1^m$. Substitute the above expansions into equation (7.27) $\sum_{n=1}^{m} \xi_n \boldsymbol{B}\boldsymbol{e}_n = \sum_{n=1}^{m} \eta_n \boldsymbol{e}_n$.

It follows from the definition of eigenvectors $\boldsymbol{e}_n$ and eigenvalues λ_n of matrix $\boldsymbol{B}$ ($\boldsymbol{B}\boldsymbol{e}_n = \lambda_n \boldsymbol{e}_n$) that in this case $\xi_n = \eta_n / \lambda_n$.

Thus we have $u = \sum_{n=1}^{m} \frac{\eta_n}{\lambda_n} e_n$. It is seen that the less are eigenvalues λ_n, the more expand the coordinates of vector $\boldsymbol{D}$ when passing to vector $\boldsymbol{u}$ and still higher becomes an effect of errors in the right-hand side of equation (7.27). Besides, additional distortion of the corresponding components $\hat{u}_n$ of the solution takes place with small λ_n, caused by computational errors in the matrix elements.

A wide spread of eigenvalues of matrix $\boldsymbol{B}$, including the emergence of very small λ_n, is observed for small values of parameter α, because in this case matrix $\boldsymbol{B}$ is close to matrix $\boldsymbol{A}_\Delta^T \boldsymbol{A}_\Delta$ corresponding to an unregularized variational form of the problem. That is why the solution obtained will have quite inadequate and oscillating form.

As value α increases, the influence of stabilizing matrix $\boldsymbol{C}$ grows, resulting in better stipulation of matrix $\boldsymbol{B}$ and gradual smoothing of solution, and the regularized approximations would appear to be the closest to a true function in the definite range of parameter variation. It is assumed that value $\alpha = \alpha_r$, chosen by the residual principle, is in this range. With further increase of α the curves $u^\alpha(\tau)$ would acquire ever more smooth outlines, but at the same time they depart from a true but less smooth solution.

Note that a solution of regularized system (7.27) can be better achieved (from the viewpoint of accuracy of results and rate of calculations) by the square root method [145]. It is possible to construct a still more efficient algorithm if transformations suggested in [58] are applied to system (7.27).

7.4 The Admissible Degree of Smoothing and Approximation Sampling Procedures

7.4.1 A residual principle

In writing a regularized form of problem (7.1) we proceeded from the assumption that we should find a more smooth in the sense of minimum of value (7.5) curve in the class of sufficiently smooth functions with a complementary constraint on a set of functions varied in the form of inequality (7.6), defined by the initial problem statement. Since the smoothness of solution becomes worse while nearing the minimum of residual $\| Au - f_\delta \|^2_{L_2}$, it is clear why the unknown and the most smooth solution lies on the frontier of set D_δ, as it follows from the problem statement given by conditions (7.10). Such a principle of sampling a regularized approximation is called a residual principle in the form of equality. Full mathematical substantiation of this approach is given in works by V.A. Morozov [103, 104]. Some of the results are shown below.

The residual principle has been used for solving the equation (7.2) under the assumption that the operator A is given exactly. But in fact, besides errors in the assignment of the right-hand side f, there will always occur the errors connected with calculation of values of operator A. It happens that these errors can be neglected in many problems.

It is stated in [104] that for a linear operator A, uniformly approximated by some operator A_Δ, i.e. when

$$\| A_\Delta - A \| \equiv \sup_{u \in U} \frac{\| A_\Delta u - Au \|_F}{\| u \|_U} \leqslant \varkappa(\Delta\tau), \quad \varkappa(\Delta\tau) \to 0, \ \Delta\tau \to 0 .$$

the residual for "disturbed" formulation of the problem

$$\rho(\alpha) = \| A_\Delta u^\alpha - f_\delta \|^2_F \tag{7.32}$$

is a continuously increasing function (see Fig. 7.2), $0 \leqslant \rho(\alpha) \leqslant \rho^*$ where $\rho^* = \| A_\Delta u^* - f_\delta \|^2_F$ (u^* is a trial solution).

The following inequalities are valid

$$\begin{aligned} \| A_\Delta u_T - f_\delta \|_F &\leqslant \| A_\Delta u_T - Au_T \|_F + \| Au_T - f_\delta \|_F \\ &\leqslant \| A_\Delta - A \| \, \| u_T \|_U + \delta \leqslant \varkappa(\Delta\tau) \, \| u_T \|_U + \delta , \end{aligned}$$

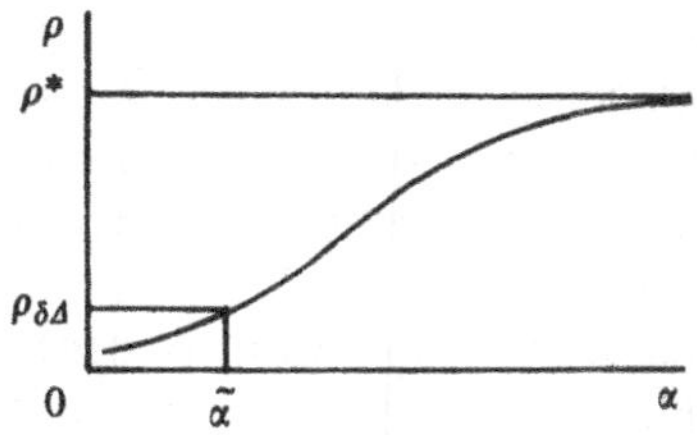

Fig. 7.2. The dependence of residual on regularization parameter

where u_T is a solution of equation (7.2), corresponding to the exact right-hand side and undisturbed operator; $\delta = \| f_T - f_\delta \|_F$ is an error of the (7.2) right-hand side.

Denote

$$\rho_{\delta\Delta} = \varkappa(\Delta\tau) \| u_T \|_U + \delta \text{ and note that for fixed } \alpha$$

$$\rho(\alpha) \leqq \| A_\Delta u_T - f_\delta \|_F^2 + \alpha \| u_T - u^* \|_U^2,$$

hence

$$\rho(\alpha) \leqslant \rho_{\delta\Delta}^2 + \alpha \| u_T - u^* \|_U^2 .$$

For sufficiently small α $(0 < \alpha < \bar{\alpha})$ $\rho(\alpha) < \rho_{\delta\Delta}^2$. Hence, we can find some value $\tilde{\alpha}$ for which equation $\rho(\tilde{\alpha}) = \rho_{\delta\Delta}^2$ is valid.

Value $\rho_{\delta\Delta}$ depends on an unknown solution u_T and that is why it is of difficulty to use it as an admissible level of residual for sampling a required value α. However, the residual principles are efficiently used when operator A is approximated through A_Δ with high accuracy, i.e. when $\varkappa(\Delta\tau) \ll \delta$. It is such a case which can be always realized for sufficiently small steps $\Delta\tau$ in the approximation of (4.1) type or under the condition of piecewise-linear approximation of the unknown function $u(\tau)$ [3].

Thus, when solving linear IHCP with constant thermophysical properties it is possible to achieve solution $\rho_{\delta\Delta} \simeq \delta$ and no additional a priori information on solution is needed for practical use of residual principle.

Pay attention to one more important point. The right-hand sides of the integral equations in some forms of IHCPs (See Ch. 3) contain, besides temperature information $T^*(\tau)$ (or $\theta^*(\tau)$) given at some point $x = d$ of the body, also integral terms, the initial temperature distribution and known boundary condition taken into account, e.g. the heat flux or temperature at points $x = b$, $b > d$. Naturally, the errors of components mentioned should also be taken into account when determining a summary error in the right-hand side. From practical standpoint the case is of most interest when these errors can be neglected. Such a situation is common case and occurs when

the initial and boundary conditions are known with high accuracy and at the same time there is a possibility to ensure high quality of approximation of the corresponding integral terms;

the smoothing effect of integral operators is great enough and the random errors in the integrand turn out to be practically suppressed; the smoothing of the initial condition will be ever stronger, the larger the time interval considered and that of the boundary condition – the farther the points a and b are from each other.

In this case we can assume that $\delta \simeq \delta_T$, where δ_T is an error in temperature data T^*.

The approximate calculation of value δ_T can be made by known estimates of the mean square error σ in the input data:

$$\delta_{L_2} \simeq \left(\int_0^{\tau_m} \sigma^2(\tau) d\tau \right)^{1/2}, \quad \delta_{E_m} \simeq \left(\sum_{n=1}^{m} \sigma_n^2 \right)^{1/2}.$$

These formulae define the level of residual in spaces L_2 and E_m respectively [see eqs. (7.12) and (7.26)].

If δ is known exactly, the residual principle produces somewhat "over-smoothed" results, the difference of the reconstructed curve from a true one being as great as stronger the mean-square error σ varies with time. Hence it may be more correct to replace the "unweighted" residual condition (7.26) by a "weighted" one, in which every current discrepancy between the calculated and given temperatures is scaled, i.e. referred to current estimates of the mean-square error:

$$\sum_{n=1}^{m} \frac{1}{\sigma_n^2} \left(\sum_{i=1}^{n} \varphi_{ni} \hat{u}_i^{\alpha} - f_{\delta n} \right)^2 = S .$$

Number $S \geqslant 0$ is sampled from the following confidence interval:

$$N - (2N)^{1/2} \leqslant S \leqslant N + (2N)^{1/2}, \quad N = m + 1 .$$

Remark 1. Along with the residual principle we can recommend an approximation sampling method, which consists in defining α from equation (see [102]) $\| A_\Delta u^\alpha - f_\delta \|_F^2 + \alpha \| u^\alpha - u^* \|_U^2 = \delta^2$. As calculations showed, this modified principle of residual yields results $\alpha = \alpha_f$ (by functional), which in fact are close to the condition of residual. Here $\alpha_r > \alpha_f$, that defining somewhat more "smoothed" functions in the case $\alpha = \alpha_r$.

This method is of particular importance for solving non-linear problems. As distinct from the residual principle, which is valid, strictly speaking, for inverse problems with linear operator, the modified residual method is also proven for nonlinear ill-posed problems [92].

Remark 2. The residual principle considered assumes that a condition $\varkappa(\Delta\tau) \ll \delta$ be satisfied. If such an assumption is not legitimate, we can make use of a generalized residual principle [66, 105] which counts errors of approximation of operator A.

7.4.2 Some Other Ways of Parameter Determination

The approximation sampling to unknown solution by the residual principle requires knowledge of the value of the initial data error δ in the appropriate metric. Sometimes this value happens to be unknown or low accuracy of its estimation does not allow to get true results. That is why we have to apply some other methods of revealing a required approximation. A method of *quasioptimal parameter* [140–142] is one of them.

With one realization of the right-hand side $f = f_\delta$ this method consists in minimization with respect to α of quantity $\|\alpha(du^\alpha/d\alpha)\|_u$. If the grid of various values of parameter α answers the geometric progression law $\alpha_{j+1} = k\alpha_j$, where $0 < k < 1$, $k \sim 1$, so $\alpha \dfrac{du^\alpha}{d\alpha} \simeq \dfrac{u^{\alpha_{j+1}} - u^{\alpha_j}}{k-1}$ and we can seek the minimum value of $\|u^{\alpha_{j+1}} - u^{\alpha_j}\|_u$.

If we assign a sufficiently broad interval of possible values of parameter $\alpha \in [\alpha_{\min}, \alpha_{\max}]$, the solution of this problem may happen to be not a single one, i.e. there are several local minima, corresponding to some values α_m. It is required to define among them the most suitable value, which would be taken as a quasioptimal parameter α_{qo}. More often it is the minimal value $\alpha_{qo} = \min_m \alpha_m$ or the next to it in increasing order.

Such an approach, consisting in defining the utmost proximity of regularized solutions (the principle of "inner" convergence), gives good results in many cases, but it carries an element of subjectivism and usually requires some qualitative and quantitative information about the unknown solution. Note also that we can encounter a situation when quantity $\|u^{\alpha_{j+1}} - u^{\alpha_j}\|_U$ has no minimum in the wanted interval of values α.

Of interest is a method of sampling the regularization parameter suggested in [85]. Let (at the minimum) two random realizations of the right-hand side of (7.1) equation be available, i.e. $f_{1\delta}(\tau)$ and $f_{2\delta}(\tau)$. Actually they can be received by using two temperature gauges or as a result of sampling different steps of process quantization as well as by "shifting" nodal points of the experimental function assignment. In any of these cases it is necessary to ensure mutual independence of these realizations.

For one of the right-hand sides, e.g. for $f_{1\delta}$, we solve the problem in minimum of the regularizing functional. Here parameter α is determined from the condition of utmost proximity of computed functions $f_1^\alpha = Au_1^\alpha$ to another random realization $f_{2\delta}$: $\min_\alpha \|f_1^\alpha - f_{2\delta}\|$. In the same way we can find u_2^α "fitting" f_2^α to realization $f_{1\delta}$ and then average the obtained results u_1^α and u_2^α. Description of other methods for parameter α choice can be found in [61, 62, 131].

Note in conclusion that in solving concrete problems it is sometimes more preferable to simultaneously use two different methods to select regularized approximations. For instance, it is useful to combine the residual principle with the quasioptimal parameter method.

7.4.3 The Automation of Approximation Sampling Procedure

In many cases when solving the inverse heat conduction problems with the help of regularized schemes a need arises in full automation of the process of sampling the reasonable (physically well-grounded) approximations to unknown boundary functions. Determination of appropriate value of parameter α from the condition of inner convergence of regularized solutions (the quasioptimal parameter method) requires a quantitative, usually visual, analysis of the obtained data. Here we use some a priori information about the solution to be expected and of the would-be nature of the exact input function. All this makes automation of the process difficult. At the same time the criterion for sampling the regularization parameter, based on the residual principle enables an algorithm of automatic search for better approximations to be constructed. An effective algorithm of such type has been suggested in [67]. Consider its application to a solution of problem (7.24).

It is shown in the above mentioned study that if we replace the residual equation $\|A_\Delta u^\alpha - f_\delta\|_{E_m} = \delta$ by equation $\|A_\Delta u^\alpha - f_\delta\|^S_{E_m} = \delta^S$, $-1 \leqslant S < 0$ and consider the latter with respect to $\lambda = 1/\alpha$, then for its solution it is possible to use an iteration method by Newton's tangents formula (owing to one-sided convexity of functions $\rho^S(\lambda) = \|A_\Delta u^{1/\lambda} - f_\delta\|^S_{E_m}$.) Besides, it is stated that the highest rate of convergence of the process in search for the root takes place at $S = -1$.

Thus, instead of (7.26) we shall further solve an equation

$$F(\lambda) \equiv \left[\sum_{n=1}^{m}\left(\sum_{i=1}^{n} \varphi_{ni}\hat{u}_i^{1/\lambda} - f_{\delta n}\right)^2\right]^{-1/2} - \delta_{E_m}^{-1} = 0 . \tag{7.33}$$

Iterative sequence is constructed by formula

$$\lambda_{j+1} = \lambda_j - F(\lambda_j)/F'(\lambda_j), \quad j = 0, 1, \ldots,$$

where $F'(\lambda) = \dfrac{1}{\lambda^2 \rho^3_{1/\lambda}} \sum_{n=1}^{m}\left[\left(\sum_{i=1}^{n} \varphi_{ni}\hat{u}_i^{1/\lambda} - f_{\delta n}\right)\sum_{i=1}^{n}\dfrac{d\hat{u}_i^{1/\lambda}}{d\alpha}\right]$.

Having differentiated (7.27), we get a system for determining $d\hat{u}^{1/\lambda}/d\alpha$:

$$\boldsymbol{B}\frac{d\boldsymbol{u}^{1/\lambda}}{d\alpha} = \boldsymbol{p} . \tag{7.34}$$

Vector-column $\boldsymbol{p}$ for matrix $\boldsymbol{C}$ and vector $\boldsymbol{g}$, defined by (7.28) and (7.29), assumes the form

$$p_1 = -c_{10} - u_1 + u_2 ;$$

$$p_k = u_{k-1} - 2u_k + u_{k+1}, \quad k = 2, 3, \ldots, m-1 ;$$

$$p_m = c_{1m} + u_{m-1} - u_m .$$

For statement (7.30) and (7.31):

$$p_1 = -u_1 + 2u_2 - u_3 + c_{20} + c_{30};\ p_2 = 2u_1 - 5u_2 + 4u_3 - u_4 - c_{20} ;$$

$$p_k = -u_{k-2} + 4u_{k-1} - 6u_k + 4u_{k+1} - u_{k+2}, \quad k = 3, 4, \ldots, m-2 ;$$

$$p_{m-1} = -u_{m-3} + 4u_{m-2} - 5u_{m-1} + 2u_m + c_{2m} ;$$

$$p_m = -u_{m-2} + 2u_{m-1} - u_m + c_{3m} - c_{2m} .$$

The process of approximation to the root of equation (7.33) (by residual) starts from some initial value λ_0 and ends by the condition

$$\left| \frac{\lambda_{j+1} - \lambda_j}{\lambda_{j+1}} \right| < \varepsilon ,$$

where ε is the prescribed relative accuracy.

As calculations show, the rate of convergence of this iterative process is high and under reasonable sampling ε we need only 4–7 iterations.

7.5 The Reconstruction Accuracy Analysis of Boundary Heat Conditions

The regularized algorithms of IHCP solution were subjected to a detailed operational examination through specially prepared methodical examples. Some of the results are given below.

Consider the problem of determining the heat flux acting upon a thermally thick wall (a semi-infinite body), which, in turn, has good heat-insulating properties: $\lambda = 1.31 \cdot 10^4$ kW/(mK), $a = 1.204 \cdot 10^{-7}$ m^2/s. Function $q(\tau) = 10.2 + 6.13\tau + 0.613\tau^2$ kW/m^2, known beforehand, has been reconstructed by values of temperature at some point inside the body (Fig. 7.3). The algebraic system (7.27)–(7.29) was used for this purpose.

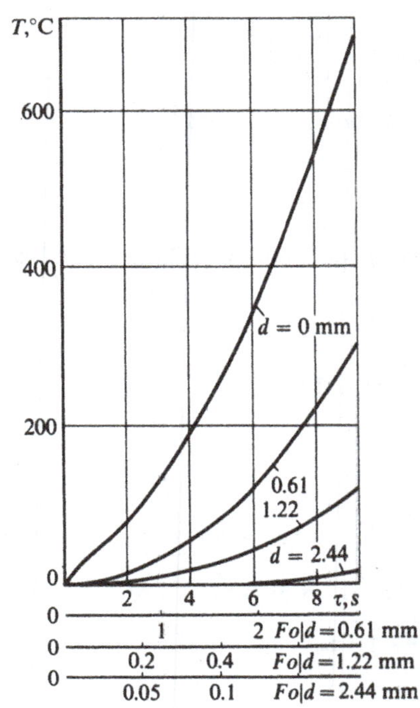

Fig. 7.3. The time dependence of temperature

The input function $T^*(\tau) = T(d, \tau)$ has been determined from the solution of a direct heat conduction problem by the given heat flux. Values T_n^*, calculated with error less than $\sim 10^{-3}$ K (10^{-3}–10^{-4}% T^*_{max}), were conventionally taken as exact data for IHCP solving.

An example of calculation for the exact data and with $\alpha = 0$ is shown in Fig. 7.4. We see that starting from $Fo_d \sim 2$ values q differ considerably from true values. If small errors are brought into T_n^* by rounding-off the input data values up to integers ($\sim 0.1\%$ T^*_{max}) there would be no possibility at all to apply the direct method with sufficiently small ΔF_0. At the same time a regularized solution is close to a true one.

In order to estimate accuracy of $q^\alpha(\tau)$ definition depending on the errors of initial data, values T_n^* were subjected to the following disturbances with the help of pseudorandom-number generators:

1) $T^*_{\delta n} = T_n^* + (-1)^n \delta_0$ are the data with saw-tooth perturbations;

2) $T^*_{\delta n} = T_n^* + (\delta_0/3)\omega_n$ are the data obtained by adding Gaussian "random" errors (a "three-sigma" rule is used);

3) $T^*_{\delta n} = T_n^* + \delta_0 \xi_n$ are the data with random errors corresponding to a uniform distribution;

4) $T^*_{\delta n} = T_n^* \pm \delta_0$ are the data with a constant systematic error.

Here δ_0 is the maximum possible error; ω_n are independent Gaussian random variables with expectation $M = 0$ and variance $\sigma^2 = 1$; ξ_n are the random variables uniformly distributed over the segment $[-1, 1]$.

Regularized approximations $\bar{q}(\tau) = q^\alpha d/\lambda$ are shown in Figs. 7.5 and 7.6. They are calculated for the normal (Gaussian) and uniform error distributions with known boundary values of first derivative $q'(0)$, $q'(\tau_m)$. Fig. 7.6 also depicts the results of $q(\tau)$ calculation, when $q'(0)$ and $q'(\tau_m)$ were assumed equalling zero.

The estimates of maximum relative error in determination of heat flux $\varepsilon = \left| \dfrac{q^\alpha_\delta - q_T}{q_{T\max}} \right| \cdot 100\%$ depending on a maximal absolute error δ_0 and a relative error $(\delta_0/T^*_{max}) \cdot 100\%$ as well as an mean-square error σ, are given in Fig. 7.7 for the first three types of disturbances. Fig. 7.8 depicts the results for the case when

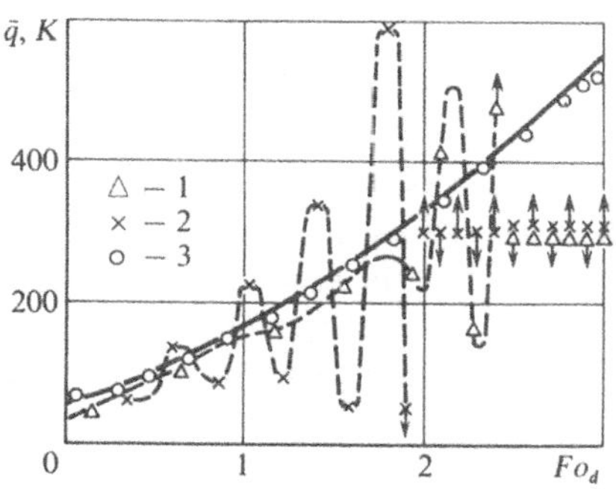

Fig. 7.4. An example for heat flux calculation using a regularized algorithm and a direct method (corresponding to condition $\alpha = 0$), $d = 0.61$ mm, $\Delta Fo_d = 0.06$ (full curve is a true solution, $\bar{q} = qd/\lambda$): 1 is a direct method, errors in temperature $\delta_0 \sim 1.5 \cdot 10^{-3} K$; 2 is a direct method, $\delta_0 \sim 0.1 \cdot 10^{-2}\ T^*_{max}$; 3 is a regularized algorithm, $\delta_0 \sim 0.1 \cdot 10^{-2}\ T^*_{max}$.

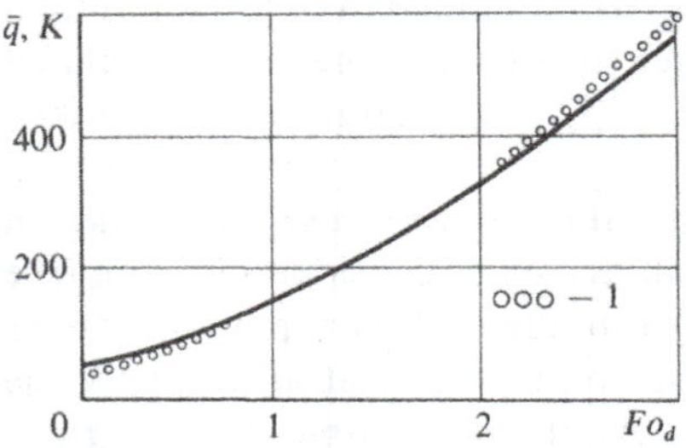

Fig. 7.5. The results of heat flux reconstruction (the normal distribution of errors in the initial temperature values), $d = 0.61$ mm, $\Delta Fo_d = 0.06$ (full curve is a true solution): 1 is a calculation for $\delta_0 = 0.25\ T^*_{max}$.

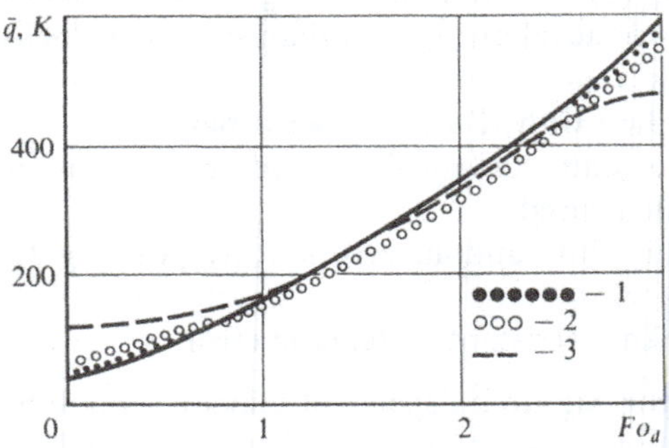

Fig. 7.6. The results of heat flux reconstruction in the case of uniform distribution of errors in the initial temperature values, $d = 0.61$ mm, $\Delta Fo_d = 0.06$ (full curve is a true solution): 1 and 2 are the calculations with exact assignment of coefficients c_{10} and c_{1m}; $\delta_0 = 0.05\ T^*_{max}$ and $0.25\ T^*_{max}$, respectively; 3 is a calculation with $c_{10} = c_{1m} = 0$, $\delta_0 = 0.25\ T^*_{max}$

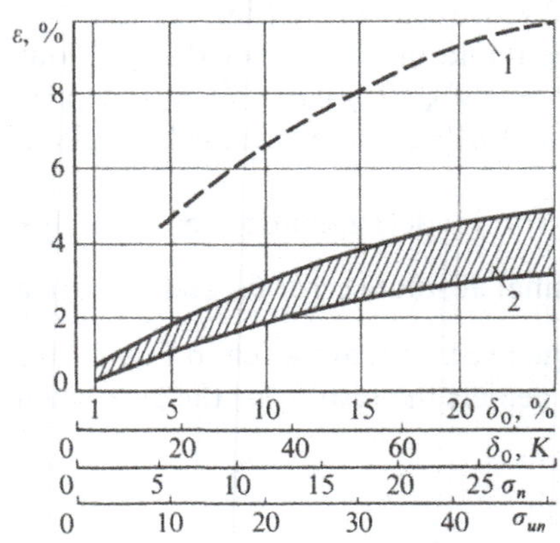

Fig. 7.7. The dependence of a relative error of heat flux determination on the temperature errors, $d = 0.61$ mm, $\Delta Fo_d = 0.06$ (σ_n, σ_{un} are the standard temperature errors for the normal and uniform distributions, respectively): 1 is the normal law of error distribution, $c_{10} = 0$; 2 is the domain (shaded) corresponding to exact values of c_{10}, c_{1m} and to all types of disturbances T_n^* except the fourth one.

a definite constant shift of temperature values (the fourth type of disturbance) is assigned, and the boundary conditions c_{10} and c_{1m} are known exactly. Here the error of heat flux determining is approximately equal to the error in the assignment of input temperature. The input data and the results obtained for $d = 2.4 \cdot 10^{-3}$ m and the normal distribution of δT^* are presented on Figs. 7.9 and 7.10. It is seen that the accuracy of reconstruction $q(\tau)$ worsens when a temperature sensor is removed from the body surface. Such a behaviour of solution is quite natural, since the initial time interval appears and then

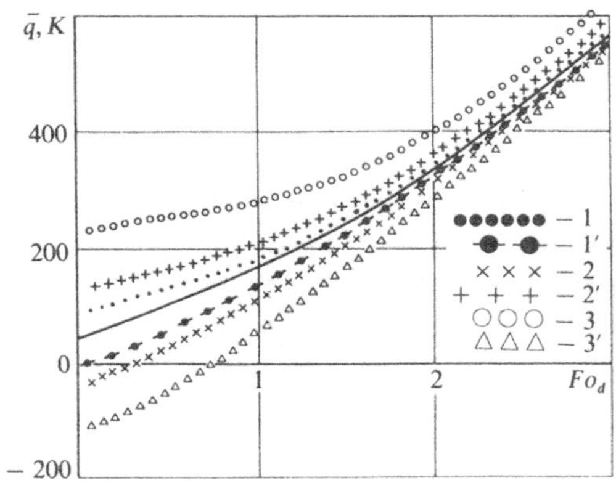

Fig. 7.8. The results of heat flux reconstruction when a systematic error is added to temperature measurements, $d = 0.61$ mm, $\Delta Fo_d = 0.06$ (full curve is a true solution): 1, 2, 3 are the calculations for $T^*_{\delta n} = T^*_n + \delta_0$; 1′, 2′, 3′ are the calculations for $T^*_{\delta n} = T^*_n - \delta_0$; 1, 1′ – with $\delta_0 = 0.05\, T^*_{max}$; 2, 2′ – with $\delta_0 = 0.1\, T^*_{max}$; 3, 3′ – with $\delta_0 = 0.25\, T^*_{max}$.

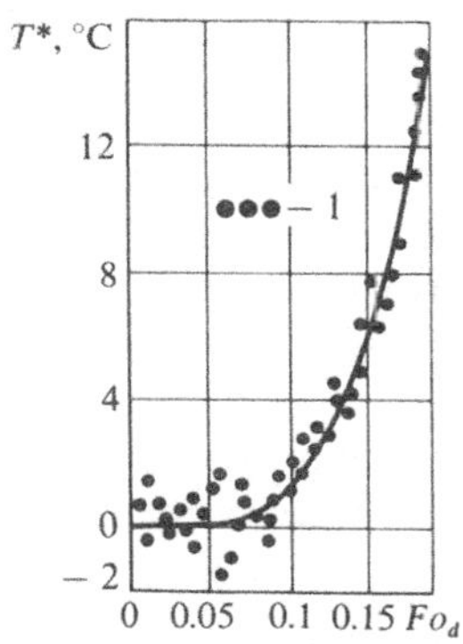

Fig. 7.9. The input data for IHCP solving, $d = 2.44$ mm (full curve is an exact function): 1 are the disturbed temperature values (random errors are distributed by the normal law with an amplitude $3\sigma = 0.1\, T^*_{max}$)

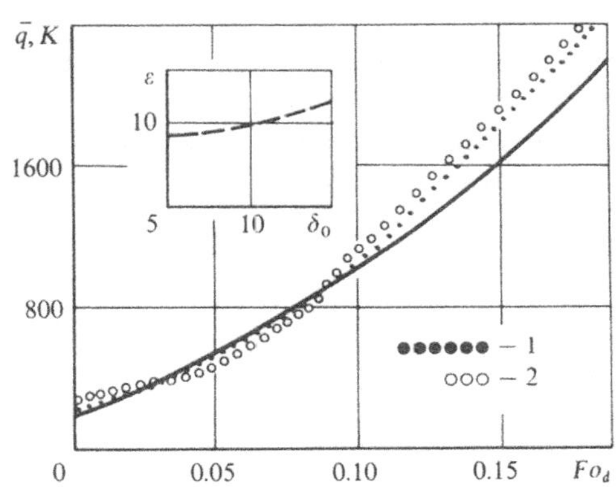

Fig. 7.10. The results of heat flux recovery with exact assignment of coefficients c_{10} and c_{1m} and the normal distribution of errors in the initial values of temperature, $d = 2.44$ mm; $\Delta Fo_d = 0.004$ (full curve is a true solution, dashed line is the estimate of reconstruction accuracy, ε, %, δ_0, %): 1 – for $\delta_0 = 0.05\, T^*_{max}$; 2 – for $\delta_0 = 0.15\, T^*_{max}$.

gradually increases, within which the input temperatures are so small in value that they are completely "jammed" by an artificially induced noise. However, the quality of reconstruction should be recognized quite satisfactory because of correct assignment of derivatives $q'(0)$ and $q'(\tau_m)$ in the boundary conditions of a regularized problem, in spite of the fact that the initial period of "uncertainty" of the input temperature ($T^* \sim 0$) in this case reaches one third of a reconstruction time.

Coefficients c_{10} and c_{1m} in a number of practical cases can be unknown and hence they are to be changed by null values. As a result, a regularized approximation deviates from a true solution in the vicinity of end points $\tau = 0$ and $\tau = \tau_m$ (see Fig. 7.6). The higher are the fluctuation errors in the initial data, the greater would be the deviation. That is why, if it is important to reduce the domain of influence of a priori assigned boundary-value conditions, the initial data could be in advance smoothed. Such an approach is also justified in those cases when an estimate of the variance of measured temperature is unknown and the condition of maximum proximity of regularized approximations is used for α sampling, which is more effectively "works" with small δ.

Another way of reducing the "end effect" assumes the use of a priori boundary conditions of higher order, e.g., in the assignment of difference analog of the second derivative of unknown function at the ends of time interval, instead of the first derivative.

The following *conclusion* can be made from the analysis of obtained curves $q^\alpha(\tau)$ for different temperature sensor locations: reconstruction of the surface heat flux by the regularization method is performed with the accuracy sufficient for use in practice in the case when a temperature sensor is far from the body surface so that it still responds to temperature changes at this measurement point.

It should be noted in addition that it is possible to noticeably increase the calculation accuracy of $q(\tau)$, $\tau \in [0, \tau_m]$ provided that the time domain of problem solving be extended up to $[0, \tau_m + \Delta\tau_m]$, where $\Delta\tau_m$ depends on thermal depth $d/\sqrt{a}$ of the temperature measurement point. Such a *technique* permits to compensate the initial uncertainty in temperature data and inaccuracy in the assignment of c_{1m}.

Consider one more example when the nominal curve $q(\tau)$ is of triangular form (Fig. 7.11). This example is chosen to check the efficiency of the algorithm treated in one of the most complicated case from methodical standpoint when continuity of the reconstructed function derivative is broken.

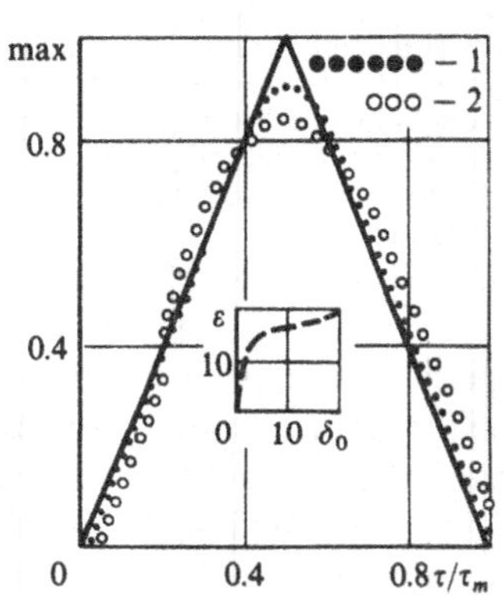

Fig. 7.11. The results of boundary condition reconstruction obtained for the normal distribution of errors in the input temperatures, providing exact assignment of coefficients c_{10} and c_{1m}, $\Delta Fo_d = 0.02$ (full curve is a true solution, dashed curve is the estimate of reconstruction accuracy, ε, %, δ_0, %): 1 – $\delta_0 = 0.01\, T^*_{max}$; 2 – $\delta_0 = 0.1\, T^*_{max}$.

The results of computations $q^{\alpha}(\tau)$ and $\varepsilon(\delta_0)$ are given in Fig. 7.12 for different temperature perturbations submitted to the normal distribution. Fig. 7.12 depicts the influence of the regularization parameter value on reconstruction of the "triangular" heat flux, provided that we have an exact assignment of the input temperature $T^*(\tau)$. It is seen that smoothing of the function peculiarity at point $\tau = \tau_m/2$ takes place and in doing so this effect gains as the input data error increases. Nevertheless, the accuracy of heat flux determination can be qualified as satisfactory.

With small errors in the initial data the influence of approximation error of the integral operator on the solution (this error increasing as ΔFo_d rises) becomes comparable with the influence of the initial data error. The operator approximation error at $m \geqslant 25$ could be considered negligible in the given examples.

Of interest is the question of possible application of the null order regularization when the smoothing functional has the following form:

$$\Phi_\alpha[q] = \| Aq - f \|^2_{L_2} + \alpha \| q \|^2_{L_2} .$$

In this case there is no uniform convergence of regularized solutions. It follows from the general Euler equation (7.20) that the regularized solutions at $r_0 = 1, r_1 = r_2 \cdots = r_k = 0$ and $u^* = 0$ will be "fixed" at the end $\tau = \tau_m$ by null

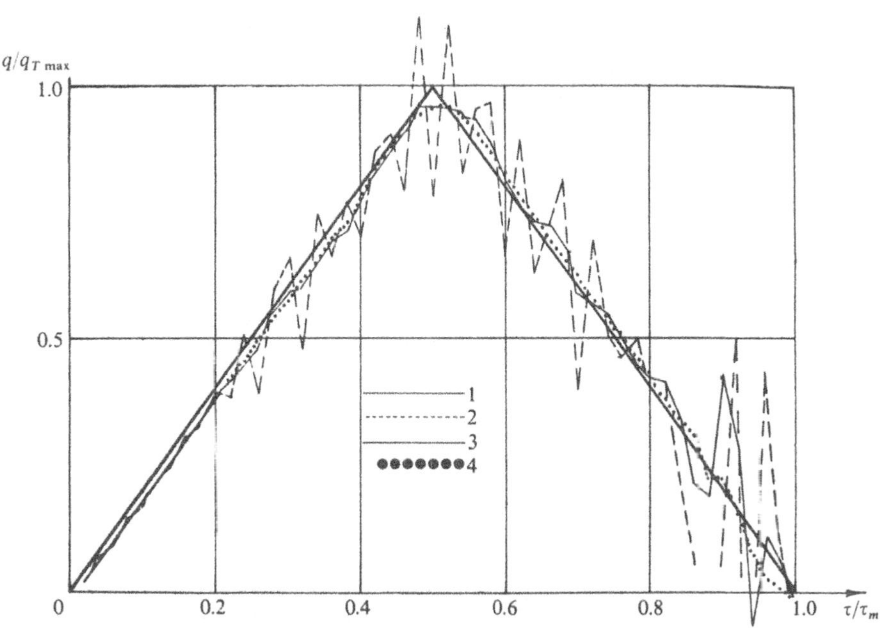

Fig. 7.12. The influence of the regularization parameter on reconstruction of the boundary condition in the case of undisturbed data: 1 is the unknown solution; 2, 3, 4 are the solutions at $\alpha = 0.1 \cdot 10^{-12}$; $0.6 \cdot 10^{-11}$; $0.3 \cdot 10^{-9}$, respectively.

value. Therefore, such a method of regularizing the IHCP solutions can yield acceptable results provided that a boundary function under reconstruction is close to zero at $\tau = \tau_m$. To illustrate this, Fig. 7.13 shows approximations to a "triangular" heat flux, which have been gained in the process of null regularization.

The sampling of approximations in all of the considered cases was made by the residual principle and according to a quasioptimal value of the regularization parameter. The value of temperature measurement error was taken equalling $\delta_{L_2} \simeq \sigma_0 \sqrt{\tau_m}$. The standard deviation σ_0 for the normal distribution was determined by a "three-sigma" rule, i.e. $\sigma_0 = \delta_0/3$; for the uniform distribution – by formula $\sigma_0 = \delta_0/\sqrt{3}$.

Calculations showed that in some vicinity of value $\alpha = \alpha_r$ (by residual) there is often a quasioptimal value α_{qo} of the parameter. It is usually to the left of α_r.

For other integral statements of one-dimensional IHCP (a slab model of finite thickness, the moving boundaries of the body, as well as other types of reconstructed and assigned boundary conditions) numerical simulation provides results which are qualitatively analogous to those considered above for the case of a thermally thick body. This proves efficiency of using the regularization method for processing and interpreting of data in various model and full-scale heat experiments. Fig. 7.14 illustrates the results of heat flux reconstruction for a slab with the inner thermally isolated surface, at which the temperature values $T^*(\tau)$ are assigned. Here the approximation steps ΔFo_b are considerably smaller

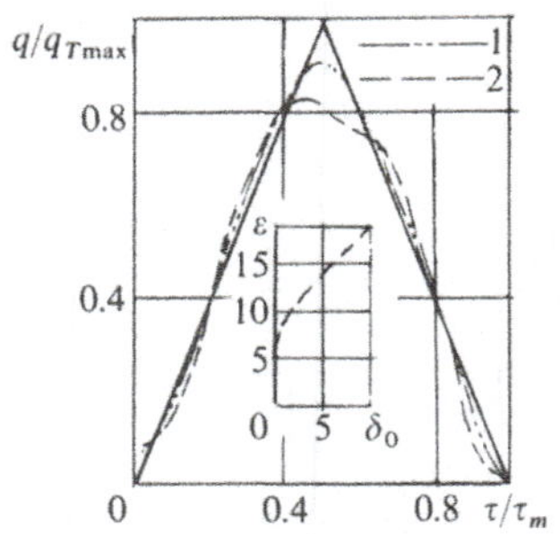

Fig. 7.13. The results of boundary condition reconstruction by the regularization method of null order with the uniform error distribution in initial temperatures, $\Delta Fo_d = 0.02$ (full curve is a true solution, dashed curve is the estimate of reconstruction accuracy, ε, %, δ_0, %): 1 – $\delta_0 = 0.01\, T^*_{\max}$; 2 – $\delta_0 = 0.1\, T^*_{\max}$.

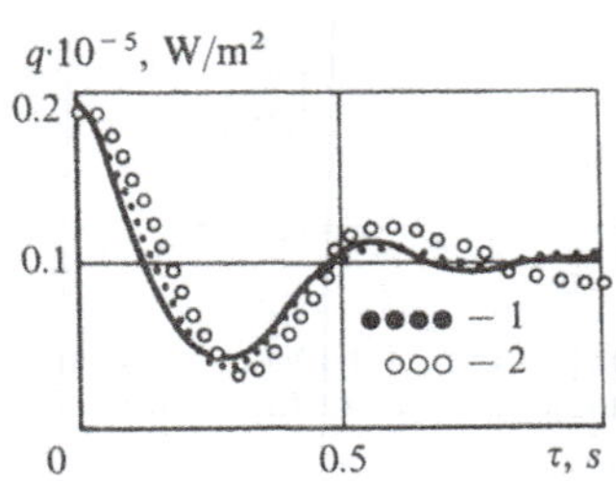

Fig. 7.14. The results of heat flux reconstruction in the case of thermally isolated slab with the normal error distribution in temperature data, $d = b$, $\Delta Fo_d = 0.01$ (full curve is a true solution): 1 is the solution with $\delta_0 = 0.01 T^*_{\max}$ 2 is the solution with $\delta_0 = 0.1 T^*_{\max}$

than $\Delta Fo_{b_{cr}}$, i.e. the critical step for a direct algebraic method of solution, and the input information being known with random errors.

7.6 By-Interval Regularization of a Non-linear Inverse Problem

Let us try to extend the regularization principles developed for the linear IHCP to include an approximate solution of the non-linear problem. Central to one of the possible approaches [10] is a linearization in time intervals of the initial problem statement with successive regularization of obtained problems according to the scheme stated above.

Consider a quasilinear equation of heat conduction in a solid body

$$C\frac{\partial T}{\partial \tau} = \frac{\partial}{\partial x}\left(\lambda \frac{\partial T}{\partial x}\right); \quad \lambda = \lambda(T),\ C = C(T).$$

Assume that the second boundary-value problem is stated for a slab with one side being exposed to the action of variable heat flux $q(\tau)$, and the other one – thermally isolated. The initial temperature profile is known.

Approximate the given problem by a suitable difference scheme (see, e.g., Sect. 5.1.1). As a result we get a computational algorithm, enabling the temperature field in the slab $T_{ln} = T(l\Delta x, n\Delta\tau)$, $l = 0, 1, \ldots, L$, $n = 1, 2, \ldots, m$ to be found from known $q(\tau)$, provided that the thermal characteristics are assigned.

Consider now an approach to be used the regularization method for solving the inverse problem, when the unknown function $q_n = q(n\Delta\tau)$, $n = 1, 2, \ldots, m$ is sought from the temperature $T_\varkappa(\tau)$ measured at some point $x_\varkappa = \varkappa\Delta x$.

In the time segment under consideration $0 \div \tau_m$ we separate an interval $\Delta\tau_N = \tau_{n=N-1} \div \tau_{n=N+k-1}$ of $k\Delta\tau$ length ($1 \leqslant N \leqslant m - k + 1$, $k > 0$ – the integer) and linearize (in it) the problem of temperature determination, assuming the thermal characteristics to be functions of coordinate and correspond to a profile of temperature at the moment $\tau = \tau_{N-1}$: $\lambda_{ln} = \lambda(T_{l,N-1})$, $C_{ln} = C(T_{l,N-1})$.

Then, according to the general principle of superposition we present temperature $T_\varkappa$ as a sum of solutions correspondingly at the zero boundary condition and the zero initial temperature distribution:

$$T_\varkappa(\tau) = T_\varkappa(\tau)\Big|_{\substack{q=0\\ \psi = T_{N-1}(x)}} + T_\varkappa(\tau)\Big|_{\substack{q=q(\tau)\\ \psi=0}}. \tag{7.35}$$

By applying the method of unit heat input at the finite time interval [118, 165], turn to a difference form of expression (7.35). As a result we obtain the following system of linear algebraic equations with triangular matrix ($j = 1$ corresponds to $n = N$) for defining the grid function q_n in the interval:

$$\sum_{j=1}^{J} q_j \varphi_j^N = f_j^N, \quad J = 1, 2, \ldots, k,$$

where

$$\varphi_j^N = T_{\varkappa j}\Big|_{\substack{q_{j=1}=1,\, q_{j>1}=0 \\ \psi_l = T_{1,N-1}}} - T_{\varkappa j}\Big|_{\substack{q_{j\geq 1}=0 \\ \psi_l = T_{1,N-1}}} ;$$

$$f_j^N = T_{\varkappa j}^* - T_{\varkappa j}\Big|_{\substack{q_{j\geq 0}=0 \\ \psi_l = T_{1,N-1}}} ;$$

$T_\varkappa^*$ is the experimental temperature.
Select a regularized residual in the following form:

$$\Phi_\alpha[q_j] = \sum_{j=1}^{k}\left(\sum_{j=1}^{J} q_j \varphi_j^N - f_j^N\right)^2 + \alpha \sum_{j=1}^{k}(q_{j+1} - q_j)^2 .$$

Find partial derivatives of Φ_α with respect to all q_j and equate them to zero. We get a system of k equations with $k + 2$ unknown quantities. To solve this problem, it is necessary to add some boundary conditions. Let it be assumed for the first series of points $n = 1, 2, \ldots, k$ $(N = 1)$ that

$$q_{j=0} = q_{j=1} \tag{7.36}$$

$$q_{j=k} = q_{j=k+1} \tag{7.37}$$

For the following series of points $(N > 1)$

$$q_{j=0} = q_{n=N-1} \tag{7.38}$$

$$q_{j=k} = q_{j=k+1} . \tag{7.39}$$

Equalities (7.36), (7.37) and (7.39) correspond to "natural boundary conditions". Value $q_{n=N-1}$ is the extreme "left" quantity of heat flux, known from a solution of the previous regularized system.

As a result, for the selected interval $\Delta\tau_N$ we get a set of algebraic equations, from the solution of which at a pre-assigned value of the regularization parameter α we obtain function $q^\alpha(\tau)$. To form the coefficients and right-hand sides of this system, it is necessary to solve the direct heat conduction problem $2k$ times with $q_n = 0$ and $q_{n=N} = 1$, $q_{n=N} = 0$, respectively $(n = N, \ldots, N + k - 1)$.

The selection of regularization parameter α is made by the condition of given residual level $\varkappa(\alpha) = \delta$, where δ is the input data error, corresponding to the time interval considered:

$$\varkappa(\alpha) = \begin{cases} \rho_\alpha = \sum_{J=1}^{k}\left(\sum_{j=1}^{J} q_j \varphi_j^N - f_j^N\right)^2 & \text{or} \\ \varphi_\alpha = \rho_\alpha + \alpha \sum_{j=1}^{k}(q_{j+1} - q_j)^2 . \end{cases}$$

Remark. Since the initial problem of boundary condition reconstruction is reduced in this algorithm to the solution of a number of individual linear problems and, in addition, it is required to find some optimal value α, we have to look for ways of reducing the computation time. This problem can be solved by automating the process of α-parameter determination (See Sect. 7.4.3). The selection of α by the condition of inner convergence for the regularized solutions is of no use in this case because of the difficulties of automation of this process.

The first M values are chosen from those obtained q_j^α ($1 \leqslant M < l$), which are taken as the final results: $q_{J \leqslant M}^\alpha = q_n$ ($N \leqslant n \leqslant N + M - 1$). The remaining values are rejected awaiting determination through further calculation.

Having computed the heat fluxes q_n one solves the direct problem for finding the temperature field T_{ln}, $N \leqslant n \leqslant N + M - 1$. The temperature profile $T_{l,N+M-1}$ is assumed to be the initial distribution while solving the analogous regularized system at the next time interval.

It is advisable to take the order k of the regularized system for the first points in time higher than for the subsequent sets of points. Since the "natural" boundary conditions $q' = 0$ are to be set at the ends of intervals (when the corresponding conditions are unknown), the solution in the result would "deviate" from a true one in some neighbourhood of these boundaries. The obtained approximate solution will be accurate enough somewhere up to the midst of the given interval. Hence, the first values of q_j till the middle of the interval could be taken as the best possible quantities; one could avoid specifying them by further calculation (the more so as changes of the thermal properties at the very initial stage are usually not large).

The interval lengths $\Delta\tau_N$, which may be unequal, as well as the number M_N should be chosen in accordance with the problem being solved and the solution to be expected. For thermal properties of the body, which change weakly with temperature, and for external heat loads, which change slowly, $\Delta\tau_N$ and M_N can be enlarged and, naturally, they should be decreased when more intensive changes of the heat fluxes occur and the thermal properties are more dependent of the temperature (especially, it concerns M).

The scheme suggested for solution of the nonlinear inverse problem can be extended to the generalized heat conduction equation

$$C\frac{\partial T}{\partial \tau} = \frac{\partial}{\partial x}\left(\lambda\frac{\partial T}{\partial x}\right) + K\frac{\partial T}{\partial x} + S\,.$$

It is assumed that only a finite-difference equivalent of the equation changes in this case, and the coefficient K is a function of x alone on the regularization interval (by analogy with λ and C). The heat source S is approximated on this time interval by a linear function of the temperature.

7.7 Regularized Continuation of the Solution of a Nonlinear Heat Conduction Equation

Regularization of numerical continuation of temperature field when the Cauchy conditions (temperature and heat flux) are given in a certain inner point of the body, can be realized by the introduction of a system of regularizing functionals, corresponding to space layers of the difference grid [12]

$$\Phi_\alpha [T_l] = \|M_l T_l - g_l\|^2_{Em} + \alpha_l \Omega_l, \tag{7.40}$$

where l is the index of a space layer; M_l is a transfer matrix from one space layer to another; $T_l = [T_{l_1}, T_{l_2}, \ldots, T_{l_m}]^T$ is an unknown vector of temperatures in lth layer; g_l is a vector of known parameters; α_l is a regularization parameter; Ω_l is a stabilizing functional. One can take $\Omega_l = k_1\|\Delta T_l\|^2_{Em} + k_2\|\Delta^2 T_l\|^2_{Em}$, $k_1 \geqslant 0$, $k_2 > 0$ which is in line with the second order regularization (Δ and Δ^2 are the first and second finite differences, respectively).

After fulfilling the operations connected with minimization of (7.40) we obtain a regularized system of algebraic equations for every lth coordinate section, and by solving of it for the given value of parameter α_l we gain T_l^α.

The transfer matrix M_l and vector g_l are defined by the form of approximation of heat conduction equation and they may correspond to the implicit or explicit finite difference schemes. These problems are investigated in [21, 22].

The choice of regularization parameters α_l is carried out according to the residual principle or by values of the regularized functional. For this purpose, the iterational search of the root $\alpha = \alpha_r$ of the following equation is arranged

$$\rho_l(\alpha) \equiv \sum_{n=1}^{m} \left(T^\alpha_{Ln} - T^*_n \right)^2 = \delta^2 \tag{7.41}$$

or the root $\alpha = \alpha_f$ of the equation

$$\varphi_l(\alpha) \equiv \rho_l(\alpha) + \alpha\Omega_l = \delta^2, \tag{7.42}$$

where T^α_{Ln} is the temperature calculated in point $x = b$; T^*_n is the experimental temperature; $\delta^2 \simeq \sum_{n=1}^{m} \sigma^2_n$ is the mean square error.

Vector T^α_L is obtained from a solution of the direct heat conduction problem in the domain $x_l \leqslant x \leqslant b$, $0 \leqslant \tau \leqslant \tau_m$ for a given condition $-\lambda(T(b,\tau)) \dfrac{\partial T(b,\tau)}{\partial x} = q^*(\tau)$ on the right-hand boundary and for temperature T^α_l of the left-hand boundary calculated for some value of α.

Equations (7.41) and (7.42) can be solved by the Newton method combined with iterations in coefficients.

This is a way to calculate the temperature field in the whole region. The unknown heat flux is determined from a finite difference approximation of the expression $q(\tau) = -\lambda(T(0,\tau) \cdot \partial T(0,\tau))/\partial x$.

Note one **important aspect** of the problem. As was pointed out in section 7.4.1, for a successful application of the regularized residual principle one should be sure that the approximation errors are smaller than the initial data errors. A characteristic feature of the residual principle is connected with this condition when the numerical solution of IHCP in the Cauchy formulation is regularized. It lies in the fact that in order to obtain a desired value of α_l, corresponding to a certain coordinate layer, one need to calculate the temperature field in the region between this layer and the point in which the Cauchy conditions are given. When going to the next layer, the region for a direct problem solution is enlarged by one step Δx_l. When a finite difference analog used for passing from one space layer to another in continuing the temperature field and one taken for solving the direct problem are not coordinated with respect to the accuracy, it is possible to encounter a large error accumulation in the calculations. As a result, a situation arises when the iterational procedure for search of the parameter α does not converge. Such a defect can become apparent for implicit schemes of the temperature field continuation, because in this case the computational formulas for direct and inverse problems are constructed according to different approximating relations.

This defect can be eliminated if *one and the same finite difference representation* is used for solving both direct and inverse problems. Such a possibility is provided by a T-type scheme of approximation (see Fig. 5.3a), which is explicit when solving a boundary-value heat conduction problem. The corresponding algorithm is constructed in [22] for the boundary condition reconstruction by solving a nonlinear IHCP in the domain with travelling boundaries. The practical well-posedness analysis of this method, performed in the process of solving model problems with different variants of simulated errors, showed efficiency of the algorithm in applications to various experimental investigations.

Following work [22], we dwell on this algorithm in more detail. Complicate the problem statement by turning to the generalized nonlinear heat conduction equation, considered in the domain with moving boundaries. Such a mathematical model can be in particular used to investigate heating of ablative composite materials in their interaction with a hot gas flow.

Let it be required to determine the temperature field in the domain $\{X_1(\tau) \leqslant x \leqslant X_2(\tau), 0 \leqslant \tau \leqslant \tau_m\}$ and heat conditions at the boundary $X_1(\tau)$ by a solution of the following generalized Cauchy problem:

$$C(T)\frac{\partial T}{\partial \tau} = \frac{\partial}{\partial x}\left(\lambda(T)\frac{\partial T}{\partial x}\right) + K(T)\frac{\partial T}{\partial x} + S(T),\ X_1(\tau) < x < X_2(\tau), 0 < \tau \leqslant \tau_m$$

$$T(x,0) = \xi(x), \quad X_1(0) \leqslant x \leqslant X_2(0);$$

$$-\lambda(T(X_2(\tau),\tau))\frac{\partial T(X_2(\tau),\tau)}{\partial x} = q^*(\tau);$$

$$T(X_2(\tau),\tau) = T^*(\tau),$$

where $C(T)$, $\lambda(T)$, $K(T)$, $S(T)$, $\xi(x)$, $q^*(\tau)$, $T^*(\tau)$ are known functions. Using new variables

$$y = \frac{x - X_1(\tau)}{X_2(\tau) - X_1(\tau)}, \quad t = \tau,$$

go to the rectangular domain of integration. Then we obtain the problem

$$C(T)\frac{\partial T}{\partial t}=\frac{1}{[X_2(t)-X_1(t)]^2}\frac{\partial}{\partial y}\left(\lambda(T)\frac{\partial T}{\partial y}\right)$$
$$+\frac{C(T)\,\{\dot{X}_1(t)+y[\dot{X}_2(t)-\dot{X}_1(t)]\}+K(T)}{X_2(t)-X_1(t)}\frac{\partial T}{\partial y}-S(T)\,.$$

$$0<y<1,\qquad 0<t\leqslant\tau_m\,; \tag{7.43}$$

$$T(y,0)=\varphi(y),\quad 0\leqslant y\leqslant 1\,; \tag{7.44}$$

$$-\frac{\lambda(T(1,t))}{X_2(t)-X_1(t)}\frac{\partial T(1,t)}{\partial y}=q^*(t)\,; \tag{7.45}$$

$$T(1,t)=T^*(t)\,, \tag{7.46}$$

where $\dot{X}_i\equiv\dfrac{\mathrm{d}X_i}{\mathrm{d}t},\qquad i=1,2\,.$

Introduce a difference grid $\{y_l=\Delta y\cdot l,\ l=1,2,\ldots,L;\ t_n=\Delta t\cdot n,\ n=1,2,\ldots,m\}$. Perform integration in the direction of the spatial coordinate. Approximate derivatives at the node $(l+1,\,n)$ by the following difference relationships:

$$\left.\frac{\partial T}{\partial t}\right|_{l+1,n}\simeq\frac{T_{l+1,n}-T_{l+1,n-1}}{\Delta t}\,,$$

$$\left.\frac{\partial T}{\partial y}\right|_{l+1,n}\simeq\frac{T_{l+2,n}-T_{ln}}{2\Delta y}\,,$$

$$\left.\frac{\partial}{\partial y}\left(\lambda\frac{\partial T}{\partial y}\right)\right|_{l+1,n}\simeq\frac{1}{\Delta y}\left[\frac{\lambda_{l+2,n}+\lambda_{l+1,n}}{2}\,\frac{T_{l+2,n}-T_{l+1,n}}{\Delta y}-\frac{\lambda_{l+1,n}+\lambda_{ln}}{2}\,\frac{T_{l+1,n}-T_{ln}}{\Delta y}\right].$$

Then a difference analog of equation (7.43) for every spatial layer with number l is

$$B_{ln}\,T_{ln}=F_{ln}\,,\qquad n=1,2,\ldots,m;\ l=L-2,L-3,\ldots,0\,, \tag{7.47}$$

where

$$B_{ln}=\frac{\lambda_{l+1,n}+\lambda_{ln}}{2\Delta y^2\,C_{l+1,n}\,(X_{2n}-X_{1n})^2}-\frac{C_{l+1,n}[\dot{X}_{1n}+\Delta y(l+1)(\dot{X}_{2n}-\dot{X}_{1n})]+K_{l+1,n}}{2\Delta y\;C_{l+1,n}(X_{2n}-X_{1n})}\,,$$

$$F_{ln} = \frac{T_{l+1,n} - T_{l+1,n-1}}{\Delta t} - \frac{\lambda_{l+2,n} + \lambda_{l+1,n}}{2\Delta y^2\, C_{l+1,n}\,(X_{2n} - X_{1n})^2}(T_{l+2,n} - T_{l+1,n})$$

$$+ \frac{\lambda_{l+1,n} + \lambda_{ln}}{2\Delta y^2\, C_{l+1,n}\,(X_{2n} - X_{1n})^2}\, T_{l+1,n}$$

$$- \frac{C_{l+1,n}[\dot{X}_{1n} + \Delta y(l+1)(\dot{X}_{2n} - \dot{X}_{1n})] + K_{l+1,n}}{2\Delta y\;C_{l+1,n}(X_{2n} - X_{1n})}\, T_{l+2,n} + \frac{S_{l+1,n}}{C_{l+1,n}}.$$

The nonlinear recurrent relation (7.47) permits the grid temperature to be restored at l-th spatial layer. The temperature values at zero time are determined from the condition (7.44): $T_{lo} = \varphi_l$, $l = \overline{1, L}$.

Relation (7.47) is a realization of one of the direct numerical methods of IHCP solution (see Chapter 5). In this case a temperature profile, received for a fixed value of index l is corrected by means of iterations through the thermal coefficients. The thermophysical characteristics computation for l-layer is carried out using values of temperature computed at the previous iteration. The grid solution, obtained for the previous layer, is used as an initial approximation (for a profile of temperatures). The iteration process ends when the condition is satisfied:

$$\max_n |T_{ln}^{(p)} - T_{ln}^{(p-1)}| \leqslant \varepsilon,$$

where p is the iteration number; $\varepsilon > 0$ is some small number. To begin the process of computation one is to know the profiles of temperatures at spatial layers with numbers L and $L-1$. The first profile is assumed by an experimental function $T^*(t)$ from the condition (7.46), the second one is computed from a finite difference analog of boundary condition (7.45). Omitting intermediate transformations, present the final recurrent formula to compute temperatures $T_{L-1,n}$, $n = \overline{1, m}$

$$T_{L-1,n}$$

$$= \frac{T_n^* + \Delta y\, q_n^*(X_{2n} - X_{1n})[2\lambda_{Ln} + \Delta y(X_{2n} - X_{1n})(C_{Ln}\dot{X}_{2n} + K_{Ln}) + (\lambda_{Ln} - \lambda_{L-1,n})]}{2\lambda_{Ln}^2}$$

$$+ \frac{\Delta y^2 (X_{2n} - X_{1n})^2}{2\lambda_{Ln}} \left[C_{Ln}\left(\frac{T_n^* - T_{n-1}^*}{\Delta t}\right) - S_{Ln}\right].$$

The numerical experiments have shown that the relations (7.47) permits obtaining a stable solution of the inverse problem providing the input temperatures T_n^* have low fluctuation errors ($\Delta T_n^* < 0.1\%\,(T_T^*)_{\max}$, T_T^* – exact temperature) and the step of integration in time Δt satisfies the approximate empirical relation

$$\Delta Fo_{\min} = \min_T\; [\lambda(T)\, \Delta t / \{C(T)\, [X_2(t) - X_1(t)]^2_{\max_\tau}\} \gtrsim 0.5\,.$$

If the input data errors and the time step do not satisfy these constraints, the described direct algorithm of solving the system (7.47) will not give any stable numerical solution. As has been already noted in chapters 4 and 5, the effect of errors in function $T^*(t)$ can be decreased performing an appropriate smoothing of the input temperatures. The constraint on the step of time integration cannot be eliminated due to ill-posedness of the problem. The value of Δt diminishing, the conditionality of the considered algorithm deteriorating steadily.

To get an algorithm for solving the algebraic problem (7.47), not bounded by the stability constraints on step Δt, let us utilize the method of regularization. Introduce a finite dimensional equivalent of the regularizing functional of the second order

$$\Phi_\alpha[T_l] = \sum_{n=1}^{m} (B_{ln} T_{ln} - F_{ln})^2 + \alpha_l k_1 \sum_{n=1}^{m} (T_{ln} - T_{l,n-1})^2$$
$$+ \alpha k_2 \sum_{n=1}^{m} (T_{l,n+1} - 2T_{ln} + T_{l,n-1})^2 , \tag{7.48}$$

where $k_1 \geqslant 0, \quad k_2 > 0$.

Assuming that thermophysical characteristics λ, C, K, S in the expression (7.48) be computed for l-th layer through temperature values obtained on the previous iteration, and minimizing $\Phi_\alpha[T_l]$ in all T_{ln}, $n = \overline{1,m}$, we receive a system of algebraic equations with a symmetric five-diagonal positive definite matrix

$$\sum_{r=n-2}^{n+2} a_{nr} T_{lr} = b_n, \quad n = 1,2,\dots,m , \tag{7.49}$$

where

$$a_{nn} = \begin{cases} B_{ln}^2 + \alpha(2k_1 + 5k_2), & n = 1;\ m-1 , \\ B_{ln}^2 + \alpha(2k_1 + 6k_2), & n = 2,3,\dots,m-2 , \\ B_{ln}^2 + \alpha(k_1 + k_2), & n = m ; \end{cases}$$

$$a_{n,n+1} = a_{n+1,n} = \begin{cases} \alpha(k_1 + 4k_2), & n = 1,2,\dots,m-2 , \\ \alpha(k_1 + 2k_2), & n = m-1 , \\ 0, & n = m ; \end{cases}$$

$$a_{n,n+2} = a_{n+2,n} = \begin{cases} \alpha k_2, & n = 1,2,\dots,m-2 , \\ 0, & n = m-1, m ; \end{cases}$$

$$b_n = B_{ln} F_{ln}, \quad n = 1,2,\dots,m-2 ;$$

$$b_{m-1} = B_{l,m-1} F_{l,m-1} - \alpha k_2 C_m ;$$

$$B_m = B_{lm} F_{lm} + 2\alpha k_2 C_m ;$$

$$C_m = \frac{d^2 T_l(\tau_m)}{dt^2} .$$

System (7.49) has the unique solution which is a regularized profile of temperatures at l-th spatial layer under definite value of parameter α_l. The problem (7.49) is better to solve by the square root method [145].

To choose a suitable value α_l of the regularization parameter use the residual principle in the form of equality (7.41), where $T^{\alpha}_{Ln} = T_{Ln}(\alpha_l)$, $n = \overline{1, m}$ is the solution of a direct heat conduction problem in the domain $\{y_l \leqslant y \leqslant l, 0 < t \leqslant \tau_m\}$ under condition (7.45) on the right-hand boundary and with given temperature on the left-hand boundary. The latter is computed by solving the system (7.49). To calculate the grid function T^{α}_{Ln} we follow an algorithm based on the same approximating relations as in system (7.47). In this case an implicit T-type scheme of approximation is used for the direct problem. The root α_l^* of equation (7.41) is determined by a chord method.

A computational algorithm for l-th spatial layer is constructed as follows:

1. Set an initial value α_l^0 of the regularization parameter (for example, assume $\alpha_l^0 = \alpha_{l-1}$).
2. Compute the regularized profile of temperatures $T_{ln}(\alpha^0)$, $n = \overline{1, m}$ at l-th layer, that is, solve the system (7.49) with value α_l^0 of the parameter.
3. Using the obtained regularized solution $T_{ln}(\alpha_l^0)$, $n = \overline{1, m}$ as the grid boundary condition, solve the direct problem of heat conduction in the domain $\{y_l \leqslant y \leqslant 1, 0 < t \leqslant \tau_m\}$ and define the grid function $T_{Ln}(\alpha_l^0)$, $n = \overline{1, m}$.
4. Compute the residual $\rho_l(\alpha_l^0)$ and solve the equation (7.37) by the chord method (steps 2 and 3 are used here).

Note that in this calculation we should successively correct the values of coefficients λ, C, K, S in the grid nodes. The computation of these coefficients must be performed by using a sufficiently smooth temperature field. Otherwise, an inadmissible distortion of the solution takes place, and the iterative process of determining root α_l^* may not converge. That is why, the correction of thermophysical coefficients is carried out only when $\rho_l(\alpha_l) > \delta^2$. In this case the regularized solution $T_{ln}(\alpha_l)$, $n = \overline{1, m}$ is in advance of a smooth nature.

In such a manner we perform computations of temperature field in the whole domain considered. The heat flux on the boundary is defined from a finite difference approximation of the expression

$$q(t) = -\frac{\lambda(T(0,t))}{X_2(t) - X_1(t)} \frac{\partial T(0,t)}{\partial y},$$

for example,

$$q_n = \frac{\lambda_{0n}}{X_{2n} - X_{1n}} \left(\frac{T_{2n} - 4T_{1n} + 3T_{0n}}{2\Delta y} \right), \quad n = 1, 2, \ldots, m.$$

A computational experiment, usually, has been constructed according to the following procedure. Boundary conditions of the second order were assumed and a direct problem of heat conduction was solved. In the result the temperatures on boundaries $X_1(t)$ and $X_2(t)$ were determined as functions of time. The temperature on boundary $X_2(t)$ after adding the fluctuating errors (the saw-tooth perturbations, as well as the normal and uniform distributions of errors) was utilized as the input temperature history in solving an inverse problem.

The direct and inverse problems have been solved using the same difference grid, that enabling to ignore the errors caused by a difference approximation of the heat conduction operator. Numerical simulation has been carried out in advance to choose parameters of the difference grid, which provided a required accuracy of solution. In the process of methodical testing of the algorithm the effect of values of parameters k_1 and k_2 was studied with respect to the accuracy of the inverse problem solution. Good results were obtained for $k_1 = 0, k_2 = 1$.

Fig. 7.15 shows the results of solving such an inverse problem when the temperature on the heated surface of the body is recovered. The input temperatures in this example were disturbed by a random error distributed at every moment of time by the normal distribution with a span of error $\pm 0.05T^*_{max}$. Dependences of the thermophysical properties on temperature corresponded to a graphite and is illustrated in Fig. 7.16. The heat flux at the right boundary of the body $q^*(\tau)$ in the given case was equal to zero. The remaining parameters $K = S = 0$, $X_1(\tau) = 0.001\tau$, $X_2(\tau) = 0.015$ m.

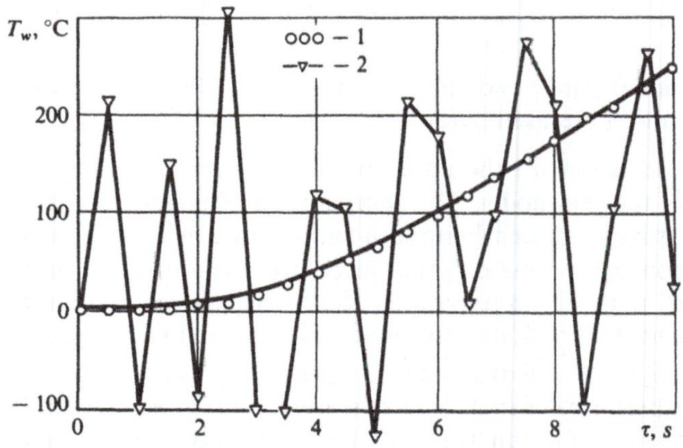

Fig. 7.15. The results of solving an inverse problem: 1 – a curve sought for; 2 – a regularized solution for disturbed data; 3 – a numerical solution with $\alpha = 0$; 4 – a solution for undisturbed data

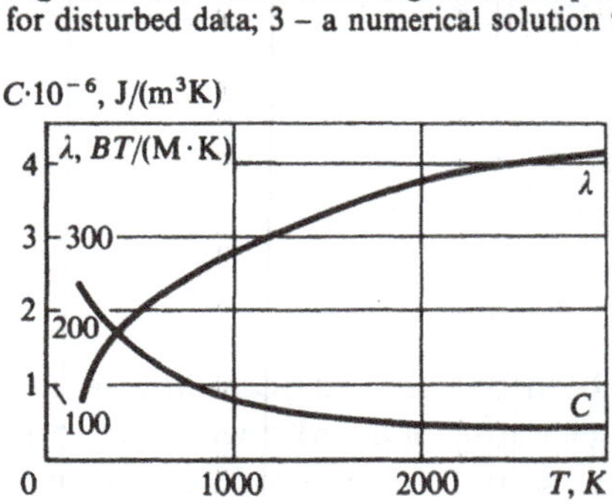

Fig. 7.16. Dependence of thermophysical properties on temperature

7.8 The Regularization of a Two-Dimensional Inverse Problem

Let us consider a possible form of regularization for a two-dimensional IHCP, represented by the integral equation (4.22). The approximation of this equation results in the matrix analog of the form (4.24)

$$A_\Delta u = f^* .$$

Choose the following regularizing functional:

$$\Phi_\alpha[u] = \int_0^{\tau_m} d\tau \int_0^b [f(x,\tau) - f^*(x,\tau)]^2 dx + \alpha \int_0^{\tau_m} d\tau \int_0^b \left(\frac{\partial^2 u}{\partial x\,\partial \tau}\right)^2 dx,$$

where $\alpha > 0$ is the regularization parameter; $f(x,\tau) = \int_0^\tau d\xi \int_0^b u(x',\xi)\, K(x,x',\tau-\xi)\, dx'$. The corresponding finite difference analog has the following form

$$\hat{\Phi}_\alpha[u] = \sum_{n=1}^{m} \sum_{l=1}^{L} (f_{ln} - f^*_{ln})^2 + \alpha \sum_{n=1}^{m} \sum_{l=1}^{L} (u_{ln} - u_{l,n-1} - u_{l-1,n} + u_{l-1,n-1})^2 .$$

The minimization of this finite dimensional form of the regularized residual, under the assumption that on the edge of the grid domain the following conditions are satisfied[1]

$$u_{on} = u_{ln},\; u_{L+1,n} = u_{Ln}, \quad n = 0, 1, \ldots, m+1 ;$$

$$u_{lo} = u_{l_1},\; u_{l,m+1} = u_{lm}, \quad l = 0, 1, \ldots, L+1 .$$

gives us the system of linear algebraic equations for determination of unknown values

$$A_\Delta^T A_\Delta u + \alpha C u = A_\Delta^T f^* . \tag{7.43}$$

Matrix C, which provides a smoothing action, has the following block form:

$$C = \begin{bmatrix} S & -S & & & & & 0 \\ -S & P & -S & & & & \\ & -S & P & -S & & & \\ & & \ddots & \ddots & \ddots & & \\ & 0 & & -S & P & -S & \\ & & & & -S & S & \end{bmatrix},$$

[1] These equalities follow from the simplest finite difference approximations of the natural boundary conditions $\partial u(0,y,\tau)/\partial x = \partial u(b,y,\tau)/\partial x = \partial u(x,y,0)/\partial \tau = \partial u(x,y,\tau_m)/\partial \tau = 0$; other conditions can be stated and the corresponding to them finite difference analogs can be assumed, if necessary.

where $j = Lm$

$$S = \begin{bmatrix} 1 & -1 & & & & & \\ -1 & 2 & -1 & & & 0 & \\ & -1 & 2 & -1 & & & \\ & \cdot & \cdot\cdot & \cdot\cdot & & & \\ & & & -1 & 2 & -1 \\ 0 & & & & -1 & 1 \end{bmatrix}_{L\times L},$$

$$P = \begin{bmatrix} 2 & -2 & & & & & \\ -2 & 4 & -2 & & 0 & & \\ & -2 & 4 & -2 & & & \\ & \cdot & \cdot\cdot & \cdot\cdot & & & \\ 0 & & & -2 & 4 & -2 \\ & & & & -2 & 2 \end{bmatrix}_{L\times L}.$$

The system (7.43) is to be solved for various values of α, and a desired element is to be chosen from the obtained family of regularized approximate solutions $\{u^{\alpha}(x, \tau)\}$. To save computer time the process of parameter choosing should be automized (See section 7.4.3).

7.9 Summary

1. Stable algorithms are developed for solving the linear and nonlinear, one-dimensional and two-dimensional boundary IHCP, based on the variational principle of regularization.
2. The numerical investigations showed a capacity for work of these algorithms and a sufficiently high accuracy of reconstruction of the heat boundary conditions. In the first place this conclusion concerns the algorithm of solving the linear one-dimensional problems. This algorithm is widely used in practical investigations. The regularized algorithms for solving the nonlinear problems are more complicated, labour-consuming; the solutions obtained by them have larger errors. Note also, that methods, considered in the given chapter and in the previous ones, can be referred only to a solution of the boundary IHCP. That is why the problem of constructing effective methods to solve nonlinear inverse problems of various types still remains urgent.

The iterative regularization offers possibilities for solving both linear and nonlinear inverse problems of heat transfer. This will be the subject of our discussion in the following chapter.

Chapter 8

Iterative Regularization of Inverse Problems

Introduction

The algorithms constructed in Chaps. 4–7 are used for the solution of boundary IHCPs. Lately, the area of practical applications of methodology, based on inverse heat-transfer problems, has expanded considerably, which necessitated solving other types of inverse problems as well. As conducted investigations show, *iterative regularization* (Chap. 6) appears to be one of the most efficient and universal approaches for the construction of stable algorithms for solving ill-posed problems. With this method algorithms convenient for practical utilization can be obtained for the solution of inverse heat-transfer problems in various formulations (linear and non-linear, one-dimensional and multi-dimensional), in domains with fixed and travelling boundaries, with minimal necessary composition of initial data and overdefinition. Also, the rigorous mathematical proof of this method for a wide class of inverse problems and modernizations of iterative algorithm to account for *qualitative and quantitative a priori information about an unknown solution* are found.

These questions are considered in the present chapter and cover the main results of studies in this area carried out by the author and his colleagues in recent years. The material is outlined as follows.

The major mathematical results on substantiation of an iterative regularization method will be given in Sect. 8.1. Section 8.2 is devoted to a derivation of analytical forms to compute gradients of functionals–residuals for the solution of various types of linear IHCP with thermophysical coefficients depending on the coordinate and time.

In Sect. 8.3 the statements of different non-linear inverse problems of one-dimensional heat conduction are formulated and a general form is obtained of the problem for an increment in temperature, necessary for derivation of a form of *an adjoint problem* statement. In Sect. 8.4 the formulations of adjoint problems are obtained and through their solution the expressions are found for a gradient of the residual for solving various IHCP, including the problems with several temperature measurements. Another universal method is described here on the formation of conditions of an adjoint problem which is illustrated on examples of boundary-coefficient IHCP and inverse problems of heat transfer in the engineering systems. Section 8.5 contains material on two different methods to allow for *qualitative* and *quantitative a priori information* concerning an unknown function both for continuous and parametrized forms of IHCP solution.

The inclusion of this information permits the reconstruction accuracy of causal characteristics to be increased. Besides, several techniques for completing iterative processes, other than a residual principle, are considered in this section when solving an ill-posed inverse problem; they do *not require knowledge of the value of the input data error*, the fact being ever such important in practice.

Examples of algorithms are presented in section 8.6 to solve linear and non-linear IHCP (a boundary problem – on determination of two unknown time-dependent heat fluxes, a coefficient one – on determination of thermal conductivity as a function of temperature and a problem of finding the source-term in heat conduction equation, dependent of two coordinates and time).

This chapter ends with section 8.7, where the results of solving a number of methodical examples are given, they illustrate efficiency of constructed algorithms. The *design of optimal experiments* is outlined in this section to obtain the greatest accuracy of required functions when solving a coefficient IHCP.

8.1 On the Rigorous Basis of the Iterative Regularization

In Chap. 6 the correct use of the iterative regularization principle was shown on the basis of heuristic considerations connected with the analysis of the behaviour of iterational processes and the possibility of their stopping according to the residual principle. Let us now consider the rigorous mathematical results related to this question.

We represent the general formulation of a certain inverse problem in the form of an operator equation of the first kind

$$Au = f, \quad u \in U, \quad f \in F \tag{8.1}$$

where U, F are Hilbert spaces, A is a linear continuous operator defined on U and u and f are unknown and observed elements respectively (u and f might be functions or vector-functions).

We shall assume further on that (8.1) has a unique solution for every element f from the image of operator A[1]. For precise operator A and a right-hand side f there exists a solution u_T of (8.1).

A well-posed statement of the problem (8.1) presumes the continuity of operator A^{-1}, inverse for A. But we shall assume this condition (see Chap. 2) to be violated, i.e. we shall assume that operator A^{-1} is not continuous.

In the same way as in Chap. 7 we introduce the following approximate initial data:

1. linear continuous operator A_h, approximating operator A under condition $\|A_h - A\| \leqslant \kappa(h)$, $\kappa(h) \to 0$, $h \to 0$;
2. right-hand side $f_\delta = f + \Delta f$, where $\Delta f \in F$ is a disturbance with the norm $\|\Delta f\|_F \leqslant \delta$.

[1] Generally this condition is not obligatory; in [37, 38] a substantiation of the method is made with no assumption of the uniqueness of problem solving.

Then the error in initial data is characterized by vector $\sigma = \{\delta, h\}$.

Suppose that a certain iterative algorithm is given which produces a sequence of approximations

$$u^{k+1} = F_A(u^k, \Delta f, h), \quad k = 0, 1, \ldots \tag{8.2}$$

beginning from given initial approximation u^0.

It is ascertained in [37, 38] that the algorithm (8.2) gives rise to the *regularizing family of operators* (see Sect. 2.6.4) with parameter of regularization k, if the following conditions are valid:

1. for every $u^0 \in U$ sequence (8.2), constructed for precise data ($\Delta f = 0, h = 0$), converges to an unknown solution u_T;
2. mapping $F_A(u, \Delta f, h)$ is continuous at all points $\{u, 0, 0\}$ except perhaps point $\{u_T, 0, 0\}$.

Thus in this case the choice of stable approximations from the corresponding iterative sequence is possible. In other words there exist such $\delta_0 > 0$, $h_0 > 0$ that for every $\delta < \delta_0, h < h_0$ an index of iteration $k = \bar{k}(\sigma)$ exists, for which $\| u^{\bar{k}(\sigma)} - u_T \|_U \xrightarrow[\sigma \to 0]{} 0$.

It was found that the above-stated requirements are satisfied, in particular, for *the steepest descent algorithm*

$$u^{k+1} = u^k - \beta_k J_u'^k, \quad \beta_k = \frac{\| J_u'^k \|_U^2}{2 \| A_h J_u'^k \|_F^2} \tag{8.3}$$

and the *conjugate gradient algorithm*

$$u^{k+1} = u^k - \beta_k S^k, \quad \beta_k = \frac{(J_u'^k, S^k)_U}{2 \| A_h S^k \|_F^2} \tag{8.4}$$

$$S^k = J_u'^k + \gamma_k S^{k-1}, \quad \gamma_0 = 0, \quad \gamma_k = \frac{\| J_u'^k \|_U^2}{\| J_u'^{k-1} \|_U^2}.$$

Here $J_u'^k = 2A_h^*(A_h u^k - f_\delta)$ is the gradient of functional $J^k = J[u^k] = \| A_h u^k - f_\delta \|_F^2$.

Thus, it is possible to construct the *regularizing algorithms* on the basis of iterative processes (8.3) and (8.4). For this purpose it is necessary to point out an admissible way for matching the index of the last iteration $\bar{k}$ to the initial data error σ. It can be constructed on the basis of the residual principle.

It was shown in [37, 38] that the algorithms of steepest descent (8.3) and conjugate gradients (8.4) with the choice of indexes of iterations respectively under conditions

$$\bar{k}(\sigma) = \min\left\{ k : \frac{J^k + J^{k+1}}{2\sqrt{J^k}} < \delta + \kappa(h) \| u_T \|_U \right\}, \tag{8.5}$$

$$\bar{k}(\sigma) = \min\left\{ k : \frac{(J^k + J^{k+1}) \sum_{i=0}^{k} \frac{1}{\| J_u'^i \|_U^2}}{2 \sum_{i=0}^{k} \sqrt{J^i} / \| J_u'^i \|_U^2} < \delta + \kappa(h) \| u_T \|_U \right\} \tag{8.6}$$

are regularizing algorithms, i.e. they produce stable approximate solutions and their accuracy would increase steadily by decreasing the initial data error – they would converge to an unknown solution by $\sigma \to 0$.

Let us analyse the condition (8.5). For every $\bar{k}$ the inequality $J^{k+1} < J^k$ holds. Hence, $\Phi^k = \dfrac{J^k + J^{k+1}}{2\sqrt{J^k}} < \sqrt{J^k}$. Alternatively, using Cauchy inequality $\frac{1}{2}(J^k + J^{k+1}) \geqslant \sqrt{J^k, J^{k+1}}$. Hence $\sqrt{J^{k+1}} \leqslant \Phi^k < \sqrt{J^k}$.

Thus, when approximation errors are not taken into consideration ($\kappa(h) = 0$) the stop of the steepest descent iterative process using the condition (8.5) is equal, within one iteration, to the stop by the condition

$$k = \bar{k}: \| Au^k - f_\delta \|_F \approx \delta \tag{8.7}$$

without crossing the level δ (Fig. 8.1).

In the case, when value δ lies between $\sqrt{J^k}$ and $\sqrt{J^{k+1}}$, the indexes of the last iterations chosen by the conditions (8.5) and (8.7) would be the same.

In a similar way it is also easy to prove for the conjugate gradient method that the stop of iterations using the condition (8.7) is almost interchangeably with the stop by the condition (8.6).

The considered results on the basis of iterative regularization are referred to as linear inverse problems. The application of such an approach for the solution of non-linear, ill-posed problems has been shown many times in computational experiments [15, 25, 26, 27, 33, 39, 41–43, 45, 47, 115].

Remark. Approximations u^{k+1}, $k = 0, 1, \ldots$ corresponding to the indexes of iterations $k \leqslant \bar{k}$ would monotonously approach the solution of Eq. (8.1). At the same time for $k > \bar{k}$ elements u^{k+1}, obtained by the steepest descent method, may no longer come near the unknown solution u_T. As for the conjugate gradient method the condition (8.6) could be used not for the stopping but for the renewal of the iterative process, if $J^{\bar{k}} > \delta^2$ [38]. Namely, after implementation of condition (8.6) one may proceed with the iterative process, taking $u_1^0 = u^{\bar{k}}$ for a new initial estimate and $J(u_1^0) = J^{\bar{k}}$ until the next implementation of condition (8.6) for certain number $\bar{k}_1$. Again if $J^{\bar{k}_1} > \delta^2$, then the renewal is carried out once more assuming that $u_2^0 = u^{\bar{k}_1}$, $J(u_2^0) = J^{\bar{k}_1}$. It continues like this

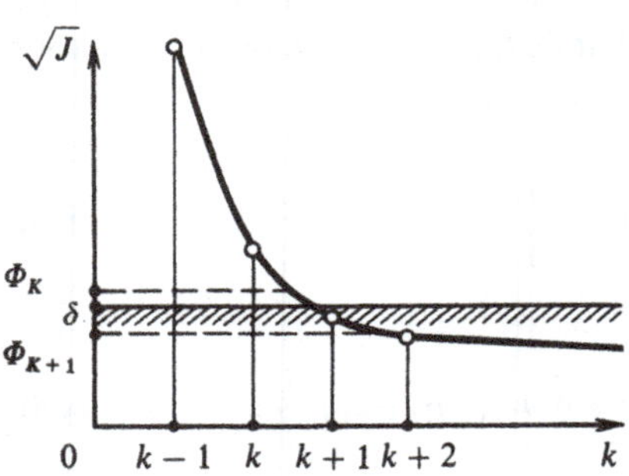

Fig. 8.1. Graphical illustration of stopping the iterative process by the residual criterion

until the condition (8.6) is fulfilled at the first iteration after the next renewal in turn. As a result accuracy of approximate solution becomes greater.

The theoretical results presented in this section imply that the regularized solutions of inverse problems using the algorithms considered in chapter 6 produce *the mean square approximations* of unknown functions. By means of iterative regularization the algorithms can also be obtained, assuring the *uniform convergence* of approximations and their derivatives (see Sect. 8.5).

8.2 General Formulation and Integral Forms of Linear Inverse Heat Conduction Problems. Gradient of the Residual Functional

As a consequence of the formulas (8.3) and (8.4) one must calculate the gradient of the corresponding residual functional for these iterative algorithms to be put into practice. In the case of linear IHCP which allow the construction of explicit forms of operators A, precise analytical expressions can be obtained for this purpose. Let us analyse this question. First of all we shall formulate a sufficiently general form of the one-dimensional inverse problem of generalized heat-conduction.

Consider the following mathematical model of heat-conduction with coefficients, depending upon coordinate and time

$$T_\tau = a_1(x,\tau)T_{xx} + a_2(x,\tau)T_x + a_3(x,\tau)T + S(x,\tau), \tag{8.8}$$

$$(x,\tau)\in Q = (0,b)\times(0,\tau_m]\,;$$

$$T(x,0) = \xi(x)\,; \tag{8.9}$$

$$[\alpha_1(\tau)T_x + \beta_1(\tau)T]_{x=0} = p_1(\tau)\,; \tag{8.10}$$

$$[\alpha_2(\tau)T_x + \beta_2(\tau)T]_{x=b} = p_2(\tau)\,. \tag{8.11}$$

Here and further in this chapter we use the following designations for partial derivatives: $z_y = \dfrac{\partial z}{\partial y}$, $z_{yy} = \dfrac{\partial^2 z}{\partial y^2}$.

The generalized form of representation of the boundary conditions (8.10) and (8.11) gives an opportunity to obtain the conditions of the first, second and third types by setting zero or nonzero coefficients $\alpha_i(\tau)$, $\beta_i(\tau)$

In this formulation we take for unknown values some of the functions $S(x,\tau)$, $\xi(x)$, $p_1(\tau)$, $p_2(\tau)$. For the uniqueness of the solution, when seeking for the source $S(x,\tau)$, one must know the whole temperature field: $f(x,\tau) = T(x,\tau)$, $(x,\tau)\in Q$. The problem of seeking for the initial temperature distribution $\xi(x)$ – retrospective IHCP – is often solved from information about the temperature in a certain moment of time: $f_m(x) = T(x,\tau_m)$. While determining functions $p_1(\tau)$ and/or $p_2(\tau)$ (boundary IHCP), and vector-functions $\{\xi(x), p_1(\tau)\}$ or $\{\xi(x), p_2(\tau)\}$ (boundary-retrospective IHCP) we suppose the temperature measurements to be given as time-functions at points N of the segment $[0,b]: f_i(\tau) = T(d_i,\tau)$, $\tau\in[0,\tau_m]$, $i=\overline{1,N}$ where $d_i < d_{i+1}$, $i=\overline{1,N-1}$, $0\leqslant d_1$ for $\alpha_1\neq 0$, $0<d_1$ for $\alpha_1=0$, $b\geqslant d_N$ for $\alpha_2\neq 0$, $b>d_N$ for $\alpha_2=0$. For the uniqueness of the solution of these problems (see Sect. 2.3) the following conditions are necessary: $N \geqslant 1$ for determination of one of the functions $p_1(\tau), p_2(\tau)$; $N \geqslant 2$ for determination of functions $\{p_1(\tau), p_2(\tau)\}$ jointly; $N \geqslant 1$, $\beta_2(\tau) \equiv 0$, $d_N = b$ while seeking vector-function $\{\xi(x), p_1(\tau)\}$; $N \geqslant 1$, $\beta_1(\tau) \equiv 0$, $d_1 = 0$ while seeking $\{\xi(x), p_2(\tau)\}$.

Operator A, produced by the boundary-value problem [see Eqs. (8.8)–(8.11)] for one or another unknown casual characteristic, may be constructed using the Green's function. Leaving aside the question of obtaining the corresponding integral representation of the solution of this boundary-value problem, we give the final results[2]:

for boundary conditions of the first kind ($\alpha_1 = \alpha_2 = 0, \beta_1 = \beta_2 = 1$) –

$$\begin{aligned} T(x,\tau) = {} & \int_0^\tau d\tau' \int_0^b S(x',\tau')G(x,\tau;x',\tau')dx' + \int_0^b \xi(x')G(x,\tau;x',0)dx' \\ & + \int_0^\tau p_1(\tau')\frac{\partial}{\partial x'}(a_1(0,\tau')G(x,\tau;0,\tau'))d\tau' \\ & - \int_0^\tau p_2(\tau')\frac{\partial}{\partial x'}(a_1(b,\tau')G(x,\tau;b,\tau'))d\tau'; \end{aligned} \tag{8.12}$$

for boundary conditions of the second kind ($\alpha_1 \neq 0, \alpha_2 \neq 0; \beta_1 = \beta_2 = 0$) and of the third kind ($\alpha_1 = \alpha_2 = 1, \beta_1 \neq 0, \beta_2 \neq 0$) –

$$\begin{aligned} T(x,\tau) = {} & \int_0^\tau d\tau' \int_0^b S(x',\tau')G(x,\tau;x',\tau')dx' + \int_0^b \xi(x')G(x,\tau;x',0)dx' \\ & - \int_0^\tau \frac{p_1(\tau')}{\alpha_1(\tau')}G(x,\tau;0,\tau')d\tau' + \int_0^\tau \frac{p_2(\tau')}{\alpha_2(\tau')}G(x,\tau;b,\tau')d\tau'; \end{aligned} \tag{8.13}$$

for combined boundary conditions ($\alpha_1 = 0, \beta_1 = 1, \alpha_2 \neq 0$) –

$$\begin{aligned} T(x,\tau) = {} & \int_0^\tau d\tau' \int_0^b S(x',\tau')G(x,\tau;x',\tau')dx' + \int_0^b \xi(x')G(x,\tau;x',0)dx' \\ & + \int_0^\tau p_1(\tau')\frac{\partial}{\partial x'}(a_1(0,\tau')G(x,\tau;0,\tau'))d\tau' \\ & + \int_0^\tau \frac{p_2(\tau')}{\alpha_2(\tau')}G(x,\tau;b,\tau')d\tau'; \end{aligned} \tag{8.14}$$

for combined boundary conditions ($\alpha_1 \neq 0, \alpha_2 = 0, \beta_2 = 1$) –

$$\begin{aligned} T(x,\tau) = {} & \int_0^\tau d\tau' \int_0^b S(x',\tau')G(x,\tau;x',\tau')dx' + \int_0^b \xi(x')G(x,\tau;x',0)dx' \\ & - \int_0^\tau \frac{p_1(\tau')}{\alpha_1(\tau')}G(x,\tau;0,\tau')d\tau' - \int_0^\tau p_2(\tau')\frac{\partial}{\partial x'}(a_1(b,\tau')G(x,\tau;b,\tau')d\tau'. \end{aligned} \tag{8.15}$$

[2] The corresponding analysis is given in: Alifanov, O.M. "Derivation of formulas for the gradient of the error in the iterative solution of inverse problems of heat conduction. I. Determination of the gradient in terms of the Green's function," J. Eng. Phys., 1987, Vol. 52, No. 3, pp. 352–359.

Here $G(x, \tau; x', \tau')$ in expression (8.12) is the Green's function of the first boundary-value problem; in expression (8.13) it is the Green's function of the second and third boundary-value problems respectively and in expressions (8.14) and (8.15) it is the Green's function of the corresponding combined boundary-value problems.

The Green's functions are defined by common means (see, for example, [154]). In general $G = G(x, \tau; x', \tau')$ satisfies the conditions

$$G_\tau = a_1 G_{xx} + a_2 G_x + a_3 G, \quad x \in (0, b), \tau > \tau';$$

$$[\alpha_1 G_x + \beta_1 G]_{x=0} = [\alpha_2 G_x + \beta_2 G]_{x=b} = 0, \quad x' \in (0, b), \tau > \tau';$$

$$\lim_{\varepsilon \to 0} \int_0^b G(x, \tau' + \varepsilon; x', \tau') f(x)\, dx = f(x') \quad \text{for } f(x) \in C[0, b], \quad \varepsilon > 0.$$

Considering any of the functions $S(x', \tau')$, $\xi(x')$, $p_1(\tau')$, $p_2(\tau')$ as unknown, we come to one or another formulation of inverse problem with linear integral operator A.

It is known that for linear operator A gradient of the residual $J[u] = \| Au - f \|_F^2$ is given by

$$J'_u = 2A^*(Au - f).$$

Indeed finding an increment of the functional by means of the equation $\| \theta - \varphi \|_F^2 = \| \theta \|_F^2 - 2(\theta, \varphi)_F + \| \varphi \|_F^2$ we obtain

$$\begin{aligned} J(u + \Delta u) - J(u) &= 2\,(Au - f, A\Delta u)_F + \| A\Delta u \|_F^2 \\ &= (2A^*(Au - f), \Delta u)_U + o(\| \Delta u \|) \end{aligned}$$

where A^* is an operator, adjoint of operator A.
This implies that $J'_u = 2A^*(Au - f)$.

Thus, to calculate J'_u it is necessary to find the operator A^* Let us consider this question examining the IHCP formulations one by one with various unknown functions assuming that all other values are given. In the case when a vector-function is unknown one should take the corresponding components of the gradient obtained below.

1. Heat source $S(x', \tau')$

In this case the residual functional is

$$J[S] = \int_0^{\tau_m} d\tau \int_0^b (AS - \tilde{f})^2\, dx\,. \tag{8.16}$$

where $AS = \int_0^\tau d\tau' \int_0^b S(x', \tau') G(x, \tau; x', \tau')\, dx'$

$$\tilde{f} = f(x, \tau) - \int_0^b \xi(x') G(x, \tau; x', 0)\, dx' - I(x, \tau)\,.$$

Function $I(x, \tau)$ is defined by the form of boundary conditions. It is an algebraic sum of the last two terms in expressions (8.12)–(8.15). For example, for boundary conditions of the first kind:

$$I(x,\tau)=\int_0^{\tau} p_1(\tau')\frac{\partial}{\partial x'}(a_1(0,\tau')G(x,\tau;0,\tau'))\,d\tau'$$
$$-\int_0^{\tau} p_2(\tau')\frac{\partial}{\partial x}(a_1(b,\tau')G(x,\tau;b,\tau'))\,d\tau'\,.$$

Considering integral operator A as a mapping from the space $L_2(Q)$ in $L_2(Q)$ we write the identity which defines the adjoint operator A^*:

$$(AS,e)_{L_2(Q)}=(S,A^*e)_{L_2(Q)}\,. \tag{8.17}$$

Here e and S are arbitrary elements from $L_2(Q)$. For our purposes it is convenient to set $e(x,\tau)=AS-f$.

By definition of the inner product in $L_2(Q)$, identity (8.17) implies

$$\int_0^{\tau_m} d\tau \int_0^{b} e(x,\tau)\,dx \int_0^{\tau} d\tau' \int_0^{b} S(x',\tau')G(x,\tau;x',\tau')\,dx'$$
$$=\int_0^{\tau_m} d\tau \int_0^{b} S(x,\tau)y^*(x,\tau)\,dx\,. \tag{8.18}$$

Function y^* in (8.18) is bound up with the gradient of functional (8.16) in the following way:

$$y^*=A^*e=\tfrac{1}{2}J'_s\,. \tag{8.19}$$

After interchanging the order of integration in the left-hand side of equality (8.18) and changing the denotions of the variables of integration $x\to x'$, $\tau\to\tau'$ in the right-hand one, we have

$$\int_0^{\tau_m} d\tau' \int_0^{b} S(x',\tau')\,dx' \int_{\tau'}^{\tau_m} d\tau \int_0^{b} e(x,\tau)G(x,\tau;x',\tau')\,dx$$
$$=\int_0^{\tau_m} d\tau' \int_0^{b} S(x',\tau')y^*(x',\tau')\,dx'\,.$$

From here by the arbitrariness of S and taking into account formula (8.19) we have

$$J'_S=2\int_{\tau'}^{\tau_m} d\tau \int_0^{b} e(x,\tau)G(x,\tau;x',\tau')\,dx'\,. \tag{8.20}$$

In many problems of practical interest the heat source $S(x,\tau)$ may be represented in the form of one of the following products:

$$S(x,\tau)=\varphi(x,\tau)\cdot v(\tau)\quad\text{or}\quad S(x,\tau)=\varphi(x,\tau)\cdot w(x)$$

where $\varphi(x,\tau)$ is a known function.

Consequently the problem of determination of the two-variable function $S(x,\tau)$ is replaced by the problem of determination of the single-variable functions $v(\tau)$ or $w(x)$.

As criterion functionals, we take the sum of the mean square residuals of the temperatures:

$$J[v] = \sum_{i=1}^{N} \int_0^{\tau_m} \rho_i(\tau)(A_{v_i}v - \tilde{f}_i)^2 \, d\tau ; \tag{8.21}$$

$$J[w] = \sum_{i=1}^{N} \int_0^{\tau_m} \rho_i(\tau)(A_{w_i}w - \tilde{f}_i)^2 \, d\tau \tag{8.22}$$

where

$$A_{v_i}v = \int_0^{\tau} d\tau' \int_0^{b} \varphi(x', \tau')v(\tau')G(d_i, \tau; x', \tau') \, dx' ,$$

$$A_{w_i}w = \int_0^{\tau} d\tau' \int_0^{b} \varphi(x', \tau')w(x')G(d_i, \tau; x', \tau') \, dx' ,$$

$$\tilde{f}_i = f_i(\tau) - \int_0^{b} \xi(x')G(d_i, \tau; x', 0) \, dx' - I(d_i, \tau) ,$$

$\rho_i(\tau)$ are the weight functions.

Functions $\rho_i(\tau)$ take into account a contribution from every experimental dependence $f_i(\tau)$, regulating, in general, these contributions by time. In the most simple case $\rho_i \equiv 1$, $i = \overline{1, N}$, i.e. all measurements are considered to be of equal significance when solving an inverse problem. Or it can be accepted that only some of the measurements are of equal significance for every moments of time and their corresponding weight functions are equal to 1 on the segment $[0, \tau_m]$. As for the other measurements they may be taken into account with the same weight on certain intervals of the segment $[0, \tau_m]$ and may be excluded in all other moments of time. Then the corresponding weights are taken as being equal to 1 on these time intervals and equal to zero for all other values of τ. Such a situation may arise in the case when temperature measurements are known not in the whole segment of time or their quality is not acceptable everywhere and their rejection is needed at certain parts of $[0, \tau_m]$.

First of all we consider the functional (8.21) assuming the linear operators A_{v_i}, $i = \overline{1, N}$ to be the transformations of the space $L_2[0, \tau_m]$.

Let $\Delta v(\tau)$ be an increment of the function $v(\tau)$. Then

$$J[v + \Delta v] = \sum_{i=1}^{N} \int_0^{\tau_m} \rho_i(\tau)[A_{vi}(v + \Delta v) - \tilde{f}_i]^2 \, d\tau . \tag{8.23}$$

Calculate the corresponding increment of the functional

$$\Delta J \equiv J[v + \Delta v] - J[v] = \sum_{i=1}^{N} \left[2 \int_0^{\tau_m} \rho_i e_{vi} A_{vi} \Delta v \, d\tau + \int_0^{\tau_m} \rho_i (A_{vi} \Delta v)^2 \, d\tau \right]$$

$$= 2 \sum_{i=1}^{N} (A_{vi} \Delta v, \rho_i e_{vi})_{L_2[0, \tau_m]} + o(\| \Delta v \|)$$

where $e_{vi} = A_{vi} v - \tilde{f}_i$.

Introducing the adjoint operators A_{vi}^* we have

$$\Delta J = 2\sum_{i=1}^{N} (\Delta v, A_{vi}^*(\rho_i e_{vi}))_{L_2[0,\tau_m]} + o(\|\Delta v\|)$$

$$= \left(\Delta v, 2\sum_{i=1}^{N} A_{vi}^*(\rho_i e_{vi})\right)_{L_2[0,\tau_m]} + o(\|\Delta v\|)\,.$$

Hence, the gradient is given by

$$J_v' = 2\sum_{i=1}^{N} A_{vi}^*(\rho_i e_{vi})\,. \tag{8.24}$$

Let us seek the values of adjoint operators A_{vi}^* at the elements $z_{vi} = \rho_i e_{vi}$. For this purpose we write the inner product of the elements z_{vi} and $A_v v$ and interchange the order of integration in it as

$$(z_{vi}, A_{vi}v)_{L_2[0,\tau_m]} = \int_0^{\tau_m} \left[z_{vi}(\tau) \int_0^{\tau} \mathrm{d}\tau' \int_0^{b} \varphi(x',\tau')v(\tau')G(d_i,\tau;x',\tau)\,\mathrm{d}x' \right] \mathrm{d}\tau$$

$$= \int_0^{\tau_m} v(\tau')\,\mathrm{d}\tau' \int_{\tau'}^{\tau_m} z_{vi}(\tau)F(d_i,\tau,\tau')\,\mathrm{d}\tau = (v, A_{vi}^* z_{vi})_{L_2[0,\tau_m]},$$

where $F(d_i,\tau,\tau') = \int_0^{b} \varphi(x',\tau')G(d_i,\tau;x',\tau')\,\mathrm{d}x'$.

Thus

$$A_{vi}^* z_{vi} = \int_{\tau'}^{\tau_m} z_{vi}(\tau)F(d_i,\tau,\tau')\,\mathrm{d}\tau\,. \tag{8.25}$$

In the same way we obtain the gradient of functional (8.22) while

$$A_{wi}: L_2[0,b] \to L_2[0,\tau_m]$$

$$J_w' = 2\sum_{i=1}^{N} A_{wi}^* z_{wi}, \tag{8.26}$$

where $z_{wi} = \rho_i e_{wi}$, $e_{wi} = A_{wi}w - \tilde{f}_i$,

$$A_{wi}^* z_{wi} = \int_0^{\tau_m} z_{wi}(\tau)K(d_i,x',\tau)\,\mathrm{d}\tau, \quad K(d_i,x',\tau) = \int_0^{\tau} \varphi(x',\tau')G(d_i,\tau;x',\tau')\,\mathrm{d}\tau'.$$

2. Initial condition $\xi(x)$

The corresponding functional for measurements at the point d_i is

$$J[\xi] = \sum_{i=1}^{N} \int_0^{\tau_m} \rho_i(\tau)(A_{\xi i}\xi - \hat{f}_i)^2\,\mathrm{d}\tau\,, \tag{8.27}$$

where $A_{\xi i}\xi = \int_0^b \xi(x')G(d_i, \tau; x', 0)\,dx', \quad A_{\xi i}: L_2[0, b] \to L_2[0, \tau_m]$,

$$\hat{f}_i = f_i(\tau) - \int_0^{\tau_m} d\tau' \int_0^b S(x', \tau')G(d_i, \tau; x', \tau')\,dx' - I(d_i, \tau).$$

The gradient of functional (8.27) is

$$J'_\xi = 2\sum_{i=1}^{N} A^*_{\xi i} z_{\xi i}, \tag{8.28}$$

where $z_{\xi i} = \rho_i e_{\xi i}, \quad e_{\xi i} = A_{\xi i}\xi - \hat{f}_i$.
Adjoint operators $A^*_{\xi i}$ are obtained in the same way as above and they are given by

$$A^*_{\xi i} z_{\xi i} = \int_0^{\tau_m} z_{\xi i}(\tau)G(d_i, \tau; x', 0)\,d\tau. \tag{8.29}$$

If the temperature measurements at the moment $\tau = \tau_m$ are used as the input data when solving the retrospective IHCP, then the residual functional may be written as

$$J_m[\xi] = \int_0^b (A_\xi \xi - \hat{f}_m)^2\,dx, \tag{8.30}$$

where $A_\xi\xi = \int_0^b \xi(x')G(x, \tau_m; x', 0)\,dx', \quad A_\xi: L_2[0, b] \to L_2[0, b]$,

$$\hat{f}_m = f_m(x) - \int_0^{\tau_m} d\tau' \int_0^b S(x', \tau')G(x, \tau_m; x', \tau')\,dx' - I(x, \tau_m).$$

The gradient of functional (8.30) is calculated by

$$J'_{m\xi} = 2\int_0^b e_\xi(x)\,G(x, \tau_m; x', 0)\,dx, \tag{8.31}$$

$$e_\xi = A_\xi \xi - \hat{f}_m.$$

3. Boundary conditions $p_1(\tau), p_2(\tau)$

In this case the criterion functionals are given by

$$J[p_1] = \sum_{i=1}^{N} \int_0^{\tau_m} \rho_i(\tau)(A_{1i}p_1 - f_{1i})^2\,d\tau; \tag{8.32}$$

$$J[p_2] = \sum_{i=1}^{N} \int_0^{\tau_m} \rho_i(\tau)(A_{2i}p_2 - f_{2i})^2\,d\tau. \tag{8.33}$$

The operators A_{1i} and A_{2i} are the mappings from $L_2[0, \tau_m]$ to $L_2[0, \tau_m]$ and they are defined as well as f_{1i}, f_{2i} according to the type of the boundary conditions to be found [see (8.12)–(8.15)]. For boundary conditions of the first kind

$$A_{1i}p_1 = \int_0^\tau p_1(\tau')\frac{\partial}{\partial x'}(a_1(0,\tau')G(d_i,\tau;0,\tau'))\,d\tau';$$

$$A_{2i}p_2 = -\int_0^\tau p_2(\tau')\frac{\partial}{\partial x'}(a_1(b,\tau')G(d_i,\tau;b,\tau'))\,d\tau'.$$

For boundary conditions of the second and third kind

$$A_{1i}p_1 = -\int_0^\tau \frac{p_1(\tau')}{\alpha_1(\tau')}G(d_i,\tau;0,\tau')\,d\tau'; \quad A_{2i}p_2 = \int_0^\tau \frac{p_2(\tau')}{\alpha_2(\tau')}G(d_i,\tau;b,\tau')\,d\tau'.$$

The functions $f_{1i} = f_{1i}(d_i, \tau)$ and $f_{2i} = f_{2i}(d_i, \tau)$ are defined, evidently, from the corresponding expression of (8.12)–(8.15). For example, if the function $p_1(\tau)$ is to be found and combined conditions are taken ($\alpha_1 \neq 0$, $\alpha_2 = 0$, $\beta_2 = 1$), then

$$f_{1i} = f_i(\tau) - \int_0^\tau d\tau' \int_0^b S(x',\tau')G(d_i,\tau;x',\tau')\,dx'$$

$$- \int_0^b \xi(x')G(d_i,\tau;x',0)\,dx' + \int_0^\tau p_2(\tau')\frac{\partial}{\partial x'}(a_1(b,\tau')G(d_i,\tau;b,\tau'))\,d\tau'.$$

The formulas for gradients are

$$J'_{p_j} = 2\sum_{i=1}^{N} A^*_{ji}z_{ji}, \quad j = 1,2 \tag{8.34}$$

where $z_{ji} = \rho_i e_{ji}$, $e_{ji} = A_{ji}p_j - f_{ji}$,

$$A^*_{1i}z_{1i} = -\frac{a_1(0,\tau')}{\alpha_1(\tau')}\int_{\tau'}^{\tau_m} z_{1i}(\tau)G(d_i,\tau;0,\tau')\,d\tau, \qquad \text{if } \alpha_1 \neq 0;$$

$$A^*_{2i}z_{2i} = \frac{a_1(b,\tau')}{\alpha_2(\tau')}\int_{\tau'}^{\tau_m} z_{2i}(\tau)G(d_i,\tau;b,\tau')\,d\tau, \qquad \text{if } \alpha_2 \neq 0;$$

$$A^*_{1i}z_{1i} = \int_{\tau'}^{\tau_m} z_{1i}(\tau)\frac{\partial}{\partial x'}(a_1(0,\tau')G(d_i,\tau;0,\tau'))\,d\tau, \qquad \text{if } \alpha_1 = 0, \beta_1 = 1;$$

$$A^*_{2i}z_{2i} = -\int_{\tau'}^{\tau_m} z_{2i}(\tau)\frac{\partial}{\partial x'}(a_1(b,\tau')G(d_i,\tau;b,\tau'))\,d\tau, \qquad \text{if } \alpha_2 = 0, \beta_2 = 1.$$

Thus, for a wide class of linear one-dimensional IHCP the analytical representations are obtained for the initial data for iterative algorithms (8.3) and (8.4): for the operators A, defined by the corresponding formulation of an inverse problem; for the criterion residual-functionals, characterizing the closeness of the calculated and measured temperatures, and finally, for the gradients of these functionals. With the use of these data the iterative algorithms are constructed which produce the mean square approximations for unknown quantities. For boundary value IHCP such examples were concluded in Chap. 6.

Remark 1. The method stated can be extended to cope with inverse problems for the regions with travelling bounds $Q_\tau = \{(x, \tau): X_1(\tau) < x < X_2(\tau), 0 < \tau < \tau_m\}$ by using the fictitious boundaries method (see Sect. 3.5) or by constructing the corresponding operators A by means of the thermal potentials technique (see Sect. 3.1).

Remark 2. The scheme considered above for the construction of the gradient of a residual can be naturally extended to cope with multidimensional IHCP. For instance, if the inverse problem from Sect. 4.5 is examined, for which the integral notation via the Green's function has the form (4.23), then the formula for the gradient of functional $J[u] = \int_0^{\tau_m} d\tau \int_0^b (Au - \bar{f})^2 \, dx$, $u = q(x, \tau)$ can, in this case, be written as

$$J'_u = 2A^*e = \frac{a}{\lambda} \int_\tau^{\tau_m} d\xi \int_0^b e(x', \xi) G(x', y^*; x, 0; \xi - \tau) \, dx' ,$$

where $e = Au - \bar{f}$.

Remark 3. When calculating the integral terms in the formulas for gradients, the stepwise approximation may be used, in particular, as in Sects. 4.1 and 4.5.

Remark 4. The examined way to determine the gradients of residuals is convenient to use in cases where a Green's function is given analytically. Whenever the Green's function is unknown and can be constructed only numerically, it is expedient to seek the gradient in a more general way – by solving an adjoint boundary-value problem (see Sect. 8.4).

Remark 5. When solving inverse problems, a priori information about the smoothness of the functions to be found needs to be taken into account. It will be shown in Sect. 8.5 that this can easily be done by carrying out the transition from the known gradient of functional in the space L_2 to the gradient of this functional in space W_2^l of functions, for which all their derivatives up to the l-th order have an integrable square. In this case formulas (8.20), (8.25), (8.26), (8.28), (8.31), (8.34) defining the gradients in L_2 of the corresponding functionals (8.16), (8.21), (8.22), (8.27), (8.30), (8.32), (8.33) give an opportunity to prepare the initial data for analytical calculation of the gradients in W_2^l.

8.3 The General Formulation of Nonlinear IHCP. The Problem for an Increment of Temperature Field

Many thermophysical processes can be approximately described by Eq. (1.1) in which subscript j is suppressed. In particular, this model is used when investigating the processes of heat conduction and thermodestruction of various materials – wood, peat, coal, polymer structural and thermoprotecting materials; when analysing the thermal effects during polymerization of organic pitches and when investigating the heat exchange in porous structures [117, 149]. The major requisite for obtaining Eq. (1.1) is an assumption about the equality of the temperatures of the solid frame of a material and of the gaseous or liquid medium filling its pores. In this case a system of equations of energy and heat conduction describing the heat transfer in such a two-phase system may be reduced to a single generalized equation of heat conduction.

In general the coefficients of this equation are the functions of temperature and time and may depend on coordinate x as well. For example, when, in the process of production of a material, an appropriate spatial variation of its properties is programmed.

According to practice already undertaken the general form of Eq. (1.1) can be simplified. Thus, for example, if the processes of thermodestruction during intensive heating of a material take place in a very thin layer near the surface, then one can ignore the distributed heat source S and filtration term $K(\partial T/\partial x)$ in the equation, taking the model of a surface destruction of a material. Then we obtain a special case of Eq. (1.1) in the form of the classical homogeneous heat conduction equation. The transition to such a simpler model of heat transfer may be also justified by the reason that thermal effects bound up with internal thermodestruction of a material may be approximately taken in certain situations by the choice of effective values of the thermal properties of the material itself.

At the same time simplification of numerical methods may be gained by excluding one or another factors within the bounds of more general inverse problems for Eq. (1.1), the formulation of which we are going to discuss now.

Consider the following boundary-value problem in the domain with travelling boundaries $\bar{Q}_\tau = \{(x, \tau): X_1(\tau) \leqslant x \leqslant X_2(\tau), 0 \leqslant \tau \leqslant \tau_m\}$

$$CT_\tau = (\lambda T_x)_x + KT_x + S, \quad (x, \tau) \in Q_\tau ; \tag{8.35}$$

$$T(x, 0) = \xi(x) ; \tag{8.36}$$

$$[\alpha_i \lambda T_x + \beta_i T]_{x = X_i(\tau)} = p_i(\tau), \quad i = 1, 2 . \tag{8.37}$$

Here numbers α_i, β_i define one or another type of boundary conditions. For $\alpha_i = 0$, $\beta_i = 1$ we obtain the conditions of the first kind and for $\alpha_i = -1$, $\beta_i = 0$ the second kind. If $\alpha_i = 1$, $\beta_i = h_i$, $p_i(\tau) = h_i T^*(\tau)$ where $h_i = \text{const}$ is a heat transfer coefficient and $T^*(\tau)$ is a temperature of the surrounding medium, then we obtain the conditions of the third kind. Any combination of the boundary conditions is also possible.

The curves $X_1(\tau)$ and $X_2(\tau)$ are assumed to be piecewise smooth functions at the segment $[0, \tau_m]$.

In the inverse problems for the Eq. (8.35) the unknown value is one or several functions from the set $u = \{C(T), \lambda(T), K(T), S(T), \xi(x), p_1(\tau), p_2(\tau)\}$. If functions $\xi(x)$, $p_1(\tau) p_2(\tau)$ are to be found, i.e. retrospective, boundary or retrospective-boundary inverse problems are to be solved, then the coefficients and the heat source in heat conduction equation are assumed to be known functions of $T(x, \tau)$, x, τ: $C(T, x, \tau)$, $\lambda(T, x, \tau)$, $K(T, x, \tau)$, $S(T, x, \tau)$.

We assume that as the additional conditions, the dependences of temperature in time, are given in certain N different points of domain $\Omega = \{X_1(\tau) \leqslant x \leqslant X_2(\tau)\}$, the two points of measurements among them possibly lying on boundaries $X_1(\tau)$ and $X_2(\tau)$ in the case of boundary conditions of the second and third kinds.

Any analysis of the problem of uniqueness of a solution for IHCP is beyond the limits of this book. In [32, 52, 59, 60, 64, 71, 77, 78, 89, 106, 107, 146, 170, 171, 198] the conditions are obtained in which the uniqueness of a solution takes place for various formulations of inverse problems. Based on these results and also on those of numerical simulation as well we can assert the following. For unique determination of any of the functions $C(T)$, $\lambda(T)$, $K(T)$, $S(T)$ one should have the minimal number of points of measurements $N_{\min} = 1$. For determination of any set of these functions, when boundary conditions of the second or third kind are given, if only one of them is non-zero, then $N_{\min}$ is equal to the number of the functions to be found. As to the minimal number of measurement points necessary for retrospective, boundary and retrospective-boundary problems, these numbers are the same as in linear IHCP (see Sect. 8.2).

Thus, we have one or another formulation of an inverse problem

$$Au = f$$

with the non-linear operator A defined by the conditions (8.35)–(8.37). The right-hand side f is a vector-function of the temperature measurements.

As has already been shown in chapter 6 the gradient methods regularized by the number of iterations can be successfully used for the solution of non-linear IHCP. The numerical algorithms for calculation of the gradient of the residual functional

$$J[u] = \| Au - f \|_F^2 \tag{8.38}$$

for boundary IHCP in the case $F = L_2$ were based on the solution of the corresponding adjoint problems, which formulations were given in chapter 6 without proof.

In the sequel we shall consider more rigorously the problem of determination of a gradient J'_u conforming to the above formulated inverse problems, including the method for obtaining the conditions for adjoint boundary-value problems.

For a Frechet differentiable operator A we have

$$J'_u = 2(A')^*(Au - f)\,, \tag{8.39}$$

where A' is a Frechet derivative of an operator A and $(A')^*$ is an adjoint operator.

It is necessary to recall that an operator A mapping a normed space U to a normed space F is called Frechet differentiable at a point u, if there exists a linear continuous operator $A'(u)$ mapping U to F so that

$$A(u + \Delta u) - A(u) = A'(u)\Delta u + \omega(u, \Delta u), \tag{8.40}$$

$$\text{where} \quad \lim_{\|\Delta u\| \to 0} \frac{\|\omega(u, \Delta u)\|}{\|\Delta u\|} = 0 . \tag{8.41}$$

Operator $A'(u)$ is called Frechet derivative of an operator A at a point u.

To obtain the expression (8.39) we shall give an increment to the vector-function u and calculate the corresponding variation of the functional:

$$\Delta J \equiv J(u + \Delta u) - J(u) = \|A(u + \Delta u) - f\|^2 - \|Au - f\|^2 .$$

By the equalities $(\theta + \varphi, z) = (\theta, z) + (\varphi, z)$, $\|\theta \pm \varphi\|^2 = \|\theta\|^2 \pm 2(\theta, \varphi) + \|\varphi\|^2$, where θ, φ, z are elements of the Hilbert space F and by the equality (8.40) after some simple transformations we have

$$\Delta J = 2(Au - f, A'\Delta u) + \eta(u, \Delta u) ,$$

where $\eta(u, \Delta u) = 2(Au - f, \omega) + \|A'\Delta u + \omega\|^2 .$

Using Cauchy inequality $|(Au - f, \omega)| \leqslant \|Au - f\| \cdot \|\omega\|$ and equality $\|A'\Delta u + \omega\|^2 = \|A'\Delta u\|^2 + 2(A'\Delta u, \omega) + \|\omega\|^2$ and the condition (8.41) as well, it is easy to see that

$$\lim_{\|\Delta u\| \to 0} \frac{\eta(u, \Delta u)}{\|\Delta u\|} = 0 . \tag{8.42}$$

Putting into consideration an adjoint operator $(A')^*$ we get

$$\Delta J = 2((A')^*(Au - f), \Delta u) + \eta(u, \Delta u) . \tag{8.43}$$

According to the definition of the gradient J'_u we have

$$\Delta J = (J'_u, \Delta u) + \eta(u, \Delta u), \tag{8.44}$$

where for η the condition (8.42) holds.

Comparing (8.43) and (8.44) we obtain the formula (8.39).

Therefore, the problem of the construction of a gradient divides into the problem of the construction of the operator A' and the problem of the determination of the adjoint operator $(A')^*$. For this purpose two auxilary problems are required – the *problem for an increment of a field* $T(x, \tau)$ *and its adjoint problem.*

The problem for an increment of temperature

Let the components of u be given increments

$$C(T) + \Delta C(T), \quad \lambda(T) + \Delta\lambda(T), \quad K(T) + \Delta K(T) ,$$
$$S(T) + \Delta S(T), \quad \xi(x) + \Delta\xi(x) ,$$
$$p_1(\tau) + \Delta p_1(\tau), \quad p_2(\tau) + \Delta p_2(\tau) .$$

By $T(x, \tau) + v(x, \tau)$ we denote the solution of problem (8.35)–(8.37) for this new set of data, and write the system (8.35)–(8.37) in a form

$$[C(T+v)+\Delta C(T+v)][T+v]_\tau=\{[\lambda(T+v)+\Delta\lambda(T+v)][T+v]_x\}_x$$
$$+[K(T+v)+\Delta K(T+v)][T+v]_x+S(T+v)+\Delta S(T+v),$$
$$(x,\tau)\in Q_\tau\,; \tag{8.45}$$
$$T(x,0)+v(x,0)=\xi(x)+\Delta\xi(x)\,; \tag{8.46}$$
$$\{\alpha_i[\lambda(T+v)+\Delta\lambda(T+v)][T+v]_x+\beta_i[T+v]\}_{x=X_i(\tau)}$$
$$=p_i(\tau)+\Delta p_i(\tau),\quad i=1,2\,. \tag{8.47}$$

Here the dependences of values C, λ, K, S on x and τ, which may take place in retrospective and boundary inverse problems are not conventionally stated.

Substract from the conditions (8.45)–(8.47) the conditions (8.35)–(8.37) respectively and for the quantities $C(T+v)$, $\lambda(T+v)$, $K(T+v)$, $S(T+v)$ use the second order Taylor formula and for the quantities $\Delta C(T+v)$, $\Delta\lambda(T+v)$, $\Delta K(T+v)$, $\Delta S(T+v)$ the first order Taylor formula. Also, take into account that coefficients C, λ, K, S in the direct problem (8.35)–(8.37) are given functions of temperature $T(x,\tau)$ which, when deriving the conditions for an increment $v(x,\tau)$, is treated as a known function. Therefore, in the sequel these coefficients are the known functions of the independent variables x and τ only. As a result after certain transformations we obtain the following formulation of the problem for determination of the quantity $v(x,\tau)$:

$$v_\tau=a_1v_{xx}+a_2v_x+a_3v+q+\omega,\quad (x,\tau)\in Q_\tau\,; \tag{8.48}$$
$$v(x,0)=\Delta\xi\,; \tag{8.49}$$
$$B_{i\tau}v(X_i(\tau),\tau)\equiv[\gamma_iv_x+\sigma_iv]_{x=X_i(\tau)}=\eta_i+\omega_i,\quad i=1,2 \tag{8.50}$$

where $a_1=\dfrac{\lambda}{C}$, $a_2=\dfrac{1}{C}(2\lambda_x+K)$, $a_3=\dfrac{1}{C}(\lambda_{xx}+S_T+K_x-C_\tau)$,

$$q=\frac{1}{C}[\Delta\lambda T_{xx}+(\Delta\lambda)_xT_x+\Delta KT_x+\Delta S-\Delta CT_\tau]\,,$$
$$\gamma_i=\alpha_i\lambda(X_i(\tau),\tau),\quad \sigma_i=\alpha_i\lambda_x(X_i(\tau),\tau)+\beta_i\,,$$
$$\eta_i=\Delta p_i(\tau)-\alpha_i\Delta\lambda(X_i(\tau),\tau)T_x(X_i(\tau),\tau)\,,$$

ω,ω_1,ω_2 are the remainders (if $\alpha_1=\alpha_2=0$, then $\omega_1=\omega_2=0$). Using the substitution $y=\dfrac{x-X_1(\tau)}{X_2(\tau)-X_1(\tau)}$, $t=\tau$ we transform the domain Q_τ in the rectangle $R=(0,1)\times(0,\tau_m)$. The values $a_i(x,\tau)$ and $q(x,\tau)$ are transformed into $\bar a_i(y,t)$ and $\bar q(y,t)$ respectively, and the system (8.48)–(8.50) transforms into

$$v_t=b_1v_{yy}+b_2v_y+b_3v+\bar q+\bar\omega,\quad (y,t)\in R\,; \tag{8.51}$$
$$v(y,0)=\Delta\bar\xi(y)\,; \tag{8.52}$$

$$B_{1t}v(0, t) \equiv [l_1 v_y + \sigma_1 v]_{y=0} = \eta_1(t) + \bar{\omega}_1 ; \tag{8.53}$$

$$B_{2t}v(1, t) \equiv [l_2 v_y + \sigma_2 v]_{y=1} = \eta_2(t) + \bar{\omega}_2 \tag{8.54}$$

where

$$b_1 = \frac{\bar{a}_1(y, t)}{[X_2(t) - X_1(t)]^2};$$

$$b_2 = \frac{1}{X_2(t) - X_1(t)} \{\bar{a}_2(y, t) + X'_1(t) + y[X'_2(t) - X'_1(t)]\}$$

$$\left(X'_i \equiv \frac{\mathrm{d}X_i}{\mathrm{d}t}\right); \quad b_3 = \bar{a}_3(y, t); \quad l_1 = \frac{\gamma_1(t)}{X_2(t) - X_1(t)}; \quad l_2 = \frac{\gamma_2(t)}{X_2(t) - X_1(t)} .$$

Thus, if we omit $\bar{\omega}, \bar{\omega}_1, \bar{\omega}_2$, i.e. if we pass to the linear approximation of the boundary-value problem for an increment $v(y, t)$ then the system (8.51)–(8.54) will be of the same form as (8.8)–(8.11) in Sect. 8.2. Moreover, by using the results of [91] one can prove that the problem obtained from Eqs. (8.48)–(8.50) and similarly from Eqs. (8.51)–(8.54) by omitting non-linear terms, represents the Frechet derivative of the operator A, which puts the quantities to be found into effect with the function $T(d(\tau), \tau)$. For this purpose it is sufficient to demand that $C(T)$, $K(T)$ and $S(T)$ are twice, and $\lambda(T)$-three times continuously differentiable functions; $p_1, p_2 \in W_2^1[0, \tau_m]$; $\xi \in W_2^1[X_1(0), X_2(0)]$, and that the conditions of coordination between initial and boundary conditions to hold.[3]

In the following we shall assume that in the problems (8.48)–(8.50) and (8.51)–(8.54) the non-linear terms are omitted.

To present the method for getting the formulation of the adjoint problem (with respect to the problem for a temperature increment), we shall use as an illustration boundary conditions of the second and third kind ($\alpha_1 \neq 0, \alpha_2 \neq 0$). Similar results for the conditions of the first kind can be obtained in the same way and they will be presented without any intermediate calculations.

For the derivation of an adjoint boundary-value problem we reduce the statement (8.51)–(8.54) to the zero boundary conditions, i.e. to the form

$$v_t = b_1 v_{yy} + b_2 v_y + b_3 v + \bar{q} + \bar{z}, \quad (y, t) \in R ; \tag{8.55}$$

$$v(y, 0) = 0 ; \tag{8.56}$$

$$[l_1 v_y + \sigma_1 v]_{y=0} = 0 ; \tag{8.57}$$

$$[l_2 v_y + \sigma_2 v]_{y=1} = 0 . \tag{8.58}$$

For the determination of the function $\bar{z} = \bar{z}(y, t)$ we write the solution of the problems (8.51)–(8.54) and (8.55)–(8.58) via the Green's function [see formula

[3] It is known that $W_2^1 \subset L_2$.

(8.13)] and equate them:

$$\int_0^1 \Delta\bar{\xi}(y')G(y,t;y',0)\,dy' - \int_0^t \frac{b_1(0,t')}{l_1(t')}\Delta p_1(t')G(y,t;0,t')\,dt'$$
$$+\int_0^t \frac{b_1(1,t')}{l_2(t')}\Delta p_2(t')G(y,t;1,t')\,dt' = \int_0^t dt' \int_0^1 \bar{z}(y',t')G(y,t;y',t')\,dy' .$$

Thereby

$$\bar{z}(y,t) = -\frac{b_1(y,t)}{l_1(t)}\eta_1(t)\delta(y) + \frac{b_1(y,t)}{l_2(t)}\eta_2(t)\delta(y-1) + \Delta\bar{\xi}(x)\delta(t),$$

where $\delta(\cdot)$ is the delta function.

For such function $\bar{z}(y,t)$ the problem (8.55)–(8.58) is identical to the statement (8.51)–(8.54), and therefore to the problem (8.48)–(8.50). Performing the reverse transition in problem (8.55)–(8.58) (from y, t to x, τ) and remembering the property of the δ-function ($\delta(a\zeta) = \frac{1}{|a|}\delta(\zeta)$, where $a \neq 0$ is a real number), we get the formulation, identical to that in Eqs. (8.48)–(8.50):

$$v_\tau = a_1 v_{xx} + a_2 v_x + a_3 v + q + z, \quad (x,\tau)\in Q_\tau; \tag{8.59}$$

$$v(x,0) = 0; \tag{8.60}$$

$$B_{i\tau}v(X_i(\tau),\tau) = 0, \quad i = 1,2, \tag{8.61}$$

where

$$z(x,\tau) = -\frac{a_1}{\gamma_1}\eta_1\delta(x - X_1(\tau)) + \frac{a_1}{\gamma_2}\eta_2\delta(x - X_2(\tau)) + \Delta\xi\delta(\tau).$$

8.4 Adjoint Problems and Gradient of a Functional

8.4.1 One-Dimensional Case

The formulation of an adjoint boundary-value problem depends on the number of the points of measurement N. Let $N = 1$. At a point $x = d(\tau)$ (see Fig. 8.2), travelling with time, the dependence $f(\tau) = T(d(\tau),\tau)$ is given. The coordinate of

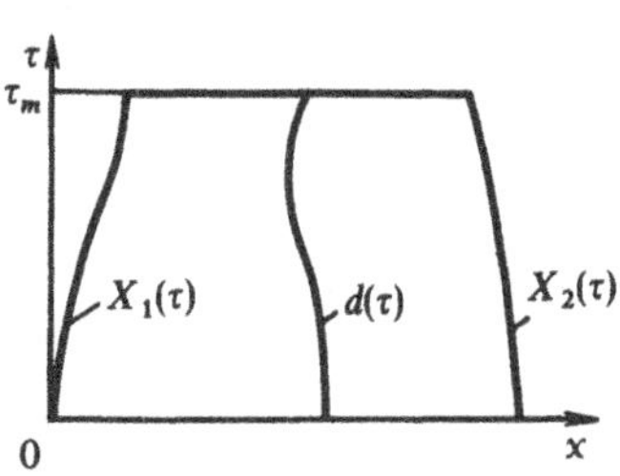

Fig. 8.2. A domain with moving boundaries and the thermosensor travelling coordinate

this point conforms to the conditions: $X_1(\tau) \leqslant d(\tau)$ for $\alpha_1 \neq 0$, $X_1(\tau) < d(\tau)$ for $\alpha_1 = 0$, $d(\tau) \leqslant X_2(\tau)$ for $\alpha_2 \neq 0$, $d(\tau) < X_2(\tau)$ for $\alpha_2 = 0$.

Function $d(\tau)$, $\tau \in [0, \tau_m]$ as well as $X_1(\tau)$ and $X_2(\tau)$, is assumed to be piecewise smooth.

In the case considered operator A assigns the temperature at a point $d(\tau)$ to the desired quantity u, i.e. $Au = T(u, d(\tau), \tau)$. We introduce a criterion functional

$$J(u) = \int_0^{\tau_m} [T(u, d(\tau), \tau) - f(\tau)]^2 \, d\tau \, . \tag{8.62}$$

Since an unknown function might be any one of the components of vector-function $u = \{\lambda(T), C(T), K(T), S(T), \xi(x), p_1(\tau), p_2(\tau)\}$ we shall find the gradient of $J[u]$ with respect to this vector-function, assuming that $\xi \in L_2[X_1(0), X_2(0)]$; $p_1, p_2, f \in L_2[0, \tau_m]$; $\lambda, C, K, S \in L_2[T_0, T_M]$ where T_0, T_M are the boundary points of a temperature interval in which the coefficients, or a source-term in Eq. (8.35), are to be found.

Denoting $L = \dfrac{\partial}{\partial \tau} - D_{x\tau}$, $D_{x\tau} = a_1 \dfrac{\partial^2}{\partial x^2} + a_2 \dfrac{\partial}{\partial x} + a_3$,

$$\chi = q - a_1 \frac{\eta_1}{\gamma_1} \delta(x - X_1(\tau) + a_1 \frac{\eta_2}{\gamma_2} \delta(x - X_2(\tau)) + \Delta\xi\delta(\tau),$$

we rewrite the linear approximation of problem (8.59)–(8.61) in a form

$$Lv = \chi, \quad v \in D_L \, . \tag{8.63}$$

Here $D_L = \{v \in G(Q_\tau)\colon\ B_{i\tau} v(X_i(\tau), \tau) = 0, i = 1, 2; v(x, 0) = 0\}$ is the domain of operator L and $G(Q_\tau)$ is a set of functions with continuous derivatives v_τ, v_x, v_{xx} in Q_τ.

Considering L as an operator, mapping $D_L \subset L_2(Q_\tau)$ to the space L_2, we write the inner product of elements $Lv, \psi \in L_2$

$$(Lv, \psi) = \iint_{Q_\tau} \psi L v \, dx \, d\tau$$

where $\psi = \psi(x, \tau)$ is sufficiently smooth in $\bar{Q}_\tau$ function.

Define an adjoint operator L^* by Lagrange identity

$$(Lv, \psi) = (v, L^*\psi) \tag{8.64}$$

and represent the inner product in the left-hand side in the following way:

$$\iint_{Q_\tau} \psi L v \, dx \, d\tau = \iint_{Q_\tau} \psi(v_\tau - a_1 v_{xx} - a_2 v_x - a_3 v) \, dx \, d\tau = I_1 - I_2 \, ,$$

where

$$I_1 = \iint_{Q_\tau} [- v\psi_\tau + (a_1\psi)_x v_x + (a_2\psi)_x v - a_3\psi v] \, dx \, d\tau \, ,$$

$$I_2 = \iint_{Q_\tau} [(a_1\psi v_x + a_2\psi v)_x - (\psi v)_\tau] \, dx \, d\tau \, .$$

For the following we need the formulas of integration by parts [90, 91], which in our case comprise

$$\iint_{Q_\tau} \theta_x \varphi \, dx \, d\tau = - \iint_{Q_\tau} \theta \varphi_x \, dx \, d\tau + \int_{\partial Q_\tau^+} \theta \varphi \, d\tau, \quad \theta, \varphi \in W_2^{1,0}(Q_\tau),$$

$$\iint_{Q_\tau} \theta_\tau \varphi \, dx \, d\tau = - \iint_{Q_\tau} \theta \varphi_\tau \, dx \, d\tau - \int_{\partial Q_\tau^+} \theta \varphi \, dx, \quad \theta, \varphi \in W_2^1(Q_\tau),$$

where ∂Q_τ^+ is the boundary of Q_τ oriented in such a way that, when rounding it, the domain Q_τ lies on the left; $W_2^{1,0}(Q_\tau)$ is the set of functions $\theta(x, \tau)$ from the space $L_2(Q_\tau)$, which have the generalized derivative θ_x, also belonging to $L_2(Q_\tau)$; $W_2^1(Q_\tau)$ is the set of functions $\theta(x, \tau)$ from $L_2(Q_\tau)$, which have generalized derivatives θ_x, θ_τ.

In our case

$$\int_{\partial Q_\tau^+} \theta \varphi \, d\tau = \int_0^{\tau_m} \theta \varphi \Big|_{x = X_2(\tau)} d\tau - \int_0^{\tau_m} \theta \varphi \Big|_{x = X_1(\tau)} d\tau \,,$$

$$\int_{\partial Q_\tau^+} \theta \varphi \, dx = \int_0^{\tau_m} \theta \varphi \Big|_{x = X_2(\tau)} \cdot X_2'(\tau) \, d\tau - \int_{X_1(\tau_m)}^{X_2(\tau_m)} \theta \varphi \Big|_{\tau = \tau_m} dx$$

$$- \int_0^{\tau_m} \theta \varphi \Big|_{x = X_1(\tau)} \cdot X_1'(\tau) \, d\tau + \int_{X_1(0)}^{X_2(0)} \theta \varphi \Big|_{\tau = 0} dx \,.$$

Therefore

$$\iint_{Q_\tau} (a_1 \psi)_x v_x \, dx \, d\tau = - \iint_{Q_\tau} (a_1 \psi)_{xx} v \, dx \, d\tau + \int_0^{\tau_m} (a_1 \psi)_x v \Big|_{X_1(\tau)}^{X_2(\tau)} d\tau \,.$$

Thus

$$I_1 = \iint_{Q_\tau} v [- \psi_\tau - (a_1 \psi)_{xx} + (a_2 \psi)_x - a_3 \psi] \, dx \, d\tau + \int_0^{\tau_m} v (c_1 \psi)_x \Big|_{x = X_1(\tau)}^{x = X_2(\tau)} d\tau \,.$$

By the same formulas of integration by parts we have for I_2

$$I_2 = \int_0^{\tau_m} (a_1 \psi v_x + a_2 \psi v) \Big|_{X_1(\tau)}^{X_2(\tau)} d\tau - \int_0^{\tau_m} \psi v \Big|_{x = X_2(\tau)} \cdot X_2'(\tau) \, d\tau$$

$$+ \int_0^{\tau_m} \psi v \Big|_{x = X_1(\tau)} \cdot X_1'(\tau) \, d\tau + \int_{X_1(\tau_m)}^{X_2(\tau_m)} \psi v \Big|_{\tau = \tau_m} dx - \int_{X_1(0)}^{X_2(0)} \psi v \Big|_{\tau = 0} dx \,.$$

The last integral is equal to zero because $v(x, 0) = 0$.

Taking into consideration the boundary conditions (8.61), we write the final expression for I_2:

$$I_2 = \int_0^{\tau_m} \left\{ v \psi \left[X_2' + a_2 - \frac{\sigma_2}{\gamma_2} a_1 \right]_{x = X_2(\tau)} \right.$$

$$\left. - v \psi \left[X_1' + a_2 - \frac{\sigma_1}{\gamma_1} a_1 \right]_{x = X_1(\tau)} \right\} d\tau$$

$$+ \int_{X_1(\tau_m)}^{X_2(\tau_m)} \psi(x, \tau_m) v(x, \tau_m) \, dx \,.$$

By the identity (8.64) we know that function $\psi(x, \tau)$ ought to satisfy the conditions of the following boundary-value problem (this problem is adjoint of the statement (8.59)–(8.61)):

$$L^*\psi(x, \tau) = \zeta(x, \tau), \quad (x, \tau) \in Q_\tau; \tag{8.65}$$

$$\psi(x, \tau_m) = 0; \tag{8.66}$$

$$B_{i\tau}^*\psi|_{x=X_i(\tau)} \equiv \left[(a_1\psi)_x - \psi\left(X_i' + a_2 - \frac{\sigma_i}{\gamma_i}a_1\right)\right]_{x=X_i(\tau)} = 0, \quad i = 1, 2, \tag{8.67}$$

where $L^* = -\dfrac{\partial}{\partial\tau} - D_{x\tau}^*, \quad D_{x\tau}^*\psi = (a_1\psi)_{xx} - (a_2\psi)_x + a_3\psi$.

Function $\zeta(x, \tau)$ will be found below.

8.4.2 Formulas for the Gradient of Residual

Let us rewrite the problem in Eqs. (8.65)–(8.67) in the form analogous to Eq. (8.63),

$$L^*\psi = \zeta, \quad \psi \in D_{L^*},$$

where

$$D_{L^*} = \{\psi \in G(Q_\tau): B_{i\tau}^*\psi|_{x=X_i(\tau)} = 0, \quad i = 1, 2; \quad \psi(x, \tau_m) = 0\}.$$

Let L^{-1} and $(L^*)^{-1}$ be inverses for L and L^* respectively. Write the inner product for elements v and ψ

$$(v, \psi) = (LL^{-1}v, \psi) = (L^{-1}v, L^*\psi) = (v, L^{-1*}L^*\psi).$$

Hence $L^{-1*}L^* = I$ in D_{L^*} (I is a unit operator). Thus

$$(L^{-1})^* = (L^*)^{-1}. \tag{8.68}$$

The increment of functional (8.62) is

$$\Delta J \equiv J(u + \Delta u) - J(u) = 2(Au - f, A'\Delta u) + o(\|\Delta u\|)$$

$$= 2\int_0^{\tau_m} [T(u, d(\tau), \tau) - f(\tau)]v(d(\tau), \tau)\,d\tau + o(\|\Delta u\|).$$

Represent ΔJ via δ-function in a form

$$\Delta J = \int_0^{\tau_m} d\tau \int_{X_1(\tau)}^{X_2(\tau)} v(x, \tau)\cdot 2e(\tau)\delta(x - d(\tau))\,dx + o(\|\Delta u\|)$$

where $e(\tau) = T(u, d(\tau), \tau) - f(\tau)$.

Observing, that $v(x, \tau) = L^{-1}\chi$ and introducing an adjoint operator we have

$$\Delta J = \iint_{Q_\tau} L^{-1}\chi\cdot 2e(\tau)\delta(x - d(\tau))\,dx\,d\tau + o(\|\Delta u\|)$$

$$= (\chi, (L^{-1})^*[2e(\tau)\delta(x - d(\tau))]) + o(\|\Delta u\|). \tag{8.69}$$

Since the equality (8.68) holds

$$(L^{-1})^*[2e(\tau)\delta(x-d(\tau))] = \psi,$$

where $\psi = \psi(x,\tau)$ is a solution of the problem in Eqs. (8.65)–(8.67) in which $\zeta(x,\tau) = 2e(\tau)\delta(x-d(\tau))$.

Substituting function χ in expression (8.69) we have

$$\Delta J = \left(\psi, q - \frac{a_1}{\gamma_1}\eta_1\delta(x-X_1) + \frac{a_1}{\gamma_2}\eta_2\delta(x-X_2) + \Delta\xi\delta(\tau)\right) + o(\|\Delta u\|). \tag{8.70}$$

The establishment of a relation between a direct problem and the adjoint one in the end results in obtaining the formula (8.70). Certainly, this requires rigorous proof, first and foremost, in the justification of application of formulas of integration by parts which raise claims of certain smoothness in the integrands. The conditions pertaining to the initial data for a direct problem, in which the smoothness required is assured in the solutions of both direct and adjoint problems, may be obtained using the results of [73]. It can be shown that it is sufficient to demand that the conditions be provided, which assure the differentiability of an operator for the direct problem.

Formula (8.70) yields the following representation of ΔJ

$$\Delta J = (\Phi, \Delta u) + o(\|\Delta u\|).$$

Hence, it is clear, that $\Phi = J'_u$.

To obtain a component J'_{u_n} of the gradient for any component u_n of vector u one needs to null all the components of vector u in formula (8.70) except the one needed. Then from equality (8.70) we obtain the representation of a form

$$\Delta J = (\Phi_{u_n}, \Delta u_n) + o(\|\Delta u_n\|), \quad \text{i.e.} \quad \Phi_{u_n} = J'_{u_n}.$$

Thus, expression (8.70) implies the formulas required for the components of the gradient in ξ, p_1, p_2:

$$J'_\xi = \psi(x, 0),$$

$$J'_{p_1} = -\psi(X_1(\tau),\tau)\frac{a_1(X_1(\tau),\tau)}{\gamma_1(\tau)}, \quad J'_{p_2} = \psi(X_2(\tau),\tau)\frac{a_1(X_2(\tau),\tau)}{\gamma_2(\tau)}.$$

Determination of components $J'_\lambda, J'_C, J'_K, J'_S$ of the gradient is a considerably more cumbersome problem. For its simplification we shall seek the coefficients and source-term of equation (8.35) in a parametrized form:

$$\lambda(T) = \sum_{j=1}^{M_1} \lambda_j\varphi_j(T), \qquad C(T) = \sum_{j=1}^{M_2} C_j\varphi_j(T),$$

$$K(T) = \sum_{j=1}^{M_3} K_j\varphi_j(T), \qquad S(T) = \sum_{j=1}^{M_4} S_j\varphi_j(T)$$

where $\{\varphi_j\}_1^M$ is a given system of basic functions; λ_j, C_j, K_j, S_j are the numerical coefficients to be found.

The increments of these functions are written in the same form

$$\Delta\lambda(T) = \sum_{j=1}^{M_1} \Delta\lambda_j \varphi_j(T), \quad \Delta C(T) = \sum_{j=1}^{M_2} \Delta C_j \varphi_j(T), \tag{8.71}$$

$$\Delta K(T) = \sum_{j=1}^{M_3} \Delta K_j \varphi_j(T), \quad \Delta S(T) = \sum_{j=1}^{M_4} \Delta S_j \varphi_j(T), \tag{8.72}$$

Substituting into Eq. (8.70) the expressions of q and η_i and taking into account the approximations in formulas (8.71) and (8.72), after simple transformations we obtain formulas for the gradients of the residual functional (8.62) with respect to vectors $\boldsymbol{\lambda} = \{\lambda_j\}_1^{M_1}, \boldsymbol{C} = \{C_j\}_1^{M_2}, \boldsymbol{K} = \{K_j\}_1^{M_3}, \boldsymbol{S} = \{S_j\}_1^{M_4}$ respectively:

$$J'_{\lambda_l} = \Phi_0 + \Phi_1 + \Phi_2, \quad l = \overline{1, M_1},$$

where $\Phi_0 = \int_0^{\tau_m} \mathrm{d}\tau \int_{X_1(\tau)}^{X_2(\tau)} \frac{\psi(x,\tau)}{C(x,\tau)} [T_{xx}(x,\tau)\varphi_l(T(x,\tau)) + T_x^2(x,\tau)\varphi_l'(T(x,\tau))]\,\mathrm{d}x$;

$$\Phi_i = \int_0^{\tau_m} \frac{\psi(x,\tau)}{C(x,\tau)} T_x(x,\tau)\varphi_l(T(x,\tau))|_{x=X_i(\tau)}\,\mathrm{d}\tau, \quad i = 1, 2;$$

$$J'_{C_l} = -\int_0^{\tau_m} \mathrm{d}\tau \int_{X_1(\tau)}^{X_2(\tau)} \frac{\psi(x,\tau)}{C(x,\tau)} T_\tau(x,\tau)\varphi_l(T(x,\tau))\,\mathrm{d}x, \quad l = \overline{1, M_2};$$

$$J'_{K_l} = \int_0^{\tau_m} \mathrm{d}\tau \int_{X_1(\tau)}^{X_2(\tau)} \frac{\psi(x,\tau)}{C(x,\tau)} T_x(x,\tau)\varphi_l(T(x,\tau))\,\mathrm{d}x, \quad l = \overline{1, M_3};$$

$$J'_{S_l} = \int_0^{\tau_m} \mathrm{d}\tau \int_{X_1(\tau)}^{X_2(\tau)} \frac{\psi(x,\tau)}{C(x,\tau)} \varphi_l(T(x,\tau))\,\mathrm{d}x, \quad l = \overline{1, M_4}.$$

8.4.3 The Transformation of an Adjoint Problem into the Form Convenient for Numerical Solution

For realization the problem in Eqs. (8.65)–(8.67), where $\zeta(x,\tau) = 2e(\tau)\delta(x - d(\tau))$, need to be transformed to a form without the singular term.

Supposing, at first, $X_1(\tau) < d(\tau) < X_2(\tau)$ we denote

$$\Psi(x,\tau) = \begin{cases} \psi_1(x,\tau), & x \in (X_1(\tau), d(\tau)) \\ \psi_2(x,\tau), & x \in (d(\tau), X_2(\tau)) \end{cases}$$

and we shall show that the following problem is identical to that in Eqs (8.65)–(8.67):

$$-\psi_{1\tau} - D^*_{x\tau}\psi_1 = 0, \ \{x,\tau\} \in Q_{1\tau} = \{X_1(\tau) < x < d(\tau), 0 < \tau < \tau_m\}; \tag{8.73}$$

$$-\psi_{2\tau} - D^*_{x\tau}\psi_2 = 0, \ (x,\tau) \in Q_{2\tau} = \{d(\tau) < x < X_2(\tau), 0 < \tau < \tau_m\}; \tag{8.74}$$

$$B^*_{1\tau}\Psi|_{x=X_1(\tau)} = B^*_{2\tau}\Psi|_{x=X_2(\tau)} = 0;$$

$$\psi_1(d(\tau),\tau) - \psi_2(d(\tau),\tau) = 0; \tag{8.75}$$

$$[(a_1\psi)_x - (a_1\psi_2)_x]_{x=d(\tau)} = 2e(\tau); \tag{8.76}$$

$$\Psi(x, \tau_m) = 0.$$

The solution of problem (8.65)–(8.67) is a distribution $\psi(x, \tau)$, $(x, \tau) \in Q_\tau$ which satisfies the system (8.65)–(8.67) in the generalized sense, i.e.

$$\psi \in \Gamma : (\varphi, L^*\varphi) = (\varphi, 2e(\tau)\delta(x - d(\tau)))$$

where Γ is the set of distributions, satisfying the initial and boundary conditions (8.66)–(8.67); $\varphi = \varphi(x, \tau)$ is a fundamental function which is defined in the following as a real function, continuous with its derivatives φ_τ, φ_x, φ_{xx} and finite in the domain Q_τ; (φ, p) is a value of the distribution p for a fundamental function φ.

The conditions, assuring differentiability of an operator A for a direct problem (see Sect. 8.3), guarantee the continuity of the coefficients of operator L^* and their similar derivatives in domain Q_τ. Therefore, differentiating $(a_1\Psi)_x$ as a distribution [57], we have with the condition (8.76)

$$(a_1\Psi)_{xx} = \{(a_1\Psi)_{xx}\} - 2e(\tau)\delta(x - d(\tau)),$$

where

$$\{(a_1\Psi)_{xx}\} = \begin{cases} (a_1\psi_1)_{xx}, & x \in (X_1(\tau), d(\tau)) \\ (a_1\psi_2)_{xx}, & x \in (d(\tau), X_2(\tau)) \end{cases}$$

is a regular part of the derivative $(a_1\Psi)_{xx}$ of distribution $(a_1\Psi)_x$.

Other derivatives Ψ_τ and $(a_2\Psi)_x$ do not possess singular terms by virtue of the condition (8.75) and continuity of a_2. Therefore, using Eqs. (8.73) and (8.74), we obtain that $\Psi(x, \tau)$ is a mild solution of Eq. (8.65) which needed to be proved.

Finally, we consider the formulation of an inverse problem when the point $x = d(\tau)$, at which the function $f(\tau)$ is given, coincides with one of the boundaries of the domain. Let, for example, $d(\tau) = X_2(\tau)$. After similar considerations, it can be shown that in this case the adjoint problem is

$$\psi_\tau + D^*_{x\tau}\psi = 0, \quad (x, \tau) \in Q_\tau;$$

$$\psi(x, \tau_m) = 0;$$

$$B^*_{1\tau}\psi|_{x=X_1(\tau)} = 0; \quad B^*_{2\tau}\psi|_{x=X_2(\tau)} = 2e(\tau).$$

8.4.4 The Case of Several Measurements

Consider an inverse problem for the equation

$$CT_\tau = (\lambda T_x)_x + KT_x + S, \ (x, \tau) \in Q_\tau = \{d_0(\tau) < x < d_{N+1}(\tau), \ 0 < \tau < \tau_m\} \tag{8.77}$$

with the boundary conditions

$$T(x, 0) = \xi(x); \tag{8.78}$$

$$[\alpha_1 \lambda T_x + \beta_1 T]_{x=d_0(\tau)} = p_1(\tau)\,; \tag{8.79}$$

$$[\alpha_2 \lambda T_x + \beta_2 T]_{x=d_{N+1}(\tau)} = p_2(\tau) \tag{8.80}$$

assuming that

$$T(d_i(\tau), \tau) = f_i(\tau), \quad i = \overline{1, N}, \ N \geqslant 1 \tag{8.81}$$

are given functions, where $d_1 < d_2 < \ldots < d_N$ (Fig. 8.3).

The value to be found might be one or several functions from the set $u = \{\xi(x), p_1(\tau), p_2(\tau), \lambda(T), C(T), K(T), S(T)\}$. While determining $\xi(x)$ and (or) $p_n(\tau)$, $n = 1, 2$ it is assumed, that $\lambda = \lambda(T, x, \tau)$, $C = C(T, x, \tau)$, $K = K(T, x, \tau)$, $S = S(T, x, \tau)$.

The problem for an increment of temperature $T(x, \tau)$ will be identical to statement (8.48)–(8.50) after substituting in them $X_1(\tau)$ and $X_2(\tau)$ by $d_0(\tau)$ and $d_{N+1}(\tau)$ respectively.

Compose the functional

$$J(u) = \sum_{i=1}^{N} \int_0^{\tau_m} \rho_i(\tau)[T(u, d_i(\tau), \tau) - f_i(\tau)]^2 \, d\tau, \tag{8.82}$$

where $\rho_i(\tau)$ are weight functions (see Sect. 8.2).

Assume that $d_i(\tau)$, $i = \overline{0, N+1}$ are piecewise-smooth functions. Let us consider first the case, when $\alpha_1 \neq 0, \alpha_2 \neq 0$ and $d_0(\tau) < d_1(\tau)$, $d_N(\tau) < d_{N+1}(\tau)$.

Following the foregoing method for obtaining an adjoint boundary-value problem and realizing the transformation to a form of it without singular terms, we obtain the following formulation of this problem:

$$\psi_{i\tau} + D^*_{x\tau}\psi_i = 0, \quad (x, \tau) \in Q_{i\tau}, \quad i = \overline{1, N+1}\,; \tag{8.83}$$

$$\psi_i(x, \tau_m) = 0, \quad i = \overline{1, N+1}\,; \tag{8.84}$$

$$\psi_i(d_i(\tau), \tau) - \psi_{i+1}(d_i(\tau), \tau) = 0, \quad i = \overline{1, N}\,; \tag{8.85}$$

$$[(a_1\psi_i)_x - (a_1\psi_{i+1})_x]_{x=d_i(\tau)} = 2\rho_i(\tau)e_i(\tau), \quad i = \overline{1, N}\,; \tag{8.86}$$

$$\left[(a_1\psi_1)_x - \psi_1\left(d_0' + a_2 - \frac{\sigma_1}{\gamma_1} a_1\right)\right]_{x=d_0(\tau)} = 0\,; \tag{8.87}$$

$$\left[(a_1\psi_{N+1})_x - \psi_{N+1}\left(d_{N+1}' + a_2 - \frac{\sigma_2}{\gamma_2} a_1\right)\right]_{x=d_{N+1}(\tau)} = 0 \tag{8.88}$$

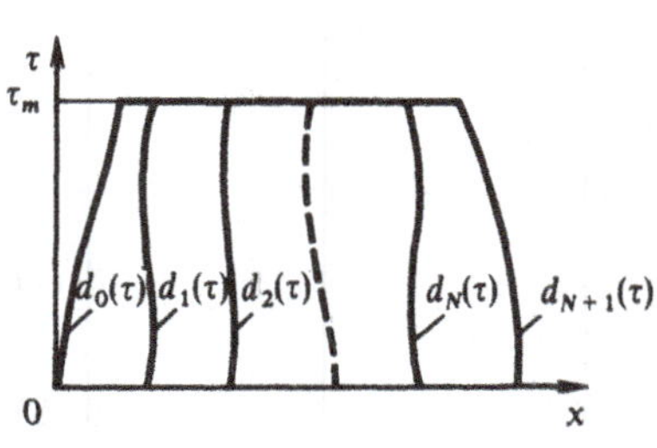

Fig. 8.3. A domain with moving boundaries and the thermosensor travelling coordinates

where $Q_{i\tau} = \{d_{i-1}(\tau) < x < d_i(\tau), 0 < \tau < \tau_m\}$; $e_i(\tau) = T(u, d_i(\tau), \tau) - f_i(\tau)$; $\gamma_1 = \alpha_1\lambda(d_0(\tau),\tau)$; $\lambda_2 = \alpha_2\lambda(d_{N+1}(\tau),\tau)$; $\sigma_1 = \alpha_1 = \alpha_1\lambda_x(d_0(\tau),\tau) + \beta_1$;
$\sigma_2 = \alpha_2\lambda_x(d_{N+1}(\tau), \tau) + \beta_2$.

Formulas for the components of the gradient of functional (8.82) in this case are

$$J'_\xi = \Psi(x, 0) \qquad d_0(0) \leqslant x \leqslant d_{N+1}(0); \tag{8.89}$$

$$J'_{p_1} = -\psi_1(d_0(\tau), \tau)\frac{a_1(d_0(\tau), \tau)}{\gamma_1(\tau)}; \tag{8.90}$$

$$J'_{p_2} = \psi_{N+1}(d_{N+1}(\tau), \tau)\frac{a_1(d_{N+1}(\tau), \tau)}{\gamma_2(\tau)}; \tag{8.91}$$

$$J'_{\lambda_l} = \Phi_0 + \Phi_1 + \Phi_{N+1}, \quad l = \overline{1, M_1}; \tag{8.92}$$

$$J'_{C_l} = -\sum_{i=1}^{N+1}\int_0^{\tau_m} d\tau \int_{d_{i-1}(\tau)}^{d_i(\tau)} \frac{\psi_i(x, \tau)}{C(x, \tau)} T_\tau(x, \tau)\varphi_l(T(x, \tau))\, dx, \quad l = \overline{1, M_2}; \tag{8.93}$$

$$J'_{K_l} = \sum_{i=1}^{N+1}\int_0^{\tau_m} d\tau \int_{d_{i-1}(\tau)}^{d_i(\tau)} \frac{\psi_i(x, \tau)}{C(x, \tau)} T_x(x, \tau)\varphi_l(T(x, \tau))\, dx, \quad l = \overline{1, M_3}; \tag{8.94}$$

$$J'_{S_l} = \sum_{i=1}^{N+1}\int_0^{\tau_m} d\tau \int_{d_{i-1}(\tau)}^{d_i(\tau)} \frac{\psi_i(x, \tau)}{C(x, \tau)} \varphi_l(T(x, \tau))\, dx, \quad l = \overline{1, M_4} \tag{8.95}$$

where $$\Psi(x, 0) = \begin{cases} \psi_1(x, 0), & x \in (d_0(\tau),\ d_1(\tau)) \\ \vdots & \vdots \qquad \vdots \\ \psi_{N+1}(x, 0), & x \in (d_N(\tau),\ d_{N+1}(\tau)); \end{cases}$$

$$\Phi_0 = \sum_{i=1}^{N+1}\int_0^{\tau_m} d\tau \int_{d_{i-1}(\tau)}^{d_i(\tau)} \frac{\psi_i(x, \tau)}{C(x, \tau)} [T_{xx}(x, \tau)\varphi_l(T(x, \tau)) + T_x^2(x, \tau)\varphi'_l(T(x, \tau))]\, dx,$$

$$\Phi_n = \int_0^{\tau_m} \frac{\psi_n(x, \tau)}{C(x, \tau)} T_x(x, \tau)\varphi_l(T(x, \tau))|_{x=d_n(\tau)}\, d\tau, \quad n = 1, N+1.$$

If to the considered N functions $f_i(\tau)$ the functions $f_0(\tau) = T(d_0(\tau), \tau)$; $f_{N+1}(\tau) = T(d_{N+1}(\tau), \tau)$ are added, then conditions (8.87) and (8.88) should be given with the right-hand sides equal to $-2\rho_0(\tau) \times [T(u, d_0(\tau), \tau) - f_0(\tau)]$ and $2\rho_{N+1}(\tau)[T(u, d_{N+1}(\tau), \tau) - f_{N+1}(\tau)]$ respectively, where $\rho_0(\tau)$ and $\rho_{N+1}(\tau)$ are weight functions.

8.4.5 The Inverse Problem with Boundary Conditions of the First Kind and in Combined Formulation

Repeating the foregoing scheme for obtaining the gradient of functional (8.82) one can show that, for the formulation of an inverse problem corresponding to conditions (8.77)–(8.81) when $\alpha_1 = \alpha_2 = 0$, $\beta_1 = \beta_2 = 1$; $d_0(\tau) < d_1(\tau)$, $d_N(\tau) < d_{N+1}(\tau)$, the following results are valid.

The problem for a temperature field variation is

$$v_\tau = a_1 v_{xx} + a_2 v_x + a_3 v + q + \omega, \quad (x, \tau) \in Q_\tau;$$

$$v(x, 0) = \Delta\xi;$$

$$v(d_0(\tau), \tau) = \Delta p_1(\tau); \quad v(d_{N+1}(\tau), \tau) = \Delta p_2(\tau),$$

where $q = \frac{1}{C}[\Delta\lambda T_{xx} + (\Delta\lambda)_x T_x + \Delta K T_x + \Delta S - \Delta C T_\tau]$;

ω is a remainder.

The adjoint problem is

$$\psi_{i\tau} + D^*_{x\tau}\psi_i = 0, \quad (x, \tau) \in Q_{i\tau}, \quad i = \overline{1, N+1};$$

$$\psi_i(x, \tau_m) = 0, \quad i = \overline{1, N+1};$$

$$\psi_i(d_i(\tau), \tau) - \psi_{i+1}(d_i(\tau), \tau) = 0, \quad i = \overline{1, N};$$

$$[(a_1\psi_i)_x - (a_1\psi_{i+1})_x]_{x=d_i(\tau)} = 2\rho_i(\tau)e_i(\tau), \quad i = \overline{1, N};$$

$$\psi_1(d_0(\tau), \tau) = \psi_{N+1}(d_{N+1}(\tau), \tau) = 0.$$

Formulas for components J'_ξ, J'_{C_l}, J'_{K_l}, J'_{S_l} of the gradient are identical to the formulas (8.89), (8.93), (8.94), (8.95) respectively. The components J'_{p_1}, J'_{p_2}, J'_{λ_l} are given by

$$J'_{p_1} = [a_1(d_0(\tau), \tau)\psi_1(d_0(\tau), \tau)]_x \tag{8.96}$$

$$J'_{p_2} = -[a_1(d_{N+1}(\tau), \tau)\psi_{N+1}(d_{N+1}(\tau), \tau)]_x \tag{8.97}$$

$$J'_{\lambda_l} = \Phi_0, \quad l = \overline{1, M_1}. \tag{8.98}$$

Finally consider the combined formulations of a boundary-value problem when $\alpha_1 = 0$, $\beta_1 = 1$, $\alpha_2 \neq 0$ and values of temperature are given at points $d_i(\tau)$, $i = \overline{1, N+1}$ and $d_0(\tau) < d_1(\tau)$. In this case the problem for a temperature field variation is defined by the conditions

$$v_\tau = a_1 v_{xx} + a_2 v_x + a_3 v + q + \omega, \quad (x, \tau) \in Q_\tau; \quad v(x, 0) = \Delta\xi;$$

$$v(d_0(\tau), \tau) = \Delta p_1(\tau); \quad [\gamma_2 v_x + \sigma_2 v]_{x=d_{N+1}(\tau)} = \Delta p_2(\tau) + \omega_2.$$

The adjoint problem is

$$\psi_{i\tau} + D^*_{x\tau}\psi_i = 0, \quad (x, \tau) \in Q_{i\tau}, \quad i = \overline{1, N+1};$$

$$\psi_i(x, \tau_m) = 0, \quad i = \overline{1, N+1};$$

$$\psi_i(d_i(\tau), \tau) - \psi_{i+1}(d_i(\tau), \tau) = 0, \quad i = \overline{1, N};$$

$$[(a_1\psi_i)_x - (a_1\psi_{i+1})_x]_{x=d_i(\tau)} = 2\rho_i(\tau)e_i(\tau), \quad i = \overline{1, N};$$

$$\psi_1(d_0(\tau), \tau) = 0;$$

$$\left[(a_1\psi_{N+1})_x - \psi_{N+1}\left(d'_{N+1} + a_2 - \frac{\sigma_2}{\gamma_2}a_1\right)\right]_{x=d_{N+1}(\tau)} = 2p_{N+1}(\tau)e_{N+1}(\tau).$$

In this case the components of the gradient are defined by formulas (8.89), (8.96), (8.91), (8.93)–(8.95) and by the formula

$$J'_{\lambda_l} = \Phi_0 + \Phi_{N+1}, \quad l = \overline{1, M_1}.$$

Remark 1. Efficient computational algorithms for solving adjoint boundary-value problems may be obtained by the transformation of the initial domain with travelling boundaries into a rectangular domain with simultaneous straightening of the curves at which initial data $f_i(\tau)$ are given. Such a transformation can be produced by substitution of the variables

$$y_i = \frac{x - d_{i-1}(\tau)}{d_i(\tau) - d_{i-1}(\tau)}, \quad i = \overline{1, N}; \quad t = \tau.$$

Remark 2. The foregoing method for obtaining the formulation of an adjoint problem and defining the gradient of the functional via its solution may be expanded to include multi-dimensional (non-linear, in general) IHCP.

Remark 3. The obtained formulas for the gradient in the space L_2 provide an opportunity to construct the directions of descent which produce approximate solutions for the required order of smoothness (see Sect. 8.5).

8.4.6 Application of the Lagrange Multipliers Method for the Construction of an Adjoint Problem

Above a method was proposed for obtaining the formulations of adjoint boundary-value problems for the inverse problems of generalized heat conduction and the formulas for the gradient were derived on their basis. There exists one more way for obtaining adjoint problems, based on the conditions necessary for stationarity of the residual functional [33, 34, 41]. The following mode is universal and can be used for various types of inverse problems in which mathematical models are those with lumped parameters and distributed ones as well.

Let us state briefly the essence of this mode by the following example.

The gradient is needed to be found of the functional

$$J(q, C) = \sum_{i=1}^{N} \int_{0}^{\tau_m} [T(q, C, d_i, \tau) - f_i(\tau)]^2 \, d\tau \tag{8.99}$$

in function $q(\tau)$ and numerical vector $C = \{C_j\}_1^M$ with the conditions

$$C(T)T_\tau = (\lambda(T)T_x)_x, \quad (x, \tau) \in Q = (0, b) \times (0, \tau_m]; \tag{8.100}$$

$$T(x, 0) = \xi(x); \tag{8.101}$$

$$-\lambda T_x|_{x=0} = q(\tau); \quad -\lambda T_x|_{x=b} = q^*(\tau), \tag{8.102}$$

where $C(T) = \sum_{j=1}^{M} C_j \varphi_j(T)$ and $\varphi_j(T)$ are the given basic functions; $p(\tau)$, $f_i(\tau) \in L_2[0, \tau_m]$ and $0 < d_1 < \ldots < d_N < b$.

Replace problem (8.100)–(8.102) by the equivalent formulation

$$C(T)T_{i\tau} = (\lambda(T)T_{ix})_x, \quad (x, \tau) \in Q_i = \{d_{i-1} < x < d_i,\ 0 < \tau \leqslant \tau_m\},$$

$$i = \overline{1, N+1}, \quad 0 = d_0 < d_1 < \ldots < d_{N+1} = b;$$

$$T_i(x, 0) = \xi_i(x), \quad d_{i-1} \leqslant x \leqslant d_i, \quad i = \overline{1, N+1};$$

$$T_i(d_i, \tau) = T_{i+1}(d_i, \tau), \quad i = \overline{1, N};$$

$$T_{ix}(d_i, \tau) = T_{i+1\,x}(d_i, \tau), \quad i = \overline{1, N};$$

$$-\lambda T_{1x}|_{x=0} = q(\tau), \quad -\lambda T_{N+1\,x}|_{x=b} = q^*(\tau),$$

where $\xi_i(x) = \xi(x)$, $d_{i-1} \leqslant x \leqslant d_i$, $i = \overline{1, N+1}$.

Transform this inverse problem into the unconstrained variational form by the introduction of augmented functional

$$I = J + I_1 + I_2 + I_3,$$

where

$$I_1 = \sum_{i=1}^{N+1} \int_{d_{i-1}}^{d_i} \mu_i(x)[T_i(x, 0) - \xi_i(x)]\,dx + \int_0^{\tau_m} \nu_0(\tau)[\lambda T_{1x}|_{x=0} + q(\tau)]\,d\tau$$

$$+ \int_0^{\tau_m} \nu_b(\tau)[\lambda T_{N+1\,x}|_{x=b} + q^*(\tau)]\,d\tau;$$

$$I_2 = \sum_{i=1}^{N} \int_0^{\tau_m} \pi_i(\tau)[T_i(d_i, \tau) - T_{i+1}(d_i, \tau)]\,d\tau$$

$$+ \sum_{i=1}^{N} \int_0^{\tau_m} \kappa_i(\tau)[T_{ix}(d_i, \tau) - T_{i+1\,x}(d_i, \tau)]\,d\tau;$$

$$I_3 = \sum_{i=1}^{N+1} \int_{d_{i-1}}^{d_i} \int_0^{\tau_m} \psi_i(x, \tau)[(\lambda(T)T_{ix})_x - C(T)T_{i\tau}]\,d\tau\,dx$$

Here $\mu_i(x)$, $i = \overline{1, N+1}$; $\nu_0(\tau)$, $\nu_b(\tau)$; $\pi_i(\tau)$; $\kappa_i(\tau)$, $i = \overline{1, N}$; $\psi_i(x, \tau)$, $i = \overline{1, N+1}$ are Lagrange multipliers.

If the conditions (8.100)–(8.102) hold, then functionals I and J are the same. Therefore the problem of minimization of functional J can be replaced by the problem of minimization of I.

The necessary condition of stationarity of $J(q, C)$ is the equality to zero of the complete variation of the functional $I(q, C)$

$$\Delta J - \Delta I_1 + \Delta I_2 + \Delta I_3 = 0. \tag{8.103}$$

For the calculation of each term in this equality, we assume that functions $C(T)$ and $q(\tau)$ were given the increments $\Delta C(T)$ and $\Delta q(\tau)$. As a result the temperature alters and its variation is denoted by $v(x, \tau)$. Following the deduction of the problem (8.48)–(8.50), we see that in linear approximation the function $v(x, \tau)$ satisfies the following boundary-value problem:

$$v_{i\tau} = a_1 v_{ixx} + a_2 v_{ix} + a_3 v_i + \frac{\Delta C}{C} T_\tau, \quad (x, \tau) \in Q_i, \quad i = \overline{1, N+1}; \tag{8.104}$$

$$v_i(x, 0) = 0, \quad d_{i-1} \leqslant x \leqslant d_i; \tag{8.105}$$

$$[-\lambda v_{1x} - \lambda_x v_1]_{x=0} = \Delta q(\tau); \tag{8.106}$$

$$v_i(d_i, \tau) = v_{i+1}(d_i, \tau), \quad i = \overline{1, N}; \tag{8.107}$$

$$v_{ix}(d_i, \tau) = v_{i+1x}(d_i, \tau), \quad i = \overline{1, N}; \tag{8.108}$$

$$[\lambda v_{N+1x} + \lambda_x v_{N+1}]_{x=b} = 0, \tag{8.109}$$

where

$$a_1 = \frac{\lambda}{C}, \quad a_2 = 2\frac{\lambda_x}{C}, \quad a_3 = \frac{1}{C}(\lambda_{xx} - C_\tau).$$

The linear part of an increment of functional is

$$\Delta J = 2 \sum_{i=1}^{N} \int_0^{\tau_m} e_i(\tau) v_i(d_i, \tau)\, d\tau \tag{8.110}$$

where $e_i(\tau) = T_i(q, C, d_i, \tau) - f_i(\tau)$.

With conditions (8.104)–(8.109) other terms in the left-hand side of equality (8.103) can be written as

$$\Delta I_1 = \sum_{i=1}^{N+1} \int_{d_{i-1}}^{d_i} \mu_i(x) v_i(x, 0)\, dx + \int_0^{\tau_m} \nu_0(\tau)[\lambda v_{1x}|_{x=0} + \lambda_x v_1|_{x=0} + \Delta q]\, d\tau$$

$$+ \int_0^{\tau_m} \nu_b(\tau)[\lambda v_{N+1x}|_{x=b} + \lambda_x v_{N+1}|_{x=b}]\, d\tau \tag{8.111}$$

$$\Delta I_2 = \sum_{i=i}^{N} \int_0^{\tau_m} \pi_i(\tau)[v_i(d_i, \tau) - v_{i+1}(d_i, \tau)]\, d\tau$$

$$+ \sum_{i=1}^{N} \int_0^{\tau_m} \kappa_i(\tau)[v_{ix}(d_i, \tau) - v_{i+1x}(d_i, \tau)]\, d\tau \tag{8.112}$$

$$\Delta I_3 = \sum_{i=1}^{N+1} \int_{d_{i-1}}^{d_i} \int_0^{\tau_m} \psi_i(x, \tau)\left(a_1 v_{ixx} + a_2 v_{ix} + a_3 v_i + \frac{\Delta C}{C} T_\tau - v_{i\tau}\right) d\tau\, dx$$

$$= \sum_{i=1}^{N+1} \int_{d_{i-1}}^{d_i} \int_0^{\tau_m} [(a_1\psi_i)_{xx} - (a_2\psi_i)_x + a_3\psi_i + \psi_{i\tau}] v_i\, d\tau\, dx$$

$$+ \sum_{i=1}^{N+1} \int_0^{\tau_m} \{a_1\psi_i v_{ix} - v_i[(a_1\psi_i)_x - a_2\psi_i]\}\Big|_{x=d_{i-1}}^{x=d_i} d\tau$$

$$- \sum_{i=1}^{N+1} \int_{d_{i-1}}^{d_i} \psi_i v_i\Big|_{\tau=0}^{\tau=\tau_m} dx\,. \tag{8.113}$$

Substituting expressions (8.111), (8.112), (8.113) into equality (8.103) we obtain on the left-hand side the summons possessing double integrals and integrals on the boundaries of domains Q_i. Since the variations v_i are arbitrary for the implementation of conditions of stationarity for I, it is necessary to equate to zero each of the groups of summons, possessing similar variations. As a result we have the formulation of an adjoint boundary-value problem:

$$-\psi_{i\tau} = (a_1\psi_i)_{xx} - (a_2\psi_i)_x + a_3\psi_i, \quad (x,\tau)\in(d_{i-1}, d_i)\times[0,\tau_m),$$

$$i = \overline{1, N+1}; \tag{8.114}$$

$$\psi_i(x,\tau_m) = 0, \quad d_{i-1} \leqslant x \leqslant d_i, \; i = \overline{1, N+1}; \tag{8.115}$$

$$\left[(a_1\psi_1]_x - a_2\psi_1 + \frac{\lambda_x}{\lambda}a_1\psi_1\right]_{x=0} = 0;$$

$$[(a_1\psi_i)_x - (a_1\psi_{i+1})_x]_{x=d_i} = 2e_i(\tau), \quad i = \overline{1, N}; \tag{8.116}$$

$$\psi_i(d_i,\tau) = \psi_{i+1}(d_i,\tau), \quad i = \overline{1, N};$$

$$\left[(a_1\psi_{N+1})_x - a_2\psi_{N+1} + \frac{\lambda_x}{\lambda}a_1\psi_{N+1}\right]_{x=b} = 0$$

which is equivalent to the formulation (8.83)–(8.88) when condition of the IHCP considered hold.

To obtain the gradient we transform the expression (8.110) with condition (8.116) taken into account, to the form

$$\begin{aligned}\Delta J &= \sum_{i=1}^{N}\int_0^{\tau_m}[(a_1\psi_i)_x - (a_1\psi_{i+1})_x]_{x=d_i}v_i(d_i,\tau)\,d\tau\\ &= \sum_{i=1}^{N+1}\int_{d_{i-1}}^{d_i}\int_0^{\tau_m}[(a_1\psi_i)_x v_i]_x\,d\tau\,dx + \int_0^{\tau_m}(a_1\psi_1)_x v_1|_{x=0}\,d\tau\\ &\quad - \int_0^{\tau_m}(a_1\psi_{N+1})_x v_{N+1}|_{x=b}\,d\tau\\ &= \sum_{i=1}^{N+1}\int_{d_{i-1}}^{d_i}\int_0^{\tau_m}[v_i(a_1\psi_i)_{xx} + v_{ix}(a_1\psi_i)_x]\,d\tau\,dx\\ &\quad + \int_0^{\tau_m}(a_1\psi_1)_x v_1|_{x=0}\,d\tau - \int_0^{\tau_m}(a_1\psi_{N+1})_x v_{N+1}|_{x=b}\,d\tau\,.\end{aligned}$$

Using the same modes as when deriving the formulas for the gradient in Sect. 6.5.2, and observing Eqs. (8.114), (8.115), (8.104), (8.105), we have the following form of a linear part of an increment of the functional

$$\Delta J = \int_0^{\tau_m}\frac{\psi_1(0,\tau)}{C(0,\tau)}\Delta q(\tau)\,d\tau - \sum_{i=1}^{N+1}\int_0^{\tau_m}\int_{d_{i-1}}^{d_i}\psi_i(x,\tau)\frac{T_{i\tau}(x,\tau)}{C(x,\tau)}\Delta C(T)\,dx\,d\tau\,. \tag{8.117}$$

From this we obtain the formula for the component J'_q

$$J'_q = \frac{\psi_1(0, \tau)}{C(0, \tau)} \tag{8.118}$$

Substituting in expression (8.117) the increment $\Delta C(T) = \sum_{j=1}^{M} \Delta C_j \varphi_j(T)$, after certain transformations we have the components of the gradient

$$J'_{C_l} = - \sum_{i=1}^{N+1} \int_0^{\tau_m} \mathrm{d}\tau \int_{d_{i-1}}^{d_i} \frac{\psi_i(x, \tau)}{C(x, \tau)} T_{i\tau}(x, \tau) \varphi_l(T(x, \tau)) \, \mathrm{d}x, \quad l = \overline{1, M} \tag{8.119}$$

The formulas (8.118) and (8.119) are equivalent to the analogous formulas for the gradient, derived in Sect. 8.4.4 for a more general case.

In the same way one can deduce the formulations of adjoint boundary-value problems and formulas for gradients for other variational forms of inverse problems. For this purpose, according to the common procedure for solving conditional extremum problems, the Lagrange functional is composed of the residual and terms corresponding to the conditions for the mathematical model: differential equations, initial and boundary conditions, conjugation conditions. For the Lagrange functional the main linear part of its increment is found corresponding to the variations of the arguments and accordingly to those of the state variables. The obtained variation of a functional is transformed using integration by parts to a form in which the integrands of the integrals, at domains where the differential equation is valid, do not contain partial derivatives of the increments of state variables. And subsequently, according to the necessary condition of stationarity for the Lagrange functional, its variation is equalled to zero. By the arbitrariness of the increments of state variables, the coefficients attached to corresponding increments are equalled to zero. The equalities obtained represent conditions for determining Lagrange multipliers which may be functions or constants depending on a corresponding condition of the mathematical model. An aggregate of these equalities gives the formulation required for an adjoint boundary-value problem.

In particular, following the above method, we can obtain an adjoint problem for the computation of a gradient of a residual for inverse heat transfer problem in any engineering system.

Let this heat exchange process be described by a system of differential equations

$$C_l \frac{\mathrm{d}T_l}{\mathrm{d}\tau} = \sum_{j=1}^{L} \lambda_{lj}(T_j - T_l) + \sum_{j=1}^{L} b_{lj}(T_j^4 - T_l^4) + Q_l, \quad l = \overline{1, L}, \quad \tau \in [0, \tau_m];$$

$$T_l(0) = T_{0l}, \quad l = \overline{1, L}, \tag{8.120}$$

where L is a number of elements in a heat model, C_l is a volumetric thermal capacity of lth element, λ_{lj} are heat exchange coefficients characterizing convection and (or) heat conduction between elements with indexes l and j, b_{lj} are coefficients of radiant heat exchange and Q_l is a total quantity of heat brought to the lth element from the outside and released in it.

Suppose that among four groups of the uniform (from the physical standpoint) parameters $\{C_l\}_1^L, \{Q_l\}_1^l, \{\lambda_{lj}\}_1^{L\times L}, \{b_{lj}\}_1^{L\times L}$, some or even all the parameters from a certain group are unknown. Denote the vector of these parameters by $\boldsymbol{u} = \{u_l\}_1^N$. Supposing further that the results of temperature measurements are known in all the elements of the heat model given in the form of time functions $T_l(\tau) = f_l(\tau)$, $l = \overline{1, L}$; we shall seek the components of vector $\boldsymbol{u}$ by minimizing the residual

$$\min_{\boldsymbol{u}} J, \quad J = \sum_{l=1}^{L} \int_0^{\tau_m} [T_l(\boldsymbol{u}, \tau) - f_l(\tau)]^2 \, d\tau,$$

where $T_l(\boldsymbol{u}, \tau)$ is a temperature of the lth element, computed by solving Eq. (8.120) for certain vector $\boldsymbol{u}$.

Omitting intermediate calculations, we state the results of deduction of an adjoint problem formulation (adjoint of the problem for temperature increments). This formulation is

$$\frac{d\psi_i}{d\tau} + \sum_{l=1}^{L} \varphi_l \frac{\partial F_l}{\partial T_i} = -2e_i(\tau), \quad i = \overline{1, L}, \quad \tau \in [0, \tau_m]; \tag{8.121}$$

$$\psi_i(\tau_m) = 0, \quad i = \overline{1, L}, \tag{8.122}$$

where

$$F_l(\tau) = \frac{1}{C_l}\left[\sum_{j=1}^{L} \lambda_{lj}(T_j - T_l) + \sum_{j=1}^{L} b_{lj}(T_j^4 - T_l^4) + Q_l\right].$$

It can be shown that gradient $J'_u = \left\{\frac{\partial J}{\partial u_i}\right\}_1^N$, is calculated via adjoint variables ψ_i by the formula

$$\frac{\partial J}{\partial u_i} = \sum_{l=1}^{L} \int_0^{\tau_m} \psi_l(\tau) \frac{\partial F_l}{\partial u_i} \, d\tau, \quad i = \overline{1, N}.$$

Using the adjoint problem (8.121)–(8.122), the form of which does not depend on the parameters considered, one can construct an effective and universal procedure for finding a descent direction when looking for causal characteristics of a heat exchange process in any engineering system.

Remark 4. The solution of inverse problems of heat exchange in engineering systems might be nonunique, being independent of the conditions of experiment (initial temperatures and external heat coercions). This question requires special consideration.

Remark 5. It is important to remember, that values of adjoint operators and those of a gradient of a functional can be found from a solution of various adjoint boundary-value problems. One can choose an adjoint problem which will be the most suitable to carry out computations. This can be seen in the example of an inverse problem in Eqs. (8.99)–(8.102). Re-write Eq. (8.104) in the problem for an increment in temperature as

$$C\vartheta_{i\tau} = \lambda\vartheta_{ixx} + 2\lambda_x\vartheta_{ix} + (\lambda_{xx} - C_\tau)\vartheta_i + \Delta CT_\tau, \quad (x, \tau)\in Q_i, \quad i = \overline{1, N+1}$$

and derive an appropriate adjoint equation. For this one needs to compute ΔI_3 once again. In the result, after some computations, we shall have the following adjoint problem:

$$- C(T(x, \tau))\psi_{i\tau} = \lambda(T(x, \tau))\psi_{ixx}, \quad (x, \tau)\in(d_{i-1}, d_i)\times[0, \tau_m),$$

$$i = \overline{1, N+1};$$

$$\psi_i(x, \tau_m) = 0, \quad d_{i-1} \leqslant x \leqslant d_i, \quad i = \overline{1, N+1};$$

$$\psi_{1x}|_{x=0} = 0;$$

$$\psi_i(d_i, \tau) = \psi_{i+1}(d_i, \tau), \quad i = \overline{1, N};$$

$$\lambda(T_i(d_i, \tau))[\psi_{ix}(d_i, \tau) - \psi_{i+1,x}(d_i, \tau)] = 2e_i(\tau), \quad i = \overline{1, N};$$

$$\psi_{N+1x}|_{x=b} = 0.$$

Following the methods described in Sect. 8.4, it is not complicated to obtain formulae for the gradient: $J'_q = \psi_1(0, \tau)$,

$$J'_{C_l} = - \sum_{i=1}^{N+1} \int_0^{\tau_m} d\tau \int_{d_{i-1}}^{d_i} \psi_i(x, \tau) T_{i\tau}(x, \tau)\varphi_l(T(x, \tau))\,dx, \quad l = \overline{1, M}.$$

Obviously, this adjoint problem is simpler for numerical solution than the boundary-value problem for Eq. (8.114).

Remark 6. As already stated, two approaches are possible for a solution of some inverse problems: (1) the problem is solved in any suitable functional space U, the selection of which depends on a priori information on the unknown characteristic; (2) the unknown function is approximated in advance (i.e. parametrized) and the solution is sought already in the finite-dimensional space E_M. Earlier in Sect. 8.4 a parametrized form was used for coefficient inverse problems (in this case a procedure of computing the gradient was simplified) which enabled us to discover more compact and economic algorithms.

Expressions for computing values of gradients of residuals with respect to boundary and retrospective inverse problems have been obtained in the frames of the first approach as being sufficient enough for many practical applications. Of course, the second approach is possible in this case too and in a number of situations could be preferable. The following considerations will explain this statement:

- With appropriate selection of finite-dimensional approximation, computational properties of algorithms can be improved, in particular, a convergence of the iterative process would be quicker in comparison with the first approach.

- Providing that we use the approximation of the unknown dependence by means of smooth basic functions, an additional property of the iterative algorithm viscosity appears that, in turn, favours using traditional techniques to halt iterations. This is especially important when the initial data error is not known or is known but with low accuracy and, hence, the residual criterion is not acceptable.

In addition, a parametrized form of any inverse problem solving is convenient from the point of view of making algorithmic-program facilities universal.

That is why, in a number of cases, it may turn out that a solution of the boundary or retrospective IHCP may also be sought as finite-dimensional approximations. In these cases the basic functions in the corresponding approximations are to be chosen so that to provide differentiability of the direct problem operator. Using the techniques described in Sect. 8.4 one can obtain appropriate expressions for gradients of functionals in solving boundary and retrospective problems (in the space E_M). Examples of the practical realization of such an approach in solving boundary IHCP will be given in Sect. 8.7.1.

8.5 Gradient Algorithms with Regard to a Priori Information

The quality of approximate solutions of inverse problems depends on taking into account completely all qualitative and quantitative a priori information about unknown quantities. Such information may be obtained from the physical peculiarities of the processes under investigation and those of the conducted experiment and from the conditions of uniqueness of solution of a problem as well. Sufficiently wide possibilities for dealing with a priori information about an unknown solution of an ill-posed problem are provided by iterational regularization.

In this section the modes are described to include the information about the smoothness of a solution and about the values of it and its derivatives at certain points of its range of definition. For this purpose in Sects. 8.5.2 and 8.5.3, when constructing an iterative sequence, a direction of descent is also chosen in the initial space L_2, but in such a way that approximations belong to a given class of smooth functions. In Sect. 8.5.4 iterative sequence is constructed directly in the space W_2^1. In both cases the corresponding directions of descent are computed via a known gradient of residual in the space L_2.

Finally, in Sect. 8.5.5 empirical modes are described for stopping the iteration when solving an ill-posed inverse problem which does not require the knowledge of the value of the initial data error, that is important in practice.

8.5.1 The Traditional Scheme for Obtaining Smooth Solutions

In Chap. 6 gradient methods have been considered for minimization of the residual functional under the constraint on its level. Such methods do not ensure the uniform convergence of approximate solutions to an unknown one and, in particular, require a good forecast of the value of an unknown function at the end point of the time interval of its definition. Therefore a necessity arises to construct algorithms which enable one to take into consideration a priori information concerning the smoothness of an unknown function and to obtain more accurate results.

The solution suggests itself straight away – to minimize the residual, regularized by the addition of the corresponding correction term. For example, in place of the criterion functional $J(u) = \| Au - f \|^2_{L_2}$ the regularized functional $\Phi_\alpha(u) = J(u) + \alpha \| u' \|^2_{L_2}$ could be considered, the rate of influence of the stabilizer $\| u' \|^2_{L_2}$ being regulated by the weight α for which choice one can use the modes given in Sect. 7.4.

Transition to the regularized residual will entail changes in algorithms for the computation of the gradient and of a step of descent as well. In particular, for a parametric optimization the corresponding expressions as applied to linear boundary IHCP are obtained in the same way as has been shown in Sect. 6.7.1. For one of the model problems of that section the results of iterative minimization of the residual with a first order stabilizer are shown in Fig. 8.4. In this

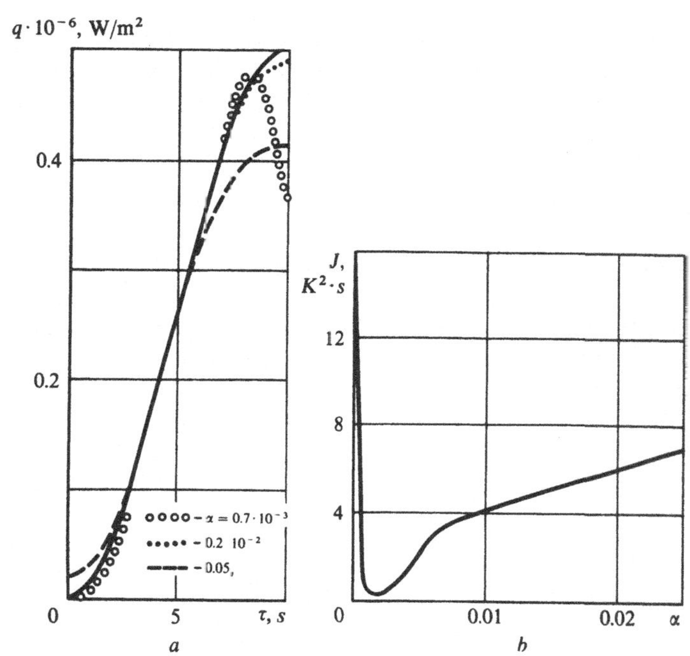

Fig. 8.4. Results of minimization of a regularized residual by the conjugate gradient method for exact input data, $a = 0.35 \cdot 10^{-6}$ m²/s, $\lambda = 0.721$ W/(m·K), $b = 0.002$ m, $d = 0.001$ m, $\Delta Fo_b = 0.017$, number of iterations $\bar{\kappa} = 100$ (with this value of κ the processes of approximations for all considered values of parameter α obviously converged): a is a time dependence of recovered heat flux for various values of the regularization parameter (full curve is an exact solution $q_T = 0.25 \cdot 10^6 \{\sin[0.1\pi(\tau - 5)] + 1\}$); b is a dependence of residual $J(q^{\kappa=100})$ on the regularization parameter

example the initial data without random errors were used and the optimal value of the parameter turned out to be $\alpha \approx 0.002$.

8.5.2 An Account of a Priori Information About an Unknown Solution (the first mode)

The idea of the first mode was suggested in [33, 36] based on the fact that the operation of integration provides an opportunity to raise the smoothness of a resultant function. The further development of this mode was produced in [48, 49, 125].

Assume that the piecewise continuous meansquare approximation is obtained of the first derivative of an unknown boundary function. Integrating it we obtain the piecewise smooth approximate solution of an inverse problem, the reconstruction of the boundary function at the end of the time interval being, in general, better than in the case of meansquare approximation. Let us consider this method, supposing that a certain one-dimensional causal characteristic $u(x)$ is to be found where $x \in [a, b]$ is a certain scalar argument (depending upon the type of inverse problem under consideration it may be time or the spatial coordinate or temperature).

Assume that the derivative $u'(x)$ is a piecewise continuous function. Let $u'(x)$ be given an increment $\Delta u'(x)$. This means that an unknown function was given a certain increment $\Delta u(x)$. The definition of a gradient of a functional in the space of functions with integrable square implies

$$\Delta J(u') \equiv J(u' + \Delta u') - J(u') = \int_a^b J'_{u'}(x) \Delta u'(x) \, \mathrm{d}x + o(\| \Delta u' \|),$$

where it is assumed that $u', \Delta u' \in L_2$ and $J'_{u'}$ is the gradient of the functional J in the derivative u'.

On the other hand, one may write $\Delta J(u) = \int_a^b J'_u(x) \Delta u(x) \mathrm{d}x + o(\| u \|)$ where Δu is an increment of an unknown function, corresponding to an increment $\Delta u'$ and J'_u is the gradient of functional J in u.

Introduce a new function

$$g(x) = \int_x^b J'_u(\xi) \, \mathrm{d}\xi \tag{8.123}$$

which derivative is $g'(x) = -J'_u(x)$.

Then

$$\Delta J(u) = -\int_a^b g'(x) \Delta u(x) \, \mathrm{d}x + o(\| \Delta u \|).$$

Integrating by parts we have

$$\Delta J(u) = \Delta u(a) g(a) + \int_a^b g(x) \Delta u'(x) \, \mathrm{d}x + o(\| \Delta u \|). \tag{8.124}$$

Since in our case $\Delta J(u') = \Delta J(u)$, the formula (8.124) implies that $g(x)$ is the gradient of the functional in u' at all points of the segment $[a, b]$ except the left one:

$$J'_{u'}(x) = g(x), \quad x \in [a, b],$$

and parameter $g(a)$ represents the value of the gradient of the functional in the constant $c = u(a)$:

$$J'_c = g(a).$$

Thus, gradient $J'_{u'}$ is defined via the known gradient J'_u.

Further we write an iterative process of the steepest descent method ($u^{0\prime}(\xi)$ is a given approximation):

$$u'^{k+1}(\xi) = u'^k(\xi) - \beta_k^0 J'^k_{u'}(\xi), \quad k = 0, 1, \ldots, \tag{8.125}$$

where

$$\beta_k^0 : J(u'^{k+1}) = \min_{\beta^0} J[u'^k - \beta^0 J'^k_{u'}].$$

Integrating expression (8.125) in ξ from a to x, we have

$$u^{k+1}(x) - u^{k+1}(a) = u^k(x) - u^k(a) - \beta_k^0 \int_a^x J'^k_{u'}(\xi)\,d\xi. \tag{8.126}$$

For the value $u^{k+1}(a) = c^{k+1}$ write a formula, corresponding to the steepest descent algorithm

$$c^{k+1} = c^k - \beta_k^1 J'^k_c \tag{8.127}$$

and substitute it in expression (8.126). As a result we have

$$u^{k+1}(x) = u^k(x) - \beta_k^0 P^k(x) - \beta_k^1 J'^k_c, \quad k = 0, 1, \ldots, \tag{8.128}$$

$$u^0(x) = z(x), \quad x \in [a, b],$$

where

$$P^k(x) = \int_a^x d\xi \int_\xi^b J'^k_u(\eta)\,d\eta, \quad J'^k_c = \int_a^b J'^k_u(x)\,dx,$$

$z(x)$ is a given approximation.

Parameters β_k^0 and β_k^1 are chosen from the condition of minimum of functional J in the present iteration:

$$\beta_k^0, \beta_k^1 : \ J(u^{k+1}) = \min_{\beta^0, \beta^1} J[u^k - \beta^0 P^k - \beta^1 J'^k_c].$$

The iterational formula (8.128) with a sufficiently smooth initial approximation $z \in W_2^l[a, b]$, $l \geq 1$ (for example, $z = \text{const}$) provides the uniform approximation of an unknown solution, if stopped according to the residual principle.

A higher speed of convergence can be obtained on the basis of a conjugate gradient method, which is given for the case under consideration in the following way:

$$u'^{k+1}(\xi) = u'^k(\xi) - \beta_k^0 J'^k_{u'}(\xi) - \gamma_k S^{k-1}(\xi), \quad k = 0, 1, \ldots; \tag{8.129}$$

$$S^{k-1}(\xi) = \beta_{k-1}^0 J'^{k-1}_{u'}(\xi) + \gamma_{k-1} S^{k-2}(\xi), \quad k \geqslant 2;$$

$$S^{-1}(\xi) = 0; \quad S^0(\xi) = \beta_0^0 J'^0_{u'}(\xi);$$

where β_k^0 and γ_k are numerical parameters.

After integrating (8.129) from a to x we have

$$u^{k+1}(x) - c^{k+1} = u^k(x) - c^k - \beta_k^0 \int_a^x J'^k_{u'}(\xi)\,d\xi - \gamma_k \int_a^x S^{k-1}(\xi)\,d\xi,$$

$$k = 0, 1, \ldots.$$

Using the relation (8.127) for the constant c we finally have

$$u^{k+1}(x) = u^k(x) - \beta_k^0 P^k(x) - \gamma_k P_1^k(x) - \beta_k^1 J'^k_c, \quad k = 0, 1, \ldots, \tag{8.130}$$

$$P^k(x) = \int_a^x d\xi \int_\xi^b J'^k_u(\eta)\,d\eta; \quad P_1^k(x) = \int_a^x S^{k-1}(\xi)\,d\xi;$$

$$S^{-1}(\xi) = 0; \quad S^0(\xi) = \beta_0^0 \int_\xi^b J'^0_u(\eta)\,d\eta;$$

$$S^{k-1}(\xi) = \beta_{k-1}^0 \int_\xi^b J'^{k-1}_u(\eta)\,d\eta + \gamma_{k-1} S^{k-2}(\xi), \quad k \geqslant 2;$$

$$J'^k_c = \int_a^b J'^k_u(x)\,dx,$$

where β_k^0, β_k^1 and γ_k are defined by the condition

$$J(u^{k+1}) = \min_{\beta^0, \beta^1, \gamma} J[u^k - \beta^0 P^k - \gamma P_1^k - \beta^1 J'^k_c].$$

The process given by formula (8.130) begins with $u^0 \in W_2^1[a, b]$, $i \geqslant 1$ and if stopped under the regularizing condition provides the *uniform convergence* of approximate solutions to the true one.

It is important to note that when the value of a function at $x = a$ is known, i.e. the constant c is given, then this a priori information can be taken into account in a natural way – by zeroing the last terms in formulas (8.128) and (8.130) and setting the initial approximation equal to the value of this constant, i.e.

$$\beta_k^1 = 0, \quad J'^k_c = 0, \quad k = 0, 1, \ldots.$$

In the case when the value of the function is known in the right-hand end of a segment (in $x = b$) and $u(a)$ is unknown, one should introduce another quantity instead of the function (8.123): $g(x) = \int_a^x J'_u(\xi)\,d\xi$, $g'(x) = J'_u(x)$. In a similar way we obtain that for $\Delta u(b) = 0$ $J'_{u'}(x) = -g(x)$. In this case integration of the sequences (8.125) and (8.129) is conducted from b to x. As a result we have

$$u^{k+1}(x) = u^k(x) - \beta_k^0 \int_x^b d\xi \int_a^\xi J_u'^k(\eta)\,d\eta, \quad k = 0, 1, \ldots$$

– for the *modified steepest descent method* and

$$u^{k+1}(x) = u^k(x) - \beta_k^0 \int_x^b d\xi \int_a^\xi J_u'^k(\eta)\,d\eta - \gamma_k \int_a^b S^{k-1}(\xi)\,d\xi\,,$$

$$S^{k-1}(\xi) = \beta_{k-1}^0 \int_a^\xi J_u'^{k-1}(\eta)\,d\eta + \gamma_{k-1} S^{k-2}(\xi), \quad k \geqslant 2\,;$$

$$S^{-1}(\xi) = 0, \quad S^0(\xi) = \beta_0^0 \int_a^\xi J_u'^0(\eta)\,d\eta$$

– for the *modified conjugate gradient method.*

Generally, when a value of function $u(x_1) = c$ is known in a certain arbitrary point x_1 of the segment $[a, b]$ represent the increment of functional in a form

$$\Delta J(u) = \Delta J_1(u) + \Delta J_2(u) + o(\|u\|)\,,$$

where

$$\Delta J_1(u) = \int_a^{x_1} J'_u(x)\Delta u(x)\,dx; \quad \Delta J_2(u) = \int_{x_1}^b J'_u(x)\Delta u(x)\,dx\,.$$

Introduce the functions $g_1(x) = \int_a^x J'_u(\xi)\,d\xi$, $x \in [a, x_1]$; $g_2(x) = \int_x^b J'_u(\xi)\,d\xi$, $x \in [x_1, b]$ for which we have $g'_1(x) = J'_u(x)$, $g'_2(x) = -J'_u(x)$. In the same way one may obtain

$$J'_{u'}(x) = \begin{cases} -g_1(x)\,, & x \in [a, x_1)\,, \\ g_2(x)\,, & x \in (x_1, b]\,. \end{cases}$$

Integrating the sequences (8.125) and (8.129) from x_1 to x and taking into account that $u^{k+1}(x_1) = u^k(x_1) = u^0(x_1)$ we obtain the computational expressions needed. In particular, for the modified steepest descent method we have

$$u^{k+1}(x) = u^k(x) - \beta_k^0 \int_{x_1}^x J_{u'}'^k(\xi)\,d\xi, \quad k = 0, 1, \ldots\,,$$

where $u^0(x) = c$.

In the following, we consider the case that the uniform convergence to the true solution should be provided with the first derivative, i.e. the smoothness of approximate solutions be raised. For this purpose one may consider, for example, a function

$$r(x) = \int_x^b d\xi \int_\xi^b J'_u(\eta)\,d\eta\,,$$

for which

$$r'(x) = -\int_x^b J'_u(\eta)\,\mathrm{d}\eta\,, \quad r''(x) = J'_u(x)\,.$$

Thus

$$\Delta J = \int_a^b r''(x)\Delta u(x)\,\mathrm{d}x + o(\|u\|)\,.$$

After integration by parts we have

$$\Delta J = -\Delta u(a)r'(a) + \Delta u'(a)r(a) + \int_a^b r(x)\Delta u''(x)\,\mathrm{d}x + o(\|\Delta u\|)\,.$$

This implies the formulas for the gradient

$$J'_{u''}(x) = r(x), \quad x \in [a, b]; \quad J'_{c_1} = r(a); \quad J'_{c_2} = -r'(a)\,,$$

where

$$c_1 = u'(a), \quad c_2 = u(a)\,.$$

The corresponding iterative formulas are constructed by the same method considered above when accounting for the condition of the first order smoothness – first of all one forms a sequence of approximations of the second derivative $u''(\tau)$ by an appropriate gradient method, then, by means of integration, one obtains the sequence of approximations to an unknown solution and conditions for the choice of numerical parameters for the iterative algorithm. In this case one can take into account not only a priori information about a value of a function at a certain point of the segment $[a, b]$, but also that of the first derivative in that or any other point as well. In particular, for the function $r(x)$ introduced and the formulas obtained for the gradient a simple way is possible to take into account information about values $u(a)$ and $u'(a)$, consisting of adding on a variation of the unknown function the restriction $\Delta u(a) = \Delta u'(a) = 0$. As a result we have an iterative formula

$$u^{k+1}(x) = u^k(x) - \beta_k^0 G_k(x), \quad k = 0, 1, \ldots\,,$$

where

$$G_k(x) = \int_a^x \mathrm{d}\xi \int_a^\xi r(\eta)\,\mathrm{d}\eta, \quad \beta_k^0\colon \ \min_{\beta^0} J[u^k - \beta^0 G_k]\,.$$

An initial approximation in this case should be chosen as sufficiently smooth $u^0(x) \in W_2^l, l \geqslant 2$, and at point $x = a$, $u^0(a) = c_2, u^{0\prime}(a) = c_1$. These conditions are valid, e.g., for function $u^0(x) = c_2 - c_1 a + c_1 x$.

The above approach can be extended to the case when it is necessary to construct regularized approximations with a higher degree of smoothness. Accordingly, if the degree of smoothness of the function to be reconstructed is l, then it is possible to take into account a priori information about the values of the derivatives up to $(l-1)$th order at certain points of a range of definition of a function. Thus, for $l=3$ the values can be assumed as the function to be found and as its first and second derivatives – it can for example, be information about extremums and points of inflexion of this function.

8.5.3 On the Construction of a Smooth Solution for a Two-dimensional Inverse Problem

Let an unknown function u depend on two variables, on spatial coordinate x and time τ. Assume, that $u(x,\tau)$ has a mixed derivative $u_{x\tau}(x,\tau) \equiv \dfrac{\partial^2 u}{\partial x\, \partial \tau}$ belonging to the space $L_2(\Gamma)$, $\Gamma = [0,b] \times [0,\tau_m]$ and the criterion functional is given by

$$J(u) = \int_0^{\tau_m} \mathrm{d}\tau \int_0^b \{A[u(x,\tau)] - f(x,\tau)\}^2 \,\mathrm{d}x\,,$$

where $f(x,\tau)$ is a known function and A is a linear or nonlinear operator, correlating to an element u an element Au.

The gradient of this functional J'_u is assumed to be known (it can be found by solving an adjoint problem).

Consider an iterative process for the derivative $u_{\xi\zeta}(\xi,\zeta)$ corresponding to the conjugate gradient method

$$\begin{aligned} u_{\xi\zeta}^{k+1} &= u_{\xi\zeta}^{k} - \beta_k^0 J'^k_{u_{\xi\zeta}} - \gamma_k^0 S^{k-1}\,, \\ S^{k-1} &= \beta_{k-1}^0 J'^{k-1}_{u_{\xi\zeta}} + \gamma_{k-1}^0 S^{k-2}, \quad k \geqslant 2\,; \\ S^{-1} &= 0, \quad S^0 = \beta_0^0 J'_{u_{\xi\zeta}}\,, \end{aligned} \tag{8.131}$$

where $J'_{u_{\xi\zeta}} = J'_{u_{\xi\zeta}}(\xi,\zeta)$ is a gradient of functional J in the derivative $u_{\xi\zeta}$ and $\beta_k^0 > 0$ and $\gamma_k^0 > 0$ are parameters of the algorithm.

After integrating both left and right-hand sides of expression (8.131) in ξ and ζ from x to b and from 0 to τ respectively we have

$$\begin{aligned} u^{k+1}(x,\tau) &= u^k(x,\tau) + \Delta u^k(x,0) + \Delta u^k(b,\tau) - \Delta u^k(b,0) \\ &\quad - \beta_k^0 \int_0^\tau \mathrm{d}\zeta \int_x^b J'^k_{u_{\xi\zeta}} \mathrm{d}\xi - \gamma_k^0 \int_0^\tau \mathrm{d}\zeta \int_x^b S^{k-1} \mathrm{d}\xi \end{aligned}$$

where $\Delta u^k(v,t) = u^{k+1}(v,t) - u^k(v,t)$.

By analogy, having at first the iterative sequences of the conjugate gradient method for the derivatives $u_\xi(\xi,0)$ and $u_\zeta(b,\zeta)$ and then passing to the sequences for $u(x,0)$ and $u(b,\tau)$ via integration we obtain

$$\Delta u^k(x,0) = \beta_k^1 \int_x^b J'^k_{u_\xi(\xi,0)} \mathrm{d}\xi + \gamma_k^1 \int_x^b S_1^{k-1}(\xi)\mathrm{d}\xi + \Delta u^k(b,0)\,.$$

$$\Delta u^k(b,\tau) = -\beta_k^2 \int_0^\tau J'^k_{u_\zeta(b,\zeta)} \mathrm{d}\zeta - \gamma_k^2 \int_0^\tau S_2^{k-1}(\zeta)\mathrm{d}\zeta + \Delta u^k(b,0)\,.$$

In these expressions values $S_1(\xi)$ and $S_2(\zeta)$ are calculated by

$$S_1^{k-1} = \beta_{k-1}^1 J'^{k-1}_{u_\zeta(\xi,0)} + \gamma_{k-1}^1 S_1^{k-2}, \quad k \geqslant 2;$$

$$S_1^{-1} = 0, \quad S_1^0 = \beta_0^1 J'^0_{u_\zeta(\xi,0)};$$

$$S_2^{k-1} = \beta_{k-1}^2 J'^{k-1}_{u_\zeta(b,\zeta)} + \gamma_{k-1}^2 S_2^{k-2}, \quad k \geqslant 2,$$

$$S_2^{-1} = 0, \quad S_2^0 = \beta_0^2 J'^0_{u_\zeta(b,\zeta)}.$$

We shall look for the value $\Delta u^k(b, 0)$ using the steepest descent formula

$$\Delta u^k(b, 0) = -\beta_k^3 J'^k_{u(b,0)}.$$

Scalar values β_k^i, $i = \overline{1, 3}$ and γ_k^i, $i = 1, 2$ are calculated from the condition of minimum of the funtional $J(u^{k+1})$.

For calculation of the components of the gradient $J'_{u_{x\tau}(x,\tau)}$, $J'_{u_x(x,0)}$, $J'_{u_\tau(b,\tau)}$, $J'_{u(b,0)}$ introduce a function $g(x, \tau)$, bound up with $J'_{u(x,\tau)}$ in the following way:

$$g(x, \tau) = \int_\tau^{\tau_m} d\zeta \int_0^x J'_{u(\xi,\zeta)} d\xi, \quad (x, \tau) \in \Gamma.$$

Therefore

$$g_\tau(x, \tau) = -\int_0^x J'_{u(\xi,\tau)} d\xi, \quad g_{x\tau}(x, \tau) = -J'_{u(x,\tau)}.$$

By the definition of the gradient we have an equation for the linear part of an increment of the functional ΔJ

$$\Delta J = \int_0^{\tau_m} d\tau \int_0^b J'_{u(x,\tau)} \Delta u(x, \tau) dx = -\int_0^{\tau_m} d\tau \int_0^b g_{x\tau} \Delta u\, dx.$$

Integrating by parts, we obtain

$$\Delta J = g(b, 0)\Delta u(b, 0) + \int_0^{\tau_m} g(b, \tau) \frac{\partial \Delta u(b, \tau)}{\partial \tau} d\tau$$

$$-\int_0^b g(x, 0) \frac{\partial \Delta u(x, 0)}{\partial x} dx - \int_0^b d\tau \int_0^{\tau_m} g(x, \tau) \frac{\partial^2 \Delta u(x, \tau)}{\partial x\, \partial \tau} d\tau.$$

Hence, the components of the gradient of the functional in components of vector $w = \{u_{x\tau}(x, \tau),\ u_x(x, 0),\ u_\tau(b, \tau),\ u(b, 0)\}$ are given by

$$J'_{u_{x\tau}(x,\tau)} = -g(x, \tau), \quad (x, \tau) \in \Gamma;$$

$$J'_{u_x(x,0)} = -g(x, 0), \quad x \in [0, b];$$

$$J'_{u_\tau(b,\tau)} = g(b, \tau), \quad \tau \in [0, \tau_m];$$

$$J'_{u(b,0)} = g(b, 0).$$

8.5.4 An Account of a Priori Information About an Unknown Solution (the second mode) [18, 20]

The construction of a smooth solution. Consider an operator equation of the first kind

$$Au = f \tag{8.132}$$

where $u = u(x)$, $x \in E_1$ is an unknown function which by assumption has an l-th generalized derivative, i.e. $u \in W_2^l[a, b]$; $f = f(x)$ is a given function from $L_2[a, b]$ and $A: W_2^l[a, b] \to L_2[a, b]$ is generally a nonlinear operator which may not have a continuous inverse.

In the sequel, we consider an algorithm for solving the problem (8.132), conforming to the steepest descent method (transition to the conjugate gradient method can be accomplished easily). In this case we have an iterative sequence

$$u^{k+1} = u^k - \beta_k J'^k_{W_2^l}, \quad k = 0, 1, \ldots, \bar{k}; \tag{8.133}$$

$$J'^k_{W_2^l} = 2(A'_{u^k})^*(Au^k - f),$$

where $J'_{W_2^l}$ is a gradient in u of the residual-functional $J(u) = \|Au - f\|^2_{L_2}$ in the space W_2^l, A'_u is the Frechet derivative of operator A at point u^k and β_k is a step length for kth iteration (β_k: $\min_\beta J(u^k - \beta J'^k_{W_2^l})$).

Initial approximation u^0 in sequence (8.133) should be taken from a class of functions with a sufficient smoothness: $u^0 \in W_2^p[a, b]$, $p \geqslant l$, in particular, one may take $u^0(x) = 0$.

As follows from the results of Sect. 8.1, when the error in right-hand side f is decreasing to zero, the approximate solution $u^{\bar{k}}$, when choosing the index of the last iteration $\bar{k}$ by the residual criteria, converges to true solution of Eq. (8.132) in the norm of the space W_2^l.

Thus, for practical application of the present algorithm we have to find the way to calculate the gradient of functional $J(u)$ in the space W_2^l. Assume that we possess an algorithm for calculating the gradient in the space $L_2[a, b]$, the latter is denoted as J'_{L_2}.

So long as the function $u(x)$ to be found belongs to the space $W_2^l[a, b]$, it is $(l-1)$ times continuously differentiable at segment $[a, b]$.

Let $u(x)$ be given a small increment $\Delta u(x) \in W_2^l$. Then the linear part of the corresponding increment of functional J is expressed via the inner product in the space W_2^l of the element Δu and the gradient $J'_{W_2^l}$:

$$\Delta J = (\Delta u, J'_{W_2^l})_{W_2^l} \tag{8.134}$$

where

$$(\Delta u, J'_{W_2^l})_{W_2^l} = \sum_{n=0}^{l} \int_a^b r_n \frac{d^n J'_{W_2^l}}{dx^n} \frac{d^n \Delta u}{dx^n} dx .$$

The values $r_n = r_n(x)$ are given non-negative continuous weight functions, r_0 and r_l being positive. By means of these functions the influence is taken into account of every derivative on the unknown function $u(x)$ at different points of the segment $[a, b]$. It is commonly set that r_n are numerical multipliers, for example, $r_0 = r_1 = \ldots = r_l = 1$.

By consequent integration by parts, the inner product in equality (8.134) can be transformed to the form without derivatives of the function $\Delta u(x)$ inside the integral (here it is assumed that $J'_{W_2^l}(x)$ and $r_n(x)$ are sufficiently smooth):

$$(\Delta u, J'_{W_2^l})_{W_2^l} = \int_a^b \Delta u(x) \sum_{n=0}^{l} (-1)^n \frac{d^n}{dx^n}\left(r_n \frac{d^n J'_{W_2^l}}{dx^n}\right) dx$$
$$+ \sum_{n=1}^{l} (-1)^{n+1} \frac{d^{n-1}\Delta u(x)}{dx^{n-1}} \sum_{i=n}^{l} (-1)^{i+1} \frac{d^{i-n}}{dx^{i-n}}\left(r_i \frac{d^i J'_{W_2^l}}{dx^i}\right)\Bigg|_{x=a}^{x=b}. \quad (8.135)$$

Consider the linear differential operator

$$D^{2l} = \sum_{n=0}^{l} (-1)^n \frac{d^n}{dx^n}\left(r^n \frac{d^n}{dx^n}\right)$$

and set the following conditions at the boundaries of the segment $[a, b]$:

$$\sum_{i=n}^{l} (-1)^{i+1} \frac{d^{i-n}}{dx^{i-n}}\left(r^i \frac{d^i J'_{W_2^l}}{dx^i}\right)\Bigg|_{x=L} = 0, \quad (8.136)$$
$$n = \overline{1, l}; \quad L = a, b.$$

Then from expression (8.135) we have

$$(\Delta u, J'_{W_2^l})_{W_2^l} = (\Delta u, D^{2l} J'_{W_2^l})_{L_2}.$$

Taking into consideration equality (8.134) we obtain as a result the differential equation

$$\sum_{n=0}^{l} (-1)^n \frac{d^n}{dx^n}\left(r_n \frac{d^n J'_{W_2^l}}{dx^n}\right) = J'_{L_2}(x), \quad x \in (a, b) \quad (8.137)$$

which is to be solved using boundary conditions (8.136).

Thus, a **convenient mode** is obtained for the calculation of the gradient of a functional in the space W_2^l via its gradient in the space L_2.

In the case $u \in W_2^1$, which is commonly used in many practical applications, the boundary-value problem (8.137)–(8.136) has the simplest form

$$r_0 J'_{W_2^1} - \frac{d}{dx}\left(r_1 \frac{dJ'_{W_2^1}}{dx}\right) = J'_{L_2}, \quad x \in (a, b); \quad (8.138)$$

$$\frac{dJ'_{W_2^1}}{dx}\Bigg|_{x=a} = \frac{dJ'_{W_2^1}}{dx}\Bigg|_{x=b} = 0 \quad (8.139)$$

If r_0 and r_1 are numbers, one may calculate a solution of the problem (8.138)–(8.139) via the Green's function:

$$J'_{W_2^1}(x) = B_1 \exp[\rho x] + B_2 \exp[-\rho x] - \frac{1}{\rho r_1} \int_a^x J'_{L_2}(\xi) \operatorname{sh}[\rho(x - \xi)] d\xi,$$
$$x \in [a, b]. \quad (8.140)$$

where

$$\rho = \sqrt{\frac{r_0}{r_1}}, \quad B_1 = B_2 \exp[-2\rho\alpha],$$

$$B_2 = \frac{\int_a^b J'_{L_2}(\xi)\mathrm{ch}[\rho(b-\xi)]\,\mathrm{d}\xi}{\rho r_1(\exp[\rho b - 2\rho a] - \exp[\rho a - 2\rho b])}.$$

Note, when computing an integral in expression (8.140) it is advisable to take advantage of the analytical integration of the function sh x In particular, the following procedure of computation $\int_a^x \ldots \mathrm{d}\xi$ turns out to be rather effective combined with the usage of the simplest step approximation of the function $J'_{L_2}(x)$, $x \in [a, b]$:

$$\int_a^x J'_{L_2}(\xi)\,\mathrm{sh}[\rho(x-\xi)]\,\mathrm{d}\xi \simeq \sum_{i=1}^{n} J'_i \int_{x_{i-1}}^{x_i} \mathrm{sh}[\rho(x_n-\xi)]\,\mathrm{d}\xi$$

$$= -\frac{1}{\rho}\sum_{i=1}^{n} J'_i\{\mathrm{ch}[\rho(x_n - x_i)] - \mathrm{ch}[\rho(x_n - x_{i-1})]\},$$

where

$$J'_i = J'_{L_2}\left(x_i - \frac{x_i - x_{i-1}}{2}\right).$$

The integral in the expression for the constant B_2 can be computed in a similar way.

In the case of space $W_2^2[0, b]$ we have a differential equation of the fourth order

$$r_0 J'_{W_2^2} - (r_1(J'_{W_2^2})_x)_x + (r_2(J'_{W_2^2})_{xx})_{xx} = J'_{L_2}, \quad x \in (0, b)$$

with boundary-value conditions

$$r_1(J'_{W_2^2})_x - (r_1(J'_{W_2^2})_{xx})_x|_{x=0;b} = 0;$$

$$r_2(J'_{W_2^2})_{xx}|_{x=0;b} = 0.$$

A solution of this problem by means of the Green's function method has real values provided that the coefficients r_0, r_1 and r_2 satisfy the condition

$$D_1 = r_1^2 - 4r_0 r_2 > 0,$$

and are written as[4]

$$J'_{W_2^2}(x) = C_1 e^{Ax} + C_2 e^{-Ax} + C_3 e^{Bx} + C_4 e^{-Bx} + S(x),$$

[4] Alifanov O.M.; Balashova I. E.: Investigating the diagnostics algorithms of the thermal effect in design. J. Eng. Phys., 1989, Vol. 56, No. 4 pp. 483–488

where

$$A = \sqrt{D_2^+/2r_2}\,; \quad B = \sqrt{D_2^-/2r_2}\,, \quad D_2^\pm = r_1 \pm \sqrt{D_1}\,;$$

$$S(x) = [r_2(B^2 - A^2)]^{-1} \int_0^x J'_{L_2}(\xi)[B^{-1}\operatorname{sh} B(x-\xi) - A^{-1}\operatorname{sh} A(x-\xi)]\,d\xi\,.$$

Coefficients in the expression are computed by solving a system of equations

$$C_1 A(1-A^2) - C_2 A(1-A^2) + C_3 B(1-B^2) + C_4 B(1-B^2) = 0$$

$$C_1 A(1-A^2)e^{bA} - C_2 A(1-A^2)e^{-bA}$$
$$+ C_3 B(1-B^2)e^{bB} - C_4 B(1-B^2)e^{-bB} = -P$$

$$C_1 A^2 - C_2 A^2 + C_3 B^2 - C_4 B^2 = 0$$

$$C_1 A^2 e^{bA} - C_2 A^2 e^{-bA} + C_3 B^2 e^{bB} - C_4 B^2 e^{-bB} = -Q,$$

where

$$P = [r_2(B^2 - A^2)]^{-1} \int_0^b J'_{L_2}(\xi)[(1-B^2)\operatorname{ch} B(b-\xi)$$
$$- (1-A^2)\operatorname{ch} A(b-\xi)]\,d\xi\,,$$

$$Q = [r_2(B^2 - A^2)]^{-1} \int_0^b J'_{L_2}(\xi)[B\operatorname{sh} B(b-\xi) - A\operatorname{sh} A(b-\xi)]\,d\xi\,.$$

A flow-chart of the computational algorithm. The sequence of solution of an inverse problem with regard to smoothness for an unknown function can be represented in the form of the following block diagram:

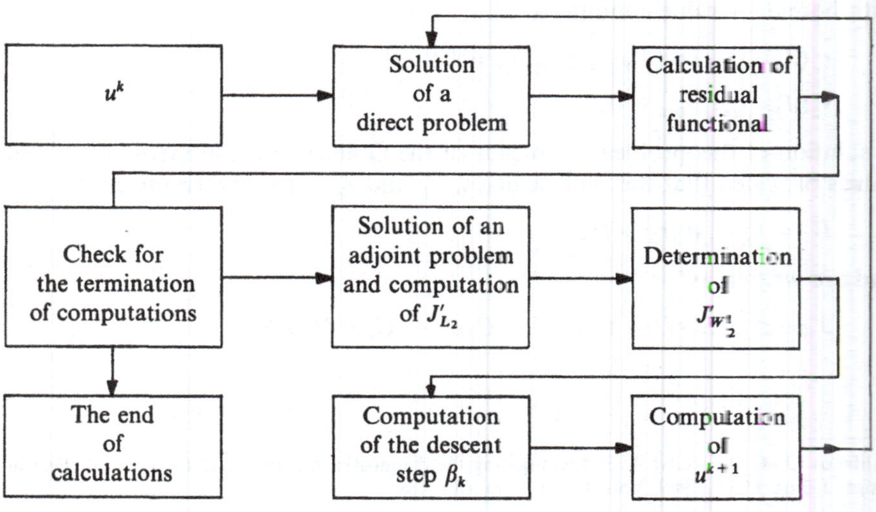

At every iteration $k = 0.1, \ldots$ the following steps of calculations are performed in succession.

1. For known approximation u^k, $k = 0.1, \ldots$ (u^0 is assumed a priori; $u^1, u^2, \ldots$ are received from computations at the previous iterations) a direct problem is being solved and a temperature field is found $T(u^k, x, \tau)$.
2. Value of a residual functional $J[u^k]$ is computed.
3. Check that the termination of an iterative process is performed; for this one can use a residual criterion or other techniques (see Sect. 8.5.5); if these conditions are satisfied, the solution of the problem ends; if not, the calculations continue.
4. By knowing the temperature field $T(u^k, x, \tau)$ one computes the coefficients and temperature residuals for the adjoint problem which is then solved and by its solution one computes the gradient J'_{L_2} in space L_2.
5. By the obtained gradient J'_{L_2}, solving the problem (8.137)–(8.136), one computes the gradient in space W^l_2 where l is a number given in advance assigned from physical considerations (usually $l \leqslant 3$). Subsequently in the result one determines the required direction of descent.
6. One computes a descent step β_k. For this one can use a linear estimate $\beta_k = \| J'^k_{W^1_2} \|^2_{W^1_2} / \| A'_{u^k} J'^k_{W^1_2} \|^2_{L_2}$ or a more exact value of the step, obtained from a solution of the problem of one-dimensional minimization $\beta_k : \min\limits_{\beta} J(u^k - \beta J'^k_{W^1_2})$; in the last case one has to solve a direct problem several times with different values of β.
7. One finds the next approximation u^{k+1}, hence it is possible to proceed with the calculations at the $k + 1$ iteration, i.e. to repeat steps 1 ...

An account of quantitative information about an unknown solution. At first we consider the situation when the values are known of a function and (or) its derivatives at boundary points of a segment $[a, b]$. These data can be easily taken into account by the appropriate choice of initial approximation to a solution and setting suitable boundary conditions for Equation (8.137). For example, let the values of the derivatives $u'(a) = \alpha_1$ and $u'(b) = \beta_1$ be known. This information should be considered when choosing the initial approximation, namely, demand that the equalities be valid $u^{0\prime}(a) = \alpha_1$, $u^{0\prime}(b) = \beta_1$. If we now make use of the solution (8.140) of the problem (8.138)–(8.139), then the conditions stated would be satisfied by equalities (8.139). In the case when the values of the function $u(a) = \alpha_0$, $u(b) = \beta_0$ are known they are taken into account when using the space W^1_2 by replacing boundary conditions (8.139) by $J'_{W^1_2}(a) = J'_{W^1_2}(b) = 0$, the initial approximation being chosen with the conditions $u^0(a) = \alpha_0$ and $u^0(b) = \beta_0$. The values of the function and its first derivatives at the ends of a segment can be chosen when solving the problem (8.132) in the space W^2_2. In this case, instead of conditions (8.136) one needs to set $\dfrac{d^n J'_{W^2_2}(L)}{dx^n} = 0$, $L = a, b$, $n = 1, 2$ and to choose appropriate $u^0(x)$. With the conditions (8.136) for $l = 2$ it is possible to assess values of second derivatives at the ends of a segment. In the same way one may consider other cases of a priori information about an unknown solution at points a and b.

The method under consideration for the construction of a smooth solution makes it possible to account for a priori information about the values of the function and its derivatives at a number of points of the segment $[a, b]$ as well. Let us clarify what kind of information it is and how to take it into account.

Functions from the space $W_2^l[a, b]$ can be represented via their generalized derivatives $u^{(l)}(x)$ in the following integral form:

$$u(x) = \sum_{n=0}^{l-1} c_n P_n(x) + \int_{x_1}^{x} d\xi_1 \int_{x_2}^{\xi_1} d\xi_2 \ldots \int_{x_l}^{\xi_{l-1}} u^{(l)}(\xi_l)\, d\xi_l, \tag{8.141}$$

where $c_n = u^{(n)}(x_{n+1})$, $n = \overline{0, l-1}$ are the values of the derivatives of function $u(x)$ up to the $(l-1)$th order at certain given points $x_{n+1} \in [a, b]$; $P_n(x)$ is a polynomial of nth degree.

The identity (8.141) is obtained in the following way. If function $u(x) \in L_2[a, b]$ has its generalized lth derivative $u^{(l)}(x) \in L_2[a, b]$ then it is continuously differentiable $(l-1)$ times at the segment $[a, b]$ and its derivative $u^{(l-1)}(x)$ is absolutely continuous at $[a, b]$. In this case for all derivatives up to the $(l-1)$th order the relation is valid

$$u^{(n-1)}(\xi_{n-1}) = \int_{x_n}^{\xi_{n-1}} u^{(n)}(\xi_n)\, d\xi_n + u^{(n-1)}(x_n), \quad n = \overline{1, l}\,.$$

Expressing $u(x)$ via $u'(\xi_1)$, then $u'(\xi_1)$ via $u''(\xi_2)$ and etc., we obtain the identity (8.141). The form of polynomials $P_n(x)$ is easily obtained for every special case.

Denote $y(x) = \int_{x_1}^{x} d\xi_1 \int_{x_2}^{\xi_2} d\xi_2 \cdots \int_{x_l}^{\xi_{l-1}} u^{(l)}(\xi_l) d\xi_l$ and substitute $u(x)$, represented in the form (8.141), in the iterative sequence (8.133). As a result we have

$$y^{k+1}(x) + \sum_{n=0}^{l-1} c_n^{k+1} P_n(x) = y^k(x) + \sum_{n=0}^{l-1} c_n^k P_n(x) - \beta_k J'^k_{W_2^l}\,. \tag{8.142}$$

Assume that values of function and their derivatives are known at points $\{x_n\}_0^{l-1}$ i.e. the numbers c_n, $n = \overline{0, l-1}$ are given. In this case the formula (8.142) yields

$$y^{k+1}(x) = y^k(x) - \beta_k J'^k_{W_2^l},$$

the initial approximation $y^0(x)$ satisfying conditions $u^{0(r)}(x_{n-1}) = c_r$, $n = \overline{0, l-1}$. In particular, one can take $u^0(x) = \sum_{n=0}^{l-1} c_n P_n(x)$, then $y^0(x) = 0$. The gradient $J'_{W_2^l}$ should satisfy the conditions $J'^{(n)}_{W_2^l}(x_{n+1}) = 0$, $n = \overline{0, l-1}$.

By this method one can satisfy l conditions for the function $u^{(n)}(x_{n+1})$, $n = \overline{0, l-1}$ at points x_{n+1} of the segment $[a, b]$ (these points, of course, may coincide). Therefore, the way of choosing the points x_{n+1} and $n = \overline{0, l-1}$ becomes clear.

Parametrized form of a solution. When solving many inverse and optimal control problems, it is convenient to represent an unknown function in the following approximate form

$$u(x) \simeq \tilde{u}(x) = \sum_{j=1}^{M} a_j \varphi_j(x), \quad x \in [a, b], \tag{8.143}$$

where $\{\varphi_j(x)\}_1^M$ is a given system of basic functions and $\boldsymbol{a} = \{a_j\}_1^M$ is a numerical vector of the coefficients to be found.

For $\varphi_j(x)$ the cubic B-splines are often taken. Then the corresponding functions in Eq. (8.143) form a subspace in the spaces $W_2^l[a, b]$, $l \leqslant 3$.

For determination of $\boldsymbol{a} \in E_M$ we utilize an iterative formula of the gradient type with stop using residual principle:

$$\begin{aligned} &\boldsymbol{a}^{k+1} = \boldsymbol{a}^k - \beta_k \boldsymbol{p}^k, \quad k = 0, 1, \ldots, \bar{k}, \\ &\bar{k}: \; J(\boldsymbol{a}^{\bar{k}}) = \| A\tilde{u}(x) - f_\delta \|_{L_2}^2 \simeq \delta^2 . \end{aligned} \tag{8.144}$$

Formulate the problem: to find such a direction of descent $\boldsymbol{p}^k = \{p_j^k\}_{j=1}^{j=M}$ for every iteration, that the convergence of approximations $\tilde{u}^k(x)$ would be ensured in the norm of the space W_2^l.

For simplicity as above, we confine our consideration to the steepest descent method only.

Let an unknown function $\tilde{u}(x) \in W_2^l[a, b]$ be given a small increment $\Delta\tilde{u}(x) \in W_2^l[a, b]$. Then a linear part of corresponding increment of the functional J may be represented in the form of an inner product in the space $W_2^l[a, b]$:

$$\Delta J = (\Delta\tilde{u}, J'_{W_2^l})_{W_2^l} \equiv \sum_{n=0}^{l} \int_a^b r_n \frac{\mathrm{d}^n J'_{W_2^l}}{\mathrm{d}x^n} \frac{\mathrm{d}^n \Delta\tilde{u}}{\mathrm{d}x^n} \, \mathrm{d}x . \tag{8.145}$$

So far as $\tilde{u}(x)$ has a form (8.143), then for an increment $\Delta\tilde{u}(x)$ we have

$$\Delta\tilde{u}(x) = \sum_{j=1}^{M} \Delta a_j \varphi_j(x) . \tag{8.146}$$

Taking into consideration that vector $\boldsymbol{p}$ in the iterative sequence (8.144) corresponds to the steepest descent method, we write a similar representation of the gradient of a functional in the space W_2^l

$$J'_{W_2^l} = \sum_{j=1}^{M} p_i \varphi_j(x) . \tag{8.147}$$

After substituting expressions (8.146) and (8.147) in the formula (8.145) and some transformations we obtain the following expression:

$$\Delta J = (\Delta \boldsymbol{a}, \boldsymbol{v})_{E_M} \tag{8.148}$$

where $\Delta \boldsymbol{a} = \{\Delta a_j\}_1^M$; $\boldsymbol{v} = \{v_j\}_1^M$; $v_j = \sum_{i=1}^{M} p_i \cdot (\varphi_j, \varphi_i)_{W_2^l}$, $(\Delta \boldsymbol{a}, \boldsymbol{v})_{E_M}$ is the inner product in the space E_M.

The formula (8.148) implies that vector v is the gradient of functional J in the space E_M therefore we obtain a system of linear algebraic equations for the calculation of component p_j when the components v_j are known:

$$\sum_{i=1}^{M} p_i b_{ji} = v_j, \quad j = \overline{1, M}, \tag{8.149}$$

where $b_{ji} = (\varphi_j, \varphi_i)_{W_2^1}$.

The matrix of this system is symmetric, positively definite and for solving Eqs. (8.149) the better-known methods in linear algebra can be utilized taking into consideration these peculiarities.

The right-hand side of the system (8.149) – the gradient of functional in $\hat{a}$ – can be calculated by solving an adjoint boundary-value problem (see Sect. 8.4).

The foregoing mode for taking into account qualitative and quantitative a priori information provides a possibility to construct effective numerical algorithms. It can be extended to multidimensional inverse problems as well.

Remark 1. In this section inverse problems have been considered to determine a single casual characteristic. In practice, the necessity may arise to find some vector-function by solving an inverse problem, for example two boundary conditions, boundary and initial conditions, several thermophysical characteristics etc. In this case, as was shown in [49], it is expedient in gradient methods to consider the step length β not as a scalar but as a *vector value* with dimensions equal to the number of values to be found. This increases appreciably the speed of operation of the algorithms and improves the quality of the approximations obtained as well. Examples of such modified gradient methods will be constructed in Sect. 8.6.

8.5.5 On the Ways of Stopping the Iteration when Initial Data Error Level is Unknown

The theoretical results, given in Sect. 8.1, and the results of computational experiments in Chap. 6 imply that the iterative process of the solution of the ill-posed inverse problem yields two stages which we conveniently call, regular and non-regular respectively. The regular stage is characterized by the steady approach of approximations to an unknown solution, during which the salient features of the reconstructed functions are revealed and specified. In the second stage one can no longer guarantee that more accurate approximations are received. In fact in successive iterations the unwanted oscillations gradually develop due to the errors in initial data.

Thus, the main problem for the practical application of the regular iterative method consists of the determination of the stop rule, i.e. an index of iteration which demarcates the stated stages.

The most widely used and mathematically proved criteria for stopping the iteration is the stop by the residual condition when the minimizing residual functional reaches the value of summary error of the experimental data (see Sect. 8.1).

Successful application of the residual criteria entails the necessity of a priori knowledge for the sufficiently accurate evalution of initial data error. But in practice this requirement is not always valid. Therefore, the necessity arises to find other ways for choosing approximations. Consider three possible empirical ways for stopping the iteration.

Stopping by additional measurement. We consider this approach as an example of one-dimensional boundary-value IHCP noting that it can be utilized for the solution of other inverse problems as well. Let the function $u(\tau)$ be found which characterizes heat exchange at one of the boundaries of a body assuming, that at the other one the boundary condition is known. In this case the minimal necessary information (regarding the uniqueness of a solution of IHCP) is the temperature measurements $T(d, \tau) = f(\tau)$ at one of the points of a body (see Sect. 2.3). Assume now that at points of a body d_1 and d_2 which may, in general, coincide the two realizations of temperature are known

$$f_{\delta_1}(\tau) = f_1(\tau) + \Delta f_1(\tau), \quad \text{and} \quad f_{\delta_2}(\tau) = f_2(\tau) + \Delta f_2(\tau),$$

when true values $f_1(\tau)$ and $f_2(\tau)$ are disturbed by random additives $\Delta f_1(\tau)$ and $\Delta f_2(\tau)$.

We construct an iterative method for solving IHCP, using realization $f_{\delta_1}(\tau)$ as initial data.

By an increase of the number of iterations, beginning with certain indexes, the functions $u^k(\tau)$ begin to deviate more from the solution to be found, "fitting" to the disturbed values of temperature $f_{\delta_1}(\tau)$. In conforming with such a course of the iteration process the residual $\Delta(u^k(\tau), f_2(\tau)) = \left\{\int_0^{\tau_m} [T(u^k(\tau), d_2, \tau) - f_2(\tau)]^2\right\}^{1/2}$ calculated at point d_2 for precise temperature $f_2(\tau)$, decreases at the beginning of the iteration process and then, after reaching its minimum, begins to increase. Naturally, the value $u^{\bar{k}}(\tau)$, corresponding to the minimum mentioned, would be a good approximation for $u_T(\tau)$.

Therefore, one may suppose taking into consideration the mutual non-correlation of disturbances $\Delta f_1(\tau)$ and $\Delta f_2(\tau)$ that the minimum of the residual $\Delta(u^k(\tau), f_{\delta_2}(\tau))$, calculated for the disturbed temperature $f_{\delta_2}(\tau)$, lies in the domain of iteration indexes, close to $\bar{k}$.

Thus, the first way for stopping the iteration is based on the determination of such an index of iteration $\tilde{k}$ for which the minimum of this residual is reached:

$$\tilde{k} : \min_k \Delta(u^k(\tau), f_{\delta_2}(\tau)) .$$

Such a way for stopping the iteration, when solving IHCP, was proposed in [16].

Remark 2. From the practical point of view two random realizations of temperatures $f_{\delta_1}(\tau)$ and $f_{\delta_2}(\tau)$ may be obtained by two different temperature sensors. In the case when $d_1 = d_2$ and there is only one gauge, one can choose out of the experimental curve $f_\delta(\tau)$ two different time series $\{f^1_{\delta_i}\}$ and $\{f^2_{\delta_i}\}$. For this purpose one may take different steps of discretization of the continuous process of measurement, and "displacement" of time grids of temperature registration as well. In both cases it is necessary, on the one hand, to avoid the correlation of data (because any valid signal is usually disturbed by nonwhite noise) and on the other hand, to exclude the effect of disgnising the frequencies of valid signal.

Traditional way of stopping by "closeness" of iterations. When solving ill-posed inverse problems, in certain cases one may use the way for stopping the iteration under the condition of closeness with accuracy ε of the neighbouring approximations, for example $\bar{k}: \| u^{\bar{k}+1} - u^{\bar{k}} \| \leqslant \varepsilon$. Such a situation arises, in particular after preliminary smoothing of initial data (see Sects. 6.8, 6.9). A similar mode may also be used in step regularization. First of all it concerns the algorithms taking into account the smoothness of the function to be reconstructed, since these algorithms possess considerable filter properties with respect to high frequency errors in initial data. After all, this approach gives good results also in the case when a number of mutual non-correlated measurements at one or several spatial points are utilized when solving inverse problems. The operation mechanism in this case is similar to the one when stopping by additional measurement.

Stopping by the increment of the functional. The present mode is based on a characteristic alteration of the residual functional in iterations (see Fig. 8.5); at the beginning of a process this value is increasing rapidly and then the speed of its alteration abruptly slows down. As a result the curve $\sqrt{J(k)}$ may have a sufficiently abrupt bend parting two regions – those of the rapid and slow alteration of the residual. For iterative algorithms with regard to a priori information on smoothness of a solution, the region of "bending" becomes particularly noticeable – for sufficiently large indexes k the curve $\sqrt{J(k)}$ has a strongly marked "shelf". The region of the values of indexes k, at which the noticeable bend of the curve occurs, commonly produces quite satisfactory approximations to an unknown solution of an ill-posed inverse problem.

Let us explain the reason for such a behaviour of the residual for a linear problem as an illustration. In this case the gradient $J'^k = A^*(Au^k - f_\varepsilon)$ $= \bar{J}'^k - \Delta J'^k$ consists of two summonds: gradient, calculated via precise data

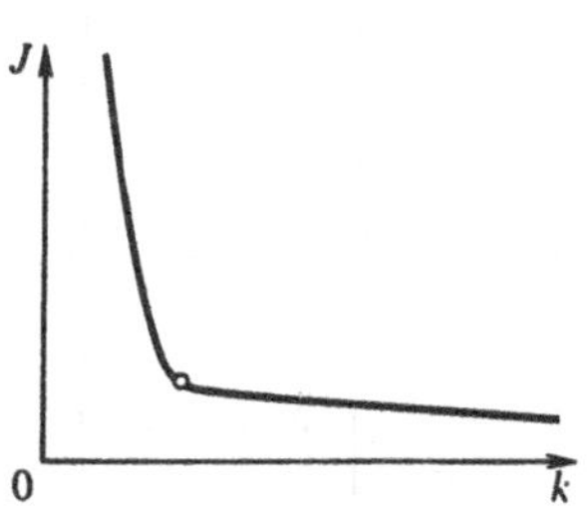

Fig. 8.5. Residual change in iterations

$\bar{J}'^k = A^*(Au^k - f)$ and additional term $A^*\Delta f$, taking into account the presence of disturbance. The disturbance Δf often represents the realization of a certain random process and oscillates with a high frequency in time. The precise function $f(\tau)$ is a smooth function, as a rule, which alters in time much more slowly as against the high-frequency disturbance $\Delta f(\tau)$. The adjoint operator A^*, as well as the operator of the direct problem A is strongly smoothing and suppresses high-frequency errors. As a result in the first iterations we have $\bar{J}'^k \gg \Delta J'^k$ and an iterative process develops almost in the same way as for $\Delta f = 0$. By approaching an unknown solution the value $\bar{J}'^k$ is decreasing rapidly, J'^k is also decreasing and the iterative process slows down.

Remark 3. For boundary-value IHCP the foregoing effect intensifies when moving off the measurement point from the boundary, where the conditions are to be reconstructed because smoothing properties of the operator A^* become stronger.

8.6 Examples of the Construction of the Algorithms for the Solution of Inverse Problems

8.6.1 Linear Boundary IHCP

Consider the problem of reconstructing the heat fluxes $q_1(\tau)$ and $q_2(\tau)$ at two boundaries of a plate, assuming that thermophysical properties of a material are constant and there are temperature measurements at N points of a body:

$$T_\tau = aT_{xx}, \quad (x, \tau) \in Q = (0, b) \times (0, \tau_m]; \tag{8.150}$$

$$T(x, 0) = 0, \quad x \in [0, b]; \tag{8.151}$$

$$-\lambda T_x(0, \tau) = q_1(\tau), \quad -\lambda T_x(b, \tau) = q_2(\tau), \quad \tau \in [0, \tau_m]; \tag{8.152}$$

$$T(d_i, \tau) = f_i(\tau), \quad \tau \in [0, \tau_m], \quad i = \overline{1, N}, \ N \geqslant 2,$$

$$0 \leqslant d_1 < \ldots < d_N \leqslant b,$$

where $a, \lambda = \text{const} > 0$ and $f_i(\tau)$ are known values.

For given approximations to the unknown vector-function $q(\tau) = \{q_1(\tau), q_2(\tau)\}$ write the solution of the direct problem in Eqs. (8.150)–(8.152) via the Green's function $G(x, x', \tau - \tau')$:

$$A_i q = A_{1i} q_1 + A_{2i} q_2,$$

where $\quad A_i q = T(q, d_i, \tau).$

$$A_{1i} q_1 = a \int_0^\tau q_1(\tau') G(d_i, 0, \tau - \tau') \, d\tau';$$

$$A_{2i} q_2 = -a \int_0^\tau q_2(\tau') G(d_i, b, \tau - \tau') \, d\tau'.$$

Compose the residual functional with weight coefficients $\rho_i(\tau)$:

$$J(u) = \sum_{i=1}^{N} \int_0^{\tau_m} \rho_i(\tau)[(A_i q)(\tau) - f_i(\tau)]^2 \, d\tau$$

and calculate its gradient in the space $L_2[0, \tau_m]$ (see Sect. 8.2)

$$J'_q = \{J'_{q1}, J'_{q2}\} .$$

The components of J'_q are given by

$$J'_{q_j} = \sum_{i=1}^{N} A^*_{ji} z_i, \quad j = 1, 2 ,$$

where $z_i = \rho_i(\tau)[(A_i q)(\tau) - f_i(\tau)]$,

$$A^*_{1i} z_i = a \int_{\tau}^{\tau_m} z_i(\tau') G(d_i, 0, \tau - \tau') \, d\tau',$$

$$A^*_{2i} z_i = -a \int_{\tau}^{\tau_m} z_i(\tau') G(d_i, b, \tau - \tau') \, d\tau' .$$

We look for a uniform approximation to the unknown solution, using the first mode to include the smoothness (see Sect. 8.5.2). In this case the *modified steepest descent algorithm* is given by (see formula in Eq. (8.128))

$$q_j^{k+1}(\tau) = q_j^k(\tau) - \beta^0_{jk} P^k_j(\tau) - \beta^1_{jk} J'^k_{cj}, \quad j = 1, 2; \; k = 0, 1, \ldots, \bar{k}, \tag{8.153}$$

$$q_j^0(\tau) = g_j(\tau) . \tag{8.154}$$

where $$P^k_j(\tau) = \int_0^{\tau} d\tau' \int_{\tau'}^{\tau_m} J'^k_{q_j}(\eta) \, d\eta, \quad J'^k_{cj} = \int_0^{\tau_m} J'^k_{q_j}(\tau) \, d\tau ; \tag{8.155}$$

$g_j(\tau) \in W^l_2[0, \tau_m]$, $l \geqslant 1$ is a given function.

Numbers β^0_{jk} and β^1_{jk} are found with the condition

$$\beta^r_{jk} : \min_{\beta^r_j} J(q^{k+1}), \quad j = 1, 2; \; r = 0, 1 ,$$

$$J(q^{k+1}) = \sum_{i=1}^{N} \int_0^{\tau_m} \rho_i(\tau) [A_{1i}(q^k_1 - \beta^0_1 P^k_1 - \beta^1_1 J'^k_{c_1}) + A_{2i}(q^k_2 - \beta^0_2 P^k_2 - \beta^1_2 J'^k_{c_2}) - f_i]^2 \, d\tau .$$

Calculating the derivatives $\partial J(q^{k+1})/\partial \beta^r_{jk}$, $j = 1, 2$; $r = 0, 1$ and equating them to zero, after simple transformations we have the system of linear algebraic equations to determine β^r_{jk}

$$\left.\begin{aligned}
a^k_{11}\beta^0_{1k} + a^k_{12}\beta^1_{1k} + a^k_{13}\beta^0_{2k} + a^k_{14}\beta^1_{2k} &= b^k_1 \\
a^k_{21}\beta^0_{1k} + a^k_{22}\beta^1_{1k} + a^k_{23}\beta^0_{2k} + a^k_{24}\beta^1_{2k} &= b^k_2 \\
a^k_{31}\beta^0_{1k} + a^k_{32}\beta^1_{1k} + a^k_{33}\beta^0_{2k} + a^k_{34}\beta^1_{2k} &= b^k_3 \\
a^k_{41}\beta^0_{1k} + a^k_{42}\beta^1_{1k} + a^k_{43}\beta^0_{2k} + a^k_{44}\beta^1_{2k} &= b^k_4
\end{aligned}\right\} \tag{8.156}$$

The coefficients and right-hand sides of the present system are given by

$$a_{nl}^k = \sum_{i=1}^{N} (\rho_i v_{ni}^k, v_{li}^k)_{L_2[0,\tau_m]}, \quad b_n^k = \sum_{i=1}^{N} (z_i^k, v_{nl}^k)_{L_2[0,\tau_m]}, \tag{8.157}$$

where $v_{1i}^k = A_{1i}P_1^k$, $v_{2i}^k = A_{1i}J_{c1}'^k$, $v_{3i}^k = A_{2i}P_2^k$,

$v_{4i}^k = A_{2i}J_{c2}'^k$, $z_i^k = \rho_i(A_i q^k - f_i)$.

The constructed algorithm is simplified naturally when one of the functions $q_1(\tau)$, $q_2(\tau)$ is known or the values $q_1(0)$ and (or) $q_2(0)$ are given.

The modified conjugate gradient algorithm can be constructed similarly.

8.6.2 Nonlinear Boundary IHCP

Let the vector-function $q(\tau) = \{q_1(\tau), q_2(\tau)\}$ of heat fluxes be reconstructed by solving the following inverse problem:

$$C(T)T_\tau = (\lambda(T)T_x)_x, (x, \tau) \in Q_\tau = \{d_0(\tau) < x < d_{N+1}(\tau), \quad 0 < \tau \leqslant \tau_m\}; \tag{8.158}$$

$$T(x, 0) = \xi(x), \quad d_0(0) \leqslant x \leqslant d_{N+1}(0); \tag{8.159}$$

$$-\lambda(T)T_x|_{x=d_0(\tau)} = q_1(\tau), \quad -\lambda(T)T_x|_{x=d_{N+1}(\tau)} = q_2(\tau), \quad \tau \in [0, \tau_m]. \tag{8.160}$$

We take the criterion functional of a form

$$J(q) = \sum_{i=1}^{N+1} \int_0^{\tau_m} \rho_i(\tau) e_i^2(\tau)\, d\tau, \tag{8.161}$$

where $\rho_i(\tau)$ are weight multipliers; $e_i(\tau) = T(q(\tau), d_i(\tau), \tau) - f_i(\tau)$.

The present formulation of IHCP is a special case of a more general problem considered in Sect. 8.4.4. Therefore, the gradient of residual (8.161) is given by formulas (8.90) and (8.91) and formulations of the problem for an increment of temperature and an adjoint problem follow from conditions (8.48)–(8.50) and (8.83)–(8.88) respectively, the zero in the right-hand side of Eq. (8.88) being replaced by the value $2\rho_{N+1}(\tau)e_{N+1}(\tau)$ since at the boundary $d_{N+1}(\tau)$ a temperature measurement is given. As a result we have the problem for an increment of the temperature field:

$$v_\tau = a_1 v_{xx} + a_2 v_x + a_3 v, \quad (x, \tau) \in Q_\tau; \tag{8.162}$$

$$v(x, 0) = 0; \tag{8.163}$$

$$\begin{aligned} &-\lambda(d_0(\tau), \tau) v_x(d_0(\tau), \tau) = \Delta q_1(\tau); \\ &-\lambda(d_{N+1}(\tau), \tau) v_x(d_{N+1}(\tau), \tau) = \Delta q_2(\tau), \end{aligned} \tag{8.164}$$

where

$$a_1 = \frac{\lambda(x, \tau)}{C(x, \tau)}, \quad a_2 = \frac{2\lambda_x(x, \tau)}{C(x, \tau)},$$

$$a_3 = \frac{1}{C(x, \tau)} [\lambda_{xx}(x, \tau) - C_\tau(x, \tau)] .$$

The adjoint problem:

$$-\psi_{i\tau} = (a_1\psi_i)_{xx} - (a_2\psi_i)_x + a_3\psi_i, \quad (x, \tau) \in Q_{i\tau}, \ i = \overline{1, N+1}; \tag{8.165}$$

$$\psi_i(x, \tau_m) = 0, \quad i = \overline{1, N+1}; \tag{8.166}$$

$$\psi_i(d_i(\tau), \tau) - \psi_{i+1}(d_i(\tau), \tau) = 0, \quad i = \overline{1, N}, \tag{8.167}$$

$$[(a_1\psi_i)_x - (a_1\psi_{i+1})_x]_{x=d_i(\tau)} = 2\rho_i(\tau)e_i(\tau), \quad i = \overline{1, N}; \tag{8.168}$$

$$\left[(a_1\psi_1)_x - \psi_1\left(d_0' + \frac{a_2}{2}\right)\right]_{x=d_0(\tau)} = 0; \tag{8.169}$$

$$\left[(a_1\psi_{N+1})_x - \psi_{N+1}\left(d_{N+1}' + \frac{a_2}{2}\right)\right]_{x=d_{N+1}(\tau)} = 2\rho_{N+1}(\tau)e_{N-1}(\tau). \tag{8.170}$$

The gradients of the functional are $J_q' = \{J_{q_1}', J_{q_2}'\}$,

$$J_{q_1}' = \frac{\psi_1(d_0(\tau), \tau)}{C(d_0(\tau), \tau)}; \quad J_{q_2}' = -\frac{\psi_{N+1}(d_{N+1}(\tau), \tau)}{C(d_{N+1}(\tau), \tau)}.$$

For the solution of formulated IHCP we utilize a gradient method which gives a solution of the first order of smoothness. In particular, we consider the algorithm given by formulas (8.153)–(8.155) in which the step lengths β_{jk}^0 and β_{jk}^1 are found, in general, by solving the problem of minimization of the residual functional at the present iteration. For this purpose, one may use well-known methods of unconditional minimization [55]. The main disadvantage of such an approach is that computer time being too great since the repeated computation of the values of the functional and therefore the repeated solution of the heat conduction problem are required for the varying values of the parameters to be found.

As shown by the computation experiment for solving various nonlinear IHCP, the problem of the determination of step lengths can often be simplified by means of linearization. Let us discuss this question.

Write an expression of the criterion functional at the $k+1$th iteration in a form

$$J(q^{k+1}) = \sum_{i=1}^{N+1} \int_0^{\tau_m} \rho_i(\tau)[T(q^k(\tau), d_i(\tau), \tau) + v(\Delta q_1^k(\tau), \Delta q_2^k(\tau), d_i(\tau), \tau) - f_i(\tau)]^2 \, d\tau, \tag{8.171}$$

where $v(\Delta q_1^k(\tau), \Delta q_2^k(\tau), d_i(\tau), \tau)$ is an increment of temperature at a point $x = d_i(\tau)$, while unknown functions were given the increments $\Delta q_1^k(\tau)$ and $\Delta q_2^k(\tau)$.

In linear approximation the function $v(\Delta q_1^k, \Delta q_2^k, x, \tau)$ is found by solving the boundary-value problem for an increment of temperature field as in Eqs. (8.162)–(8.164). Since this problem is linear one can use the principle of superposition writing its solution as a sum of solutions corresponding to the separate components of coercion:

$$
\begin{aligned}
v(\Delta q_1^k, \Delta q_2^k, x, \tau) &= v(\Delta q_1^k, 0, x, \tau) + v(0, \Delta q_2^k, x, \tau) \\
&= -\beta_{1k}^0 v(P_1^k, 0, x, \tau) - \beta_{1k}^1 (J_{c1}^{\prime k}, 0, x, \tau) \\
&\quad - \beta_{2k}^0 v(0, P_2^k, x, \tau) - \beta_{2k}^1 v(0, J_{c2}^{\prime k}, x, \tau)\,.
\end{aligned} \tag{8.172}
$$

Substituting expression (8.172) into formula (8.171) and using the necessary conditions of extremum $\partial J(q^{k+1})/\partial \beta_{jk}^r = 0$, $j = 1, 2$, $r = 0, 1$, we have a system of algebraic equations of the form shown in Eqs. (8.156) for the computation of step lengths β_{jk}^r. In this system the coefficients and right-hand sides are given by the expressions of the form (8.157) with the limits of summation $N + 1$, while values v_{ni}^k and z_i^k are defined by the corresponding increments of temperature

$$
\begin{aligned}
&v_{1i}^k = v(P_1^k, 0, d_i, \tau), \quad v_{2i}^k = v(J_{c1}^{\prime k}, 0, d_i, \tau)\,, \\
&v_{3i}^k = v(0, P_2^k, d_i, \tau), \quad v_{4i}^k = v(0, J_{c2}^{\prime k}, d_i, \tau)\,, \\
&z_i^k = \rho_i(\tau)[T(q^k(\tau), d_i, \tau) - f_i(\tau)]\,.
\end{aligned}
$$

Thus to obtain linear evaluation of the vectors for step lengths it is necessary to solve the boundary-value problem in Eqs. (8.162)–(8.164) four times using the boundary conditions $\{\Delta q_1, \Delta q_2\} = \{P_1^k(\tau), 0\}, \{J_{c1}^{\prime k}, 0\}, \{0, P_2^k(\tau)\}, \{0, J_{c2}^{\prime k}\}$ respectively.

Naturally if some of the values $q_1(\tau)$, $q_2(\tau)$, $q_1(0)$, $q_2(0)$ are known then the dimension of the algebraic system is correspondingly reduced.

Remark 1. The above scheme of approximate determination of the vector for step lengths could be utilized when solving other non-linear multiparametric inverse problems.

For the numerical solution of inverse problems considered here it is convenient to pass on to some new spatial coordinates y_i bound up with the travelling boundaries. Introduce new variables

$$
y_i = \frac{x - d_{i-1}(\tau)}{d_i(\tau) - d_{i-1}(\tau)} \quad \text{and} \quad i = \overline{1, N+1}, \quad t = \tau\,.
$$

Then the subdomains $\bar{Q}_{i\tau} = \{(x, \tau)\colon\ d_{i-1}(\tau) \leqslant x \leqslant d_i(\tau),\ 0 \leqslant \tau \leqslant \tau_m\}$, $i = \overline{1, N+1}$ are transformed onto subdomains $\bar{Q}_i = \{(y_i, t)\colon 0 \leqslant y_i \leqslant 1,\ 0 \leqslant t \leqslant \tau_m\}$, $i = \overline{1, N+1}$, i.e. the boundaries of the corresponding layers become straight. In the new variables every function $w(x, \tau)$ is given by

$$
w(x, \tau) = w(d_{i-1}(\tau) + y_i[d_i(\tau) - d_{i-1}(\tau)], \tau) = \bar{w}(y_i, t)\,. \tag{8.173}
$$

The derivatives in x and τ are connected with the derivatives in y_i and t in the following way:

$$\left.\begin{aligned}&\frac{\partial w}{\partial x}=\frac{\partial \bar{w}}{\partial y_i}\frac{\partial y_i}{\partial x}=\frac{1}{h_i(\tau)}\frac{\partial \bar{w}}{\partial y_i};\quad \frac{\partial^2 w}{\partial x^2}=\frac{1}{h_i^2(\tau)}\frac{\partial^2 \bar{w}}{\partial y_i^2};\\ &\frac{\partial w}{\partial \tau}=\frac{\partial \bar{w}}{\partial t}+\frac{\partial \bar{w}}{\partial y_i}\frac{\partial y_i}{\partial \tau}=\frac{\partial \bar{w}}{\partial t}-\frac{1}{h_i(\tau)}[d'_{i-1}(\tau)+y_i h'_i(\tau)]\frac{\partial \bar{w}}{\partial y_i},\end{aligned}\right\}\tag{8.174}$$

where $h_i(\tau)=d_i(\tau)-d_{i-1}(\tau)$.

The boundary-value problems (8.158)–(8.160), (8.162)–(8.164) and (8.165)–(8.170) written in the new variables, in view of equalities (8.173) and (8.174) are given by (bars over corresponding values are suppressed):

The direct problem:

$$C(T)T_{it}=\frac{1}{h_i^2(t)}(\lambda(T)T_{iy_i})_{y_i}+\frac{C(T)}{h_i(t)}[d'_{i-1}(t)+y_i h'_i(t)]T_{iy_i},$$

$$0<y_i<1,\quad i=\overline{1,N+1},\quad 0<t\leqslant\tau_m;$$

$$T_i(y_i,0)=\xi_i(y_i),\quad 0\leqslant y_i\leqslant 1,\quad i=\overline{1,N+1};$$

$$-\frac{\lambda(T)}{h_1(t)}T_{1y_1}|_{y_1=0}=q_1(t);\quad -\frac{\lambda(T)}{h_{N+1}(t)}T_{N+1\,y_{N+1}}|_{y_{N+1}=1}=q_2(t);$$

$$T_i(1,t)=T_{i+1}(0,t),\quad i=\overline{1,N};$$

$$h_{i+1}(t)T_{iy_i}|_{y_i=1}=h_i(t)T_{i+1\,y_{i+1}}|_{y_{i+1}=0},\quad i=\overline{1,N};$$

$$T_i(1,t)=f_i(t),\quad i=\overline{1,N+1},$$

where $T_i=T_i(y_i,t)$ is a temperature field in the layer $0\leqslant y_i\leqslant 1$.

The problem for an increment of the temperature field:

$$v_{it}=a_1^i(y_i,t)v_{iy_iy_i}+a_2^i(y_i,t)v_{iy_i}+a_3^i(y_i,t)v_i,$$

$$0<y_i<1,\quad 0<t\leqslant\tau_m,\quad i=\overline{1,N+1};$$

$$v_i(y_i,0)=0,\quad 0\leqslant y_i\leqslant 1,\ i=\overline{1,N+1};$$

$$-\frac{1}{h_1(t)}[\lambda_{1y_1}v_1+\lambda_1 v_{1y_1}]_{y_1=0}=\Delta q_1(t);$$

$$-\frac{1}{h_{N+1}(t)}[\lambda_{N+1\,y_{N+1}}v_{N+1}+\lambda_{N+1}v_{N+1\,y_{N+1}}]_{y_{N+1}=1}=\Delta q_2(t);$$

$$v_i(1,t)=v_{i+1}(0,t),\quad i=\overline{1,N};$$

$$h_i(t)v_{iy_i}|_{y_i=1}=h_{i+1}(t)v_{i+1\,y_{i+1}}|_{y_{i+1}=0},\quad i=\overline{1,N},$$

where

$$a_1^i(y_i, t) = \frac{1}{h_i^2(t)} \frac{\lambda_i(y_i, t)}{C_i(y_i, t)};$$

$$a_2^i(y_i, t) = \frac{2}{h_i^2(t) C_i(y_i, t)} \lambda_{iy_i}(y_i, t) + \frac{1}{h_i(t)} [d'_{i-1}(t) + y_i h'_i(t)];$$

$$a_3^i(y_i, t) = \frac{1}{h_i^2(t) C_i(y_i, t)} \lambda_{iy_iy_i}(y_i, t)$$

$$+ \frac{1}{h_i(t) C_i(y_i, t)} C_{iy_i}(y_i, t)[d'_{i-1}(t) + y_i h'_i(t)] - \frac{1}{C_i(y_i, t)} C_{it}(y_i, t).$$

The adjoint problem:

$$-\psi_{it} = (a_1^i \psi_i)_{y_i y_i} - (a_2^i \psi_i)_{y_i} + a_3^i \psi_i, \quad 0 < y_i < 1,$$

$$i = \overline{1, N+1}, \ 0 \leqslant t < \tau_m;$$

$$\psi_i(y_i, \tau_m) = 0, \quad 0 \leqslant y_i \leqslant 1, \ i = \overline{1, N+1};$$

$$\psi_i(1, t) = \psi_{i+1}(0, t), \quad i = \overline{1, N};$$

$$\frac{1}{h_i(t)} (a_1^i \psi_i)_{y_i}|_{y_i=1} - \frac{1}{h_{i+1}(t)} (a_1^{i+1} \psi_{i+1})_{y_{i+1}}|_{y_{i+1}=0}$$

$$= 2\rho_i(t)[T_i(1, t) - f_i(t)], \quad i = \overline{1, N};$$

$$\left[\left(a_1^1 \frac{\lambda_{1y_1}}{\lambda_1} - a_2^1\right)\psi_1 + (a_1^1 \psi_1)_{y_1}\right]_{y_1=0} = 0;$$

$$\left[\left(a_1^{N+1} \frac{\lambda_{N+1\,y_{N+1}}}{\lambda_{N+1}} - a_2^{N+1}\right)\psi_{N+1} + (a_1^{N+1} \psi_{N+1})_{y_{N+1}}\right]_{y_{N+1}}$$

$$= 2\rho_{N+1}(t)[T_{N+1}(1, t) - f_{N+1}(t)].$$

The implemented transformations of the boundary-value problems give an opportunity to simplify their approximation. Since subdomains $\bar{Q}_i$ are rectangular, the rectangular finite-difference grids can be introduced in them and there are no specific difficulties in approximating the equations in the neighbourhood of curvilinear boundaries, although this convenience entails a certain complication of the equations to be approximated with respect to initial ones.

A parametrized form of the problem. Now consider a solution of boundary IHCP in the finite-dimensional space E_M. There are two statements of this problem – one as not overdefined (A) and the other as overdefined (B).

Statement A.

$$C(T)T_\tau = (\lambda(T)T_x)_x, \quad x\in(0, b), \ \tau\in(0, \tau_m],$$
$$T(x, 0) = \xi(x),$$
$$-\lambda(T)T_x|_{x=0} = q(\tau),$$
$$-\lambda(T)T_x|_{x=b} = q^*(\tau).$$

Here $q(\tau)$ is the unknown value; $C(T)$, $\lambda(T)$, $\xi(x)$, $q^*(\tau)$ are the given functions and the temperature measurements are known at some point $x = d$, $d\in[0, b]$ in the form of a dependence $f(\tau)$.

A finite-dimensional approximation of the unknown function can be represented by the formula

$$q(\tau) \simeq \sum_{j=1}^{M} q_j\varphi_j(\tau),$$

where $\{\varphi_j\}_1^M$ is a set of some base functions and $\boldsymbol{q} = \{q_j\}_1^M$ is a vector for the parameters sought.

Following one of the approaches stated in Sect. 8.4, for example the Lagrange multipliers method, and taking into account Remark 5 of Sect. 8.4.6 it is possible to demonstrate that the gradient of the functional $J(\boldsymbol{q})= \int_0^{\tau_m} [T(\boldsymbol{q}, d, \tau) - f(\tau)]^2 \,d\tau$ in space E_M is expressed through an adjoint variable $\psi(x, \tau)$ in the form of

$$J'(\boldsymbol{q}) = \{J'_j\}_1^M, \quad J'_j = \int_0^{\tau_m} \psi(0, \tau)\varphi_j(\tau)\,d\tau.$$

Function $\psi(x, \tau)$ is a solution of the adjoint boundary-value problem:

$$-C(T(x, \tau))\psi_\tau = \lambda(T(x, \tau))\psi_{xx}, \quad 0 < x < b, \quad 0 \leqslant \tau < \tau_m;$$
$$\psi(x, \tau_m) = 0, \quad 0 \leqslant x \leqslant b;$$
$$-\lambda(T(x, \tau))\psi_x|_{x=0} = \begin{cases} 0, & \text{if } 0 < d \leqslant b \\ 2[T(0, \tau) - f(\tau)], & \text{if } d = 0; \end{cases}$$
$$\lambda(T(x, \tau))\psi|_{x=b} = \begin{cases} 0, & \text{if } 0 \leqslant d < b \\ 2[T(b, \tau) - f(\tau)], & \text{if } d = b; \end{cases}$$
$$\psi(d - 0, \tau) = \psi(d + 0, \tau), \quad \text{if } 0 < d < b;$$
$$-\lambda(T(x, \tau))\psi_x|_{x=d+0} + \lambda(T(x, \tau))\psi_x|_{x=d-0} = 2[T(d, \tau) - f(\tau)], \text{ if } 0 < d < b.$$

The coefficients of this problem are calculated in advance by the temperature field $T(\boldsymbol{q}^k, x, \tau)$, found for approximation $q^k(\tau)$. Knowing J'_q, construct a procedure to search vector $q \in E_M$ in accordance with the gradient method

$$q_j^{k+1} = q_j^k - \beta_k S_j^k, \quad k = 1, 2, \ldots, k^*, \quad j = \overline{1, M};$$
$$S_j^k = J_j'^k + \gamma_k S_j^{k-1}$$
$$\gamma_k = (J'^k, J'^k - J'^{k-1})_{E_M} / \| J'^{k-1} \|_{E_M}^2, \quad \gamma_0 = 0.$$

Providing that a descent step β_k is obtained from the condition of minimum of the functional at the given iteration: $\beta_k = \underset{\beta > 0}{\operatorname{argmin}} J(q^k - \beta S^k)$, this algorithm will correspond to the conjugate gradient method. A steepest descent algorithm is obtained from it by nullifying all γ_k, $k = 1, 2, \ldots, k^*$. A linear estimate of the step can be computed by the formula

$$\beta_k = \int_0^{\tau_m} [T(d, \tau) - f(\tau)] \vartheta(d, \tau) d\tau / 2 \int_0^{\tau_m} \vartheta^2(d, \tau) d\tau .$$

Here $\vartheta(x, \tau)$ is used for the solution of a linear boundary-value problem for temperature increment:

$$(C(T(x, \tau))\vartheta)_\tau = (\lambda(T(x, \tau))\theta)_{xx}, \quad 0 < x < b, \quad 0 < \tau \leqslant \tau_m;$$
$$\theta(x, 0) = 0;$$
$$-(\lambda(T(x, \tau))\theta)_x|_{x=0} = \sum_{j=1}^{M} S_j^k \varphi_j(T);$$
$$-(\lambda(T(x, \tau))\theta)_x|_{x=b} = 0 .$$

The coefficients of this problem are also computed by means of $T(x, \tau) \equiv T(q^k, x, \tau)$. The number of the last iteration $\bar{k}$ is determined by a regularizing condition of the stop, for example, by a residual criterion.

To complete the description of this algorithm, one needs to choose the basic functions $\varphi_j(\tau)$. Good results are obtained using the B-splines. The function

$$B^{(l-1)} = B^{(l-1)}(\tau_j, \tau_{j+1}, \ldots, \tau_{j+l}, \tau) = l \sum_{s=j}^{j+l} \frac{(\tau_s - \tau)_+^{l-1}}{W_j'(\tau_s)},$$

where

$$W_j' = (\tau - \tau_j)(\tau - \tau_{j+1}) \cdots (\tau - \tau_{j+l})$$
$$(\tau_s - \tau)_+^{l-1} = \max\{0, (\tau_s - \tau)^{l-1}\},$$

is called the B-spline of degree $(l - 1)$ for such nodes as $\tau_j, \tau_{j+1}, \ldots, \tau_{j+l}$ [130].

When using the cubic B-splines a segment $[0, \tau_m]$ is divided into m equal portions and a uniform grid is introduced:

$$\omega = \{\tau_j = j\Delta\tau, j = -2, -1, \ldots, m + 3; \Delta\tau = \tau_m/m\} .$$

In solving practical problems one usually uses B-splines with so-called "natural" boundary conditions [61, 130]

$$q''(0) = q''(\tau_m) = 0 .$$

In this case an unknown function is represented as follows:

$$q(\tau) \simeq \sum_{j=1}^{M} q_j \varphi_j(\tau) ,$$

$$\varphi_1 = 2B_0(\tau + \Delta\tau) + B_0(\tau) ,$$

$$\varphi_2 = - B_0(\tau + \Delta\tau) + B_0(\tau) ,$$

$$\dots\dots\dots\dots\dots\dots$$

$$\varphi_j = B_{j-1}(\tau), \quad j = \overline{3, M-2},$$

$$\varphi_{M-1} = B_0(\tau - (M-3)\Delta\tau) - B_0(\tau - (M-1)\Delta\tau) ,$$

$$\varphi_M = B_0(\tau - (M-2)\Delta\tau) + 2B_0(\tau - (M-1)\Delta\tau) ,$$

where

$$B_j(\tau) = B_0(\tau - j\Delta\tau) ,$$

$$B_0(\tau) = \frac{1}{6(\Delta\tau)^3} [(\tau + 2\Delta\tau)_+^3 - 4(\tau + \Delta\tau)_+^3 + 6(\tau)_+^3 - 4(\tau - \Delta\tau)_+^3 + (\tau - 2\Delta\tau)_+^3] ,$$

$$M = m + 1 .$$

Function $B_0(\tau)$ has a property

$$B_0(\tau) \begin{cases} > 0, & |\tau| < 2\Delta\tau \\ = 0 & |\tau| \geqslant 2\Delta\tau , \end{cases}$$

which sufficiently simplifies the algorithm.

Note that functions $q(\tau)$ constructed in this manner form a subspace in space $W_2^2[0, \tau_m]$. This fact certainly does not change the meansquare convergence nature of the algorithm considered.

To ensure a convergence of approximations $q^k(\tau)$ in the norm of space $W_2^\alpha[0, \tau_m]$, $\alpha = 1$ or 2, one needs to find a corresponding vector of the descent direction p^k instead of S^k at every iteration (see Sect. 8.5.4).

Statement B. This differs from problem A only in that instead of one measurement $f(\tau)$ several measurements $f_i(\tau)$, $i = \overline{1, N}$; $N \geqslant 1$ are given at various points d_i inside the body, with $0 < d_1 < \dots < d_N < b$.

For the unknown function $q(\tau)$ use the same finite-dimensional representation via the cubic B-splines as for the solution of problem A, and as a criterion functional take the value

$$J(\bar{q}) = \sum_{i=1}^{N} \int_{0}^{\tau_m} [T(q, d_i, \tau) - f_i(\tau)]^2 \, d\tau .$$

A gradient of this functional in space E_M has the components

$$J'_j = \int_0^{\tau_m} \psi(0,\tau)\varphi_j(\tau)\mathrm{d}\tau .$$

An adjoint variable $\psi(x,\tau)$ in the case given can be found from a solution of the following problem:

$$-C(T(x,\tau))\psi_\tau = \lambda(T(x,\tau))\psi_{xx}, \quad 0 < x < b, \quad 0 \leqslant \tau < \tau_m ;$$

$$\psi(x,\tau_m) = 0, \quad 0 \leqslant x \leqslant b ;$$

$$\psi_x|_{x=0} = 0; \quad \psi_x|_{x=b} = 0 ;$$

$$\psi(d_i - 0,\tau) = \psi(d_i + 0,\tau), \quad i = \overline{1,N} ,$$

$$-\lambda(T(x,\tau))\psi_x|_{x=d_i+0} + \lambda(T(x,\tau))\psi_x|_{d_i-0} = 2[T(d_i,\tau) - f_i(\tau)],\ i = \overline{1,N} .$$

The problem for temperature increment remains unchanged. Its solution is used for the determination of a linear approximation to a descent step

$$\beta_k = \sum_{i=1}^{N} \int_0^{\tau_m} [T(d_i,\tau) - f_i(\tau)]\vartheta(d_i,\tau)\mathrm{d}\tau \Big/ \sum_{i=1}^{N} \int_0^{\tau_m} \vartheta^2(d_i,\tau)\mathrm{d}\tau .$$

In other respects a computation procedure is similar to that considered in solving problem A.

8.6.3 Coefficient IHCP

As a function to be found we consider the dependence of thermal conductivity $\lambda(T)$ on temperatures having the following conditions:

$$C(T)T_\tau = (\lambda(T)T_x)_x, \ (x,\tau) \in Q = (0,b) \times (0,\tau_m] ; \tag{8.175}$$

$$T(x,0) = \xi_0, \quad x \in [0,b] ; \tag{8.176}$$

$$T(0,\tau) = T_1(\tau),\ T(b,\tau) = T_2(\tau), \quad \tau \in [0,\tau_m] ; \tag{8.177}$$

$$T(d_i,\tau) = f_i(\tau), \quad i = \overline{1,N},\ \tau \in [0,\tau_m], \quad 0 < d_1 < \ldots < d_N < b , \tag{8.178}$$

where $C(T)$; $T_1(\tau)$; $T_2(\tau)$; $f_i(\tau)$, $i = \overline{1,N}$ are given functions and ξ_0; b; τ_m; d_i, $i = \overline{1,N}$ are given numbers.

The range of definition of $\lambda(T)$ is a segment $[T_0, T_M]$ where T_0 and T_M are the minimal and the maximal values of temperature in domain Q respectively. As Eq. (8.175) is homogeneous then, by the maximum principle, the temperatures T_0 and T_M are reached at $x = 0$ or $x = b$. Therefore, when solving this inverse problem T_0 and T_M are known beforehand: they are defined by the minimal and maximal values of functions $T_1(\tau)$ and $T_2(\tau)$.

According to the method of iterative regularization and assuming that boundary conditions $T_1(\tau)$ and $T_2(\tau)$ are given with a high degree of accuracy, we consider a variational problem for finding $\lambda(T)$ within the scope of the model

[Eqs. (8.175)–(8.178)] and with the condition

$$\lambda(T)\colon \min_{\lambda(T)} |J(\lambda) - \delta_J^2| ,$$

where $J(\lambda) = \sum_{i=1}^{N} \int_0^{\tau_m} \rho_i(\tau)[T(\lambda, d_i, \tau) - f_i(\tau)]^2 \, d\tau$ is a residual functional and δ_J^2 is a minimal permissible level of the residual, defined by the errors in functions $f_i(\tau)$.

We write the parametric representation of the unknown function $\lambda(T)$ taking the *cubic B-splines* [130] for basic functions. For this purpose we take at the segment $[T_0, T_M]$ the uniform grid $\{T_j = T_0 + jH\}$, $j = \overline{0, M}$ with the step $H = (T_M - T_0)/M$ and supplement it with points $T_{-3} = T_0 - 3H$, $T_{-2} = T_0 - 2H$, $T_{-1} = T_0 - H$, $T_{M+1} = T_0 + (M+1)H$, $T_{M+2} = T_0 + (M+2)H$, $T_{M+3} = T_0 + (M+3)H$. Introduce on this supplemented grid a sequence of cubic B-splines given by

$$B_j(\bar{T}) = B_0(\bar{T} - jH), \quad j = -1, 0, 1, \ldots, M+1 ;$$

$$B_0(\bar{T}) = \frac{1}{6H^3} [(\bar{T} - 2H)_+^3 - 4(\bar{T} - H)_+^3 + 6(\bar{T})_+^3 - 4(\bar{T} + H)_+^3 + (\bar{T} + 2H)_+^3] ,$$

where $\bar{T} = T - T_0$; $(\bar{T} - g)_+^3 = \begin{cases} (\bar{T} - g)^3 & \text{when } \bar{T} \geq g \\ 0 & \text{when } \bar{T} < g . \end{cases}$

The present sequence is a basis in the subspace $\tilde{S}$ of functions $S = W_2^2[T_0, T_M]$ such, that

(a) $S(T)$ is a cubic polynomial at every interval $[T_0, T_1)$, (T_1, T_2), $\ldots, (T_{M-1}, T_M]$;
(b) the derivatives $S^{(\nu)}(T)$, $\nu = 0, 1, 2$ are continuous at points $T = T_j$, $j = 1, \ldots, M-1$:

$$S^{(\nu)}(T_j - 0) = S^{(\nu)}(T_j + 0) .$$

Approximation of solution $\lambda(T)$ by functions from subspace $\tilde{S}$ gives an opportunity to obtain the approximate solution $\tilde{\lambda}(T)$ of second order smoothness. For this purpose represent $\tilde{\lambda}(T)$ in a form of linear combination

$$\tilde{\lambda}(T) = \sum_{j=-2}^{M+1} \lambda_j B_j(\bar{T}) .$$

where λ_j are numerical coefficients to be found.

Then variational formulation of IHCP consists of determining the vector $\boldsymbol{\lambda} = [\lambda_{-1}, \lambda_0, \ldots, \lambda_{M+1}]^T$ having the condition

$$\min_{\boldsymbol{\lambda}} |\hat{J}(\boldsymbol{\lambda}) - \delta_J^2| ,$$

where

$$\hat{J}(\lambda) = \sum_{i=1}^{N} \int_{0}^{\tau_m} \rho_i(\tau)[T(\tilde{\lambda}, d_i, \tau) - f_i(\tau)]^2 \mathrm{d}\tau .$$

The iterative algorithm for the determination of λ is given by

$$\lambda^{k+1} = \lambda^k - \beta_k p^k, \quad k = 0, 1, \ldots, \bar{k},$$

where $\bar{k}$: $\hat{J}(\lambda^{\bar{k}}) \simeq \delta_J^2$; β_k: $\min_{\beta} \hat{J}(\lambda^k - \beta p^k)$.

A direction of descent $p = [p_{-1}, p_0, \ldots, p_{M+1}]^T$, corresponding to one or another gradient method, can be chosen in such a way that the uniform convergence of approximations can be assured. When the meansquare approximation is sought, for the conjugate gradient method, we have

$$p^k = J'(\lambda^k) + \gamma_k p^{k-1},$$

$$\gamma_0 = 0, \quad \gamma_k = \frac{\sum_{j=-1}^{M+1} J'^k_{\lambda_j}(J'^k_{\lambda_j} - J'^{k-1}_{\lambda_j})}{\sum_{j=-1}^{M+1} (J'^k_{\lambda_j})^2}, \quad k > 0 .$$

The components of gradient $J'(\lambda)$ are calculated by the formula (8.98), which in this case has a form

$$J'_{\lambda_j} = \sum_{i=1}^{N+1} \int_{0}^{\tau_m} \mathrm{d}\tau \int_{d_{i-1}}^{d_i} \frac{\psi_i(x, \tau)}{C(x, \tau)} [T_{xx}(x, \tau) B_j(\bar{T}) + T_x^2(x, \tau) B'_j(\bar{T})] \,\mathrm{d}x ,$$

$$j = -1, 0, \ldots, M+1$$

The computation of functions ψ_i, $i = \overline{1, N+1}$ by solving the corresponding adjoint problem, is carried out with the conditions $Q_\tau = Q$, $u = \lambda$,

$$D_{x\tau}\psi_i = (a_1\psi_i)_{xx} - (a_2\psi_i)_x + a_3\psi_i ,$$

$$a_1 = \frac{1}{C} \sum_{j=-1}^{M+1} \lambda_j B_j(\bar{T}), \quad a_2 = \frac{2}{C} T_x \sum_{j=-1}^{M+1} \lambda_j B'_j(\bar{T}),$$

$$a_3 = \frac{1}{C}\left[T_{xx} \sum_{j=-1}^{M+1} \lambda_j B'_j(\bar{T}) + T_x^2 \sum_{j=-1}^{M+1} \lambda_j B''_j(\bar{T}) = C_T T_\tau \right].$$

The linear approximation of the step length β_k is derived by the scheme considered in Sect. 8.6.2 and is given by

$$\beta_k = \frac{\sum_{i=1}^{N} \int_{0}^{\tau_m} [T(\lambda^k, d_i, \tau) - f_i(\tau)] v(\Delta\tilde{\lambda}^k, d_i, \tau) \mathrm{d}\tau}{\sum_{i=1}^{N} \int_{0}^{\tau_m} v^2(\Delta\tilde{\lambda}^k, d_i, \tau) \mathrm{d}\tau},$$

where $v(\Delta\tilde{\lambda}^k, d_i, \tau)$ is an increment of temperature, determined by solving the following problem (see Sect. 8.4.5)

$$v_\tau = a_1 v_{xx} + a_2 v_x + a_3 v + \frac{1}{C}(\Delta\tilde{\lambda}^k T_{xx} + \Delta\tilde{\lambda}^k_T T_x^2), \quad (x,\tau)\in Q; \tag{8.179}$$

$$v(x,0) = 0; \tag{8.180}$$

$$v(0,\tau) = v(b,\tau) = 0; \tag{8.181}$$

$$\Delta\tilde{\lambda}^k = \sum_{j=-1}^{M+1} p_j^k B_j(\bar{T}).$$

Remark 2. In the foregoing construction of a spline-approximation of an unknown function $\lambda(T)$ the boundary conditions have not been used. In certain cases it is expedient to "fasten" function $\tilde{\lambda}(T)$ or one of its derivatives $\tilde{\lambda}^{(l)}(T)$ by certain values at one or both ends of the segment $[T_0, T_M]$. Such a situation is quite usual, when there is an applicable a priori information on the behaviour of function $\lambda(T)$ in the neighbourhood of the points T_0 and T_M. Also, accounting for similar boundary conditions reduces the dimension of vector λ and improves the conditionality of the problem to be solved. For example, the so-called natural boundary conditions can be set

$$\tilde{\lambda}''(T_0) = \tilde{\lambda}''(T_M) = 0 \tag{8.182}$$

Calculating $\tilde{\lambda}''(T)$ and using the conditions (8.182) with regard to $B_j''(\bar{T}) = 0$ when $j \geqslant 2$ and $j \leqslant m-2$, we have

$$\tilde{\lambda}(T) = \sum_{j=0}^{M} \lambda_j \varphi_j(\bar{T}),$$

where

$$\varphi_0(\bar{T}) = B_0(\bar{T}) + 2B_{-1}(\bar{T}); \quad \varphi_1(\bar{T}) = B_1(\bar{T}) - B_{-1}(\bar{T});$$

$$\varphi_i(\bar{T}) = B_j(\bar{T}), \quad j = \overline{2, M-2};$$

$$\varphi_{M-1}(\bar{T}) = B_{M-1}(\bar{T}) - B_{M+1}(\bar{T}); \quad \varphi_M(\bar{T}) = B_M(\bar{T}) + 2B_{M+1}(\bar{T});$$

$\{\lambda_j\}_0^M$ are parameters to be found.

Remark 3. A number of segments in the spline-approximation may considerably influence the quality of the reconstructed solution. When the number of basic functions is too small, an approximation $\tilde{\lambda}(T)$ is "oversmoothed" and the minimal value of residual $J_{\min}$ becomes greater than the initial data error δ^2. On the other hand, when a basis consists of a great number of functions the problem becomes too ill-conditioned and the computation time grows. In this case, if the error δ^2 is known to be inaccurate and is taken as being understated, then the noises in temperature measurements may result in objectionable oscillations of approximate solution. Therefore, the number M should be chosen to be compatible with the smoothness of the function to be reconstructed. If M cannot be set a priori, it is expedient to set a small number of segments of spline-approximation at first, and then successively increase M in each one, every time minimizing a functional (without stopping from the residual) and so on till the level δ^2 is reached. A similar result $\tilde{\lambda}(T)$ is taken for approximate solution.

Remark 4. The quality of reconstruction of $\lambda(T)$ can be improved, if the restriction of $\lambda(T) > 0$ (in evidence) is taken into account. To solve this problem one may utilize, in particular, the conjugate gradient method combined with the method of penalty functions.

The other coefficient IHCP may be solved in a similar way.

8.6.4 The Problem of Reconstruction of a Source

The reconstruction of a distributed source-term in a heat-conduction equation may be associated with various practical applications. Let us discuss one possible application in transient heat measurements – the diagnostics of heat transfer in thin constructions. Consider the following example in which one must find a spatial-time distribution of heat flux $q(x, y, \tau)$ at a plate with thickness b (Fig. 8.6). The function $q(x, y, \tau)$ is a result of summary heat coercion $q_1(x, y, \tau) = -\lambda \left.\frac{\partial T}{\partial z}\right|_{z=0}$ and $q_2(x, y, \tau) = -\lambda \left.\frac{\partial T}{\partial z}\right|_{z=b}$ at both sides of a plate. If one of the sides of a plate is heat-insulated, for example, $q_2 = 0$, then the problem consists in finding a distribution of the heat flux at the other side.

Assuming that thermophysical properties are constant, we write an equation of heat conduction in the present body

$$CT_\tau = \lambda(T_{xx} + T_{yy} + T_{zz}) .$$

We integrate both sides of this equality in z from 0 to b, assuming that the temperature T does not depend on z owing to the smallness of thickness b. As a result we have

$$a_1 T_\tau = a_2(T_{xx} + T_{yy}) + q(x, y, \tau) , \tag{8.183}$$

where $q = q_1 - q_2$; $a_1 = Cb$; $a_2 = \lambda b$.

The determination of $q(x, y, \tau)$ can be carried out uniquely only if the whole temperature field is given. Therefore, we assume the following function to be known.

$$f(x, y, \tau) = T(x, y, \tau) . \tag{8.184}$$

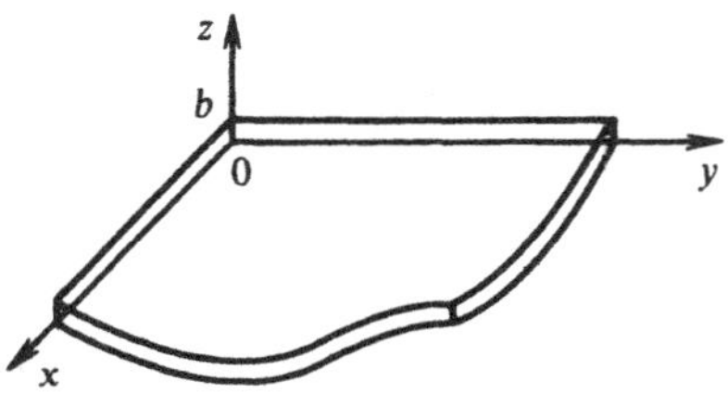

Fig. 8.6. A flat plate

Complete the formulation of IHCP by assuming that the plate occupies on the plane a certain domain Ω with a piecewise-smooth boundary $\partial\Omega$ and that the boundary conditions are given

$$T(x, y, 0) = \xi(x, y); \tag{8.185}$$

$$T|_{\partial\Omega} = g(\omega, \tau), \quad \omega \in \partial\Omega. \tag{8.186}$$

Introduce a residual functional

$$J(q) = \int_0^{\tau_m} d\tau \iint_\Omega [T(q, x, y, \tau) - f(x, y, \tau)]^2 \, dx\, dy \tag{8.187}$$

and seek a function $q(x, y, \tau)$ which minimizes the criterion functional (8.187) under constraint $J(q) \geqslant \delta_J^2$. When the residual criteria is utilized δ_J^2 is given by

$$\delta_J^2 = \int_0^{\tau_m} d\tau \iint_\Omega \sigma^2(x, y, \tau) dx\, dy$$

where σ^2 is a variance of function f.

We construct an iterative process in the space $L_2(Q)$, $Q = \Omega \times [0, \tau_m]$ by the conjugate gradient method

$$q^{k+1} = q^k - \beta_k S^k, \quad k = 0, 1, \ldots, \bar{k}, \tag{8.188}$$

where

$$S^k = J_q'^k + \gamma_k S^{k-1}, \quad \gamma_0 = 0, \quad \gamma_k = \|J_q'^k\|^2 / \|J_q'^{k-1}\|^2,$$

$$\beta_k = (T^k - f, v^k) / \|v^k\|^2.$$

In these formulas $T^k = T(q^k, x, y, \tau)$ is a temperature field in the k-th iteration; $v^k = v(\Delta q^k, x, y, \tau)$ is a field of an increment of temperature in the k-th iteration for the source increment Δq^k;

$$(u, w) = \iiint_Q u(x, y, \tau) w(x, y, \tau) \, dx\, dy\, d\tau, \quad \|u\| = \sqrt{(u, u)}$$

are inner products of elements u and w and the norm of element u in the space $L_2(Q)$ respectively.

An increment of temperature field v^k is computed by solving the following boundary-value problem (in the following index "k" is suppressed for simplicity)

$$a_1 v_\tau = a_2(v_{xx} + v_{yy}) + \Delta q(x, y, \tau), \quad (x, y, \tau) \in Q; \tag{8.189}$$

$$v(x, y, 0) = 0; \tag{8.190}$$

$$v|_{\partial\Omega} = 0. \tag{8.191}$$

The gradient of the criterion functional will be calculated via the adjoint variable $\psi(x, y, \tau)$.

Introduce an operator $Lv = a_1 v_\tau - a_2(v_{xx} + v_{yy})$ and rewrite the problem in Eqs. (8.189)–(8.191) in a form

$$Lv = \Delta q, \quad v \in H_L. \tag{8.192}$$

Here H_L is a range of definition of the operator L, consisting of the functions $v = v(x, y, \tau)$, which possess continuous derivatives $v_\tau, v_x, v_y, v_{xx}, v_{yy}$ in Q and satisfy the conditions (8.190) and (8.191).

Calculate an increment of the functional

$$\Delta J \equiv J(q + \Delta q) - J(q) = (v, 2e) + o(\|\Delta u\|),$$

where $e = T(x, y, \tau) - f(x, y, \tau)$.

Since Eq. (8.192) implies $v = L^{-1}\Delta q$, then

$$\Delta J = (L^{-1}\Delta q, 2e) + o(\|\Delta u\|).$$

By means of operator L^{-1*}, adjoint of L^{-1},

$$\Delta J = (\Delta q, L^{-1*}(2e)) + o(\|\Delta u\|).$$

Thus, the gradient of functional in $L_2(Q)$ is given by

$$J'_q = L^{-1*}(2e). \tag{8.193}$$

Let us find the operator L^{-1*}. For this purpose consider the inner product of elements v and ψ and carry out the following transformations, assuming that ψ is a sufficiently smooth function

$$(v, \psi) = (LL^{-1}v, \psi) = (L^{-1}v, L^*\psi) = (v, L^{-1*}L^*\psi).$$

Hence, the operator $L^{-1*}L^*$ is unique which implies

$$L^{-1*} = L^{*-1}.$$

This fact provides an opportunity to introduce the formula (8.193) in a form

$$J'_q = L^{*-1}(2e). \tag{8.194}$$

In the sequel we shall use the formula of integration by parts [90]:

$$\iiint\limits_Q \frac{\partial u}{\partial \eta_i} w \, dx\,dy\,d\tau = -\iiint\limits_Q u \frac{\partial w}{\partial \eta_i} dx\,dy\,d\tau + \iint\limits_{\partial Q} uw\cos(\widehat{n_Q, \eta_i})\,ds, \tag{8.195}$$

where $\eta_1 = \tau$; $\eta_2 = x$; $\eta_3 = y$; ∂Q is a boundary of domain Q; dS is an element of the area of the surface ∂Q; n_Q is an outer normal and u and w are some sufficiently smooth functions.

Using the formula (8.195) we transform the inner product

$$(Lv, \psi) = \iiint\limits_Q \psi(a_1 v_\tau - a_2 v_{xx} - a_2 v_{yy})\,dx\,dy\,d\tau$$

to a form

$$\begin{aligned}(Lv, \psi) = & -(v, a_1\psi_\tau + a_2\psi_{xx} + a_2\psi_{yy}) + a_1 \iint\limits_\Omega \psi v|_{\tau=0}^{\tau=\tau_m}\,dx\,dy \\ & + a_2 \int_0^{\tau_m} d\tau \int\limits_{\partial\Omega} \{v[\psi_x \cos(\widehat{n_\omega, x}) + \psi_y \cos(\widehat{n_\omega, y})] \\ & - \psi[v_x \cos(\widehat{n_\omega, x}) + v_y \cos(\widehat{n_\omega, y})]\}\,d\omega,\end{aligned} \tag{8.196}$$

where $d\omega$ is an element of $\partial\Omega$ and n_ω is an outer normal for $\partial\Omega$.

Using the representation of the derivative of some function $\theta(x, y)$ in direction n via its derivatives in x and y

$$\theta_n = \theta_x \cos(\hat{n}, x) + \theta_y \cos(\hat{n}, y),$$

we have

$$(Lv, \psi) = -(v, a_1\psi_\tau + a_2\psi_{xx} + a_2\psi_{yy}) + a_1 \iint_\Omega \psi v|_{\tau=0}^{\tau=\tau_m} dxdy$$

$$+ a_2 \int_0^{\tau_m} d\tau \int_{\partial\Omega} (v\psi_{n_\omega} - \psi v_{n_\omega}) d\omega.$$

With the conditions (8.190) and (8.191) we finally obtain

$$(Lv, \psi) = -(v, a_1\psi_\tau + a_2\psi_{xx} + a_2\psi_{yy})$$

$$+ a_1 \iint_\Omega \psi v|_{\tau=\tau_m} dxdy - a_2 \int_0^{\tau_m} d\tau \int_{\partial\Omega} \psi v_{n_\omega} d\omega. \tag{8.197}$$

The expression (8.197) and identity $(Lv, \psi) = (v, L^*\psi)$ yield the conditions of an adjoint problem:

$$L^*\psi = \zeta, \quad (x, \tau) \in Q; \tag{8.198}$$

$$\psi(x, \tau_m) = 0; \tag{8.199}$$

$$\psi|_{\partial\Omega} = 0, \tag{8.200}$$

where $L^*\psi = a_1\psi_\tau + a_2\psi_{xx} + a_2\psi_{yy}$; $\zeta = \zeta(x, y, \tau)$ is a certain function.

Introduce the problem (8.198)–(8.200) in a form

$$L^*\psi = \zeta, \quad \psi \in H_{L^*},$$

where $H_{L^*} = \{\psi \in G(Q): \ \psi(x, \tau_m) = 0, \psi|_{\partial\Omega} = 0\}$ is the range of definition of the operator L^* and $G(Q)$ is a set of functions $\psi(x, y, \tau)$ which possess continuous derivatives $\psi_\tau, \psi_x, \psi_y, \psi_{xx}, \psi_{yy}$ in Q.

So far as $L^{*-1}\zeta = \psi$, the formula (8.194) for the gradient may be finally given in a form

$$J'_q = \psi, \quad (x, y, \tau) \in Q,$$

where $\psi = \psi(x, y, \tau)$ is a solution of the adjoint problem in Eqs. (8.198)–(8.200) in which $\zeta = 2e$.

In order to take into consideration of the smoothness of a function reconstructed, one may draw on the mode proposed in Sect. 8.5.3.

For simplicity we restrict our consideration to the rectangular domain

$$\Omega = \{(x, y); 0 \leqslant x \leqslant b_1, \quad 0 \leqslant y \leqslant b_2\}.$$

In the present case we introduce an auxiliary function

$$g(x, y, \tau) = \int_\tau^{\tau_m} d\tau' \int_0^y dy' \int_0^x J'_q(x', y', \tau') dx',$$

$$0 \leqslant x \leqslant b_1, \quad 0 \leqslant y \leqslant b_2, \quad 0 \leqslant \tau \leqslant \tau_m.$$

By integration by parts of the expression defining the increment of functional it is easy to obtain the corresponding components of the gradient

$$J'_{p_1} = -g(x, y, \tau), \quad (x, y, \tau) \in \bar{Q};$$
$$J'_{p_2} = -g(x, y, 0), \quad (x, y) \in \bar{\Omega};$$
$$J'_{p_3} = g(x, b_2, \tau), \quad x \in [0, b_1], \tau \in [0, \tau_m];$$
$$J'_{p_4} = g(b_1, y, \tau), \quad y \in [0, b_2], \tau \in [0, \tau_m];$$
$$J'_{p_5} = g(x, b_2, 0), \quad x \in [0, b_1];$$
$$J'_{p_6} = g(b_1, y, 0), \quad y \in [0, b_2];$$
$$J'_{p_7} = -g(b_1, b_2, \tau), \quad \tau \in [0, \tau_m];$$
$$J'_{p_8} = -g(b_1, b_2, 0),$$

where

$$p_i = q_{xy\tau}(x, y, \tau); \quad p_2 = q_{xy}(x, y, 0); \quad p_3 = q_{x\tau}(x, b_2, \tau);$$
$$p_4 = q_{y\tau}(b_1, y, \tau); \quad p_5 = q_x(x, b_2, 0); \quad p_6 = q_y(b_1, y, 0);$$
$$p_7 = q_\tau(b_1, b_2, \tau); \quad p_8 = q(b_1, b_2, 0).$$

8.7 Computational Experiments

8.7.1 Boundary IHCP

One-Dimensional Formulation. In Sects. 6.8 and 6.9 it has been ascertained, when considering the characteristic properties of the iterative solution of boundary IHCP in the space L_2, that general regularities exist when solving both linear and non-linear problems. As the results of computational experiments showed the common salient features are also observed when solving linear and nonlinear IHCP by the gradient algorithms taking into account the smoothness of functions reconstructed. In the sequel, we discuss the results of solving model inverse problems by algorithms in which a direction of descent is calculated via the adjoint variable (Sect. 8.6.2). Similar computations were performed using the modified conjugate gradient method. The problems in Eqs. (8.158)–(8.160), (8.162)–(8.164) and (8.165)–(8.170) were solved numerically. For this purpose the difference grid was introduced by dividing every segment $0 \leqslant y_i \leqslant 1$, $i = \overline{1, N+1}$ in l_i subsegments with the length $\Delta y_i = 1/l_i$. In the time coordinate the division of the segment $[0, \tau_m]$ was carried out with the constant step $\Delta\tau = \tau_m/m$, where m is a number of steps. As a result we obtain in every subdomain $\bar{Q}_i$ a grid of points (y_{i_k}, τ_m), $k = \overline{0, l_i}$; $n = \overline{0, m}$. To approximate the boundary-value problems, an implicit finite-difference scheme was utilized with the calculation of coefficients at the same time layer in which the unknown values have to be computed, i.e. at the layer $\tau = \tau_n$, assuming that for $\tau = \tau_{n-1}$

the values T, v and ψ have already been found. Hence, the finite-difference system of equations for a nonlinear direct heat conduction problem is also non-linear. Therefore, to compute the values of temperatures at the n-th layer the "run" method was utilized with iterations in coefficients of the equation. When solving the problem numerically for an increment of temperature field and the adjoint problem the iterations are not needed since these problems are linear. The steps Δy_i and $\Delta\tau$ were chosen to provide the neglected small errors in computation of functions $T(x, \tau)$, $\psi(x, \tau)$ and $v(x, \tau)$. This gave an opportunity to neglect approximation errors when solving IHCP.

The results of solving the problem for finding the heat flux at one of the boundaries of a plate [27, 34, 35, 99] are presented below. An account of the smoothness of a solution was conducted according to the first mode (see Sect. 8.5.2) with the assumption that $q \in W_2^1[0, \tau_m]$ and the heat flux $q(0)$ is given for the initial moment. The precise temperature data at different distances from the surface of the heat exchange were obtained by solving the direct heat-conduction problem for the given boundary conditions. The computed dependences of temperature in time at one or several internal points were used as initial data for the solution of the inverse problem – to find the heat flux $q(\tau) = q_1(\tau)$ and temperature $T(\tau)$ at one of the sides of a plate. On the other side of a plate the heat flux $q_2(\tau)$ was taken as being equal to zero (heat insulated wall). Initial approximation was taken as $q^0(\tau) = 0$. The evaluation of the accuracy of a solution of the problem was carried out by the value of the integral error in computation of the boundary function $u(\tau)$ which was $q(\tau)$ or $T(\tau)$:

$$\varepsilon_u = \left[\int_0^{\tau_m} [u_T(\tau) - u(\tau)]^2 \, d\tau \bigg/ \int_0^{\tau_m} u_T^2(\tau) \, d\tau \right]^{1/2},$$

where $u_T(\tau)$ is a precise function and $u(\tau)$ is a function, computed by solving the inverse problem.

In Table 8.1 the obtained values of ε_q and ε_T are given as functions of the distance between the point of measurement and heat exchange surface in the case of one measurement ($N = 1$) with constant thermophysical properties and fixed boundaries of a body and points of measurement.

The data in Table 8.2 depend on the number of temperature sensors to be taken into account when solving IHCP. In these computations the stopping of

Table 8.1

No. of measuring points	$\bar{d}_i = \dfrac{d_i}{\sqrt{a\tau_m}}$	$\Delta Fo = \dfrac{a\Delta\tau}{d_i^2}$	ε_q	ε_T
1	0.058	5.93	0.14	0.045
2	0.116	1.48	0.17	0.05
3	0.232	0.37	0.20	0.07
4	0.348	0.16	0.25	0.09
5	0.464	0.09	0.29	0.12
6	0.580	0.06	0.53	0.25

the iteration was carried out with the condition of reaching the number of iterations being equal to dimension m of the vector of unknown parameters. In the present case $\bar{k} = m = 50$. This condition was accepted due to the fact that when solving well-posed problems the number of iterations sufficient for obtaining the minimum of quadratic function by the conjugate gradient method is theoretically equal to m. Since the step in time $\Delta\tau$ remained invariable for all distances d_i then the different steps in Fourier number corresponded to different distances. The results, represented in Tables 8.1 and 8.2, may be interpreted as an indirect evaluation of the speed of convergence of the iterative process depending upon the location of points of temperature measurement and the number of temperature sensors.

The results obtained show that with distance of the measurement point from the heat exchange surface, the speed of convergence of the iterative process decreases. The reason is that the variational problem becomes more ill-conditioned. On the other hand, the increase of the number of spatial points of temperature measurements taken into account when seeking the boundary condition results in an increase of convergence speed. This circumstance provides an opportunity of ensuring the required accuracy of determination for boundary conditions with a smaller number of iterations than in the case $N = 1$.

An account of temperature measurements at points at the distance $\Delta d = [d_i(\tau) - d_1(\tau)]/(a\tau_m)^{1/2} > 0.5$ (a is a thermal diffusivity) from the front point of measurement does not in practice increase the accuracy of solving the variational problem and the speed of convergence. It is apparently related to the fact that the characteristic properties of the solution of boundary IHCP for given time τ_m are defined by the information about temperature field only up to a certain depth (see Sects. 2.1.6 and 4.2).

For the illustration in Fig. 8.7 the results of the solution of the inverse problem for precise initial data are represented when the results are given of temperature measurements at one and two points.

In the presence of random errors in measured temperatures $f_i(\tau)$, $i = \overline{1, N}$, caused by limited accuracy in measurement and registration of experimental data, the principal question, as has been demonstrated, is to determine the conditions for stopping the iteration. The main difficulty in applying the stop-

Table 8.2

Total number of measuring points	Coordinates of measuring points, $\bar{d}_i = d_i/\sqrt{a\tau_m}$							
	1	2	3	4	5	6	ε_q	ε_T
1	0.058	–	–	–	–	–	0.140	0.045
2	0.058	0.116	–	–	–	–	0.056	0.011
3	0.058	0.116	0.232	–	–	–	0.039	0.009
4	0.058	0.116	0.232	0.348	–	–	0.025	0.006
5	0.058	0.116	0.232	0.348	0.464	–	0.015	0.003
6	0.058	0.116	0.232	0.348	0.464	0.58	0.016	0.004

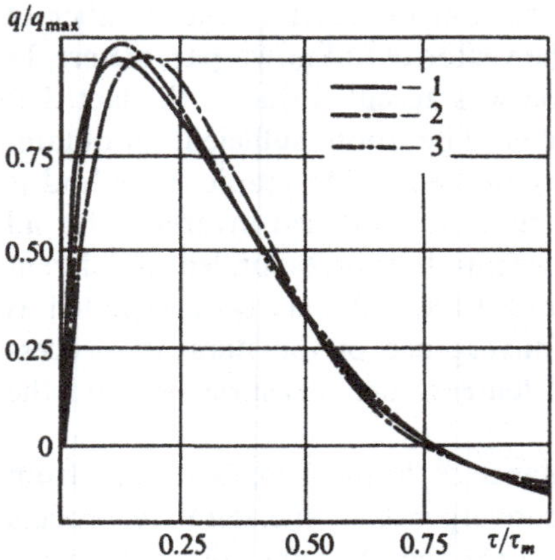

Fig. 8.7. The results of recovering the heat flux with errorless input data ($\bar{\kappa} = 50$): 1 is a true solution; 2 is a solution of IHCP obtained with measurements in one point ($\bar{d}_1 = 0.058$); 3 is a solution of IHCP obtained with measurements in two points ($\bar{d}_1 = 0.058$; $\bar{d}_2 = 0.116$)

ping rule under residual criteria is that of the availability of reliable evaluation of the initial data error δ.

Accordingly, the investigation was carried out to evaluate the influence of the number of measurement points for accuracy and stability of approximate solutions in the case where initial temperatures $f_i(\tau)$ are given with errors. The levels of maximum possible errors have varied in the range of $5 \div 10\%$ of f_i^{max} $(i = \overline{1, N})$ when modelled by the random numbers generator.

Figures 8.8 and 8.9 show the alterations by iterations of the errors in the determination of the heat flux ε_q and the temperature of a surface ε_T for various numbers of measurement points and different distribution laws of random errors.

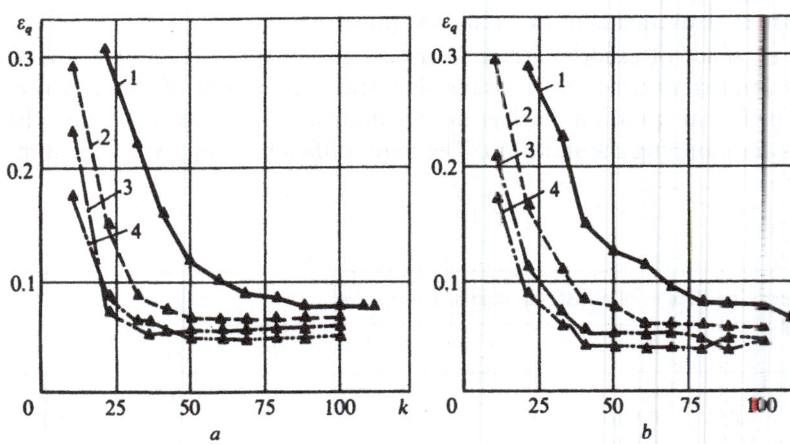

Fig. 8.8. Change of error of the heat flux recovery in iterations (maximum level of errors in the input data $\delta_{0i}^{max} = 0.05 f_i^{max}$): *a* – a uniform law of distribution; *b* – a normal law of distribution; 1 – $N = 1$; 2 – $N = 2$; 3 – $N = 3$; 4 – $N = 4$

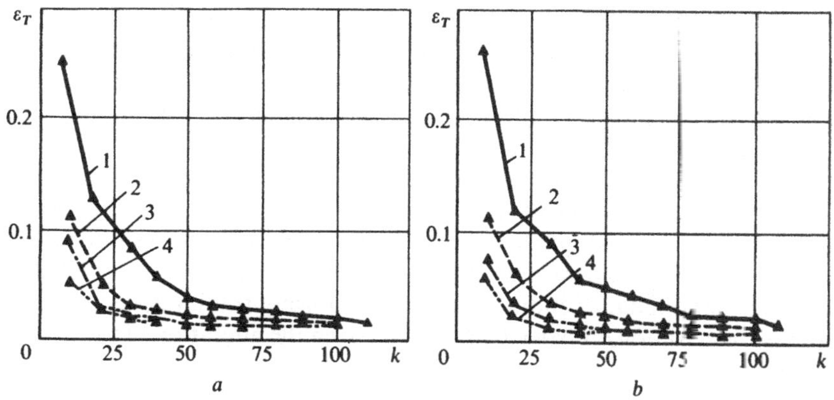

Fig. 8.9. Change of error of the surface temperature recovery in iterations, $\delta_{0i}^{max} = 0.05 \, f_i^{max}$: a – a uniform law of distribution; b – a normal law of distribution; $1 - N = 1$; $2 - N = 2$; $3 - N = 3$; $4 - N = 4$

The results presented show that by increasing the number of measurement points the stability of the iterative process with respect to random errors in initial temperatures increases as well. This is revealed in the fact that after rapid lowering at the initial stage of the process, the error of approximate solutions stabilizes at a certain, practically constant level. As an example, Figure 8.10 illustrates the unknown function $q(\tau)$, the function obtained under the residual criterion and the function obtained at the 100th iteration in the case of three

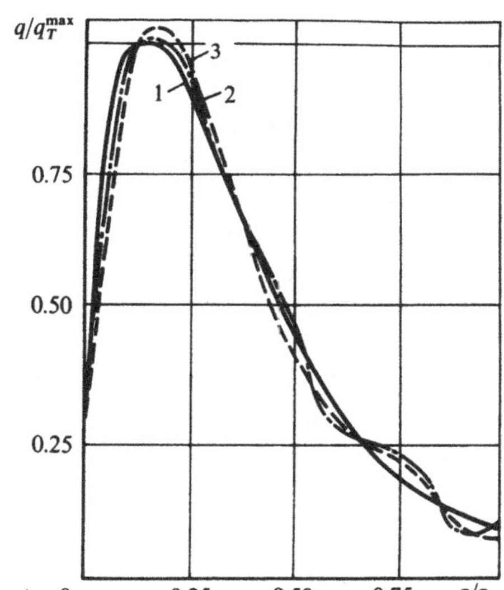

Fig. 8.10. The comparison of approximate solutions with the exact one: 1 is the unknown function; 2 is an approximation chosen by the residual criterion; 3 is an approximation corresponding to $\bar{k} = 100$

measurements and the uniform distribution of errors when $\varepsilon_i^{\max} = 0{,}1 f_i^{\max}$, $i = \overline{1,3}$. As expected, when continuing the iterative process after fulfilment of condition $J(q^k) \simeq \delta^2$ a certain deviation of the approximate solutions from the exact one is observed. However, this deviation is small and commensurable with errors of measurements.

Thus, in the absence of estimates of errors in initial temperatures, the utilization of results of independent measurements at several points increases the stability of the iterative method, when solving IHCP, with respect to the disturbances in initial temperatures. Also, the possibility of sufficiently reliable determination of boundary conditions arises when using traditional conditions for stopping the iteration, in particular, by analysing the alteration of the values of the minimizing functional at neighbouring iterations.

In addition, the installation of additional temperature sensors provides an opportunity to reduce the errors in determination of boundary conditions. But it should be noted that the increase of the number of temperature sensors over certain number N^* does not result in any further improvement of the quality of reconstruction. This number depends on the thermophysical properties of a body and the spatial location of the measurement points – usually $N^* \leqslant 4 \div 5$.

Remark 1. Since the stability properties of boundary-value IHCP are best when measurements are carried out at the same boundary of a body where an unknown function is to be reconstructed, it is expedient then to organize an experiment in such a way that the points d_i are installed as close as possible to this boundary and even at the boundary itself (with the condition that unidimensionness of the temperature field is accomplished with high accuracy and random disturbances in measured functions $f_i(\tau)$ are mutually non-correlated).

Numerous computational experiments have also been conducted to investigate the stopping of the iteration by additional measurement (see Sect. 8.5.5). The results obtained allow the conclusion to be drawn that such a mode produces the solutions close to the ones obtained under residual criteria. For two types of reconstructed functions this conclusion is illustrated by Fig. 8.11, where the computation data are represented for the initial temperatures, disturbed under normal law with an amplitude of errors $3\sigma = 0.05 f^{\max}$ ($N = 1$).

Finally the utilization of an algorithm constructed by means of the second mode to take proper account of the smoothness operation (see Sect. 8.5.4), produced results when reconstructing $q \in W_2^1[0, \tau_m]$ close to the results of the solution of IHCP using the first mode. Figure 8.12 illustrates the results of the solution of a boundary IHCP with constant coefficients for $N = 1$ and $\Delta F_{0d} = 0.2$. For the information on smoothness assumed was the unknown function $q(\tau)$ belonging to the space $W_2^1[0, \tau_m]$ and the values of derivative $q'(0) = q'(\tau_m) = 0$ at the ends of time segment were supposed to be known as well. For initial approximation the function $q^0(\tau) = 0$ was taken. The gradient of the residual was computed by the formula (8.140). The errors in the disturbed temperature $f_\delta(\tau)$ were distributed according to normal law with $3\sigma = 0.05 f^{\max}$.

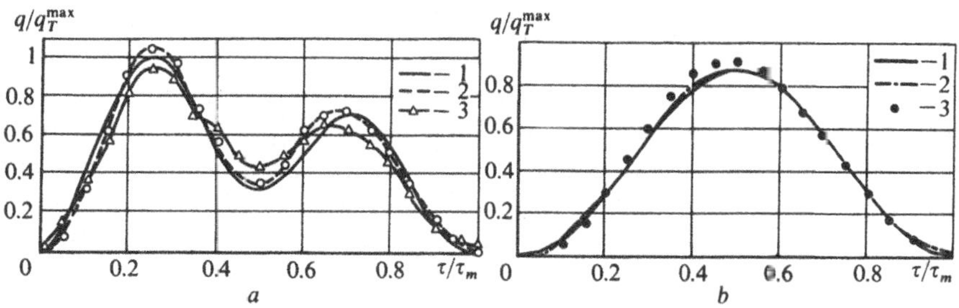

Fig. 8.11. The results of solving a nonlinear inverse problem by the conjugate gradient method in the space L_2: a – $d_1 = 0.42b$; $0.06 \leqslant \Delta Fo_{d_1} \leqslant 0.1$; 1 is the unknown function given as $q_T = \gamma(\frac{B-i}{B-1})^2\{1 - \cos(4\pi\frac{i-1}{m-1}) + 0.5[1 - \cos(2\pi\frac{i-1}{m-1})]\}$, $i = \overline{1, m}$, $B = \frac{\gamma}{\beta(m-1)}$, $\gamma > 0$, $\beta > 0$, in this case it was assumed that $\gamma = 0.833 \cdot 10^6$, $\beta = 3$, $m = 20$; 2 – a halt by the additional measurement at point $d_2 = 0.08b$ ($\bar{\kappa} = 10$); 3 – a halt by the residual criterion ($\bar{\kappa} = 7$); b – $d_1 = b$; $0.01 \leqslant \Delta Fo_{d_1} \leqslant 0.018$; 1 is the unknown function given as $q_T = \gamma[1 - \cos(2\pi\frac{i-1}{m-1})]$, $i = \overline{1, m}$, $\gamma > 0$, in this case it was assumed that $\gamma = 0.833 \cdot 10^6$, $m = 20$; 2 is an iterative solution for exact input data ($\bar{\kappa} = 18$); 3 is an iterative solution for disturbed input data with a halt by the additional measurement at point $d_2 = 0.08b$ ($\bar{\kappa} = 5$)

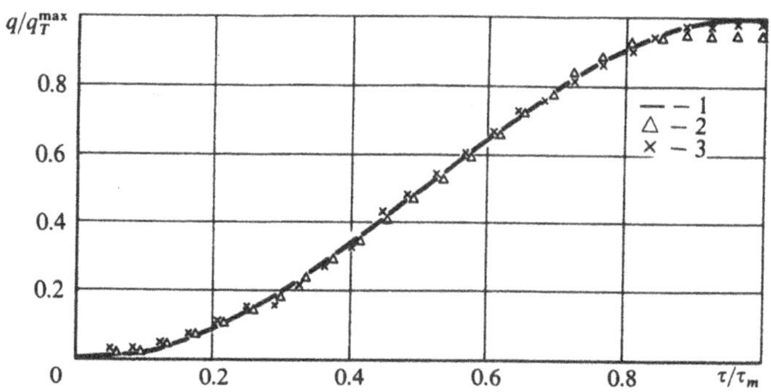

Fig. 8.12. Results of the heat flux recovery allowing for smoothness of the unknown function, $\Delta Fo_d = 0.1$, $u^0(\tau) = 0$: 1 is a true solution; $q_T = \frac{q_{T\max}}{2}\{\sin[\frac{2\pi}{\tau_m}(\tau - \frac{\tau_m}{4})] + 1\}$; 2 is an approximation obtained for the input data disturbed by the normal law of error distribution with an amplitude $3\sigma = 0.05 f^{\max}$ (the halt by the residual criterion $\bar{\kappa} = 7$); 3 is an approximation for the exact input data ($\bar{\kappa} = 50$)

As one can see in the figure, the reconstructed solutions are sufficiently close to the true one.

Remark 2. As model computations showed the damping properties of the algorithms with regard to the smoothness of the function to be reconstructed are higher than for the algorithms producing mean square approximations. In general, this peculiarity reduces considerably the requirements as to the accuracy of determination of the initial data error when using the residual criteria for stopping the iteration.

In Figs. 8.13–8.14 are shown the results on restoration of the heat flux $q(\tau)$ and surface temperature $T_w(\tau)$ of the slab by precise and disturbed temperature data on its other boundary for non-linear boundary IHCP formulation with thermophysical properties $\lambda(T)$ and $C(T)$, corresponding to type St. 20 steel.

In these computations, performed by means of algorithms of conjugate gradient type, it was assumed that $d = b = 0.01$ m, $\Delta x = 0.4 \cdot 10^{-4}$ m, $\Delta\tau = 0.02$ s, $\xi(x) = T_0 = \text{const}$. With this, the parameter ΔF_0 computed by formula

$$\Delta F_{0\,\min} = \min_T \left[\lambda(T)\Delta\tau/(C(T)d^2)\right],$$

was within $0.012 \leqslant \Delta F_0 \leqslant 0.038$.

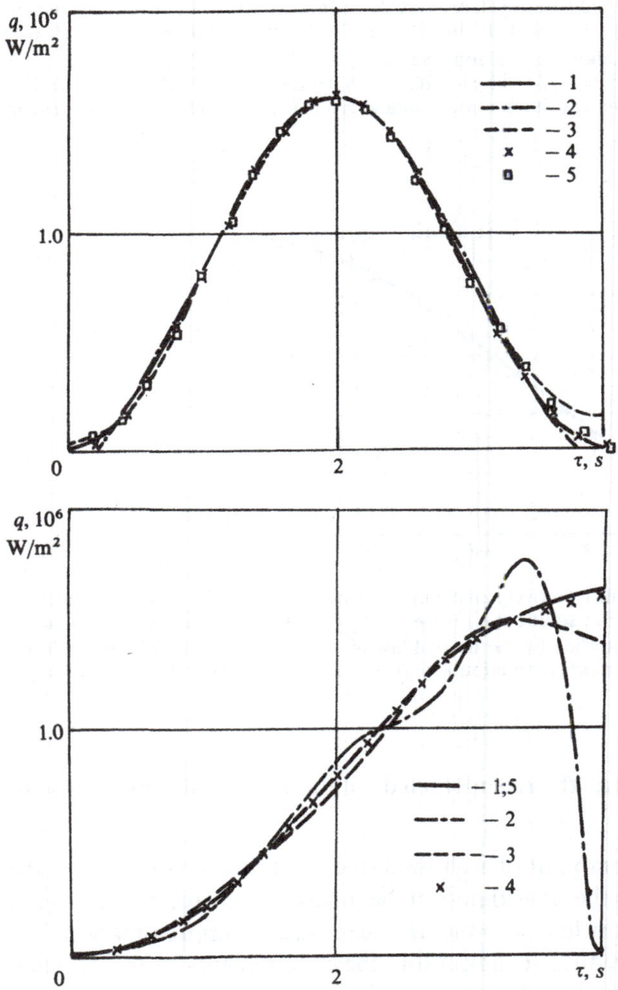

a

Fig. 8.13 continued

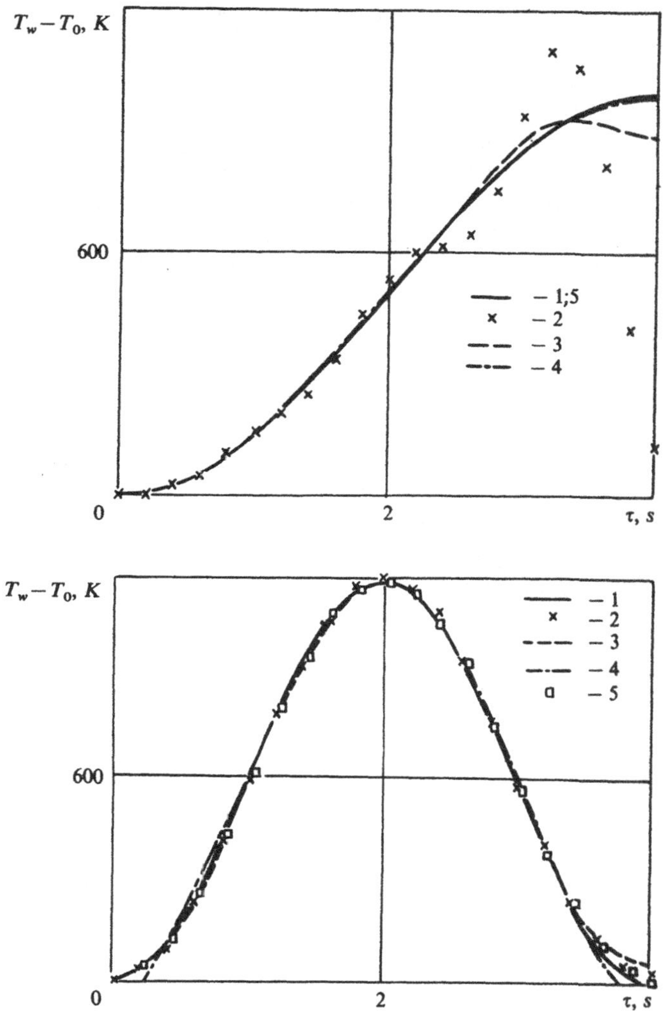

b

Fig. 8.13. Results of the iterative solution of the inverse problem with the exact input data: 1 are true solutions: $q_T = 0.833 \cdot 10^6[1 - \cos(2\pi \frac{i-1}{m-1})]$, $i = \overline{1, m}$ – the upper graph (a); $q_T = 0.833 \cdot 10^6(1 - \cos \frac{\pi i}{m})$, $i = \overline{1, m}$ – the lower graph (a); $T_W - T_0 = 500\ (1 - \cos \frac{\pi i}{m})$, $i = \overline{1, m}$ – the upper graph (b); $T_W - T_0 = 600[1 - \cos(2\pi \frac{i-1}{m-1})]$, $i = \overline{1, m}$ – the lower graph (b); 2 is a solution in L_2; 3 and 4 – the second mode of allowing for smoothness: (3 is a solution in W_2^1 with conditions $q'(0) = q'(\tau_m) = 0$ (a) or $T'_W(0) = T'_W(\tau_m) = 0$ (b); 4 is a solution in W_2^2); 5 is the first mode of allowing for smoothness with unknown value $q(\tau_m)$ (a) or $T_W(\tau_m)$ (b)

An error in the input data was simulated by the dependence including the instrumental error and error of "manual" decoding:

$$f(\tau) = T(d, \tau) + \frac{\delta_0 \omega(\tau)}{3} \max_{\tau \in [0, \tau_m]} T(d, \tau) + \frac{\delta_1 \omega(\tau)}{3},$$

where $T(d, \tau)$ was the temperature computed from a solution of the direct heat conduction problem for the assumed functions $q(\tau)$ or $T_w(\tau)$, ω was the random value, distributed by normal law with unit-equal variance and zero expectation and $\delta_0 = 0.045$ and $\delta_1 = 12.5$ K.

The data presented in these figures are obtained for the case when the smoothness of dependences was not taken into account (solution in the space $L_2[0, \tau_m]$) and for the cases when this information was taken into account. The account for smoothness was considered by means of the first and the second modes. The uniform approximations to unknown functions were constructed

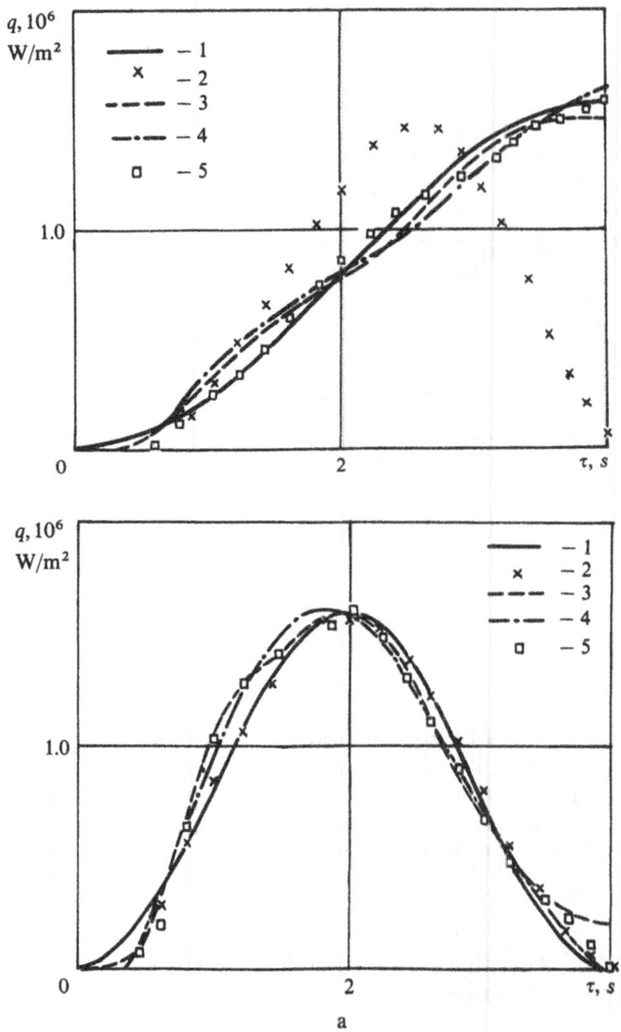

Fig. 8.14 continued

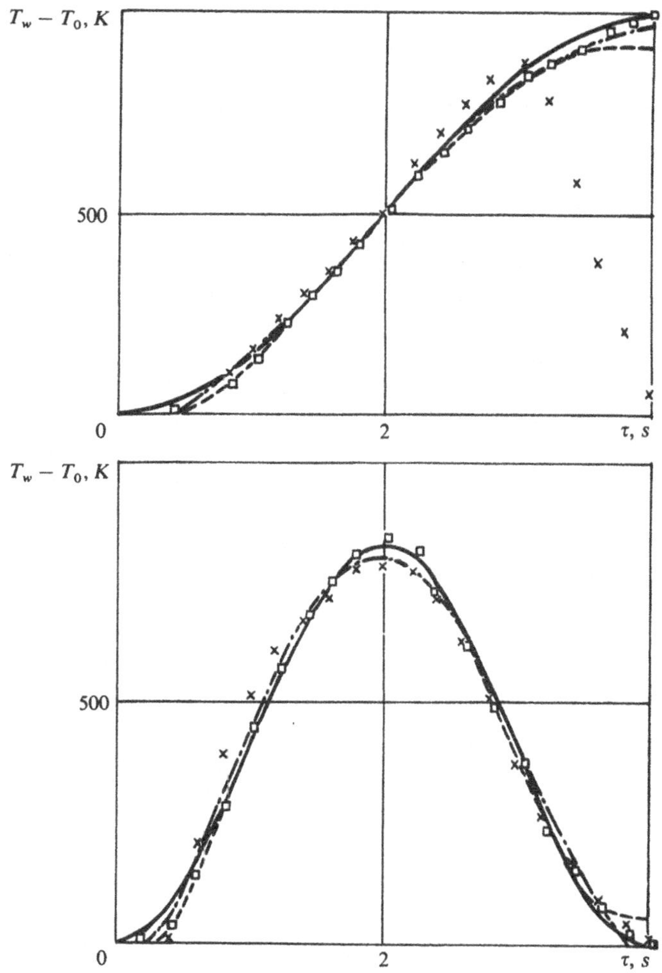

b

Fig. 8.14. Results of the iterative solution of the inverse problem with the disturbed input data: 1 are true solutions: $q_T(\tau)$ – see Fig. 8.13a; $T_W - T_0 = 500(1 - \cos\frac{\pi i}{m})$, $i = \overline{1, m}$ – the upper graph (b); $T_W - T_0 = 415[1 - \cos(2\pi\frac{i-1}{m-1})]$, $i = \overline{1, m}$ – the lower graph (b); 2 is a solution in L_2; 3 and 4 – the second mode of allowing for smoothness (3 is a solution in W_2^1 with conditions $q'(0) = q'(\tau_m) = 0$ (a) or $T'_W(0) = T'_W(\tau_m) = 0$ (b); 4 is a solution in W_2^2); 5 – the first mode of allowing for smoothness with unknown value $q(\tau_m)$ (a).

following the first mode, solutions in the spaces $W_2^1[0, \tau_m]$ and $W_2^2[0, \tau_m]$ were sought by the second mode. Different boundary-value conditions were assumed for dependences under restoration.

A mathematical simulation in advance enabled "good" relations of coefficients to be established which when using the computer BESM-6 were taken for the curves in Figs. 8.13a and 8.14a to be $r_0 = r_1 = 1$ (in the case W_2^1), and $r_0 = 1$,

$r_1 = 2.1$, $r_2 = 1$ (in the case W_2^2) and for the curves in Figs. 8.13b and 8.14b $r_0 = 6.25$, $r_1 = 1$ (in the case W_2^1) and $r_0 = 5$, $r_1 = 7$, $r_2 = 1$ (in the case W_2^2).

The general conclusion, which can be made from the results of this computational experiment, is that by considering the smoothness and quantitative a priori information the accuracy in defining thermal boundary conditions is increased.

Analysis of a parametrized form of the problem. Here are the results presented of a mathematical simulation with respect to a non-linear boundary IHCP in statements A and B (see Sect. 8.6.2) when using a finite-dimensional approximation of the unknown heat flux $q(\tau)$.[5] To solve three boundary-value problems (direct, adjoint and temperature increment) which are necessary to perform one iterative step, an implicit four-point difference scheme has been used and all differential operators have been approximated on one and the same difference grid with a constant step in time and with a constant step in spatial coordinate inside every layer, to the left and to the right of the measurement point. A conjugation of numerical solutions at the points of measurements has been performed with the use of difference representations for the conditions of energy conjunction. Such an approach disturbs the uniformity of the difference grid but allows increased accuracy of a numerical solution when there is a discontinuity of the first kind in the derivative of the solution on the boundary between layers (adjoint problems). The initial direct boundary-value problems are non-linear hence their solutions have undergone successive specifications at every time step using the method of simple iteration until the approximate solutions or the adjacent iterations coincide with a relative accuracy 0.001 given in advance.

The values of steps of difference grid were selected in preliminary mathematical simulations by solving a direct problem with a successive increase of the step numbers along the spatial coordinate and in time until the fields of temperatures coincided with the relative accuracy of 0.001.

At first, consider the results of computational experiments for a problem in *statement* A when its solution is sought using the conjugate gradient method in space E_M regardless of the unknown function smoothness. Mathematical simulation was performed with the conditions $d = b = 1$, $\xi(x) = 0$, $q^*(\tau) = 0$. Thermophysical properties were considered to be constant $\lambda(T) = 1$, $C(T) = 1$.

All computations were made with $\Delta Fo_d = \dfrac{\lambda}{C}\dfrac{\Delta\tau}{d^2} = 0.04$ ($\Delta\tau$, step of a difference grid in time). Computational experiments have been conducted for answers to questions given below.

---

[5] The results of a computational experiment are taken for the most part from Alifanov, O.M.; Artyukhin, E.A.; Nenarokomov, A.V: Spline-approximation of the solution of the inverse heat-conduction problem, taking account of the smoothness of the desired function. High Temperature, 1987, Vol. 25, No. 4, pp. 520–526. Alifanov, O.M.; Nenarokomov, A.V: Effect of different factors on the accuracy of the solution of a parametrized inverse problem of heat conduction. J. Eng. Phys., 1989, Vol. 56, No. 3, pp. 308–312.

1. How do the conditionality and the accuracy characteristics of the algorithm of IHCP solution depend on the number of approximation parameters (M) and on the input data error?

To answer this question, calculations have been conducted on the restoration of four relations $q(\tau)$ which kept a special feature in that they could be *exactly* approximated by the cubic B-splines with $M = 3, 5, 7, 11$, correspondingly. These dependences are depicted in Fig. 8.15a–d. Calculations have been made for exact and disturbed temperature data $f(\tau)$.

A computational experiment was constructed using a traditional technique as described earlier. At first a direct problem was solved with the known value $q(\tau)$ and a resultant temperature $T(d, \tau)$ was used to form function $f(\tau)$. Namely, the dependence $T(d, \tau)$ was superpositioned by either errors to model some real experimental situation. Then an inverse problem on the restoration of function $q(\tau)$ was solved (at this stage of computations $q(\tau)$ was unknown).

An exact representation $q(\tau)$ by the corresponding B-splines as well as a solution of the direct problem, both at the stage of preparation of the input data and at the stage of IHCP solution with the use of one and the same difference grid have practically eliminated the effect of the approximation error δ_A (see Sect. 6.2) on the accuracy of determination $q(\tau)$.

The error of thermosensor indications was simulated as follows:

$$f(\tau) = T(d, \tau)(1 + \gamma\omega),$$

where $\gamma \in [0; 0.25]$ and ω was a random variable, distributed by normal law with

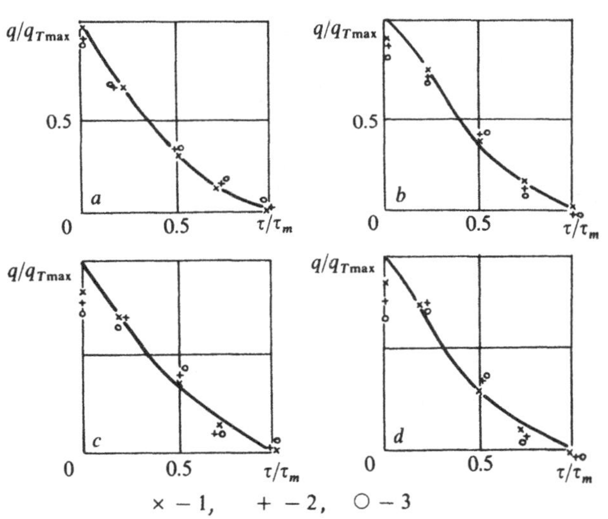

Fig. 8.15. Computational results on restoration of dependences $q(\tau)$, (full curves are unknown solutions); 1, 2, 3 are iterative solutions for $\delta_f = 0$; 0.05 and 0.1, respectively; $a - M = 3$, $b - M = 5$, $c - M = 7$, $d - M = 11$

a zero mean value and unit-equal variance. The results of these computations, performed with stopping the iterations using the residual criterion, are shown in Fig. 8.15a–d. A definite deterioration of restoration accuracy is observed with any increase in the dimension of the problem. To estimate this situation quantitatively, relative errors of IHCP solution in two metrics have been computed:

$$\varepsilon_{L_2} = \left\{ \int_0^{\tau_m} [q_T(\tau) - q(\tau)]^2 \, d\tau \Big/ \int_0^{\tau_m} q_T^2(\tau) \, d\tau \right\}^{1/2},$$

$$\varepsilon_{C^{(0)}} = \max_\tau |q_T(\tau) - q(\tau)| / \max_\tau |q_T(\tau)|,$$

where under $q_T(\tau)$ one assumes a true solution. The corresponding data depending on the dimensions of the problem are given in Fig. 8.16a, b which show a lower accuracy of determination of $q(\tau)$ with the growth of number M, especially when estimating the accuracy in space $C^{(0)}[0, \tau_m]$.

The main conclusion which can be made from this analysis is in the expediency of using the least number of basic functions of approximation from

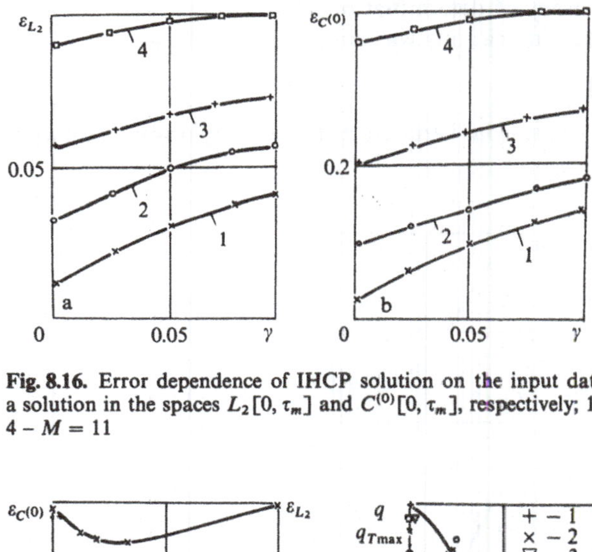

Fig. 8.16. Error dependence of IHCP solution on the input data error: a, b – error estimates of a solution in the spaces $L_2[0, \tau_m]$ and $C^{(0)}[0, \tau_m]$, respectively; 1 – $M = 3$, 2 – $M = 5$, 3 – $M = 7$, 4 – $M = 11$

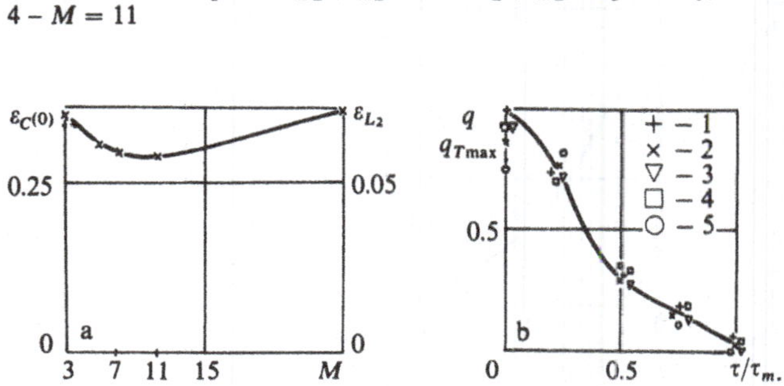

Fig. 8.17. The influence of approximation dimension on the IHCP solution error: *a* is a dependence of solution error on the number M; *b* are the results of inverse problem solution (1, 2, 3, 4, 5 – restored values $q(\tau)$ with $M = 3$, 5, 7, 11 and 31, respectively

which one can obtain fulfillment of the condition of stopping the iteration process, i.e. when a value of the minimized functional J becomes equal to or will be slightly lower than the error of temperature measurements δ_f (metric-matched with J). This conclusion is proved by the results obtained in the variation of the number of unknown parameters (Fig. 8.17). In this case the function sought $q(\tau)$ was exactly approximated by the B-splines with $M = 7$. In Fig. 8.17a one can see the minimum of solution error ε_{L_2} in the vicinity of this value.

2. How does an uncertainty in the knowledge of the value of error of the input data influence the accuracy of IHCP solution?

This is a very important question when using a residual criterion to choose the number of the last iteration, since a wrong assignment of value δ_f leads either to an earlier halt of the iterative process (if δ_f is overestimated) or to its ungrounded continuation (if δ_f is underestimated). What errors might appear at the end in the solution of an inverse problem, one can judge from the results of mathematical simulation, given in Fig. 8.18. The following estimate of error in the assignment of δ_f is accepted here:

$$\Delta\delta_f = (\tilde{\delta}_f - \delta_f)/\delta_f ,$$

where δ_f is the exact value of error and $\tilde{\delta}_f$ is the value of error used in the computation to stop the iterations. It can be seen that any overestimation of the residual level δ_f leads to more considerable errors in IHCP solution in the space metric $C^{(0)}[0, \tau_m]$ compared to underestimation.

3. How much does the accuracy of IHCP solution depend on the error in the definition of the thermosensor's location?

An answer to this question gives data of computation, presented in Fig. 8.19, where a relative error in the determination of the unknown function $\varepsilon_{C(0)}$ is

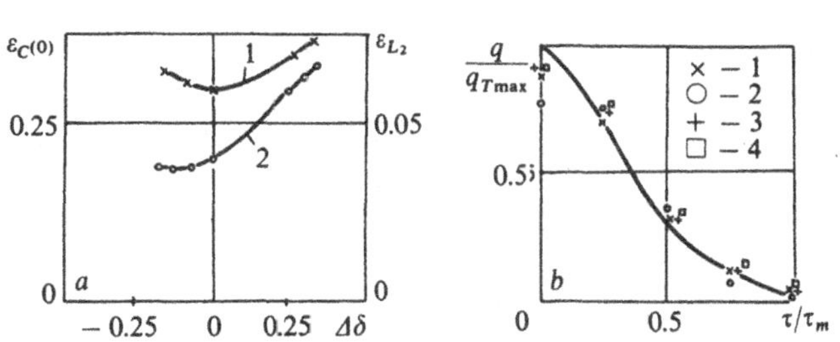

Fig. 8.18. The influence of errors in the assignment of admissible residual level on the error in the inverse problem solution: *a* is an error dependence of the solution on the error in the assignment of temperature data error (1, 2 – solution error estimates in the spaces $L_2[0, \tau_m]$ and $C^{(0)}[0, \tau_m]$, respectively; *b* are results of IHCP solution (1, 2, 3, 4 – restored values of $q(\tau)$ with $\Delta\delta_f = 0$; 0.25; −0.1; −0.15, respectively)

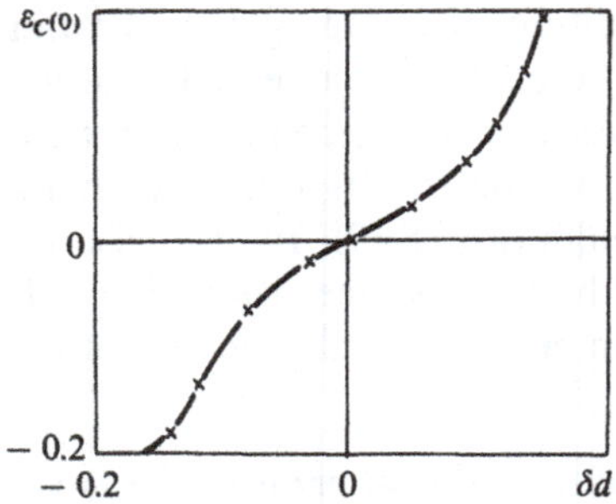

Fig. 8.19. Error dependence of the heat flux computation on the error in determining the thermosensor location coordinate

shown depending on a relative error in the thermosensor's location $\delta d = (\tilde{d} - d)/d$, where d and $\tilde{d}$ are exact and approximately known co-ordinates of the temperature sensor, respectively. Thus, inaccurate definition of thermosensor positioning, giving rise to a systematic error in the input data, results in large errors in the solution of an inverse problem. Hence, one must aim to achieve a sufficiently accurate measurement of the co-ordinate of the thermosensor's positioning.

4. What is the degree of the impact of errors in the assignment of thermophysical characteristics $\lambda(T)$ and $C(T)$ on the IHCP solution?

Of course, this impact will be different depending on the initial data of the inverse problem and in particular on the type of nominal curves $\lambda(T)$ and $C(T)$. However, even one example of computation can be very instructive. So, for constant thermophysical properties the error of restoring $q(\tau)$ is shown in Fig. 8.20 due to the uniform offset of these properties

$$\delta\lambda = \int_T [\tilde{\lambda}(T) - \lambda(T)]\,\mathrm{d}T / \int_T \lambda(T)\,\mathrm{d}T,$$

$$\delta C = \int_T [\tilde{C}(T) - C(T)]\,\mathrm{d}T / \int_T C(T)\,\mathrm{d}T,$$

where $\tilde{\lambda}$ and $\tilde{C}$ are the "shifted" values of the thermal conductivity and heat capacity per unit volume, respectively. It is seen that the accuracy of definition $C(T)$ is of great importance.

With this we complete the analysis of a numerical solution of problem A. It should be noted that the conclusions made herein will not change in principle for the problem in statement B. The only exception, perhaps, can be in the evaluation of the error in the assignment of a minimal permissible level of the residual and its influence on the accuracy of restoring $q(\tau)$. In this case when indications from sevral thermosensors are taken into account, similar to that already shown in analysing a solution of the boundary IHCP in functional space, the requirements for accuracy regarding assignment δ_f can be lowered.

In the example of an inverse problem in *statement B* a comparison of iterative algorithms will be carried out constructed with and withut the smoothness of $q(\tau)$ in order to answer the question:

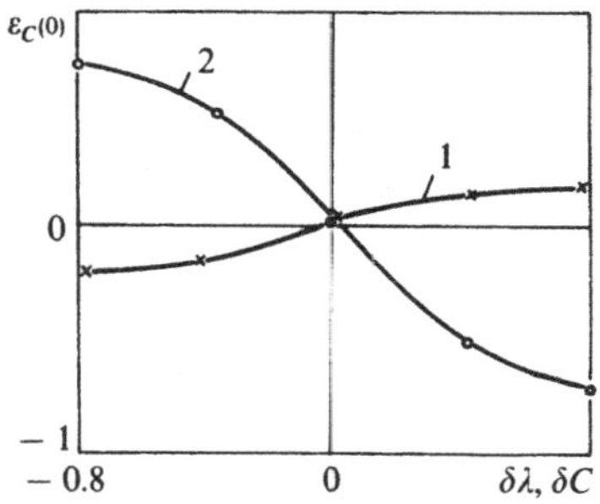

Fig. 8.20. Errors in computations of $q(\tau)$ against the error in the assignment of thermophysical properties: 1 is $\varepsilon_{C(0)}(\delta\lambda)$, 2 is $\varepsilon_{C(0)}(\delta C)$

5. How does an account of a priori information concerning the smoothness of function being defined influence the character and accuracy of iterative solution?

An appropriate investigation has been conducted for the following conditions: an unknown heat flux $q(\tau)$ was assigned as a sign-variable function as given in Fig. 8.21. The initial distribution of temperture is constant and equals 300 K. A thickness of the plate b equals 0.024 m and duration of the process $\tau_m = 100$ s. Two thermosensors with co-ordinates $d_1 = 0.006$ m and $d_2 = 0.012$ m are positioned in the plate. The inner boundary is thermally insulated ($q^*(\tau) = 0$). Thermophysial properties are given in Table 8.3.

A numerical solution was carried out on the difference grid $L \times m = 40 \times 40$, the required accuracy for temperature field computation having been reached.

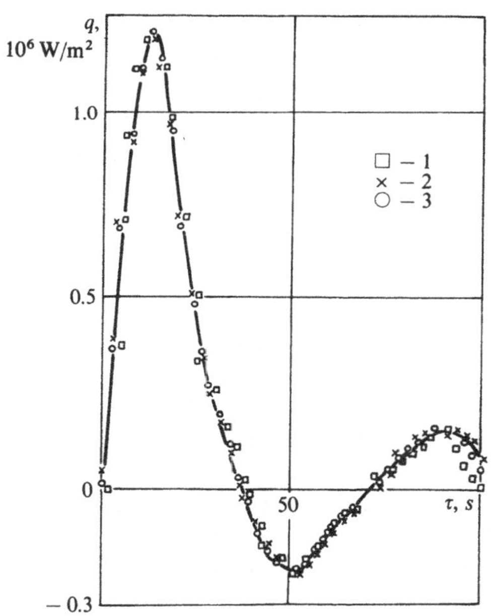

Fig. 8.21. Results of IHCP iterative solution (full curve is the unknown solution); 1 is a solution in the space $L_2[0, \tau_m]$; 2 is a solution with the B-spline approximation; 3 is a solution with the B-spline approximation, smoothness taken into account

Table 8.3. Thermophysical properties of the material of the plate

T, K	C, 10^7 J/($m^3 \cdot$ K)	λ, W/(m $\cdot$ K)
300	0.2120	46.10
450	0.2123	50.12
600	0.3201	55.03
750	0.2956	58.05
900	0.2783	59.54
1050	0.2823	61.01
1200	0.2892	63.11

Iterative processes were constructed based on the steepest descent method, i.e. it was assumed that $\gamma_k \equiv 0$.

With the information on the smoothness of unknown function, the required directions of descent at every iteration were computed by solving the algebraic system (8.149). It was assumed that $l = 2$, i.e. a convergence of approximations $q_k(\tau)$ was provided by the norm of space $W_2^2\ [0, \tau_m]$. Values of gradient components

$$J'_j = \int_0^{\tau_m} \psi(0, \tau)\varphi_j(\tau)\mathrm{d}\tau, \quad j = \overline{1, M}$$

were used as the right sides of Eqs. (8.149).
The results of IHCP solution obtained by means of these two algorithms (with and without smoothness) for non-disturbed temperature data are given on Fig. 8.21. In addition, a solution of the given IHCP in space $L_2[0, \tau_m]$ is included which was found by using the algorithm described in Sect. 8.6.2 under the piecewise–constant approximation $q(\tau)$ when the number of approximation intervals was equal to the number of time steps m in the difference grid. In these computations the initial approximation was assumed as $q^0(\tau) = 0$. The fact that this solution does not "come" to zero in point $\tau = \tau_m$ can be explained by the use of the following technique. In solving an adjoint problem one introduces a fictitious time layer $\tau \in [\tau_m, \tau_m + \Delta\tau]$ and the condition $\psi(x, \tau_m) = 0$ is substituted by the condition $\psi(x, \tau_m + \Delta\tau) = 0$. An adjoint problem is hereby solved in domain $\{0 \leqslant x \leqslant b, \Delta\tau \leqslant \tau \leqslant \tau_m + \Delta\tau\}$, i.e. this domain is shifted with respect to the domain for solving a direct problem of heat conduction for one time step. Such a techique, as computational experiments showed, improves solution in the vicinity of the point $\tau = \tau_m$ provided that the initial approximaion is not equal to an unknown function in this point. Of course, one should remember that as $\Delta\tau$ decreases, the computed values $q(\tau_m)$ will be closer and closer to the value $q^0(\tau_m)$.

A comparison of algorithms used for their convergence can be based on an indirect estimate, i.e. on the value of the minimized funcional obtained while continuing an iterative process until the relative functional values for adjacent iterations differ by a small value ε. These data for $\varepsilon = 0.0001$ are given in Table

8.4, where the following numbering of algorithms is accepted: 1, solution in space $L_2[0, \tau_m]$; 3, 2, parametrized form of solution with and without smoothness of $q(\tau)$, respectively.

Thus, we can point out definite advantages of a parametrized form of the solution, especially when it includes the property of smoothess of the dependence being recovered.

By including the property of smoothness a problem in the choice of weight r_n appears. For this parametric calculations have been carried out with a variation of values r_0, r_1 and r_2 to establish an optimal ratio between them. These results are shown in Table 8.5. We can conclude that the equivalent weights of the function, the first and second derivatives give better results.

As outlined when analysing the solution of problem A, there is an optimal number of intervals in partitioning the domain of definiion of approximating B-splines. For the problem considered $M_{\mathrm{opt}} = 14$, and all results were obtained having this number of intervals.

Computations have also been made for disturbed input data in which the random errors were set by normal law with their maximum spread from nominal curves $f_1(\tau)$ and $f_2(\tau)$, equalling $3\sigma = 0.05 \max_{\tau\in[0;\tau_m]} f_i(\tau)$. The corresponding results with the use of residual criterion for halt are given in Fig. 8.22.

So the computational experiments conducted show that a priori account of smoothness of a solution leads to a better convergence of the iterative process. In this case this manifested itself in the decrease of error of the heat flux recovery $\varepsilon^2_{L_2}$, by approximately 30%, compared with a parametrized form of the solution which does not take this factor into account.

Table 8.4. The obtained values of a minimized functional

Algorithm of solution	J, $K^2 \cdot c$
1	106.35
2	35.49
3	6.23

Table 8.5. The obtained values of a minimized functional depending on the weight coefficients

r_0	r_1	r_2	J, $K^2 \cdot c$
1.0	1.0	1.0	6.23
1.0	0	0	118.0
1.0	1.0	0	97.0
1.0	0	1.0	150.0
$0.1 \cdot 10^{-5}$	1.0	0	$0.1 \cdot 10^5$
0.1	1.0	0	200.0
0.1	0	1.0	190.0
$0.1 \cdot 10^{-5}$	0	1.0	$0.1 \cdot 10^4$

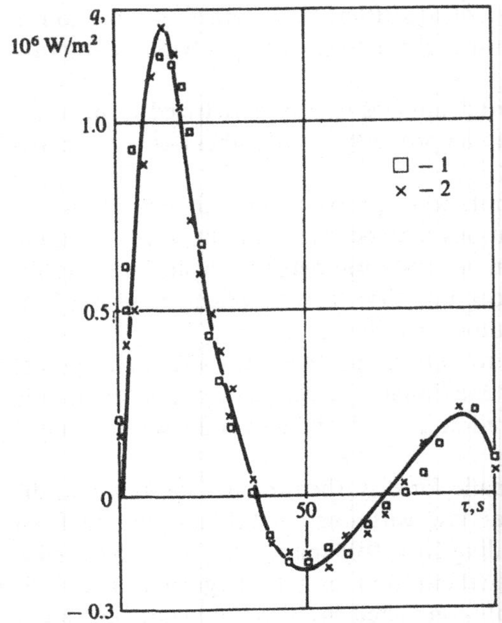

Fig. 8.22. Results of IHCP solution using disturbed data (full curve is the unknown solution); 1 is a solution obtained without regard to smoothness; 2 is a solution obtained with allowance for smoothness

Another conclusion which was obtained during these calculations and proven by corresponding theoretical prediction is that as the order of smoothness increases the stability of the iterative process grows with respect to fluctuation errors in the input temperatures. That is, the stability of the iterative solution for ill-posed inverse problem providing a convergence in the norm of space W_2^1 is higher than the stability of solution with a convergence in the norm $W_2^0 = L_2$. Also the algorithm giving a convergence in the norm W_2^2 is more stable than the algorithm with a convergence in W_2^1.

This property, first of all, decreases the required accuracy of assignment of the temperature measurement error when used to stop iterations of the residual condition. Second, it enables the traditional stopping criteria to be used, e.g. by closeness of two approximations or two values of the minimized functional at adjacent iterations (see Sect. 8.5.5). This is explained by a strengthening (as l grows) of the effect of curve's "fracture" $\sqrt{J(k)}$, after a sharp decreasing of functional values at the initial iterations its values change very little. We must remember, however, that the increase of the considered order of smoothness makes the algorithm more and more "rigid" and, hence, impairs its capability for restoring the specific thin structural features of an unknown function. That is why in the majority of cases we should confine ourselves with values of $l \leqslant 2$ or 3. In the process of mathematical simulation the dependences $q(\tau)$ have also been considered which differ from those shown in Fig. 8.21. The results obtained coincided qualitatively with those presented above.

Thus, considering the parametrized form of IHCP we can come again to a conclusion on expediency of a priori account for smoothness when restoring heat loads and, in addition, on the efficiency of the given method in obtaining smooth solutions.

Remark 3. Basically, to this point we consider and compare only those methods for solving boundary IHCP which have been suggested by the author. But there are other and quite effective methods, in particular, a functional specification method, shown by Beck. In Figs. 8.23 and 8.24 for the case of non-disturbed temperature data concerning problem A these are the results obtained when using a method described in the present book for IHCP solution in space E_M, to simultaneously compare them with the results from Beck, Wolf [165] and Beck, Murio.[6] Unfortunately, the information available does not allow an exact quantitative comparison, but the data presnted do illustrate the method analysed as being competitive.

Two dimensional formulation. The corresponding results were obtained when solving the two dimensional IHCP [30, 31, 76]. In particular, we present the results of computations for the formulation of inverse problem, given in Sect. 6.11. An account of smoothness was carried out according to the mode described in Sect. 8.5.3. The gradient of functional in the space L_2 was computed via the adjoint problem [Eqs. (6.71)–(6.74)].

The solutions of the direct problem $T(x, 0, \tau)$, the adjoint problem $\psi(x, c, \tau)$ and the problem for an increment of temperature $\Delta T(x, 0, \tau)$ which are necessary for organization of iterative process, were computed using the Green's function and approximate-analytical computational mode, set forth in Sect. 4.5. Instead of initial integral operator A we have in this case a matrix A_Δ. For a suitable choice of parameter of finite-difference grid $x_l = l\Delta x$, $l = \overline{0, L}$;

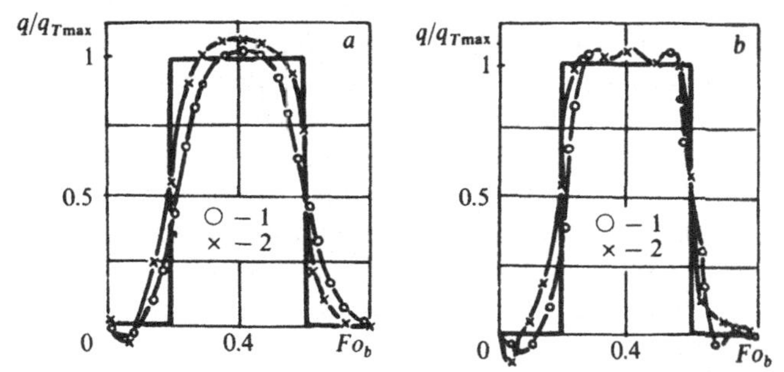

Fig. 8.23. Results of the heat flux recovery (full curves are the unknown dependences): *a*, *b* – computations with $\Delta Fo_d = 0.05$ and 0.01, respectively; 1 are the results, obtained by Beck and Murio; 2 are the results, obtained at 11-th iteration by means of the present method with solution approximation by the B-splines, $M = 20$ (a) and $M = 30$ (b)

[6] Beck, J.V., Murio, D.A., Combined function specification regularization procedure for solution of inverse heat conduction problem, *AIAA Journal*, 1986, **24**, pp. 180–185.

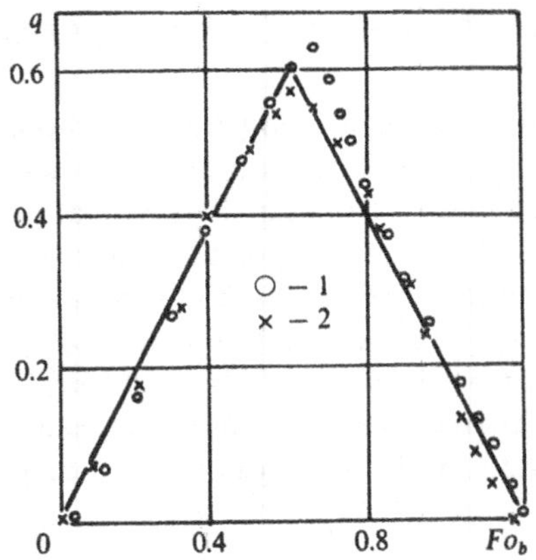

Fig. 8.24. Results of the heat flux recovery (full curve is the unknown function): 1 are the results obtained by Beck and Wolf; 2 are the results, obtained at 15-th iteration by means of the present method with solution approximation by the B-splines, $M = 20$

$\tau_j = j\Delta\tau$, $j = \overline{0, m}$ a sufficiently accurate approximation of an operator A is obtained which provides an opportunity to neglect the corresponding errors, when constructing an algorithm for the solution of inverse problem.

The calculation of the elements of matrix A_Δ entails the summation of infinite series (see, for example, the Eq. (4.23)). The truncated series give rise to additional errors, and as a result the matrix $A_{\Delta H} = A_\Delta + H$ is computed instead of matrix A_Δ, where H is a matrix of errors. To economize on computational time, one should sum only the number of elements of the series, sufficient to solve the problem with a predetermined accuracy.

As shown in [30], an error due to truncated series may be sufficiently small as compared with the error in experimental data, if the condition $h \leqslant \Delta f^{\max}/(mLq^{\max})$ is valid, where h is an element of matrix H (assuming that all elements of matrix H are equal to some positive number h); $\Delta f^{\max}$ is a maximal error in temperature measurements and $q^{\max}$ is a maximal value of a heat flux.

Figure 8.25a presents the results of a solution of one of the model problems using a modified conjugate gradient algorithm with regard to the smoothness of $q(x, \tau)$ (see Sect. 8.5.3) for three spatial points: $x_1 = 0.17 \cdot 10^{-2}$ m, $x_3 = 0.5 \cdot 10^{-2}$ m and $x_5 = 0.83 \cdot 10^{-2}$ m. In these computations it was assumed that $\Delta x = 0.17 \cdot 10^{-2}$ m, $\Delta\tau = 0.05c$, $a = 1.1 \cdot 10^{-4}$ m²/s, $\lambda = 45$ w/(m·K). Initial temperatures were taken to be undisturbed. The stopping of the iteration was carried out having the condition

$$k = \bar{k}: \quad |\bar{J}(q^{\bar{k}+1}) - \bar{J}(q^{\bar{k}})| \leqslant 10^{-8},$$

where

$$\bar{J}(q) = J(q) \Big/ \int_0^{\tau_m} \mathrm{d}\tau \int_0^b f^2(x, \tau)\mathrm{d}x .$$

If the values of function at the boundaries of a domain are given then there is no need to calculate the corresponding increments and the accuracy of reconstruc-

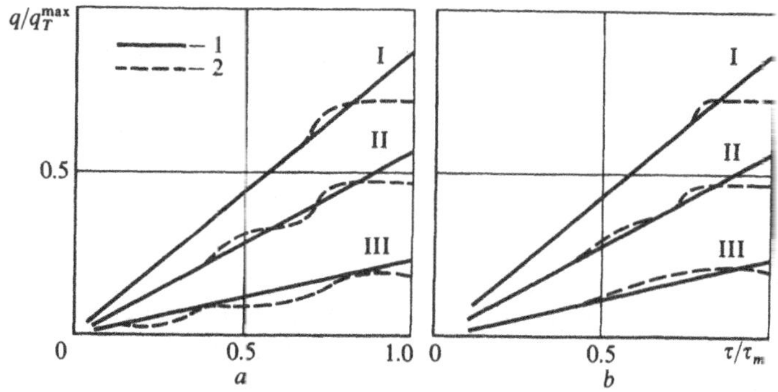

Fig. 8.25. Results of solving a two-dimensional IHCP by a modified method of conjugate gradients allowing for smoothness of the unknown function for the case when all variations are unknown (a) and when the variations in $q_\kappa(x, 0)$, $q_\kappa(b, \tau)$ and $q_\kappa(b, 0)$ are not determined (b) ($c = 0.02$ m, $b = 0.01$ m): 1 is a true solution; 2 – 30th (a) and 20th (b) iterations; I is $q(x_1, \tau)$; II is $q(x_3, \tau)$; III is $q(x_5, \tau)$

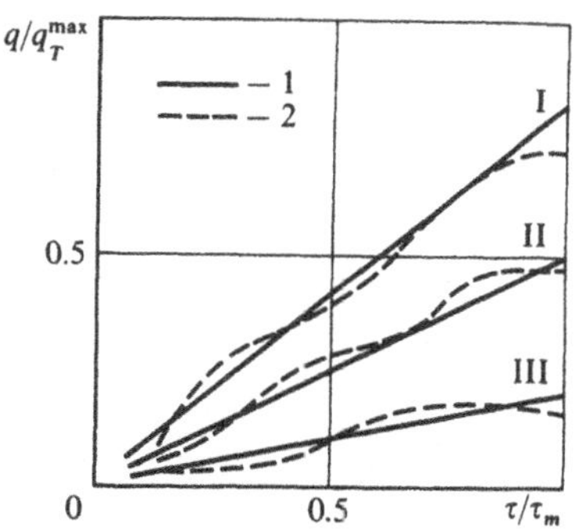

Fig. 8.26. Reconstruction of the heat flux by a modified method of conjugate gradients allowing for smoothness of the solution for disturbed data: 1 is a true solution; 2 are the results of calculations with temperature values including the random errors distributed by normal law with an amplitude $3\sigma = 0.01 f^{max}$; I is $q(x_1, \tau)$; II is $q(x_3, \tau)$; III is $q(x_5, \tau)$

tion of a function increases slightly. Figure 8.25b depicts the corresponding results assuming that $q(x, 0) = q(b, \tau) = 0$.

The influence of the perturbances in initial data for the solution of IHCP has also been analysed. The disturbance was summed to the function $f(x, \tau)$ which was modelled by the generator of random numbers under normal distribution law with amplitude $3\sigma = 0.01 f^{max}$. The stopping of the iteration was carried out by the residual criterion $\bar{k}$: $J(q^{\bar{k}}) \simeq \delta^2$, where $\delta^2 = \int_0^{\tau_m} d\tau \int_0^b \sigma^2(x, \tau) dx$, σ^2 is a variance of function $f(x, \tau)$. The results of the solution of one of the examples are presented in Fig. 8.26 and testify the satisfactory quality of the reconstruction of $q(x, \tau)$.

8.7.2 The Coefficient IHCP and the Experiment Design

Based on the results of [25, 42, 47], we present the results of the solution of model IHCP using the algorithm, set out in Sect. 8.6.3. In the computations the differential operators and functions in boundary-value problems for temperature $T(x, \tau)$, increment $v(x, \tau)$ and adjoint variable $\psi_i(x, \tau)$, $i = \overline{1, N+1}$ were substituted by the corresponding finite-difference analogues, the implicit scheme of approximation being utilized. The parameters of finite difference grids were chosen in such a way that the errors of approximation could be neglected.

First we discuss the solution of the problems with the minimal necessary points of measurements ($N = 1$).

Figure 8.27 depicts the results of the reconstruction of thermal conductivity $\lambda(T) = 0.5 + 2T^2$ as a function of temperature having the following conditions $C(T) = 1$, $b = 1$, $d_1 = 0.5$, $\tau_m = 1$ and for precise initial data. The temperatures $T_1(\tau)$, $T_2(\tau)$ and $f_1(\tau)$ were determined by solving the direct problem using boundary conditions of the second kind $q_1(\tau) = 1$, $q_2(\tau) = 0$.

To approximate the unknown function $\lambda(T)$ by splines, the segment $[T_0, T_M]$ was divided into three parts. For the initial approximation $\lambda^0 = 1.75$ was taken. The iterations were stopped at $\bar{k} = 25$, when the process of improving the approximate solution almost stopped.

The computational experiments have shown that the accuracy of the reconstruction of $\lambda(T)$ depends critically upon the location of a temperature sensor. Figure 8.28 shows one of the examples where the results of the solution of IHCP for precise initial data and for different locations of a point of temperature measurement are presented in dimensionless form. In this case the function to be found was given by

$$\lambda(T) = 0.217 - 6.08 \cdot 10^4\, T + 6.7 \cdot 10^{-7}\, T^2, \quad \frac{W}{m \cdot k}. \tag{8.201}$$

The number of the segments for spline-approximation was taken $M = 3$.

Therefore, when solving the coefficient IHCP the *experiment design*, i.e. the determination of a number of temperature sensors and coordinates of their locations in a body $\boldsymbol{d} = \{d_i\}_1^N$, becomes very important [44, 50]. As shown in [43, 44], the D-optimal design can be successfully utilized for this purpose. In

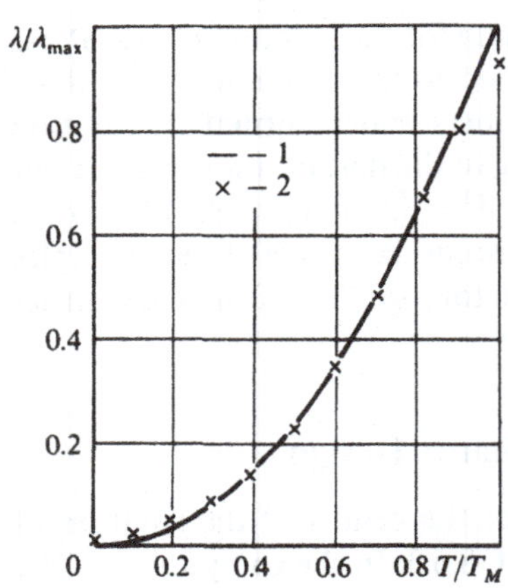

Fig. 8.27. Estimation of thermal conductivity for exact input data: 1 is the unknown function: 2 is the IHCP solution

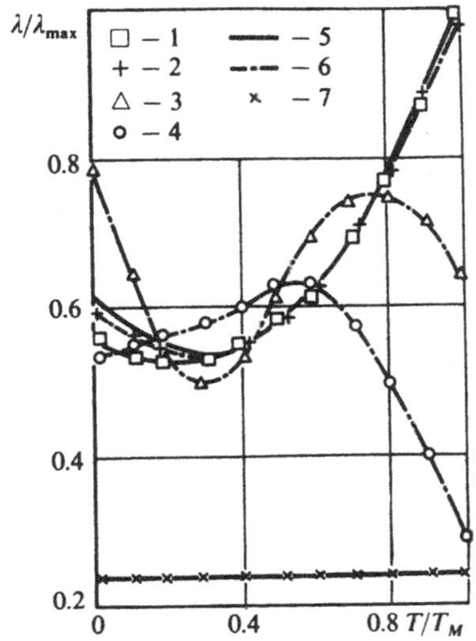

Fig. 8.28. Results of an IHCP solution with the use of exact input data for different locations of the thermosensor $d\cdot 10^3$ m, $b = 49\cdot 10^{-3}$ m, $\bar{\kappa} = 25$; $1 - d = 0.7$; $2 - d = 1.4$; $3 - d = 2.1$; $4 - d = 4.2$; 5 is the unknown function; 6 are the calculated dependences; 7 is the initial approximation

this case the experiment design $\varepsilon = \{N, d\}$ is determined having the condition of maximum of determinant D of the normed Fisher information matrix $F(\varepsilon)$

$$\varepsilon = \varepsilon^*: \max_{\varepsilon} D[F(\varepsilon)] ;$$

$$F(\varepsilon) = \frac{1}{N} \phi_{jn}, \quad j, n = \overline{1, M} ,$$

where

$$\phi_{jn} = \sum_{i=1}^{N} \int_0^{\tau_m} \theta_j(d_i, \tau)\theta_n (d_i, \tau) \mathrm{d}\tau ,$$

$\theta_j(d_i, \tau)$ are the sensitivity functions calculated at the points of location of the temperature sensors: $\theta_j(d_i, \tau) = \partial T(d_i, \tau)/\partial \lambda_j$, $j = \overline{-1, M + 1}$.

Since the values ϕ_{jn} depend on the vector of unknown parameters λ, then the optimal design may be constructed using the method of successive approximations beginning with a certain initial evaluation of vector λ and conducting a series of experiments.

Figure 8.29 illustrates the results of the experiment design for one of the problems of determination of the coefficient $\lambda(T)$ with one temperature sensors. For comparison the same figure shows the dependences of the error in reconstructing $\lambda(T)$ at the location of the measurement point d. The results of this design are in good agreement with the results obtained by the repeated solving of the inverse problem for various values of d. It is clear that there exists a rather narrow domain in which the temperature sensor should be installed. As has been pointed out in [44], this domain narrows as the number of the segments of spline-approximation M increases. At the same time by increasing the number of points of temperature measurement this domain can be widened.

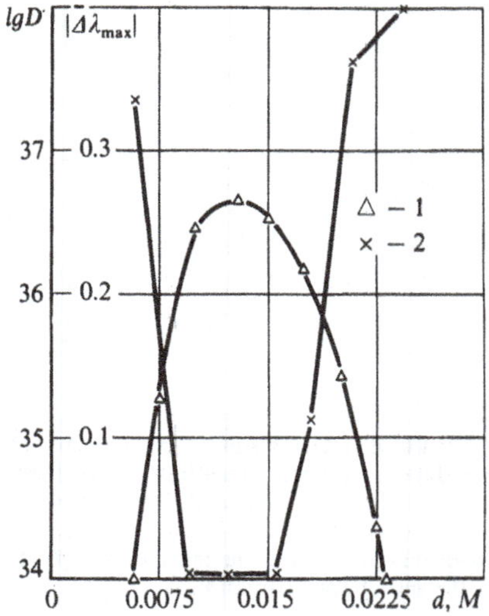

Fig. 8.29. Results of determining the optimal experimental design and the dependence of maximum relative error of thermal conductivity estimate ($\lambda(T) = 0.39 \cdot 10^{-1} + 0.17 \cdot 10^{-3}\, T$) on the location of the thermosensor: 1 is lg D; 2 is $|\Delta\lambda_{max}|$

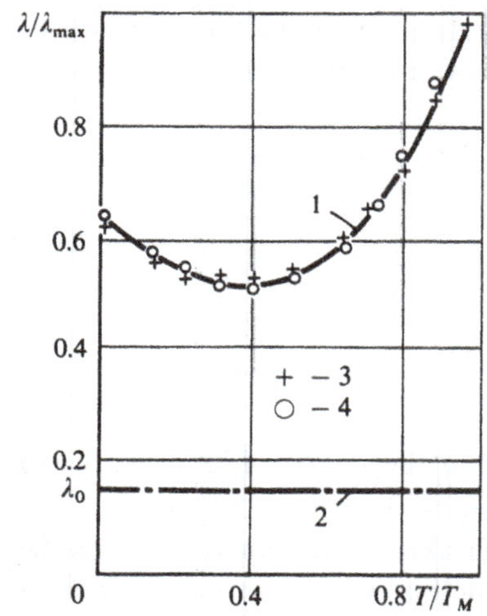

Fig. 8.30. Estimation of thermal conductivity: 1 is the exact solution; 2 is the initial approximation; 3, 4 are the solutions of inverse problem for exact and disturbed data, respectively

When initial temperatures were given with errors the stopping was carried out using the residual criterion

$$\sum_{i=1}^{N} \int_{0}^{\tau_m} [T_i(\lambda, d_i, \tau) - f_i(\tau)]^2 \mathrm{d}\tau \simeq \delta^2 ,$$

where $\delta^2 = \sum_{i=1}^{N} \int_{0}^{\tau_m} \sigma_i^2 \, \mathrm{d}\tau$ is an estimate of the summary error in initial data and $\sigma_i(\tau)$ is a meansquare deviation of temperature at a point d_i at a moment τ.

Figure 8.30 shows the results of the reconstruction of $\lambda(T)$ for the dependence in Eq. (8.201) for $N = 1$. In these computations the initial temperatures were disturbed by the errors with the uniform law of distribution and an amplitude of up to 3% of the maximal values of temperature. In the present example the utilization of measurements by two sensors slightly improved the results.

The general result of computational experiments implies that a sufficiently accurate estimation of thermophysical properties for transient heat flow in the body can be obtained by solving coefficient IHCP.

Remark 4. An example was considered above to define one thermophysical property $\lambda(T)$. Good results were also obtained when defining a vector-function of thermophysical properties. Calculations on simultaneous restoration of dependence on temperature of the thermal conductivity and distributed heat source have been given by Alifanov, Balakovsky and Klibanov[7]. Results on simultaneous determination of the thermal conductivity, heat capacity per unit volume and internal heat source as functions of temperature have been given by Artyukhin and Nenarokomov[8].

8.7.3 The Reconstruction of a Source in the Heat Conduction Equation

Let us discuss the results of the solution of inverse problem considered in Sect. 8.6.4 in the case of a thin rectangular plate $\bar{\Omega} = \{(x, y): 0 \leqslant x \leqslant b_1, 0 \leqslant y \leqslant b_2\}$.

To organize the iterative process given by formula (8.188) one must calculate the temperature $T(x, y, \tau)$, the increment of the temperature field $v(x, y, \tau)$ and the adjoint variable $\psi(x, y, \tau)$ in each iteration. Since an adjoint problem may be transformed into the common form of a boundary-value heat conduction problem by introducing the variable $t = \tau_m - \tau$, all three problems in Eqs. (8.183), (8.185), (8.186); (8.188)–(8.191) and (8.198)–(8.200) have the same form and may be solved by a single algorithm. For the construction of this algorithm we utilize the alternating-direction method [176, 196].

Approximate the Eq. (8.183) in the domain $Q = \Omega \times (0, \tau_m]$ at the grid $\{x_l = \Delta x \cdot l, l = \overline{0, L}; y_j = \Delta y \cdot j, j = \overline{0, k}; \tau_n = \Delta\tau \cdot n, n = \overline{0, m}\}$ by means of the pattern containing an intermediate time layer $\bar{\tau} = \tau + \dfrac{\Delta\tau}{2}$ (Fig. 8.31). For this purpose we use the finite-difference scheme of the alternating direction method depending on which transition from the n-th layer to the $(n+1)$-th one is carried out in the two stages having the steps $0.5 \cdot \Delta\tau$

$$\frac{a_1}{0.5 \cdot \Delta\tau}(T_{lj}^{n+1/2} - T_{lj}^{n}) = \Lambda_1 T_{lj}^{n+1/2} + \Lambda_2 T_{lj}^{n} + q_{lj}^{n+1/2} \tag{8.202}$$

$$\frac{a_1}{0.5 \cdot \Delta\tau}(T_{lj}^{n+1} - T_{lj}^{n+1/2}) = \Lambda_1 T_{lj}^{n+1/2} + \Lambda_2 T_{lj}^{n+1} + q_{lj}^{n+1/2} \tag{8.203}$$

---

[7] Alifanov, O.M.; Balakovskii, S.L.; Klibanov, M.V.: Restoration of causal characteristics of the thermal conductivity process from the solution of the combined inverse problem. *J. Eng. Phys.*, 1987, **52**, pp. 617–622.

[8] Artyukhin, E.A.; Nenarokomov, A.V.: Coefficient inverse heat-conduction problem. *J. Eng. Phys.*, 1987, **53** (3), pp. 1085–1090.

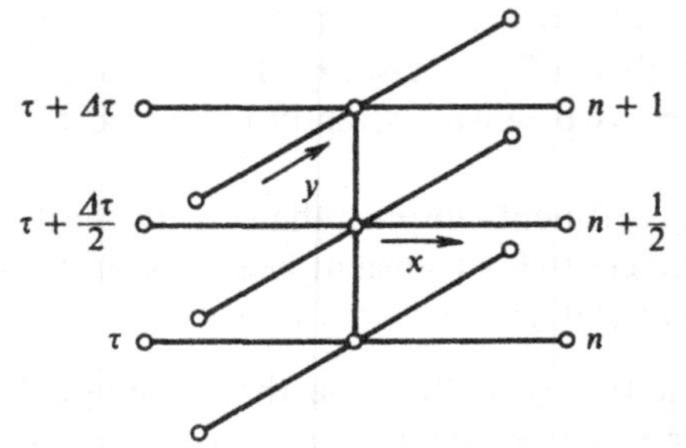

Fig. 8.31. A pattern of the alternating-direction method

where

$$T_{lj}^{\kappa} = T(x_l, y_j, \tau_\kappa); \quad q_{lj}^{n+1/2} = q(x_l, y_j, \tau_{n+1/2}) .$$

The finite-difference operators Λ_1 and Λ_2 are given by

$$\Lambda_1 T_{lj} = \frac{a_2}{\Delta x^2}(T_{l+1,j} - 2T_{lj} + T_{l-1,j}) ,$$

$$\Lambda_2 T_{lj} = \frac{a_2}{\Delta y^2}(T_{l,j+1} - 2T_{lj} + T_{l,j-1}) .$$

We introduce the initial condition

$$T_{lj}^0 = \xi(x_l, y_j), \quad l = \overline{0, L}, \quad j = \overline{0, K}$$

and boundary conditions

$$T_{l0}^{n+1} = g_1(x_l, \tau_{n+1}), \quad T_{lK}^{n+1} = g_2(x_l, \tau_{n+1}), \quad l = \overline{1, L-1} ;$$

$$T_{0j}^{n+1/2} = \bar{g}_{3j}, \quad T_{Lj}^{n+1/2} = \bar{g}_{4j}, \quad j = \overline{1, K} ,$$

where $\bar{g}_{\gamma j} = \frac{1}{2}[g_\gamma(y_j, \tau_{n+1}) + g_\gamma(y_j, \tau_n)] - \frac{\Delta\tau}{4}\Lambda_2[g_\gamma(y_j, \tau_{n+1}) - g_\gamma(y_j, \tau_n)]$,

$$\gamma = 3, 4 .$$

The numbers $\bar{g}_{\gamma j}$ are the intermediate values of functions $g_\gamma(y, \tau)$, $\gamma = 3, 4$ at the layer $n + \frac{1}{2}$ corresponding to the intermediate values $T_{lj}^{n+1/2}$ and $l = \overline{0, L}$ which arise when Eq. (8.203) is subtracted from Eq. (8.202).

The first finite-difference problem in Eq. (8.202) is solved for any fixed index j by the one-dimensional "run" method in the direction x. As a result, the intermediate values of function $T_{lj}^{n+1/2}$, $l = \overline{1, L-1}$ are found. The problem in Eq. (8.203) is solved by the one-dimensional "run" method by the coordinate y and as a result the values T_{lj}^{n+1}, $j = \overline{1, K}$ may be calculated.

Thus the transition from the nth time layer to the $(n+1)$th one may be carried out using the one-dimensional "run" method first in the longitudinal direction and then in the transverse.

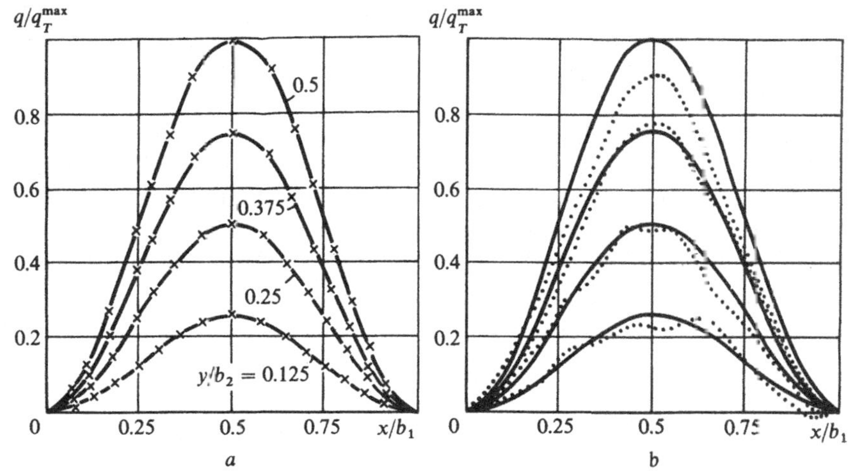

Fig. 8.32. Recovery of the source-term for exact (a) and disturbed (b) input data

It is known that this finite-difference scheme of the alternating-direction method is stable and has an accuracy of $O(\Delta\tau^2 + \Delta x^2 + \Delta y^2)$.

Figure 8.32,a presents the results of one of the examples of computation of a mean square approximation to an unknown source-term as a function of variable x at four cuttings along coordinate y and at one time point $\tau = \frac{\tau_m}{2}$. In the present case the undisturbed initial data were taken in a form of the grid function $f(x_l, y_j, \tau_n)$, obtained by the numerical solution of the direct problem. Note a good accuracy of the solution obtained. But even comparatively small random errors noticeably worsen the quality of the reconstructed function which is clear from Fig. 8.32,b. Here the results of the solution of IHCP are presented while using the function $f_\delta(x_l, y_j, \tau_n)$, which was disturbed by the errors with uniform distribution and with amplitude up to 1% of f^{max}. Such a change for the worse in the results emanates from the high level of instability of IHCP under consideration which, when utilizing an algorithm providing mean square convergence only, results in the appearance of wave-like deviations in the reconstructed function.

Therefore, when solving similar inverse problems one should take special care about the accuracy of temperature measurements and utilize the algorithms appropriate with regard to smoothness.

8.8 Summary

1. Mathematical results are presented which rigorously substantiate the method of iterative regularization for solving different types of linear inverse problems.

2. When applied to the inverse problem of generalized heat conduction two effective approaches are developed for constructing the gradient of residual functional, based on the use of the Green's function and adjoint problem, respectively.
3. Two new techniques are described for including quantitative and qualitative a priori information on the unknown solution in the gradient algorithms.
4. Examples are given on iterative algorithms for solving boundary and coefficient inverse problems of heat conduction. The results of computational experiments illustrate their high efficiency and broad possibilities for solving different linear and non-linear inverse problems.

Conclusions

The methods based on solving inverse heat transfer problems provide an opportunity to devise general and universal approaches for identification of thermal processes in various fields of science and technology, to make thermophysical investigations more reliable and informative while designing and testing technical objects and to carry out efficient heat diagnostics on the machinery and on the equipment in operation.

Among methods for the solution of the inverse problem the most universal is the iterative regularization. This method is widely used at present for the solution of various types and formulations of inverse problems, but does not exclude the successful application of other approaches which might be more effective in certain special cases. In particular, the direct methods (algebraic and numerical) for the solution of boundary IHCP proved to be especially useful and convenient for the express-processing of the experimental data and in cases where it is necessary to carry out the data processing of measurements in the real time.

The areas and conditions of the effective practical application of the methods developed for boundary IHCP are illustrated in Table A, in which “yes” means that the use of the given method is possible and “no” means impossible or inexpedient.

Table A The conditions of effective application of the methods for the solution of boundary IHCP

NN	Distinctions of application of methods	Direct approximate-analytical method (Chap. 4)	Direct numerical method (Chap. 5)	Iterative regularization method (Chaps. 6, 8)	Regularized algebraic method (Chap. 7)	Regularized numerical method (Chap. 7)
1	2	3	4	5	6	7
	The peculiarities of a formulation of IHCP					
1.	Linear with constant TPP	yes	yes	yes	yes	yes
2.	Nonlinear or linear with variable TPP	no	yes	yes	no	yes
3.	Homogeneous heat-conduction equation	yes	yes	yes	yes	yes
4.	Generalized heat-conduction equation	no	yes	yes	no	yes
5.	Fixed boundaries of a body	yes	yes	yes	yes	yes
6.	Travelling boundaries of a body	in one-dimensional case	yes	yes	in one-dimensional case	yes
7.	Fixed temperature sensors	yes	yes	yes	yes	yes
8.	Travelling temperature sensors	no	yes	yes	no	yes
9.	One-dimensional	yes	yes	yes	yes	yes
10.	Two-dimensional	for the domains of simple shape	yes	yes	for the domains of simple shape	yes
11.	Overdefined	no	no	yes	no	no
	General conditions of application					
12.	The determination of heat loads	yes	yes	yes	yes	yes
13.	The computation of temperature fields	no	yes	yes	no	yes
14.	The slowly alterating heat-exchange processes (HP)	yes	yes	yes	yes	yes

15.	The quickly alterating and short-term HP	no	no	yes	yes	yes
16.	Low-temperature HP	yes	yes	yes	yes	yes
17.	High-temperature HP with essential variable intensity	no	yes	yes	no	yes
18.	Small heat-depth of the location of temperature sensors	yes	yes	yes	yes	yes
19.	Big heat-depth of the location of temperature sensors	no	no	yes	yes	yes
20.	Metallic structural elements	yes	yes	yes	yes	yes
21.	Thermal protection and thermal insulation	no	for materials with high thermal conductivity	yes	yes	yes
	The conditions of practical application					
22.	The complexity of the algorithms	simple	medium complexity	medium complexity	heightened complexity	complex
23.	Computation time expenses	small	small	medium	medium	large
24.	Computation in real time	possible	possible	in a few isolated cases	in a few isolated cases	impossible
25.	The restrictions on the values of the steps of approximation	very strict	strict	no restriction	no restriction	no restriction
26.	Preliminary smoothing of initial data for high levels of errors	necessary	necessary	not necessary	not necessary	not necessary
27.	Obtaining the required degree of smoothness of the results	impossible	impossible	possible	possible	possible
28.	An account of the quantitative a priori information about an unknown solution.	impossible	impossible	possible	hard to implement	hard to implement

Bibliography

1. Alifanov, O.M.; Zantsev, V.K.; Pankratov, B.M. et al. (Ed.: Academician V.P. Mishin): Algorithms for diagnostics of heat loads of flying vehicles. Moscow, Mashinostroenie, 1983, (in Russian).
2. Aldoshin, G.T.; Golosov, A.S.; Zhuk, V.I.: Solution of an inverse problem of transient heat conduction for a slab. In: "Heat and mass transfer" (in Russian). Minsk, Nauka i Tekhnika, Vol. 8, 1968, pp. 186–199.
3. Alifanov, O.M.: Solution of the problem of transient heat conduction and its application to studying heat shield materials. In: "Investigations of transient convective heat and mass transfer" (in Russian). Minsk, Nauka i Tekhnika, 1971, pp. 322–333.
4. Alifanov, O.M.: Rugularization of solutions of inverse problem of heat conduction. In: "Heat Transfer-Soviet Research" (USA), Vol. 5, No. 4, 163, 1973.
5. Alifanov, O.M.: Inverse problem of heat conduction. J. Eng. Phys., 1973, Vol. 25, No. 3, pp. 1187–1192.
6. Alifanov, O.M.: Direct methods of solving linear reverse heat-conduction problems. J. Eng. Phys., 1973, Vol. 24, No. 6, pp. 791–797.
7. Alifanov, O.M.: Regularization of the scheme for solving reverse heat-conduction problems. J. Eng. Phys., 1973, Vol. 24, No. 2, pp. 235–241.
8. Alifanov, O.M.: Numerical solution of a nonlinear reverse problem of heat conduction. J. Eng. Phys., 1973 Vol. 25, No. 2, pp. 1070–1076.
9. Alifanov, O.M.: Determination of boundary thermal regime from the solution of an inverse heat-conduction problem. J. Eng. Phys., 1974, Vol. 26, No. 2, pp. 246–253.
10. Alifanov, O.M.: Regularization schemes for approximate solution of nonlinear reverse problem of heat-conduction. J. Eng. Phys., 1974, Vol. 26, No. 1, pp. 90–94.
11. Alifanov, O.M.: Solution of an inverse problem of heat-conduction by iterative methods. J. Eng. Phys., 1974, Vol. 26, No. 4, pp. 471–476.
12. Alifanov, O.M.: Inverse boundary-value problems of heat-conduction. J. Eng. Phys., 1975, Vol. 29, No. 1, pp. 821–830.
13. Alifanov, O.M.: Gradient methods of reconstructing boundary heat conditions. In: "Heat and mass transfer -V" (in Russian), Vol. 9, Minsk, 1976, pp. 85–93.
14. Alifanov, O.M.: Inverse heat-transfer problems in the investigation of thermal processes and the design of engineering systems. J. Eng. Phys., 1977, Vol. 33, No. 6, pp. 1402–1409.
15. Alifanov, O.M.: Determination of heat loads from a solution of the nonlinear inverse problem. High Temperature, 1977, Vol 15, No. 3, pp. 498–504.
16. Alifanov, O.M., Some questions involved in the solution of inverse problems in heat conduction and automated data processing in thermophysical investigations. J. Eng. Phys., 1980, Vol. 39, No. 2, pp. 831–837.
17. Alifanov, O.M.: Identification of processes of heat and mass transfer by methods of inverse problems. In: Modern experimental methods of studying the processes of heat and mass transfer (in Russian). Proc. of Int. school-seminar. Part 2, Minsk, Izd. ITMO Akad. Nauk BSSR, 1981, pp. 133–147.
18. Alifanov, O.M.: Methods of solving ill-posed problems. J. Eng. Phys., 1983, Vol. 45, No. 5, pp. 1237–1245.
19. Alifanov, O.M.: Identification of physical processes and inverse problems. J. Eng. Phys., 1985, Vol. 49, No. 6, pp. 1380–1386.
20. Alifanov, O.M.: A method of taking account of a priori information in solving incorrect inverse problems. J. Eng. Phys., 1985, Vol. 49, No. 6, pp. 1412–1418.

21. Alifanov, O.M.; Artyukhin, E.A.: Regularized numerical solution of nonlinear inverse heat-conduction problem. J. Eng. Phys., 1975, Vol. 29, No. 1, pp. 934–938
22. Alifanov, O.M.; Artyukhin, E.A.: Determination of boundary conditions during thermal gasdynamic testing. High Temperature, 1978, Vol. 16, No. 4, pp. 699–704.
23. Alifanov, O.M.; Artyukhin, E.A.; Loginov, S.N.; Malozemov, V.V.: Solution of inverse problems of heat conduction by the method of dynamic filtration. J. Eng. Phys., 1981, Vol. 41, No. 5, pp. 1260–1264.
24. Alifanov, O.M.; Artyukhin, E.A.; Pankratov, B.M.: Solution of the nonlinear inverse problem for a generalized heat-conduction equation in a region with moving boundaries. J. Eng. Phys., 1975, Vol. 29, No. 1, pp. 928–933.
25. Alifanov, O.M.; Artyukhin, E.A.; Rumyantsev, S.V.: Solution of boundary and coefficient inverse heat conduction problems by iterative methods. In: "Heat and mass transfer -VII" (in Russian), Vol. 9, Minsk, 1980, pp. 106–112.
26. Alifanov, O.M.; Artyukhin, E.A.; Tryanin, A.P.: Determining the heat-flux density at the boundary of a porous body from the solution of the inverse problem. High Temperature, 1983, Vol. 21, No. 6, pp. 888–895.
27. Alifanov, O.M.; Balashova, I.E.: Selection of approximate solutions of inverse heat-conduction problems. J. Eng. Phys., 1985, Vol. 48, No. 5, pp. 626–632.
28. Alifanov, O.M.; Batura, N.I.; Bespalov, A.M.; Gorshkov, M.I.; Kuz'min, N.A.; Maiorov, A.I.: New method for nonsteady-heat-transfer investigations in a thermal aerodynamic tube. J. Eng. Phys., 1977, Vol. 33, No. 6, pp. 1413–1416.
29. Alifanov, O.M.; Gorshkov, M.I.; Zantsev, V.K.; Pankratov, B.M.: Transient heat transfer between a solid and a plasma stream. J. Eng. Phys., 1975, Vol. 29, No 1, pp. 831–834.
30. Alifanov, O.M.; Egorov, Yu.V.: Algorithms and results of solving the inverse heat-conduction boundary problem in a two-dimensional formulation. J. Eng. Phys., 1985, Vol. 48, No. 4, pp. 489–496.
31. Alifanov, O.M.; Kerov, N.V.: Determination of external thermal load parameters by solving the two-dimensional inverse heat-conduction problem. J. Eng. Phys., 1981, Vol. 41, No. 4, pp. 1049–1053.
32. Alifanov, O.M.; Klibanov, M.V.: Uniqueness conditions and method of solution of the coefficient inverse problem of thermal conductivity. J. Eng. Phys., 1985, Vol. 48, No. 6, pp. 730–735.
33. Alifanov, O.M.; Mikhailov V.V.: Solution of the nonlinear inverse thermal conductivity problem by the iteration method. J. Eng. Phys., 1978, Vol. 35, No. 6, pp. 1501–1506.
34. Alifanov, O.M.; Mikhailov, V.V.: Determining thermal loads from the data of temperature measurements in a solid. High temperature, 1983, Vol. 21, No. 5, pp. 724–730.
35. Alifanov, O.M.; Mikhailov, V.V.: Solution of the overdetermined inverse problem of thermal conductivity involving inaccurate data. High Temperature, 1985, Vol. 23, No. 1, pp. 112–117.
36. Alifanov, O.M.; Rumyantsev, S.V.: One method of solving incorrectly stated problems. J. Eng. Phys., 1978, Vol. 34, No. 2, pp. 223–226.
37. Alifanov, O.M.; Rumyantsev, S.V.: On the stability of iterative methods for the solution of linear ill-posed problems. Soviet Math. Dokl., 1979, Vol. 20, No. 5, pp. 1133–1136.
38. Alifanov, O.M.; Rumyantsev, S.V.: Regularising gradient algorithms for inverse thermal-conduction problems. J. Eng. Phys., 1980, Vol. 39, No. 2, pp. 858–861.
39. Alifanov, O.M.; Tryanin, A.P.: Determination of the coefficient of internal heat exchange and the effective thermal conductivity of a porous solid on the basis of a nonstationary experiment. J. Eng. Phys., 1985, Vol. 48, No. 3, pp. 356–365.
40. Artyukhin, E.A.: Determination of thermal diffusivity from experimental data. J. Eng. Phys., 1975, Vol. 29, No. 1, pp. 878–881.
41. Artyukhin, E.A.: Reconstruction of the thermal conductivity coefficient from the solution of the nonlinear inverse problem. J. Eng. Phys., 1981, Vol. 41, No. 4, pp. 1054–1058.
42. Artyukhin, E.A.: Recovery of the temperature dependence of the thermal conductivity coefficient from the solution of the invérse problem. High Temperature, 1981, Vol. 19, No. 5, pp. 698–702.
43. Artyukhin, E.A.: Sensitivity analysis and experiment design in the problems of identification of processes of generalized heat conduction (in Russian). In: "Heat- and Mass Transfer-VII". Minsk, v. 7, 1984, pp. 81–84.
44. Artyukhin, E.A.: Experimental design of measurements for the solution of coefficient-type inverse heat-conduction problem. J. Eng. Phys., 1985, Vol. 48, No. 3, pp. 372–376.
45. Artyukhin, E.A.; Killikh, V.E.; Okhapkin A.S.: Reconstructing the effective coefficient of

thermal conductivity of asbestos-textolite from the solution of the inverse problem. J. Eng. Phys., 1983, Vol. 45, No. 5, pp. 1281–1286.

46. Artyukhin, E.A.; Okhapkin, A.S.: Determination of the parameters in the generalized heat-conduction equation from transient experimental data. J. Eng. Phys., 1982, Vol. 42, No. 6 pp. 693–698.
47. Artyukhin, E.A.; Okhapkin, A.S.: Parametric analysis of the accuracy of solution of a nonlinear inverse problem of recovering the thermal conductivity of a composite material. J. Eng. Phys., 1983, Vol. 45, No. 5, pp. 1275–1281.
48. Artyukhin, E.A.; Rumyantsev, S.V.: A gradient method of defining smooth solutions to inverse boundary-value problems in thermal conduction. J. Eng. Phys., 1980, Vol. 39, No. 2, pp. 862–865.
49. Artyukhin, E.A.; Rumyantsev, S.V.: Optimal choice of descent steps in gradient methods of solution of inverse heat conduction problems. J. Eng. Phys., 1980, Vol. 39, No. 2, pp. 865–869.
50. Budnik, S.A.: The problem of planning thermal measurements. J. Eng. Phys., 1980, Vol. 39, No. 2, pp. 840–844.
51. Budak, B.M.; Samarsky, A.A.; Tikhonov A.N.: Problem Exercises for mathematical physics (in Russian). Nauka, Moscow, 1972.
52. Bulychev, E.V.; Glasko, V.B.; Fedorov, S.M.: Reconstructing initial temperature from value measured on surface. USSR Comput. Math. and Math. Phys., 1983, Vol. 23, No. 6.
53. But, E.N.: Spline identification of heat fluxes. J. Eng. Phys., 1977, Vol. 33, No. 6, pp. 1085–1089.
54. Vanichev, A.P.: An approximate method of solving heat conduction problems in solids (in Russian). Izv. Akad. Nauk SSSR, Ser. OTN, 1946, No. 12, pp. 1767–1774.
55. Vasil'ev, F.P.: Lectures on the methods of solving extremal problems (in Russian). Izd. Moscow State Univ., Moscow, 1974.
56. Vasil'ev, F.P.: Methods of solving extremal problems (in Russian). Nauka, Moscow, 1981.
57. Vladimirov, V.S.: Equations of mathematical physics (in Russian). Nauka, Moscow, 1976.
58. Voevodin, V.V.: The method of regularization. USSR Comput. Math. and Math. Phys., 1969, Vol. 9, No. 3, pp. 228–232.
59. Volkov, V.M.: An inverse problem for quasi-linear equation of parabolic type (in Russian). In: "A study of inverse problems correctness and some operator equations", Vichisl. Centre, Sibirsk. Otdel. Akad. Nauk SSSR, Novosibirsk, 1981, pp. 27–36.
60. Volkov, V.M.: An inverse problem for quasi-linear equation of parabolic type (in Russian). In: "Ill-posed problems of Mathematical physics and analysis", Nauka, Novosibirsk, 1984, pp. 227–228.
61. Voskoboinikov, Yu. E.; Preobrazhensky, N.G.; Sedelnikov A.I.: Mathematical processing of an experiment in molecular gas dynamics (in Russian). Nauka, Novosibirsk, 1984.
62. Voskoboinikov, Yu. E.; Thomsons, Ya. Ya.: Construction of a regularized solution to one inverse heat-conduction problem with random errors in the initial data. J. Eng. Phys., 1977 Vol. 33, No. 6, pp. 1489–1493.
63. Geraschenko, O.A.: Fundamentals of heat thermometry (in Russian). Naukova Dumka, Kiev 1971.
64. Glasko, V.B.: Inverse problems of mathematical physics. Izd. Moscow State Univ., Moscow 1984. English transl. in American Institute of Physics, New York, 1984.
65. Goltsov, A.S.; Simbirsky, D.F.: Identification of an nonstationary nonlinear heat object using the Kalmann filter (in Russian). Avtometria, 1975, No. 1, pp. 36–42.
66. Goncharsky, A.V.; Yagola, A.G.; Leonov, A.G.: A generalized discrepancy principle, USSR Comput. Math. and Math. Phys., 1973, Vol. 13, No. 2, pp. 25–38.
67. Gordonova V.I., Morozov, V.A.: Numerical algorithms for the selection of the parameter in the regularization method. USSR Comput. Math. and Math. Phys., 1973, Vol. 13, No. 3, pp. 1–9
68. Goryachev, A.A.; Yudin, V.M.: Solution of the inverse coefficient problem of heat conduction. J. Eng. Phys., 1982, Vol. 43, No. 4, pp. 1148–1154.
69. Ditkin, V.A.; Prudnikov, A.P.: Operational calculus (in Russian). Vysshaya Shkola, Moscow, 1975.
70. Zhuk, V.I.; Golosov, A.S.: Engineering methods of determining thermal boundary conditions by means of temperature measurement data. J. Eng. Phys., 1975, Vol. 29, No. 1, pp 845–849.
71. Iskenderov, A.D.: An inverse problem for quasilinear parabolic equations. Differential equations, 1974, Vol. 10, No. 5, pp. 687–692.
72. Kalinnikov, L.D.; Shumakov, N.V.: The inverse problem of thermal conductivity for a plate. High Temperature, 1971, Vol. 9, No. 4, pp. 733–736.
73. Kamyinin, L.I.: On the existence of solution of boundary-value problems for a parabolic

equation with discontinuous coefficients (in Russian). Izv. Akad. Nauk SSSR, ser. Mat., 1964, Vol. 28, pp. 721–744.

74. Karmanov, V.G.: Mathematical programming (in Russian). Nauka, Moscow, 1980.
75. Kartashov, E.M.: Analytical methods in heat conduction of solids (in Russian). Vysshaya Shkola, Moscow, 1979.
76. Kerov, N.V.: Solution of the two-dimensional inverse heat-conduction problem in a cylindrical coordinate system. J. Eng. Phys., 1983, Vol. 45, No. 5, pp. 1245–1249.
77. Klibanov, M.V.: Uniqueness in the large of solutions of inverse problems for a class of differential equations. Differential equations, 1984, Vol. 20, No. 11, pp. 1390–1395.
78. Klibanov, M.V.: On a class of inverse problems for nonlinear parabolic equations. Soviet Math. Dokl., 1985, Vol. 31, No. 1, pp. 104–107.
79. Kover'yanov, V.A.: Inverse problem of nonsteady-state thermal conductivity. High Temperature, 1967, Vol. 5, No. 1, pp. 121–127.
80. Kozdoba, L.A.: Solution of nonlinear problems of heat conduction (in Russian). Naukova Dumka, Kiev, 1976.
81. Kozdoba, L.A.; Krukovsky, P.G.: Method of solving inverse problems of heat transfer (in Russian) Naukova Dumka, Kiev, 1982.
82. Kozdoba, L.A.; Krivoshey, F.A.: Determination of thermal properties of solids by electric simulation of solutions of inverse problems of transient heat conduction (in Russian). In: "Thermophysical properties of solids", Nauka, Moscow, 1973, pp. 28–31.
83. Kolmogorov, A.N.; Fomin, S.V.: The elements of function theory and functional analysis (in Russian). Nauka, Moscow, 1972.
84. Kondratiev, G.M.: Regular heat regime (in Russian). Gostekhizdat, Moscow, 1954.
85. Kochetov, I.I.: A new way of choosing the regularization parameter. USSR Comput. Math. and Math. Phys., 1976, Vol. 16, No. 2, pp. 215–219.
86. Kudryavtsev, E.V.; Chakalev, K.N.; Shumakov, N.V.: Transient heat transfer (in Russian). Izd. Akad. Nauk SSSR, Moscow, 1961.
87. Kuzmin, M.P.: Electric simulation of transient processes of heat transfer (in Russian). Energiya, Moscow, 1974.
88. Lavrent'ev, M.M.: Some improperly posed problems of mathematical physics (in Russian). Izd. Sib. Otd. Akad. Nauk SSSR, Novosibirsk, 1962. English transl. in Springer-Verlag. New York, 1967.
89. Lavrent'ev, M.M.; Romanov, V.G.; Shishatskii, S.P.: Ill-posed problems of mathematical physics and analysis (in Russian). Nauka, Moscow, 1980. English transl. in Trans. of Math. Monographs., Vol. 64, American Mathematical Society, Providence, Rhode Island, 1986.
90. Ladyzhenskaya, O.A.: Boundary-value problems of mathematical physics (in Russian). Nauka, Moscow, 1973.
91. Ladyzhenskaya, O.A.; Solonnikov, V.A.; Uraltseva, N.N.: Linear and quasilinear equations of parabolic type (in Russian). Nauka, Moscow, 1967.
92. Liscovec, O.A.: A method of choosing the regularization parameter to solve nonlinear ill-posed problem. Soviet Math. Dokl., 1976, Vol. 17, No. 4, pp. 1013–1017.
93. Lyikov, A.V.: The theory of heat conduction (in Russian). Visshaya Shkola, Moscow, 1967.
94. Lyikov, A.V.: Heat and mass transfer (in Russian). Reference book. Energiya, Moscow, 1972.
95. Matsevity, Yu. M.: Electric simulation of nonlinear problems of engineering thermophysics (in Russian). Naukova Dumka, Kiev, 1977.
96. Matsevity, Yu. M.; Multanovsky, A.V.: Identification in the problems of heat conduction (in Russian). Naukova Dumka, Kiev, 1982.
97. "Methods of determination of heat conduction and thermal diffusivity" A.G. Shaskhov, G.M. Volokhov, T.N. Abramenko, V.P. Kozlov. – Ch. ed. A.V. Lyikov, Energiya, Moscow, 1973 (in Russian).
98. Mikhailov, V.V.: Optimization of multilayer thermal insulation. J. Eng. Phys., 1980, Vol. 39, No. 2, pp. 880–884.
99. Mikhailov, V.V.: Question of the convergence of iteration methods of solving the inverse heat-conduction problem. J. Eng. Phys., 1983, Vol. 45, No. 5, pp. 1263–1266.
100. Mishin, V.P.; Alifanov, O.M.: Inverse problems of heat exchange – fields of application in the planning and testing technical objects. J. Eng. Phys., 1982, Vol. 42, No. 5, pp. 115–123.
101. Simulation of heat conditions of a space vehicle and environment. L.V. Kozlov, M.D. Nusinov, A.I. Akishin, et al., Ch. ed. Academician G.I. Petrov, Mashinostroenie, Moscow, 1971 (in Russian).

102. Morozov, V.A.: Regularization of incorrectly posed problems and the choice of regularization parameter, USSR Comput. Math. and Math. Phys., 1966, Vol. 6, No. 1, pp. 242–25[illegible].
103. Morozov, V.A.: Methods of solving unstable problems (in Russian). Izd. Vychisl. Centre, Moscow State Univ., Moscow, 1967.
104. Morozov, V.A.: The error principle in the solution of operational equations by the regularization method. USSR Comput. Math. and Math. Phys., 1968, Vol. 8, No. 2, pp. 63–87.
105. Morozov, V.A.: Regular methods of solving ill-posed problems (in Russian). Izd. Moscow State Univ., Moscow, 1974.
106. Muzylev, N.V.: Uniqueness theorems for some inverse problems of heat transfer. USSR Comput. Math. and Math. Phys., 1980, Vol. 20, No. 2, pp. 120–134.
107. Muzylev, N.V.: Uniqueness of simultaneous determining of coefficients of thermal conductivity and volumetric heat capacity, Zh. Vychisl. Mat. Mat. Fiz., 1983, Vol. 23, No. 1, pp. 102–108. – Engl. transl. in USSR Comput. Math. and Math. Phys., 1983, Vol. 23, No. 1.
108. Myuntz, G.: Integral equations (in Russian). Gostekhisdat, Moscow-Leningrad, 1934, P.I.
109. Koshkin, V.K.; Kalinin, E.K.; Dreytser, G.A. et al.: (in Russian). Transient heat transfer. Mashinostroenie, Moscow, 1973.
110. Nikitenko, N.I.: Study of nonstationary heat- and mass-transfer processes by the grid method (in Russian). Naukova Dumka, Kiev, 1978.
111. Omelchenko, K.G.; Volodina, M.V.: Analysis of inverse heat-conduction problems by means of similarity solutions. J. Eng. Phys., 1975, Vol. 29, No. 1, pp. 860–862.
112. Omelchenko, K.G.; Pchyolkina, V.G.: Solution of an inverse problem of nonlinear heat conduction to determine thermophysical characteristics. J. Eng. Phys., 1975, Vol. 29, No. 1, pp. 885–887.
113. Omelchenko, K.G.; Savelov, M.V.; Timoshenko, V.P.: Investigation of heat- and mass-transfer processes in decomposing porous materials, High Temperature, 1974, Vol. 12, No. 4, pp. 664–670.
114. The fundamentals of heat transfer in aviation and rocketry. V.S. Avduevsky, D.M. Galitseisky, G.A. Glebov, et al., ch. ed. V.K. Koshkin, Mashinostroenie, Moscow, 1975 (in Russian).
115. Okhapkin, A.S.: Study of the heat-transfer characteristics of a thermally protective composite materials. J. Eng. Phys., 1985, Vol. 49, No. 6, pp. 1469–1473.
116. Pankratov, B.M.: Thermal design of units of flying vehicles (in Russian). Mashinostroenie, Moscow, 1981.
117. Pankratov, B.M.; Polezhaev, Yu. V.; Rud'ko, A.K.: Materials interaction with gas flows (in Russian). Mashinostroenie, Moscow, 1975.
118. Pekhovich, A.I.; Zhidkih, V.M.: Analysis of heat regime of solids (in Russian). Energiya, Moscow, 1976.
119. Petrovsky, I.G.: Lectures on equations with partial derivatives (in Russian). Fizmatgiz, Moscow, 1961.
120. Platunov, E.S.: Thermophysical calculations for monotonous regime (in Russian). Energiya, Leningrad, 1973.
121. Polezhaev, Yu.V.; Killikh, V.E.; Narozhnyi, Yu.G.: Problems regarding the transient heating of heat shielding materials. J. Eng. Phys., 1975, Vol. 29, No. 1, pp. 840–844.
122. Polezhaev, Yu.V.; Narozhny, Yu.G.; Safonov, V.E.: A method for determining the thermal conductivity coefficient of high-temperature materials during their unsteady heating. High Temperature, 1973, Vol. 11, No. 3, pp. 543–548.
123. Polezhaev, Yu.V.; Yurevich, F.B.: Heat protection (in Russian). Energiya, Moscow, 1976.
124. Prosuntsov, P.V.; Reznik, S.V.: Determination of the thermophysical properties of translucent materials. J. Eng. Phys., 1985, Vol. 49, No. 6, pp. 1458–1462.
125. Rumyantsev, S.V.: Ways of allowing for a priori information in regularizing gradient algorithms, J. Eng. Phys., 1985, Vol. 49, No. 6, pp. 1418–1421.
126. Samarsky, A.A.: A theory of difference schemes (in Russian). Nauka, Moscow, 197[illegible].
127. Simbirsky, D.F.: The temperature diagnostics of engines (in Russian). Tekhnika, Kiev, 1976.
128. Simbirsky, D.F.; But E.N.: The measurement of heat fluxes by one-dimensional temperature detectors using the Kalmann filter and spline-approximation (in Russian). In: "Experimental methods of investigation of gas-turbine engines", Kharkov, 1975, Iss. 2, pp. 33–43.
129. Smolskyi, B.M.; Sergeeva, L.A.; Sergeev, V.L.: Transient heat transfer (in Russian). Nauka i Tekhnika, Minsk, 1974.
130. Stechkin, S.B.; Subotin, Yu. N.: Splines in computational mathematics (in Russian). Nauka, Moscow, 1976.

131. Strahov, V.N.; Valjashko, G.M.: On the problem of choosing the regularization parameter when solving ill-posed linear problems. Soviet Math. Dokl., 1976, Vol. 17, No. 3, pp. 664–668.
132. Tyomkin, A.G.: Determination of time dependent external heat actions by methods of heat conduction (in Russian). Izv. Vyssh. Uchebn. Zaved., Ser. Energetika, 1961, No. 5, pp. 60–71.
133. Tyomkin, A.G.: Inverse methods of heat conduction (in Russian). Energiya, Moscow, 1973.
134. Heat- and mass- exchange in the flow of heated gas, A.G. Shashkov, F.B. Yurevich, V.L. Sergeev, et al., ch. ed. A.G. Shashkov, Nauka i Tekhnika, Minsk, 1974.
135. Tikhonov, A.N.: On a heat conduction equation for several variables (in Russian). Byull. Moscow State Univ., Section A, 1938, Vol. 1, Iss. 9, pp. 1–49.
136. Tikhonov, A.N.: On the stability of inverse problems (in Russian). Dokl. Akad. Nauk SSSR, 1943, Vol. 39, No. 5, pp. 195–198.
137. Tihonov (Tikhonov), A.N.: Solution of incorrectly formulated problems and the regularization method. Soviet Math. Dokl., 1963, Vol. 4, No. 4, pp. 1035–1038.
138. Tikhonov, A.N.: Regularization of incorrectly posed problems. Soviet Math. Dokl., 1963, Vol. 4, No. 6, pp. 1624–1627.
139. Tikhonov, A.N.: Inverse problems in heat conduction. J. Eng. Phys., 1975, Vol. 29, No. 1, pp. 816–820.
140. Tikhonov, A.N.; Arsenin, V. Ya.: Methods of solving ill-posed problems (in Russian). Nauka, Moscow, 1986.
141. Tikhonov, A.N.; Glasko, V.B.: The approximate solution of Fredholm integral equations of the first kind. USSR Comput. Math. and Math. Phys., 1964, Vol. 4, No. 3, pp. 236–247.
142. Tikhonov, A.N.; Glasko, V.B.: Use of the regularization method in nonlinear problems, USSR Comput. Math. and Math. Phys., 1965, Vol. 5, No. 3, pp. 93–107.
143. Tikhonov, A.N.; Samarsky, A.A.: Equations of mathematical physics (in Russian). Nauka, Moscow, 1972.
144. Faddeyev, D.K.; Faddeyeva, V.N.: On ill-conditioned systems of linear equations. USSR Comput. Math. and Math. Phys., 1961, Vol. 1, No. 3, pp. 447–454.
145. Faddeyev, D.K.; Faddeyeva, V.N.: Computational methods of linear algebra (in Russian). Fizmatgiz, Moscow-Leningrad, 1963.
146. Frolov, V.V.: Uniqueness theorems for the solution of the inverse problem of heat conduction. J. Eng. Phys., 1975, Vol. 29, No. 1, pp. 923–927.
147. Cherpakov, P.V.: A theory of regular heat transfer (in Russian). Energiya, Moscow, 1975.
148. Shashkov A.G.; Kreychi, L.; Sergeev, V.L.: Transient heat transfer in plasma yet (in Russian). In: "Heat- and mass- transfer". Minsk, 1972, Vol. 1, P. 1, pp. 373–378.
149. Shlensky, O.F.; Shashkov, A.G.; Aksyonov, L.N.: Thermal physics of decomposing materials (in Russian). Energoatomizdat, Moscow, 1985.
150. Shumakov, N.V.: A method for the experimental study of the process of heating a solid body. – Teplofiz. Zh., 1957, Vol. 27, No. 4, pp. 844–855. – Engl. transl. in Soviet Physics (translated by American Institute of Physics), Vol. 2, 1957, p. 771.
151. Shumakov, N.V.: A method of successive intervals in thermometry of transient processes (in Russian). Atomizdat, Moscow, 1979.
152. Yudin, V.M.: Distribution of heat in glass-reinforced plastics (in Russian). Proc. of TsAGI, Iss. 1267, Izd. TsAGI, Moscow, 1970.
153. Yaryishev, N.A.: Theoretical basis of measuring transient temperatures (in Russian). Energiya, Leningrad, 1967.
154. Arima R.: On General Boundary Value Problem for Parabolic Equation. J. Math. Kyoto Univ., 1964, Vol. 4, No. 1, pp. 207–244.
155. Audley, D.R.; Hodge, J.K.: Identifying the Aerothermodynamic Environment of the Space Shuttle Orbiter, Columbia. Proc. of 6-th IFAC Simp. Identification and System Parameter Estimation, Arlington, Virginia, USA, June 7–11, 1982, pp. 1208–1213
156. Bartllett, E.P.; Curry D.M.: Angevaluation of Flight Data for the Apollo Thermal Protection System. In: "Space Simul. Proc. Symp., New York, 1972". Washington, D.C., 1972, pp. 749–768.
157. Bass, B.R.: Applications of the Finite Element to the Inverse Heat Conduction Problem Using Beck's Second Method. J. Eng. Ind., 1980, Vol. 102, pp. 168–176.
158. Beck, J.V.: Calculation of Surface Heat Flux from an Internal Temperature History. ASME Paper 62-HT-46, 1962.
159. Beck, J.V.: Surface Heat Flux Determination Using an Integral Method. Nucl. Eng. Des., 1968, Vol. 7, No. 2, pp. 170–178.

160. Beck, J.V.: Nonlinear Estimation Applied to the Nonlinear Inverse Heat Conduction Problem. Int. J. Heat and Mass Transfer, 1970, Vol. 13, No. 4, pp. 703–716.
161. Beck, J.V.; Arnold, K.J.: Parameter Estimation in Engineering and Science. Wiley, New York, 1977.
162. Beck, J.V.; Blackwell, B.; St. Clair, C.R., Jr.: Inverse Heat Conduction, Ill-posed Problems. A Wiley-Interscience Publication, New York, 1985.
163. Beck, J.V.; Litkouhi, B.; St. Clair C.R., Jr: Efficient Sequential Solution of the Nonlinear Inverse Heat Conduction Problem. Numer. Heat Transfer, 1982, Vol. 5, pp. 275–286.
164. Beck, J.V.; Murio, D.: Combined Function Specification-Regularization Procedure for Solution of Inverse Heat Conduction Problem. AIAA Paper No. AIAA-84-0491, January 1984.
165. Beck, J.V.; Wolf, H.: The Nonlinear Inverse Heat Conduction Problem. ASME Paper 65-HT-40, 1965.
166. Bell, J.B.: The Noncharacteristic Cauchy Problem for Class of Equations with Time Dependence. -I. Problems in One Space Dimensions, SIAM J. Math. Anal., 1981, Vol. 12, pp. 759–777.
167. Betrin, J.J.: Effects of Surface Temperature and Reynolds Number of Leeward Shuttle Heating, J. Spacecraft, 1976, Vol. 13, No. 8, pp. 473–480.
168. Blackwell, B.F.: An Efficient Technique for the Numerical Solution of the One-Dimensional Inverse Problem of Heat Conduction. Numer. Heat Transfer, 1981, Vol. 4, pp. 229–239.
169. Burggraf, O.R.: An Exact Solution of the Inverse Problem in Heat Conduction Theory and Applications. J. Heat Transfer, 1964, Vol. 86c., pp. 373–382.
170. Cannon, J.R.; Du Chateau, P.: An Inverse Problem for an Unknown Source in a Heat Equation. J. of Math. Anal. and App., 1980, No. 75, pp. 465–485.
171. Cannon, J.R.; Du Chateau, P.: An Inverse Problem for an Nonlinear Diffusion Equation. SIAM J. Appl. Math., 1980, Vol. 39, No. 2, pp. 272–289.
172. Carslaw, H.S.; Jaeger, J.C.: Conduction of Heat in Solids. 2-nd ed., Oxford Univ. Press, New York, 1959.
173. Cordero, J.; Diederich, F.W.: Hurwicz, H.: Aerothermodynamic Test Techniques for Re-entry Structures and Materials. Aerospace Engineering, 1963, Vol. 22, No. 1, pp. 170–191.
174. Davis Don., D.; Wood, C.: A Pyrotechnic-activated Calorimeter for Flight Research at Lunar Return Velocity. In: "Utilis. Elements. Pyrotechn. et. Explosifs Systems Spat." Paris, 1969, pp. 369–378.
175. Deverall, L.I.; Channapragada, R.S.: A New Integral Equation for Heat Flux in Inverse Heat Conduction. J. Heat Transfer, 1966, Vol. 88 c, p. 327.
176. Douglas, J.: On the Numerical Integration of $\frac{\partial^2 u}{\partial x^2} + \frac{\partial^2 u}{\partial y^2} = \frac{\partial u}{\partial t}$ by Implicit Methods. J. Soc. Industr. Appl. Math., 1955, Vol. 3, No. 1, pp. 42–65.
177. Fire I. The Reentry Heating Spacecraft. NASA Facts, 1965, Vol. 2, No. 11, pp. 1–8.
178. Forsythe, G.; Moler, C.B.: Computer Solutions of Linear Algebraic Systems. Prentice-Hall, Inc. Englewood Cliffs, New York, 1967.
179. Friedman, A.: Partial Differential Equations of Parabolic Type. Prentice-Hall, Inc. Englewood Cliffs, New York, 1964.
180. Hadamard, J.: Lectures on Cauchy's Problem in Linear Partial Differential Equations. Yale University Press, New Haven, CT, 1923.
181. Heat Transfer and Spacecraft Thermal Control. Ed. by J.W. Lucas. Ser. "Progress in Astronautics and Aeronautics", Vol. 24, Jet Propulsion Laboratory, Pasadena, California, 1971.
182. Hills, R.G.; Mulholand, G.P.: Accuracy and Resolving Power of One-Dimensional Transient Inverse Heat Conduction Theory as Applied to Discrete and Inaccourate Measurements. Int. J. Heat and Mass Transfer, 1979, Vol. 22, pp. 1221–1229.
183. Imber, M.: Two-Dimensional Inverse Conduction Problem Further Observations. AIAA J., 1975, Vol. 13, pp. 114–115.
184. Imber, M.: Nonlinear Heat Transfer in Planar Solids: Direct and Inverse Applications. AIAA J., 1979, Vol. 17, pp. 204–212.
185. Imber, M.; Khan, J.: Prediction of Transient Temperature Distributions with Embedded Thermocouples. AIAA J., 1972, Vol. 10, pp. 784–789.
186. Joerg, P.: Determination of Back Side Wall Temperatures of Ablative Rocket Nozzles. Heat Transfer-Cleveland. Chemical Engineering Progress Symposium series, 1965, Vol. 61, No. 59, pp. 99–107.
187. Lanczos, C.: Applied Analysis. Prentice Hall, Inc., 1956.

188. Langford, D.: New Analytic Solutions of the One-Dimensional Heat Equation for Temperature and Heat Flow Rate Both Prescribed at the Same Fixed Boundary. Quart. Appl. Math., 1967, Vol. 24, No. 4, pp. 315–322.
189. Legendre, P.J., Chase, G.T.: The Operational Performance of Reentry Vehicle Heat Shield Thermodynamic Instrumentation. Instrum. Aerospace Ind. Vol. 19, Pittsburg, 1973, pp. 85–96.
190. Lorenzi, A.: An Inverse Problem for a Semilinear Parabolic Equation. Annali di Mat. Pura ed Applicata, 1982, Vol. 31, pp. 145–166.
191. Lotkin, M.: The Numerical Integration of Heat Conduction Equations. J. Math. and Phys., 1958, Vol. 37, No. 2, pp. 178–187.
192. Mirsepassi, T.J.: Heat Transfer Charts for Time-Variable Boundary Conditions. Part I- Semi-Infinite Solid. ASME Paper 58-SA-7, 1958.
193. Murio, D.A.: The Modification Method and the Numerical Solution of an Inverse Heat Conduction Problem. SIAM J. Sci. Stat. Comput., 1981, Vol. 2, pp. 17–34.
194. Murphy, J.D.: Flight Test Aerodynamic Heating Data for the Afterbody of the Project Mercury Spacecraft with Comparisons to Available Prediction Methods, NASA Contractor Report CR-649, Washington, 1967, p. 127.
195. Murphy, J.D.; Rubesin, M.W.: A Re-evalution of Heat-Transfer Data Obtained in Flight Tests of Heat-sink Shielded Re-entry Vehicles. AIAA Entry Technol. Conf. Williamsburg-Hampton, Va, 1964. New York. Amer. Inst. Aeronaut. and Astronaut, pp. 77–92.
196. Peaceman, D.W.; Rachford, H.H.: The Numerical Solution of Parabolic and Elliptic Differential Equations. J. Soc. Industr. Appl. Math., 1955, Vol. 3, No. 1, pp. 28–42.
197. Polak, E.: Computational Methods in Optimization. A Unified Approach. Mathematics in Science and Engineering. Vol. 77. Department of Electrical Engineering and Computer Sciences. University of California. Berkley, California, Academic Press. New York-London. 1971.
198. Protter, M.H.: Properties of Solutions of Parabolic Equations and Inequalities. Canad. J. Math., 1961, Vol. 13, pp. 331–345.
199. Reinsch, C.H.: Smoothing by Spline Functions. Numerische Mathematic, Bd. 10, 1967, pp. 177–183.
200. Sparrow, E.M.; Haji-Sheikh, A.; Lundgren, T.S.: The Inverse Problem in Transient Heat Conduction. J. Appl. Mech., Trans. ASME, series E, 1964, Vol. 86, pp. 369–375.
201. Speedy, C.B.; Brown, R.F.; Goodwin, G.C.: Control theory: Identification and Optimal Control. Oliver and Boyd Edinburgh, 1970.
202. Stefan, J.: On the Theory of Ice Formation, Especially on Ice Formation in Polar Seas (Über die Theorie der Eisbildung, insbesondere über die Eisbildung im Polarmeere) (In German). Sitzungsberichte der Kaiserlichen Akademie Wiss. Wien. Math.-naturwiss. Kl., 1890, Vol. 98(2a), pp. 965–973.
203. Stolz, G.: Numerical Solutions to an Inverse Problem of Heat Conduction for Simple Shapes. J. Heat Transfer, 1960, Vol. 82, No. 1, pp. 20–26.
204. Vojvodich, N.S.: PAET Entry Heating and Heat Protection Experiment. J. Spacecraft, 1973, Vol. 10, No. 3, pp. 181–189.
205. Weber, C.F.: Analysis and Solution of the Ill-posed Inverse Heat Conduction Problem. Int. J. Heat and Mass Transfer, 1981, Vol. 24(11), pp. 1783–1792.
206. Wetmore, W.C.: Project Fire Design Nearing Completion. Aviation Week and Space Technology, December 1962, pp. 53–57.

An Additional Bibliography on Inverse Heat Transfer and Ill-Posed Problems (it is added after finalization of editing English text of the manuscript)

1. Alifanov, O.M.: Application of the regularization principle to the formulation of approximate solution of inverse heat-conduction problems. J. Eng. Phys., 1972, Vol. 23, No. 6 pp. 1566–1571.
2. Alifanov, O.M.: Inverse-problem methods in thermophysical investigations and thermal design of technical systems (in Russian). In: "Heat- and Mass Transfer-VI", Problem papers of the VI-th All-Union Conference on Heat- and Mass Transfer, part 2, Izd. ITMO Akad. Nauk BSSR, Minsk, 1981, pp. 166–178.
3. Alifanov, O.M.: Methods for the solution of problems of identification of heat- and mass transfer processes. In: "Mathematical Models, Analytical and Numerical Methods in the Theory of Transfer" (in Russian). Ed. A.A. Samarskii, Izd. ITMO Akad. Nauk BSSR, Minsk, 1982, pp. 79–89.
4. Alifanov, O.M.: Diagnostics and identification of heat-exchange processes (in Russian). In: "Heat- and Mass Transfer-VII", Problem papers of the VII-th All-Union Conference on Heat- and Mass Transfer, part 2, Izd. ITMO Akad. Nauk BSSR, Minsk, 1985, pp. 117–127.
5. Alifanov, O.M.: Derivation of formulas for the gradient of the error in the iterative solution of inverse problems of heat conduction. I. Determination of the gradient in terms of the Green's function. J. Eng. Phys., 1987, Vol. 52, No. 3, pp. 352–359.
6. Alifanov, O.M.: Extreme methods in heat transfer problems (in Russian). In "Heat/Mass Transfer-MIF", Proceedings of the Minsk International Heat and Mass Transfer Forum, Minsk, May 24–27, 1988, Problem papers, Minsk, Izd. ITMO AN BSSR, Minsk, 1988.
7. Alifanov, O.M.: Inverse heat transfer problems (in Russian). Mashinostroenie, Moscow, 1988.
8. Alifanov, O.M.: Inverse problems in the design, modeling and testing of engineering systems. In: Proceeding of the 3-rd International Conference on Inverse Design Concepts and Optimiz. in Eng. Sci. (ICIDES-III), ed. G.S. Dulikravich, Washington, D.C. October 23–25, 1991.
9. Alifanov, O.M.; Artyukhin, E.A.; Pankratov, B.M.: Solution of nonlinear inverse problem of heat conduction. In: "Heat Transfer-Soviet Research" (USA), Vol. 10, No. 3, pp. 63–71, 1978.
10. Alifanov, O.M.; Artyukhin, E.A.: Identification of mathematical model of transient heat transfer processes. In: "Numerical Methods in Thermal Problems". Proc. 4th Int. Conf., Swansea, July 15–18, Swansea, 1985.
11. Alifanov, O.M.; Artyukhin, E.A.; Nenarokomov, A.V.: Spline approximation of the solution of the inverse heat-conduction problem, taking account of the smoothness of the desired function. High Temp. 1987, Vol. 25, No. 4, pp. 520–526.
12. Alifanov, O.M.; Artyukhin, E.A., Rumyantsev, S.V.: Extremal methods of solving ill-posed problems and their applications to inverse heat transfer problems (in Russian). Nauka, Moscow, 1988.
13. Alifanov, O.M.; Balaboshko, N.G.; Desner, I.G.; Kerov, N.V.; Kuznetsov, V.S.; Pankratov, B.M.: Hardware and software structures for processing thermophysics experiment data. J Eng. Phys., 1983, Vol. 45, No. 5, pp. 1223–1226.
14. Alifanov, O.M.; Balakovskii, S.L.; Klibanov, M.V.: Reconstruction of causal characteristics of the thermal conductivity process from the solution of the combined inverse problem. J. Eng. Phys., 1987, Vol. 52, No. 5, pp. 617–622.
15. Alifanov, O.M.; Balashova, I.E.: Solving inverse problems of heat transfer with consideration of data on the smoothness of the reconstructed relation (in Russian). Izv. Sib. Otdel. Akad. Nauk SSSR, Ser. Tech. Nauk, 1987, Vol. 7, No. 2, pp. 5–11.
16. Alifanov, O.M.; Balashova, I.E.: Investigating the diagnostics algorithms of the thermal effect in design. J. Eng. Phys., 1989, Vol. 56, No. 4, pp. 483–488.

17. Alifanov, O.M.; Mikhailov, V.V.: Solution of the inverse boundary-value problem of heat conduction in an overdefined formulation. J. Eng. Phys., 1983, Vol. 45, No. 5, pp. 1270–1274.
18. Alifanov, O.M.; Nagoga, G.P.; Sapozhnikov, V.M.: Problem of determining internal boundary conditions in measuring the heat of gas-turbine blades undergoing cooling. J. Eng. Phys., 1986, Vol. 51, No. 3, pp. 1040–1045.
19. Alifanov, O.M.; Nenarokomov, A.V.: Effect of different factors on the accuracy of the solution of a parameterized inverse problem of heat conduction. J. Eng. Phys., 1989, Vol. 56, No. 3, pp. 308–312.
20. Alifanov, O.M.; Nenarokomov, A.V.: A three-dimensional inverse heat conduction problem in the extremal statement (in Russian). Dokl. Russian Akad. Nauk, 1992, Vol. 325, No. 5, pp. 950–954.
21. Alifanov, O.M.; Rumyantsev, S.V.: Formulas for the discrepancy gradient in the iterative solution of inverse heat-conduction problems. II. Determining the gradient in terms of a conjugate variable. J. Eng. Phys., 1987, Vol. 52, No. 4, pp. 489–495.
22. Alifanov, O.M.; Rumyantsev, S.V.: Gradient of the discrepancy in the iterative solution of inverse heat-conduction problems. III. Calculation of the gradient using a conjugate boundary problem. J. Eng. Phys., 1987, Vol. 52, No. 6, pp. 708–713.
23. Alifanov, O.M.; Rumyantsev, S.V.: Application of iterative regularization for the solution of incorrect inverse problems. J. Eng. Phys., 1987, Vol. 53, No. 5, pp. 1335–1342.
24. Anger, G.: Inverse problems in differential equations. Plenum Press, New York, 1990.
25. Artyukhin, E.A.: Optimum planning of experiments in the identification of heat-transfer processes. J. Eng. Phys., 1989, Vol. 56, No. 3, pp. 256–260.
26. Artyukhin, E.A.; Baranov, V.V.; Ganchev, B.G.; Nenarokomov, A.V.: Nonsteady heat-transfer in the wetting of heated surfaces. High Temp., 1987, Vol. 25, No. 5, pp. 719–723.
27. Artyukhin, E.A.; Guseva, L.I.; Tryanin, A.P.; Shibin, A.G.: Data processing and planning of nonstationary thermophysical experiments. J. Eng. Phys., 1989, Vol. 56, No. 3, pp. 286–290.
28. Artyukhin, E.A.; Killikh, V.E.; Nenarokomov, A.V.; Repin, I.V.: Investigation of the thermal interaction of material with two-phase flows by the inverse problem method. High Temp. 1990, Vol. 28, No. 1, pp. 94–99.
29. Artyukhin, E.A.; Mamolov, V.A.; Nenarokomov, A.V.: Evaluating the effect of shrinkage on the effective thermal conductivity of a glass-reinforced plastic. J. Eng. Phys., 1989, Vol. 56, No. 6, pp. 707–712.
30. Artyukhin, E.A.; Nenarokomov, A.V.; Inverse problem of deriving the integral emissivity of a solid. High Temp., 1986, Vol. 24, No. 5, pp. 725–729.
31. Artyukhin, E.A.; Nenarokomov, A.V.: Coefficient inverse heat-conduction problem. J. Eng. Phys., 1987, Vol. 53, No. 3, pp. 1085–1090.
32. Balakovskii, S.L.: Resolving power of the iteration method of solving inverse heat conduction boundary value problems. J. Eng. Phys., 1987, Vol. 52, No. 6, pp. 713–717.
33. Balakovskii, S.L.: Solution of inverse heat transfer problems by a two-model method. J. Eng. Phys., 1990, Vol. 57, No. 3, pp. 1118–1122.
34. Balk, P.I.: Practically achievable accuracy and reliability of the solution of inverse heat conduction problems. J. Eng. Phys., 1987, Vol. 52, No. 2, pp. 243–249.
35. Banks, H.T.; Kunisch, K.: Estimation Techniques for Distributed Parameter Systems, Birkhäuser, Boston, 1989.
36. Baumeister, J.: Stable solution of inverse problems. Vieweg & Sohn, Braunschweig/Wiesbaden, 1987.
37. Baumeister, J., Reinhardt, H.J.: Inverse heat conduction problems and their numerical solution. In: "Inverse and Ill-posed Problems", H.W. Engl, C.W. Groetsch (eds.), Academic Press 1987, pp. 325–344.
38. Beck, J.V.: Criteria for comparison of methods of solution of the inverse heat conduction problem. Nucl. Eng. Des., 1979, 5, pp. 11–12.
39. Beck, J.V.; Litkouhi, B.; St. Clair, C.R., Jr.: Efficient sequential solution of the nonlinear inverse heat conduction problem. Numer. Heat Transfer, 1982, 5, pp. 275–286.
40. Ben-Haim, Y.; Elias, E.: Indirect measurements of surface temperature and heat flux: optimal design using convexity analysis. International J. of Heat and Mass Transfer, Vol. 1987, 30, No. 8, pp. 1673–1683.
41. Blackwell, B.F.: An efficient technique for the numerical solution of the one-dimensional inverse problem of heat conduction. Numer. Heat Transfer, 1981, 4, pp. 229–239.
42. Blackwell, B.F.: Some comments of Beck's solution of the inverse problem of heat conduction through the use of Duhamel's theorem. Int. J. Heat and Mass Transfer, 1983, 26, pp. 302–305.

43. Budnik, S.A.; Guseva, L.T.; Shibin, A.G.: Analysis of the method of temperature measurements used to determine the set of characteristics of a thermally protective coatings. J. Eng. Phys., 1989, Vol. 56, No. 3, pp. 301–308.
44. Cannon, J.; Hornung, U. (Eds.): Inverse Problems. Birkhäuser, Basel, 1986.
45. Cannon, J.R.; DuChateau, P.: An inverse problem for an unknown source in a heat equation. J. Math. Anal. Appl., 1980, 75, No. 2, pp. 465–485.
46. Cannon, J.R.; Zachmann, D.: Parameter determination in parabolic partial-differential equations from overspecified boundary data. Internat. J. Eng. Sci., 1982, 20, No. 6, pp. 779–788.
47. Carasso, A.S.: Impulse response acquisition as an inverse heat conduction problem. SIAM J. Appl. Math., 1990, Vol. 50, No. 1, pp. 75–90.
48. Carasso, A.S.: Corrigendum: 'Impulse response acquisition as an inverse heat conduction problem'. SIAM J. Appl. Math., 1990, Vol. 50, No. 3, pp. 942.
49. Chen, H.; Chang, S.: Application of the hybrid method to inverse heat conduction problems. International J. of Heat and Mass Transfer, 1990, Vol. 33, No. 4, pp. 621–628.
50. Denisov, A.M.: Uniqueness of the solutions of some inverse problems for the heat equation with a piecewise-constant coefficient. Zh. Vychisl. Mat. i Mat. Fiz., 1982, 22, No. 4, pp. 858–864. Engl. transl. in: USSR Comput. Math. and Math. Phys., 1982, Vol. 22, No. 4.
51. Denisov, V.N.: Stabilization of the solution of the Cauchy problem for the heat equation. Soviet Math. Dokl., 1988, Vol. 37, No. 3, pp. 688–692
52. Elden, L.: Hyperbolic approximations for a Cauchy problem for the heat equation. Inverse Problems, 1988, 4, pp. 59–70.
53. Engl, H.W., Groetsch, C.W. (Eds.): Inverse and Ill-posed Problems. Academic Press, Orlando, 1987.
54. Engl, H., Langthaler, T.: Numerical solution of an inverse problem connected with continuous casting of steel. ZOR, 29 (1985); B, pp. 185–199.
55. Engl, H.; Lanthaler, T.; Manselli, P.: On an inverse problem for a nonlinear heat equation connected with continuous casting of steel. In: "Optimal control with partial differential equations II", K-H. Hoffman, W. Krabs, (eds.), Birkhauser, Basel, 1987, pp. 67–89.
56. Engl, H.; Manselli, P.: Stability estimates and regularization for an inverse heat conduction problem. Numer. Func. Anal. Optim. 10 (1989), pp. 517–540.
57. Flach, G.P., Ozisik, M.N.: Whole domain function-specification method for linear inverse heat conduction. Annals of Nuclear Energy, 1986, Vol. 13, No. 6, pp. 325–336.
58. Flach, G.P., Ozisik, M.N.: Inverse heat conduction of periodically contacting surfaces. Journal of Heat Transfer, Transactions ASME, 1988, Vol. 110, No. 4, pp. 821–829.
59. Flach, G.P., Ozisik, M.N.: Inverse heat conduction problem of simultaneously estimating spatially varying thermal conductivity and heat capacity per unit volume. Numerical Heat transfer, 1989, Vol. 16, No. 2, pp. 249–266.
60. Gentili, G.: Dissipativity condition and inverse problems for heat conduction with linear memory: Inverse Problems, 1991, Vol. 7, No. 1, pp. 77–84.
61. Gillian, D.S.; Mair, B.A., Martin, C.F.: An inverse convolution method for regular parabolic equations. SIAM J. Control Optim., 1991, Vol. 29, No. 1, pp. 71–88.
62. Gillian, D.S., Mair, B.A., Martin, C.F.: A convolution method for inverse heat conduction problems. Math. Systems Theory, 1988, Vol. 21, No. 1, pp. 49–60.
63. Glasko, V.B.; Kulik, N.N.; Tikhonov, A.N.; Shkliano, I.N.: An inverse problem of heat conduction. USSR Comput. Math. and Math. Phys., 1980, Vol. 19, pp. 223–230.
64. Hensel, E.: Inverse Theory and Applications for Engineers. Prentice Hall. Englewood Cliffs New Jersey. 1991.
65. Hensel, E.; Hills, R.: Steady-state two-dimensional inverse heat conduction. Numerical Heat Transfer, Part B, 1989, Vol. 15, pp. 227–240.
66. Jarny, Y.; Ozisik, M.N., Bardon, J.P.: General optimization method using adjoint equation for solving multidimensional inverse heat conduction. International J. of Heat and Mass Transfer, 1991, Vol. 34, No. 11, pp. 2911–2919.
67. Kammerer, W.J.; Nashed, M.Z.: On the convergence of the conjugate gradient method for singular linear operator equations. SIAM J. Numer. Anal., 1972, Vol. 9, No. 1, pp. 165–181
68. Klibanov, M.V.: A uniqueness theorem for a class of inverse thermal-conduction problems involving temperature measurements at internal points. J. Eng. Phys., 1985, Vol. 49, No. 6, pp. 1484–1486.
69. Knabner, P.: Regularizing the Cauchy problem for the heat equation by sign restrictions. In: "Improperly posed problems and their numerical treatment", G. Hammerlin, K.-H. Hoffman (eds.), Birkhäuser 1983, pp. 165–177.

70. Knabner, P.: Regularization of the Cauchy problem for the heat equation by norm bounds. Appl. Anal., 1984, Vol. 17, No. 4, pp. 295–312.
71. Kuroyanagi, T.: Surface temperature and surface heat flux determination of the inverse heat conduction problem. Bulletin of the JSME, 1986, Vol. 29, No. 255, pp. 2961–2969.
72. Kurpisz, K.: Method for determining steady state temperature distribution within blast furnace hearth lining by measuring temperature at selected points (in Japanese). Transactions of the Iron and Steel Institute of Japan, 1988, Vol. 28, No. 11, pp. 926–929.
73. Kurpisz, K.: Numerical solution of one case of inverse heat conduction problems. Journal of Heat Transfer, Transactions ASME, 1991, Vol. 113, No. 2, pp. 280–286.
74. Liu, G.L.: A novel variational formulation of inverse problem of heat conduction with free boundary on an image plane. Numerical Methods in Thermal Problems, 1989, Vol. VI, Part 1, 2, Pineridge, Swansea, pp. 1712–1720.
75. Maillet, D.; Degiovanni, A., Pasquetti, R.: Inverse heat conduction applied to the measurement of heat transfer coefficient on a cylinder. Comparison between an analytical and a boundary element technique. Journal of Heat Transfer, Transactions of ASME, 1991, Vol. 113, No. 3, pp. 549–557.
76. Markin, A.D.; Pyatyshkin, G.G.: Regularization of the solution of the inverse heat conduction problem in a variational formulation. J. Eng. Phys., 1987, Vol. 52, No. 6, pp. 717–721.
77. Marquardt, W.; Auracher, H.: Observer-based solution of inverse heat conduction problems. International J. of Heat and Mass Transfer, 1990, Vol. 33, No. 7, pp. 1545–1562.
78. Mehta, R.C.: Extension of the solution of inverse conduction problem. Int. J. Heat Mass Transfer, 1979, Vol. 22, pp. 1149–1150.
79. Mishin, V.P.; Alifanov, O.M.: Quality improvement in final design of thermally stressed structures, and inverse heat-transfer problems. 1. General problems of the theory. Soviet Mashine Science, 1986, No. 5, pp. 16–25.
80. Mishin, V.P.; Alifanov, O.M.: Improving final designs of thermally stressed structures, and inverse heat-transfer problems. 2. Practical applications. Soviet Mashine Science, 1986, No. 6, pp. 8–16.
81. Morozov, V.A.: Methods for Solving Incorrectly Posed Problems. Springer-Verlag, New York, 1984.
82. Murio, D.A.: Parameter selection by discrete mollification and the numerical solution of the inverse heat conduction problem. J. Comp. Appl. Math., 1988, Vol. 22, pp. 25–34.
83. Murio, D.A.: The mollification method and the numerical solution of the inverse heat conduction problem by finite differences, Computers Math. Applic., 1989, Vol. 17, No. 10, pp. 1385–1396.
84. Murio, D.A.; Guo, L.; A stable marching finite difference algorithm for the inverse heat conduction problem with no initial filtering procedure. Computers Math. Applic., 1990, Vol. 19, No. 10, pp. 35–50.
85. Muzylev, N.V.: Uniqueness of the determination of the temperature dependence of thermophysical properties in piecewise monotonic processes. J. Eng. Phys., 1985, Vol. 49, No. 6, pp. 1487–1490.
86. Nashed M.Z.: Steepest descent for singular linear operator equations. SIAM J. Numer. Anal., 1970, Vol. 7, No. 3, pp. 358–362.
87. Nashed, M.Z.: On nonlinear ill-posed problems 1: Classes of operator equations and minimization of functionals. In: "Nonlinear Analysis and Applications" (V. Lakshmikantham, Ed.), Dekker, New York, 1987, p. 351–373.
88. Palancz, B.; Szabolcs, G.: Solution of inverse heat conduction problem by orthogonal collocation techniques. Acta Technica, Budapest, 1988, Vol. 101, No. 3, pp. 323–337.
89. Prosuntsov, P.V.; Reznik, S.V.: Problems of heat transfer in semitransparent scattering materials. 1. The theoretical fundamentals of the absolute stationary method for determination of thermal conductivity (in Russian). Sib. Fiz.-Tekhn. Zh. (Izv. Sib. Otd. Russ. Akad. Nauk), 1991, Iss. 1, pp. 51–56.
90. Prosuntsov, P.V.; Reznik, S.V.; Tyutyunnik, V.E.: Problems of heat transfer in semitransparent scattering materials. 2. The experimental facility and results of investigating thermal conductivity by the absolute stationary method (in Russian). Sib. Fiz.-Tekhn. Zh. (Izv. Sib. Otd. Russ. Akad. Nauk), 1992, Iss. 4, pp. 27–32.
91. Prosuntsov, P.V.; Reznik, S.V.: Problems of heat transfer in semitransparent scattering materials. 3. The theoretical fundamentals of determination of thermal conductivity by the method of monotonous heating (in Russian). Sib. Fiz.-Tekhn. Zh. (Izv. Sib. Otd. Russ. Akad. Nauk), 1992, Iss. 4, pp. 33–38.

92. Raynaud, M.: Comparison of space marching finite difference technique and function minimization technique for the estimation of the front location in nonlinear melting problems. Fifth Symposium on Control of Distributed Parameter Systems, A. El Jai and M. Amouroux (eds.), Université de Perpignan, France, 1989, pp. 215–220.
93. Raynaud, M., Bransier, J.: A new finite difference method for nonlinear inverse heat conduction problem. Numerical Heat Transfer, 1986, Vol. 9, No. 1, pp. 27–42.
94. Raynaud, M.; Beck, J.V.: Methodology for comparison of inverse heat conduction methods. Transactions of the ASME, 1988, Vol. 110, pp. 30–37.
95. Reemtsen, R.; Kirsch, A.: A method for the numerical solution of the one-dimensional inverse Stefan problem. Numer. Math., 1984, Vol. 43, pp. 253–273.
96. Reinhardt, H.J.: Numerical method for the solution of two-dimensional inverse heat conduction problem. International J. for Numerical Methods in Engineering, 1991, Vol. 32, No. 2, pp. 363–383.
97. Reinhardt, H.-J.; Valensia, L.: The numerical solution of inverse heat conduction problems with application to reactor technology. In: "Structural Mechanics in Reactor Technology (9th SMLPL-Conf., Lausanne). 1987, Ed. F.H. Wittman, Vol. B, pp. 25–29.
98. Scott, E.P.; Beck, J.V.: Analysis of order of the sequential regularization solutions of inverse heat conduction problems. Journal of Heat Transfer, Transactions ASME, 1989, Vol. 111 No. 2, pp. 218–224.
99. Shoji, M.; Ono, N.: Application of the boundary element to the inverse problem of heat conduction (in Japanese). Nippon Kikai Gakkai Ronbunshu, B Hen, 1988, Vol. 54, No. 506 pp. 2893–2900.
100. Sun, X., Jaggard, D.L.: The inverse black body radiation problem: a regularized solution Journal of Applied Physics, 1987, Vol. 62, No. 11, pp. 4382–4386.
101. Talenti, G. (Ed.): Inverse Problems. LNM 1225, Springer-Verlag, New York, 1986.
102. Tandy, D.F.; Trujillo, D.M.; Busby, H.R.: Solution of inverse heat conduction problems using an eigenvalue reduction technique. Numerical Heat Transfer, 1986, Vol. 10, No. 6, pp. 597–617.
103. Tikhonov, A.N.; Goncharsky, A.V. (Eds): Ill-posed Problems in the Natural Sciences. MIR, Moscow, 1989.
104. Weiland, E.; Babary, J.P.: Comparative study for a new solution to the inverse heat conduction problem. Comm. in Applied Numerical Methods, 1988, Vol. 4, No. 5, pp. 687–689.
105. Williams, S.D.; Curry, D.M.: An analytical experimental study for surface heat flux determination. J. Spacecraft, 1977, Vol. 14, pp. 632–637.
106. Woodbury, K.: Effect of thermocouple sensor dynamic on surface heat flux predictions obtained via inverse heat transfer analysis. Int. J. of Heat and Mass transfer 1990, Vol. 33, No. 12, pp. 2641–2649.
107. Woodbury, K.A.: Determination of Surface Heat Fluxes During Spray Quenching of Aluminum Using an Inverse Technique. ASME Paper No. 91-WA-HT-12, Winter Annual Meeting, 1991, Atlanta, Georgia.
108. Yoshimura, T.; Ikuta, K.: Inverse heat conduction problem by finite element formulation. Int. J. Systems Sci., 1985, Vol. 16, pp. 1365–1376.
109. Zabaras, N.: Inverse Finite element techniques for the analysis of solidification processes. International J. for Numerical Methods in Engineering, 1990, Vol. 29, No. 7, pp. 1569–1587.
110. Zabaras, N.; Liu, J.C.: An analysis of two-dimensional linear inverse heat transfer problems using an integral method. Numer. Heat Transfer, 1990, Vol. 13, pp. 527–533.
111. Zabaras, N.; Mukherjee, S., Richmond D.: An analysis of inverse heat transfer problems with phase changes using an integral method. ASME J. of Heat Transfer, 1988, Vol. 110, pp. 554–561.

Subject Index

SPRINGER NATURE

GPSR Compliance

The European Union's (EU) General Product Safety Regulation (GPSR) is a set of rules that requires consumer products to be safe and our obligations to ensure this.

If you have any concerns about our products, you can contact us on ProductSafety@springernature.com

In case Publisher is established outside the EU, the EU authorized representative is:

Springer Nature Customer Service Center GmbH
Europaplatz 3
69115 Heidelberg, Germany

The manufacturer's authorised representative in the EU is Springer Nature Customer Service Centre GmbH, Europaplatz 3, 69115 Heidelberg, Germany. If you have any concerns regarding our products, please contact ProductSafety@springernature.com

Printed and bound by CPI Group (UK) Ltd, Croydon, CR0 4YY
22/07/2026
02174188-0006